SMART SENSORS

FOR INDUSTRIAL APPLICATIONS

Devices, Circuits, and Systems

Series Editor
Krzysztof Iniewski
CMOS Emerging Technologies Inc., Vancouver, British Columbia, Canada

FORTHCOMING TITLES:

SMART SENSORS

FOR INDUSTRIAL APPLICATIONS

Edited by
Krzysztof Iniewski

CRC Press
Taylor & Francis Group
Boca Raton   London   New York

CRC Press is an imprint of the
Taylor & Francis Group, an **informa** business

MATLAB® is a trademark of The MathWorks, Inc. and is used with permission. The MathWorks does not warrant the accuracy of the text or exercises in this book. This book's use or discussion of MATLAB® software or related products does not constitute endorsement or sponsorship by The MathWorks of a particular pedagogical approach or particular use of the MATLAB® software.

CRC Press
Taylor & Francis Group
6000 Broken Sound Parkway NW, Suite 300
Boca Raton, FL 33487-2742

First issued in paperback 2017

© 2013 by Taylor & Francis Group, LLC
CRC Press is an imprint of Taylor & Francis Group, an Informa business

No claim to original U.S. Government works
Version Date: 20130401

ISBN 13: 978-1-138-07764-5 (pbk)
ISBN 13: 978-1-4665-6810-5 (hbk)

Library of Congress Cataloging-in-Publication Data

Smart sensors for industrial applications / editor, Krzysztof Iniewski.
    pages cm. -- (Devices, circuits, and systems)
    Summary: "This book assembles the latest research in the field of smart sensors technology in one place. It exposes readers to myriad applications that smart sensors technology has enabled. The book is meant for advanced graduate research work or for academicians and researchers. This way the book has a widespread appeal, including practicing engineers, academicians, research scientists and senior graduate students"-- Provided by publisher.
    Includes bibliographical references and index.
    ISBN 978-1-4665-6810-5 (hardback)
    1. Detectors--Industrial applications. I. Iniewski, Krzysztof, 1960-

TA165.S56 2013
681'.2--dc23   2013008836

Visit the Taylor & Francis Web site at
http://www.taylorandfrancis.com

and the CRC Press Web site at
http://www.crcpress.com

Contents

PART I   *Photonic and Optoelectronics Sensors*

PART II   Infrared and Thermal Sensors

PART III   Magnetic and Inductive Sensors

Contents

PART IV Sound and Ultrasound Sensors

PART V Piezoresistive, Wireless, and Electrical Sensors

List of Figures

Preface

Sensor technologies are a rapidly growing topic in science and product design, embracing developments in electronics, photonics, mechanics, chemistry, and biology. Their presence is widespread in everyday life; they sense sound, movement, optical, or magnetic signals. The demand for portable and lightweight sensors is relentless, filling various needs in several industrial environments.

The book is divided into five parts. Part I deals with photonics and optoelectronics sensors. Various developments in optical fibers, Brillouin detection, and Doppler effect analysis are described. Oxygen detection, directional discrimination, and optical sensing are some key technological applications. Part II deals with infrared and thermal sensors. Bragg gratings, thin films, and microbolometers are described. Temperature measurements in industrial conditions, including sensing inside explosions, are widely covered. Part III deals with magnetic and inductive sensors. Magnetometers, inductive coupling ferro-fluidics are described. Magnetic field and inductive current measurements in various industrial conditions, including airplanes, are covered in detail. Part IV deals with sound and ultrasound sensors. Underwater acoustic modem, vibrational spectroscopy, and photoacoustics are described. Finally, Part V deals with piezo-resistive, wireless, and electrical sensors.

With such a wide variety of topics covered, I am hoping that the reader will find something stimulating to read and discover the field of sensor technologies to be both exciting and useful in industrial practice. Books like this one would not be possible without many creative individuals meeting together in one place to exchange thoughts and ideas in a relaxed atmosphere. I would like to invite you to attend the CMOS Emerging Technologies Research events that are held annually in beautiful British Columbia, Canada, where many topics covered in this book are discussed. See http://www.cmosetr.com for presentation slides from the previous meeting and announcements about future ones. If you have any suggestions or comments about the book, please email me at kris.iniewski@gmail.com.

Kris Iniewski
Vancouver, British Columbia, Canada

MATLAB® is a registered trademark of The MathWorks, Inc. For product information, please contact:

The MathWorks, Inc.
3 Apple Hill Drive
Natick, MA 01760-2098 USA
Tel: 508-647-7000
Fax: 508-647-7001
E-mail: info@mathworks.com
Web: www.mathworks.com

Editor

Dr. Krzysztof (Kris) Iniewski manages R&D at Redlen Technologies Inc., a start-up company in Vancouver, British Columbia, Canada. Redlen's revolutionary production process for advanced semiconductor materials enables a new generation of more accurate, all-digital, radiation-based imaging solutions. Kris is also a president of CMOS Emerging Technologies Research (www.cmo-set.com), an organization of high-tech events covering communications, microsystems, optoelectronics, and sensors.

During his career, Dr. Iniewski held numerous faculty and management positions at the University of Toronto, the University of Alberta, Simon Fraser University, and PMC-Sierra Inc. He has published more than 100 research papers in international journals and conferences and holds 18 international patents granted in the United States, Canada, France, Germany, and Japan. He is a frequent invited speaker and has consulted for multiple organizations internationally. He has also written and edited several books for IEEE Press, Wiley, CRC Press, McGraw Hill, Artech House, and Springer. His personal goal is to contribute to healthy living and sustainability through innovative engineering solutions. In his leisurely time, Kris can be found hiking, sailing, skiing, or biking in beautiful British Columbia. He can be reached at kris.iniewski@gmail.com.

Contributors

Ehad Akeila
Department of Electrical and Computer
  Engineering
University of Auckland
Auckland, New Zealand

Nélia Jordão Alberto
Institute of Telecommunications
and
Centre for Mechanical Technology and
  Automation
University of Aveiro
Aveiro, Portugal

Maan M. Alkaisi
Department of Electrical and Computer
  Engineering
The MacDiarmid Institute for Advanced
  Materials and Nanotechnology
University of Canterbury
Christchurch, New Zealand

Mahmoud Almasri
Department of Electrical and Computer
  Engineering
University of Missouri
Columbia, Missouri

B. Andò
Department of Electrical, Electronic and
  Computer Engineering
University of Catania
Catania, Italy

Udo Ausserlechner
Sense and Control
Infineon Technologies AG
Villach, Austria

S. Baglio
Department of Electrical, Electronic and
  Computer Engineering
University of Catania
Catania, Italy

Fabricio G. Baptista
Faculdade de Engenharia de Bauru
Departamento de Engenharia Elétrica
Universidade Estadual Paulista
Sao Paulo, Brazil

Egidio De Benedetto
Department of Engineering for Innovation
University of Salento
Lecce, Italy

A. Beninato
Department of Electrical, Electronic and
  Computer Engineering
University of Catania
Catania, Italy

Bridget Benson
Department of Electrical Engineering
California Polytechnic State University
San Luis Obispo, California

Lúcia Maria Botas Bilro
Institute of Telecommunications
University of Aveiro
Aveiro, Portugal

Richard J. Blaikie
Department of Physics
University of Otago
Dunedin, New Zealand

Chris J. Bleakley
Complex & Adaptive Systems Laboratory
School of Computer Science and Informatics
University College Dublin
Dublin, Ireland

François Boussu
Engineering and Textile Materials Laboratory
  (GEMTEX)
National Higher School of Arts and Textile
  Industries (ENSAIT)
Roubaix, France

and

University of Lille Nord de France
Lille, France

Christian-Alexander Bunge
Leipzig Deutsche Telekom AG
University for Telecommunication
Leipzig, Germany

Dean Callaghan
Photonics Research Centre
Dublin Institute of Technology
Dublin, Ireland

Giuseppe Cannazza
Department of Engineering for Innovation
University of Salento
Lecce, Italy

Antoni J. Canós
Microwave Division (DIMAS)
ITACA Research Institute
Universidad Politécnica de Valencia
Valencia, Spain

Jose M. Catalá-Civera
Microwave Division (DIMAS)
ITACA Research Institute
Universidad Politécnica de Valencia
Valencia, Spain

Andrea Cataldo
Department of Engineering for Innovation
University of Salento
Lecce, Italy

Qi Cheng
Department of Electrical and Computer
    Engineering
University of Missouri
Columbia, Missouri

A. Conesa-Roca
Department of Electronic Engineering
Universitat Politècnica de Catalunya,
    BarcelonaTech (UPC)
Barcelona, Spain

Irina Cristian
Faculty of Textiles—Leather Engineering and
    Industrial Management
Gheorghe Asachi Technical University of Iasi
Iasi, Romania

A.P.J. van Deursen
Department of Electrical Engineering
Eindhoven University of Technology
Eindhoven, the Netherlands

Gerald Farrell
Photonics Research Centre
Dublin Institute of Technology
Dublin, Ireland

Jozue Vieira Filho
Faculdade de Engenharia de Ilha Solteira
Departamento de Engenharia Elétrica
Universidade Estadual Paulista
Sao Paulo, Brazil

Yusaku Fujii
Department of Electronic Engineering
Gunma University
Gunma, Japan

B. García-Baños
Microwave Division (DIMAS)
ITACA Research Institute
Universidad Politécnica de Valencia
Valencia, Spain

Juan Ramon Gonzalez
Complex & Adaptive Systems Laboratory
School of Computer Science and Informatics
University College Dublin
Dublin, Ireland

Asaf Grosz
Department of Electrical and Computer
    Engineering
Ben-Gurion University of the Negev
Beer-Sheva, Israel

Jinseok Heo
Department of Chemistry
Buffalo State College
State University of New York
Buffalo, New York

Ellen L. Holthoff
Sensors and Electron Devices
United States Army Research Laboratory
Adelphi, Maryland

Jonathan F. Holzman
School of Engineering
University of British Columbia
Kelowna, British Columbia, Canada

Hendrik Husstedt
Measurement and Actuators Division
Vienna University of Technology
Vienna, Austria

Daniel J. Inman
Department of Aerospace Engineering
University of Michigan
Ann Arbor, Michigan

Paweł Jamróz
Laboratory of Flow Metrology
Strata Mechanics Research Institute
Polish Academy of Sciences
Krakow, Poland

Xian Jin
School of Engineering
University of British Columbia
Kelowna, British Columbia, Canada

Manfred Kaltenbacher
Measurement and Actuators Division
Vienna University of Technology
Vienna, Austria

Ryan Kastner
Department of Computer Science and
    Engineering
University of California, San Diego
La Jolla, California

Chang-Soo Kim
Department of Electrical & Computer
    Engineering
and
Department of Biological Sciences
Missouri University of Science and Technology
Rolla, Missouri

H.S. Kim
Department of Mechanical Engineering
University of Minnesota
Minneapolis, Minnesota

Vladan Koncar
Engineering and Textile Materials Laboratory
    (GEMTEX)
National Higher School of Arts and Textile
    Industries (ENSAIT)
Roubaix, France

and

University of Lille Nord de France
Lille, France

Fred Lacy
Department of Electrical Engineering
Southern University and A&M College
Baton Rouge, Louisiana

Bernhard Lendl
Institute of Chemical Technologies and
    Analytics
Vienna University of Technology
Vienna, Austria

Hugo Filipe Teixeira Lima
Institute for Nanostructures, Nanomodelling
    and Nanofabrication
and
Department of Physics
University of Aveiro
Aveiro, Portugal

Mariusz Litwa
Division of Metrology and Optolectronics
Institute of Electrical Engineering and
    Electronics
Poznań University of Technology
Poznań, Poland

Alayn Loayssa
Department of Electrical and Electronic
    Engineering
Public University of Navarra
Pamplona, Spain

Merlin L. Mah
Department of Electrical and Computer
    Engineering
University of Minnesota
Minneapolis, Minnesota

V. Marletta
Department of Electrical, Electronic and
    Computer Engineering
University of Catania
Catania, Italy

Koichi Maru
Department of Electronics and Information
    Engineering
Kagawa University
Kagawa, Japan

Sari Merilampi
Research and Competence Center (Welfare
    Technology)
Satakunta University of Applied Sciences
Pori, Finland

Jerzy Nabielec
Faculty of Electrical Engineering, Automatics,
    Computer Science and Electronics
Department of Measurement and
    Instrumentation
AGH University of Science and
    Technology
Krakow, Poland

Saad Nauman
Department of Materials Science and
    Engineering
Institute of Space Technology
Islamabad, Pakistan

Holger Neumann
Cryogenics Division
Institute of Technical Physics
Karlsruhe Institute of Technology
Eggenstein-Leopoldshafen, Germany

Volker Nock
Department of Electrical and Computer
    Engineering
The MacDiarmid Institute for Advanced
    Materials and Nanotechnology
University of Canterbury
Christchurch, New Zealand

Rogério Nunes Nogueira
Institute for Telecommunications
University of Aveiro
Aveiro, Portugal

Eugene Paperno
Department of Electrical and Computer
    Engineering
Ben-Gurion University of the Negev
Beer-Sheva, Israel

Suzanne Paradis
Defence Research and Development
Micro Systems Group
Toronto, Ontario, Canada

Paul M. Pellegrino
Sensors and Electron Devices
United States Army Research Laboratory
Adelphi, Maryland

Felipe L. Peñaranda-Foix
Microwave Division (DIMAS)
ITACA Research Institute
Universidad Politécnica de Valencia
Valencia, Spain

João de Lemos Pinto
Institute for Nanostructures, Nanomodelling
    and Nanofabrication
and
Department of Physics
University of Aveiro
Aveiro, Portugal

Hans Poisel
POF Application Center
Ohm-Hochschule Nürnberg
Nürnberg, Germany

Stefan Radel
Institute of Chemical Technologies and Analytics
Vienna University of Technology
Vienna, Austria

R. Rajamani
Department of Mechanical Engineering
University of Minnesota
Minneapolis, Minnesota

Ginu Rajan
School of Electrical Engineering and
    Telecommunications
The University of New South Wales
Sydney, New South Wales, Australia

Rajini Kumar Ramalingam
Cryogenics Division
Institute of Technical Physics
Karlsruhe Institute of Technology
Eggenstein-Leopoldshafen, Germany

M. Román-Lumbreras
Department of Electronic Engineering
Universitat Politècnica de Catalunya,
    BarcelonaTech (UPC)
Barcelona, Spain

Mohamed Saad
Complex & Adaptive Systems Laboratory
School of Computer Science and Informatics
University College Dublin
Dublin, Ireland

Mikel Sagues
Department of Electrical and Electronic
    Engineering
Public University of Navarra
Pamplona, Spain

Zoran Salcic
Department of Electrical and Computer
    Engineering
University of Auckland
Auckland, New Zealand

Johannes Schnöller
Institute of Water Quality, Resource and Waste
    Management
Vienna University of Technology
Vienna, Austria

Yuliya Semenova
Photonics Research Centre
Dublin Institute of Technology
Dublin, Ireland

A.S. Sezen
Department of Mechanical Engineering
University of Minnesota
Minneapolis, Minnesota

S. Sivaramakrishnan
Department of Mechanical Engineering
University of Minnesota
Minneapolis, Minnesota

Akshya Swain
Department of Electrical and Computer
    Engineering
University of Auckland
Auckland, New Zealand

Joseph J. Talghader
Department of Electrical and Computer
    Engineering
University of Minnesota
Minneapolis, Minnesota

G. Velasco-Quesada
Department of Electronic Engineering
Universitat Politècnica de Catalunya,
    BarcelonaTech (UPC)
Barcelona, Spain

Tao Wang
The City College of New York
The City University of New York
New York, New York

Huaxiang Yi
State Key Laboratory of Advanced Optical
    Communication Systems and Networks
School of Electronics Engineering and
    Computer Science
Peking University
Beijing, China

Zhiping Zhou
State Key Laboratory of Advanced Optical
    Communication Systems and Networks
School of Electronics Engineering and
    Computer Science
Peking University
Beijing, China

Zhigang Zhu
The City College of New York
The City University of New York
New York, New York

Ander Zornoza
Department of Electrical and Electronic
    Engineering
Public University of Navarra
Pamplona, Spain

Part I

Photonic and Optoelectronics Sensors

1 Optical Fiber Sensors
Devices and Techniques

*Rogério Nunes Nogueira, Lúcia Maria Botas Bilro,
Nélia Jordão Alberto, Hugo Filipe Teixeira Lima,
and João de Lemos Pinto*

CONTENTS

1.1 INTRODUCTION

The research on optical fiber sensors produced and still continues to give life to a variety of measurement techniques for different applications. The technology is now in a mature state, with different applications already using commercial optical fiber sensors as a standard. This includes not only massive deployment for real-time structural health monitoring in airspace, civil, and oil industry but also more specific applications, such as environment monitoring, biochemical analyses, or gas leak monitoring in hazardous environments.

The success of this technology relies on the intrinsic flexibility, low weight, immunity to electromagnetic interference, passive operation, and high dynamic range, associated with remote monitoring and multiplexing capabilities, which allow these optical fiber sensors to succeed in difficult measurement situations where conventional sensors fail.

Optical fiber sensors operate by modifying one or more properties of the light passing through the sensor, when the parameter to be measured changes. An interrogation scheme is then used to evaluate the changes in the optical signal by converting them to a signal that can be interpreted. In this way, depending on the light property that is modified, optical fiber sensors can be divided into three main categories: intensity-, wavelength-, and phase-based sensors.

1.2   INTENSITY-BASED SENSORS

Intensity-based sensors represent one of the earliest and perhaps the simplest type of optical fiber sensors. They offer the advantages of ease of fabrication, low price–performance ratio, and simplicity of signal processing. These make them highly attractive, particularly in applications where the cost of implementation frequently excluded more expensive optical fiber systems. Grating and interferometric sensors allow high-resolution measurements, but this is not always necessary and less costly intensity-based sensing methods may offer an option in industry.

A wide number of intensity-based sensors are being presented and developed using different schemes; still they can be grouped in two major classes: intrinsic and extrinsic. In the extrinsic sensors, the optical fiber is used as a means to transport light to an external sensing system. In the intrinsic scheme, the light does not leave the optical fiber to perform the sensing function. Here, the fiber plays an active role in the sensing function, and this may involve the modification of the optical fiber structure.

1.2.1   TRANSMISSION AND REFLECTION SCHEMES

An additional classification scheme is related to the way the optical signal is collected. If the receiver and emitter are in opposite ends of the fiber(s), the sensor is of a transmission kind, otherwise it is of a reflection kind. An example of the first situation is the dependence of the power transmitted from one fiber to another on their separation [1]. With respect to reflection, most methods use reflecting surfaces to couple the light again in the fiber [2]. Other sensors are based on Fresnel reflection mechanisms [3,4] or special geometries of the fiber tip [5].

Intensity variation detection can also be performed through bending. If the bend radius is reduced below a critical value, the loss in the transmitted signal increases very rapidly, allowing the construction of a macrobending optical fiber sensor. These devices can be used to measure parameters such as deformation [6], pressure [7], and temperature [8].

Among the transmission and reflection systems reported, there are several transduction mechanisms, namely, evanescent wave–based sensors and spectrally based sensors.

1.2.2   SPECTRALLY BASED SENSORS

Spectroscopic detection has been a reliable method for the design of optical fiber sensors and is popularly used for chemical, biological, and biochemical sensing [9]. When a properly designed sensor reacts to changes in a physical quantity like refractive index (RI), absorption, or fluorescence intensity, a simple change of optical signal can be correlated to the concentration of a measurand [10].

Generally, the design of the sensors can simply comprise optical fibers with a sample cell, for direct spectroscopic measurements, or be configured as fiber optrodes, where a chemical selective layer comprising chemical reagents in suitable immobilizing matrices is deposited onto the optical fiber (Figure 1.1).

In its simplest form, the technique involves confining a sample between two fibers and the quantification of the light transmitted through the sample [11,12]. The fiber can play an active role acting as a sensing probe. The activation can be accomplished replacing the original cladding material, on a small section or end of the fiber, by a chemical agent or an environmentally sensitive material. There are a wide number of sensors that make use of this technique. Goicoechea et al. [13], using the reflection method, developed an optical fiber pH sensor based on the indicator neutral red. Regarding humidity sensing, most spectroscopic-based configurations are based on moisture-sensitive reagents (such as cobalt chloride, cobalt oxide, rhodamine B) attached to the tip of the fiber, usually with the aid of a polymeric material to form the supporting matrix [10]. Also, SiO_2 nanoparticles were pointed as a possible humidity reagent, since they are superhydrophilic [14].

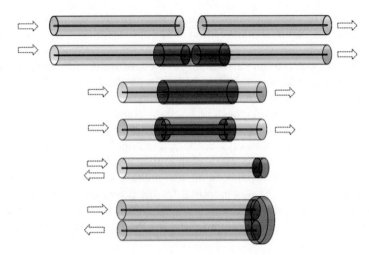

FIGURE 1.1   Schematic diagrams of different sensing methods for spectrally based sensors.

In the case of fluorescent materials, the properties that can be used for sensing are intensity, decay time, anisotropy, quenching efficiency, and luminescence energy transfer [15]. A common measurand in fluorescent methods is oxygen using mainly Ru [16] and Pt complexes [17]. These dye complexes are easily excited using low-cost light emission diode (LED) light sources, and their phosphorescence wavelengths are well separated from the excitation LED wavelength. A glucose sensor was designed by Scully et al. [18], using the oxygen-consuming enzymatic conversion of glucose to gluconic acid. In order to compensate temperature-induced variations in the luminescence intensity, Lo et al. [17] proposed a variable attenuator design using a negatively thermal expansion material as a temperature compensation method for gaseous oxygen measurements. On the other hand, Ganesh and Radhakrishnan [19] coated the sensitive area with black silicone, avoiding interferences due to changes in RI, turbidity, ambient light, and background fluorescence.

1.2.3   EVANESCENT WAVE–BASED SENSORS

Aside from the large portion of light guided in the core, there is a small component, known as the evanescent field, which decays exponentially away from the core surface. The evanescent field strength is a function of discontinuities of the interface, refractive indices, launching angle of the light beam, and dimensions of the fiber core. The methods used to increase the evanescent field strength are total or partial removal of the cladding, tapering, or side polishing [20].

The three main methods of removing the fiber cladding are thermal tapering, side polishing, and chemical etching. The chemical etching method is, both technically and economically, advantageous over the first two. This method allows a uniform production of long (up to meters) sections of fiber. A recent method comprises also the use of commercial CO_2 laser systems for rapid construction of discontinuities in silica and polymer fibers [20].

With tapered fibers, the evanescent field not only extends beyond the cladding, but its magnitude is also enhanced in that tapered region. When a liquid medium is placed at the tapered region, mode coupling changes the magnitude of the evanescent field, which can be monitored through sensor output. The evanescent field strength is also determined by the diameter and the taper geometry. Regarding tapers as biosensors, Souza et al. [21] performed a chemical treatment of a taper allowing the attachment of the covalent protein (isolated from *Staphylococcus aureus* cell surface). It is common to combine the use of tapers and surface plasmon resonance (SPR) methods. Leung et al. [22] presented a biosensor that consists of a taper coated with gold and is housed in a flow cell. For the detection of single-stranded DNA (ssDNA), authors showed that it was feasible to directly detect the hybridization of ssDNA to its complementary strand immobilized on the sensor surface. They also presented a similar study for the continuous detection of various concentrations of bovine serum albumin (BSA) and of the target BSA in the presence of a contaminating protein, ovalbumin (OVA) [23].

Side polishing with core exposure also enhances the sensitivity of an optical fiber sensor to a certain physical parameter. This method is valuable for physical, chemical, and biological sensing. With respect to physical devices, the most frequent is the curvature sensors. Bilro et al. [24] presented a curvature sensor for development of a wearable and wireless system to quantitatively evaluate the human gait. Regarding sensors based on RI variations at the polished interface, there are some interesting applied works, namely, a multipoint liquid level measurement sensor [25], spectroelectrochemical characterization [26], and resin cure monitoring [11]. Another variant of the method is the incorporation of SPR technique. An immunosensor using gold-coated side polished optical fiber was proposed for the detection of *Legionella pneumophila* with 850 nm LED and halogen light source–sensing system [27]. Another study presented an optical fiber ammonia sensor using ZnO nanostructure grown on the side polished section [28].

1.2.4 SELF-REFERENCE TECHNIQUES

The major disadvantage of intensity-based optical fiber sensors is the stability of light sources. To overcome this problem, several referencing techniques can be used. The simplest form uses an optical element that splits the optical signal in different paths that are equally affected by power fluctuations (ratiometric method). Another self-referencing method is given by Lo et al. [17] that used two different fluorescent indicators immobilized in the same matrix: one designed for sensing and the other for reference. Another practice consists in the transmission of multiwavelengths, where only one signal is attenuated, or excites the fluorescent dye, according to the measurand quantity [29].

1.3 PHASE-BASED SENSORS

Phase-modulated sensors, also known as interferometric sensors, offer sensitivities high as 10^{-13} m by comparing the phase difference of coherent light traveling along two different paths. The light is provided by a coherent laser source and is generally injected in two single-mode fibers and recombined latter. If one fiber is perturbed relative to the other, a phase shift occurs and can be detected with high precision using an interferometer.

1.3.1 PHASE DETECTION

When a lightwave of a given wavelength λ propagates inside an optical fiber of length L, the phase angle Φ at the end of the fiber is given by

$$\Phi = \frac{2\pi L}{\lambda} = \frac{2\pi n_{\text{core}}}{\lambda_0} \tag{1.1}$$

where
  n_{core} is the RI of the fiber core
  λ_0 is the wavelength of the light in vacuum

If an external perturbation causes a change in the RI (Δn_{core}) or in the length of the fiber, a phase change occurs and can be defined by

$$\Delta\Phi = \frac{2\pi}{\lambda}(n_{\text{core}}\Delta L + L\Delta n_{\text{core}}) \tag{1.2}$$

Assuming the RI remains constant, a length variation will induce a phase change ($\Delta\Phi$) of

$$\Delta\Phi = \frac{2\pi}{\lambda} n_{\text{core}} \Delta L \qquad (1.3)$$

When a lightwave is injected into two equal single-mode fibers, the power is split, but the phase remains the same. If the two optical fibers experience the same conditions, the lightwaves will recombine at the same phase angle and constructive interference will occur, giving the maximum intensity output. However, if the fibers are subject to different thermal or mechanical strains, they will recombine with a phase difference proportional to the different lengths the lightwaves traveled, and destructive interference will occur, causing the output intensity to decrease.

Some interferometer configurations that can be used to detect the phase shifts are Mach–Zehnder, Michelson, Fabry–Perot, and Sagnac interferometers. These will now be discussed.

1.3.2 MACH–ZEHNDER

The Mach–Zehnder interferometer (Figure 1.2a) consists of two 3 dB couplers, separated by two optical fibers, where one is the reference and the other is the sensing fiber. In the first coupler, 50% of the power is injected to each fiber, and the fibers recombine at the second coupler. If both fiber lengths are the same, or differ by an integral number of wavelengths, the lightwaves will recombine in exact phase and the output intensity will be at its maximum value. However, if the fiber lengths differ by ½ wavelength, the recombined beams will be in opposite phase and the output intensity will be minimum.

1.3.3 MICHELSON

The Michelson interferometer configuration may be considered as an inverted over itself version of the Mach–Zehnder interferometer. This configuration (Figure 1.2b) uses a single optical fiber coupler and fibers with mirrored ends that back reflect the laser beams, which are recombining at the coupler and directed to the detector.

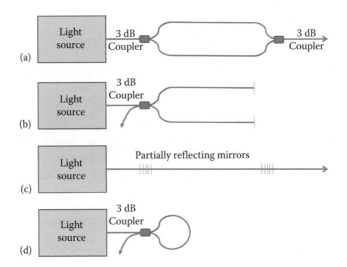

FIGURE 1.2 Illustration of the four main interferometer configurations: (a) Mach–Zehnder, (b) Michelson, (c) Fabry–Perot, and (d) Sagnac.

1.3.4 FABRY–PEROT

The Fabry–Perot interferometer differs from the previously presented interferometers by not requiring a reference fiber. In this configuration (Figure 1.2c), the fiber presents two partially reflective mirrors. The partially transmitting mirrors cause the light to travel multiple passes inside the cavity transmitted at the second mirror and reaching the detector, which magnifies the phase difference, doubling the sensitivity to phase differences when compared to other interferometer configurations.

1.3.5 SAGNAC

The Sagnac interferometer uses counterpropagating light beams in a ring path. This is achieved by connecting the laser source into a 3 dB optical fiber coupler injecting light into both ends of the same optical fiber in a coiled configuration (Figure 1.2d). This causes light to travel along the fiber in both directions, and both directions are the sensing fibers. If the fiber coil is rotated in an axis perpendicular to the coil plane, the light propagation time in one direction will be shortened, while the light in the other path requires more time because light will need to travel a longer distance, resulting in a relative phase shift (Sagnac effect). On the other hand, if the coil is kept stationary, light travels the same distance in both directions and no phase shift occurs. This configuration allows to measure rotation with high precision, such as the rotation of earth around its axis.

1.4 WAVELENGTH-BASED SENSORS

Most wavelength-modulated sensors are based on fiber Bragg gratings (FBGs). An FBG can be described as a periodic modulation of the RI of the fiber core. It is generally obtained when a photosensitive optical fiber is exposed to an UV radiation in a periodic pattern. For sensor applications, the typical modulation of the RI is $\Delta n \approx 10^{-4}$ with a few millimeters length.

When an FBG is illuminated by a broadband light source, the spectral component that satisfies the Bragg condition is reflected by the grating. In the transmission spectrum, this component is missing (Figure 1.3). The Bragg condition is given by the following expression [30]:

$$\lambda_B = 2n_{\text{eff,core}}\Lambda \tag{1.4}$$

where
λ_B is the wavelength of the back-reflected light (Bragg wavelength)
$n_{\text{eff,core}}$ is the RI of the core
Λ is the periodicity of the RI modulation ($\approx 0.5 \, \mu m$)

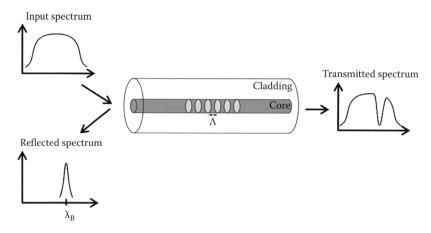

FIGURE 1.3 Schematic representation of an FBG.

Equation (1.4) denotes that the signal reflected by the grating is dependent on the FBG physical parameters. When the grating is subjected to mechanical deformation or temperature variation, a shift in wavelength of the Bragg signal is obtained:

$$\Delta\lambda_B = 2\left(\Lambda\frac{\partial n_{eff,core}}{\partial\varepsilon} + n_{eff,core}\frac{\partial\Lambda}{\partial\varepsilon}\right)\Delta\varepsilon + 2\left(\Lambda\frac{\partial n_{eff,core}}{\partial T} + n_{eff,core}\frac{\partial\Lambda}{\partial T}\right)\Delta T \qquad (1.5)$$

The first term in Equation 1.5 represents the mechanical perturbation on the Bragg wavelength given by the alteration of the grating pitch and the changes in the RI caused by the strain-optic effect.

The second term on Equation 1.5 is related to the thermal perturbation on the Bragg wavelength. The resulting wavelength shift is caused by changes in the grating pitch, due to thermal expansion, and by changes in the RI. For a Bragg grating written in a germanium-doped optical fiber, strain and thermal sensitivities of 1.2 pm/µε and 13.7 pm/°C are expected, respectively [31]. Although the main application of FBGs as sensors is in temperature and strain monitoring, it is possible to monitor other parameters, such as pressure, acceleration, or the presence of certain chemicals, by adapting the FBG to a structure that can translate these parameters into changes of temperature and/or strain. Reducing the diameter of the fiber, it becomes sensitive to the environmental medium, and RI can be monitored [32]. Other designs of optical fiber sensors, as for instance tilted FBGs (TFBGs) and long-period gratings (LPGs), can also be used in RI sensing. TFBGs are FBGs where the RI modulation is purposely tilted to the fiber axis in order to enhance the coupling of the light from the forward propagating core mode to cladding modes [33]. The reflected Bragg wavelength λ_{TFBG} and the cladding mode resonances λ^i_{clad} are determined by the phase-matching condition through the following equations:

$$\lambda_{TFBG} = \frac{2n_{eff,core}\Lambda}{\cos\theta}; \quad \lambda^i_{clad} = \frac{\left(n^i_{eff,core} + n^i_{eff,clad}\right)\Lambda}{\cos\theta} \qquad (1.6)$$

where
$n_{eff,core}$ is the effective index of the core mode at λ_{TFBG}
$n^i_{eff,core}$ is the effective index of the core mode at λ^i_{clad}
$n^i_{eff,clad}$ is the effective index of the ith cladding mode at λ^i_{clad}

The grating period along the axis of the fiber, Λ_{TFBG}, is given by $\Lambda_{TFBG} = \Lambda/\cos\theta$, where θ is the tilt angle. The cladding modes attenuate rapidly, being only observed in the transmission spectrum as numerous resonances (Figure 1.4).

In an LPG, the RI is periodically modulated to produce a grating; the periodicity of this modulation is typically in the range 100 µm to 1 mm, instead of ≈0.5 µm as in FBGs. Normally, the grating length is between 2 and 4 cm. The small grating wave vector, $2\pi/\Lambda_{LPG}$, where Λ_{LPG} is the periodicity of the RI modulation, promotes the coupling of light from the guided fundamental core mode (the LP_{01} mode) to different forward-propagating cladding modes (LP_{0i} mode with $i = 2, 3, 4, ...$). The cladding modes are quickly attenuated as they propagate along the fiber axis, due to scattering

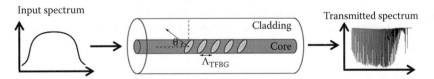

FIGURE 1.4 Schematic representation of a TFBG.

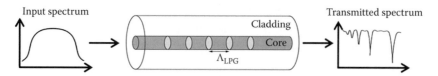

FIGURE 1.5   Schematic representation of an LPG.

losses at the cladding–air interface [34]. As a result, the transmission spectrum of an LPG has several loss bands (Figure 1.5), at different wavelengths, given by

$$\lambda_{LPG,i} = \left(n_{\text{eff,core}} - n^i_{\text{eff,clad}} \right) \Lambda_{LPG} \tag{1.7}$$

where
λ_i is the coupling wavelength
$n_{\text{eff,core}}$ is the effective index of the core mode
$n^i_{\text{eff,clad}}$ is the RI of the ith cladding mode

The most typical techniques to produce LPGs include the use of UV lasers, CO_2 lasers, electrical discharges, and irradiation by femtosecond pulses in the infrared. However, it is also possible to obtain LPGs by diffusion of dopants into the fiber core, ion implantation, and deformation of the fiber [35].

In both LPGs and TFBGs, the attenuation bands obtained in the transmission spectrum have been widely explored in the development of sensors, since they are sensitive to a range of parameters. The response of the resonances to the measurand is dependent on the cladding mode order. This feature offers the possibility of development of multiparameter sensing systems, using a single sensor element [36,37].

All gratings that were introduced before were written in silica fibers. However, a major drawback of using silica is that only 3% of elongation is achieved. This fact motivated the scientific community to think in new fiber for FBG inscription, namely, in polymer optical fiber (POF). The first Bragg grating in step-index fiber was reported in 1999 [38] and the first in microstructured POF only in 2005 [39]. The properties of POF are quite different to those of silica and offer some significant potential advantages. The Young's modulus of poly[methyl methacrylate] (PMMA) is ∼25 times less than that of silica and POF can undergo much higher strains. The most common method for FBG inscription in POF is the phase mask method using a 325 nm HeCd laser [40].

1.4.1 Multiparameter Sensors

Multiparameter measurement using fiber sensors is a challenging topic, while it enables the minimization of size, cost, and complexity of sensing systems. FBG sensors are intrinsically sensitive to strain and temperature; however, they present cross-sensitivity between these two parameters, because changes in both the parameters are encoded in the peak wavelength shift. The majority of FBG applications handle this cross-sensitivity using a pair of FBGs, where one grating is isolated from strain, measuring only temperature variations, while the other FBG measures both strain and temperature. However, the first multiparameter sensing head, proposed by Xu et al. [41] in 1994, used a different principle of operation. The reported scheme (Figure 1.6) was based on two superimposed FBGs written at two very distant wavelengths, 850 and 1300 nm, and explored the wavelength dependence of the photoelastic and thermo-optic coefficients to obtain four different strain and temperature coefficients.

The concept of using different strain and temperature sensitivities was then widely explored. In 1996, James et al. [42] spliced two FBGs written in fibers with different diameters achieving distinct

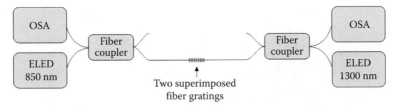

FIGURE 1.6 Experimental setup for strain–temperature discrimination using a dual-wavelength FBG.

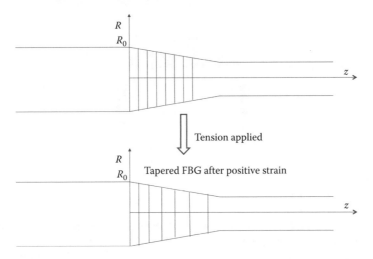

FIGURE 1.7 Illustration of a tapered FBG (up) and tapered FBG after positive strain (down).

strain coefficients. The distinct temperature sensitivities of FBG sensors written in the splice region of fibers with different dopants was later presented [43]. Also, a sensing head was accomplished based on fibers doped with different concentrations [44] or formed by splicing different FBG types [45].

In 2010, Lima et al. [46] designed a sensing head and presented the necessary interrogation parameters to perform strain and temperature discrimination. They used a single FBG written in an optical fiber taper with a linear diameter variation, as represented in Figure 1.7. When subjected to tension and due to the different cross sections of the fiber along its length, different values of strain arise, causing the broadening of the FBG signal and allowing the use of the information contained in both peak wavelength and spectral width.

Using a single TFBG, Chehura et al. [47] discriminated these two parameters based on the fact that core mode resonance and cladding mode resonance exhibit different thermal sensitivities but approximately equal strain sensitivities. The application of Hi-Bi fibers and more complex configurations based on Fabry–Perot cavities with FBG mirrors [48], sampled FBGs [49], superstructured FBGs [50], or the combination of single-mode/multimode fibers [51], holey fibers [52], and photonic crystal fibers [53], as well as many other sensing configurations involving fiber grating devices, were also proposed.

Beyond temperature and strain discrimination, research on FBG sensors evolved toward measurement of other parameters, namely, displacement/temperature [54], pressure/temperature [55], transverse load/temperature [56], stress/cracks [57], and RI/strain [58].

1.4.2 FBG INSCRIPTION METHODS

The most common FBG inscription methods include the phase mask, interferometric, and point-by-point techniques. In the first, a phase mask is used; this is an optical diffractive device that is designed to diffract light under normal incidence. The superposition of the diffracted light occurs

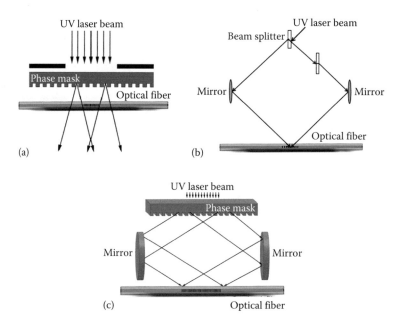

FIGURE 1.8 Schematic representation of grating inscription techniques: (a) phase mask, (b) interferometric with amplitude beam splitter, and (c) interferometric with phase mask.

in front of the phase mask, where the photosensitive optical fiber needs to be placed, creating an interference pattern on the core of the optical fiber that will be responsible for the RI modulation (Figure 1.8a). In the interferometric method, generically, the UV beam is split into two beams and recombined in the core of the fiber after being reflected by two UV mirrors, creating an interference pattern. Adjusting the angle of incidence, by changing the angle of the mirrors, is possible to control the wavelength of the inscribed grating, from wavelengths close to the UV source to virtually infinity. The separation of the beams is usually made with an amplitude beam splitter (Figure 1.8b) or a phase mask (Figure 1.8c).

In the point-by-point method, each grating plane is produced individually by a focused single pulse from a UV laser. After the first grating plane inscribed, the fiber is then translated and a new point is written. The distance between two planes corresponds to the grating pitch. More information regarding the FBGs inscription methods can be found in [31].

1.4.3 INTERROGATION OF FBG SENSORS

The principle of operation of a wavelength-based sensor system consists in monitoring the resonance wavelength shift caused by a perturbation. An interrogation system is required to measure this change.

A common setup to interrogate FBGs, especially in laboratory environment, is composed by a broadband light source, such as a superluminescent LED, an optical spectrum analyzer (OSA), and an optical circulator or optical coupler, as shown in the scheme of Figure 1.9. However, conventional spectrometers/OSAs have typical resolutions of 0.1 nm, so they are normally used for inspection of the optical properties of the FBGs rather than for high precision wavelength shift detection.

Agile tunable lasers and simple photodiode detectors allow obtaining the transmission and reflection spectrum of an FBG, offering an improvement of several orders of magnitude in both output power and signal linewidth, when compared to broadband sources and spectrometers. These advantages are recognized for some time, but the high cost of these systems has discouraged their use.

One alternative approach to interrogate an FBG signal is based on the use of a tunable passband filter. This is used to scan the wavelength range of interest, where the FBG signal is located and the

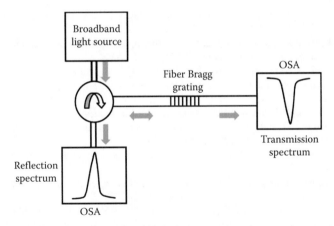

FIGURE 1.9 Schematic representation of an FBG interrogation setup.

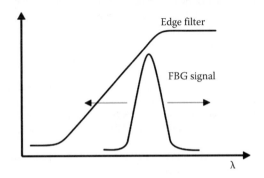

FIGURE 1.10 Principle of operation of the edge filter interrogation method.

output is the convolution of both the spectrum of the FBG and the tunable filter. When the spectra of both tunable filter and FBG match, the maximum output occurs. This method has a relatively high resolution over a large working range, allowing to interrogate, simultaneously, several FBGs. The resolution is mainly dependent on the linewidth of the tunable filter and on the signal-to-noise ratio of the FBG signal. The most common tunable filters used in this configuration are Fabry–Perot filters, acousto-optic filters, and FBG-based filters.

The methods described earlier can be used for both analysis of the optical properties of FBGs and wavelength shift detection, since they allow to obtain the complete spectrum of one or more FBGs, from which the peak wavelengths are easily obtained. The edge filter technique only permits the identification of the peak wavelength of one FBG but is a cost-efficient method when high sampling rates are demanded. This is based on the use of an edge filter with a linear dependence between wavelength shifts and the output intensity variations of the filter. Using this setup, as represented in Figure 1.10, a wavelength change can be converted to an intensity change, easily detected using a photodetector.

REFERENCES

1. Kuang, K., Quek, S., and Maalej, M., Assessment of an extrinsic polymer-based optical fibre sensor for structural health monitoring, *Measurement Science and Technology*, 15; 2133–2141; 2004.
2. Binu, S., Pillai, V., Pradeepkumar, V., Padhy, B., Joseph, C., and Chandrasekaran, N., Fibre optic glucose sensor, *Materials Science and Engineering C*, 29(1); 183–186; 2009.
3. Chen, J. and Huang, X., Fresnel-reflection-based fiber sensor for on-line measurement of ambient temperature, *Optics Communications*, 283(9); 1674–1677; 2010.

4. Zhou, A., Liu, Z., and Yuan, L., Fiber-optic dipping liquid analyzer: Theoretical and experimental study of light transmission, *Applied Optics*, 48(36); 6928–6933; 2009.

5. Nath, P., Singh, H., Datta, P., and Sarma, K., All-fiber optic sensor for measurement of liquid refractive index, *Sensors and Actuators A: Physical*, 148(1); 16–18; 2008.

6. Vijayan, A., Gawli, S., Kulkarni, A., Karekar, R., and Aiyer, R., An optical fiber weighing sensor based on bending, *Measurement Science and Technology*, 19; 105302(8pp.); 2008.

7. Regez, B., Sayeh, M., Mahajan, A., and Figueroa, F., A novel fiber optics based method to measure very low strains in large scale infrastructures, *Measurement*, 42(2); 183–188; 2009.

8. Rajan, G., Semenova, Y., and Farrell, G., All-fibre temperature sensor based on macro-bend single mode fibre loop, *Electronics Letters*, 44(19); 1123–1124; 2008.

9. McDonagh, C., Burke, C., and MacCraith, B., Optical chemical sensors, *Chemical Reviews*, 108(2); 400–422; 2008.

10. Yeo, T., Sun, T., and Grattan, K., Fibre-optic sensor technologies for humidity and moisture measurement, *Sensors and Actuators A: Physical*, 144(2); 280–295; 2008.

11. Bilro, L., Alberto, N., Pinto, J., and Nogueira, R., Simple and low-cost cure monitoring system based on side-polished plastic optical fibre, *Measurement Science and Technology*, 21(11); 117001; 2010.

12. Yokota, M., Okada, T., and Yamaguchi, I., An optical sensor for analysis of soil nutrients by using LED light sources, *Measurement Science and Technology*, 18(7); 2197–2201; 2007.

13. Goicoechea, J., Zamarreno, C., Matias, I., and Arregui, F., Optical fibre pH sensor based on layer-by-layer electrostatic self-assembled Neutral Red, *Sensors and Actuators B*, 132(1); 305–311; 2008.

14. Corres, J., Matias, I., Hernaez, M., Bravo, J., and Arregui, F., Optical fiber humidity sensors using nano-structured coatings of SiO_2 nanoparticles, *IEEE Sensors Journal*, 8(3); 281–285; 2008.

15. Borisov, S. and Wolfbeis, O., Optical biosensors, *Chemical Reviews*, 108; 423–461; 2008.

16. Chu, F., Yang, J., Cai, H., Qu, R., and Fang, Z., Characterization of a dissolved oxygen sensor made of plastic optical fibre coated with ruthenium-incorporated sol gel, *Applied Optics*, 48; 338–342; 2009.

17. Lo, Y.-L., Chu, C.-H., Yur, J., and Chang, Y., Temperature compensation of fluorescence intensity-based fiber-optic oxygen sensors using modified Stern–Volmer model, *Sensors and Actuators B: Chemical*, 131(2); 479–488; 2008.

18. Scully, P., Betancor, L., Bolyo, J., Dzyadevych, S., Guisan, J., Fernández-Lafuente, R., Jaffrezic-Renault, N., Kuncová, G., Matějec, V., O'Kennedy, B., Podrazky, O., Rose, K., Sasek, L., and Young, J., Optical fibre biosensors using enzymatic transducers to monitor glucose, *Measurement Science and Technology*, 18(10); 3177–3186; 2007.

19. Ganesh, A. and Radhakrishnan, T., Fiber-optic sensors for the estimation of oxygen gradients within biofilms on metals, *Optics and Lasers in Engineering*, 46; 321–327; 2008.

20. Irawan, R., Chuan, T., Meng, T., and Ming, T., Rapid constructions of microstructures for optical fiber sensors using a commercial CO_2 laser system, *Biomedical Engineering Journal*, 2; 28–35; 2008.

21. Souza, N., Beres, C., Yugue, E., Carvalho, C., Neto, J., Silva, M., Werneck, M., and Miguel, M., Development of a biosensor based in polymeric optical fiber to detect cells in water and fluids, *Proceedings of the 18th International Conference on Plastic Optical Fibre*, Sydney, Australia; 4pp.; 2009.

22. Leung, A., Shankar, P., and Mutharasan, R., Label-free detection of DNA hybridization using gold-coated tapered fiber optic biosensors (TFOBS) in a flow cell at 1310 nm and 1550 nm, *Sensors and Actuators B: Chemical*, 131(2); 640–645; 2008.

23. Leung, A., Shankar, P., and Mutharasan, R., Model protein detection using antibody-immobilized tapered fiber optic biosensors (TFOBS) in a flow cell at 1310 nm and 1550 nm, *Sensors and Actuators B: Chemical*, 129(2); 716–725; 2008.

24. Bilro, L., Oliveira, J., Pinto, J., and Nogueira, R., Gait monitoring with a wearable plastic optical sensor, *Proceedings of IEEE Sensors Conference*, Lecce, Italy; pp. 787–790; 2008.

25. Lomer, M., Arrue, J., Jauregui, C., Aiestaran, P., Zubia, J., and Lopez-Higuera, J., Lateral polishing of bends in plastic optical fibres applied to a multipoint liquid-level measurement sensor, *Sensors and Actuators A: Physical*, 137; 68–73; 2007.

26. Beam, B., Armstrong, N., and Mendes, S., An electroactive fiber optic chip for spectroelectrochemical characterization of ultra-thin redox-active films, *Analyst*, 134; 454–459; 2009.

27. Lin, H.-Y., Tsao, Y.-C., Tsai, W.-H., Yang, Y.-W., Yan, T.-R., and Sheu, B.-C., Development and application of side-polished fiber immunosensor based on surface plasmon resonance for the detection of *Legionella pneumophila* with halogens light and 850 nm-LED, *Sensors and Actuators A: Physical*, 138(2); 299–305; 2007.

28. Dikovska, A., Atanasova, G., Nedyalkov, N., Stefanov, P., Atanasov, P., Karakoleva, E., and Andreev, A., Optical sensing of ammonia using ZnO nanostructure grown on a side-polished optical-fiber, *Sensors and Actuators B: Chemical*, 146(1); 331–336; 2010.

29. Benjamin, V., Satish, J., and Madhusoodanan, N., Fiber optic sensor for the measurement of concentration of silica in water with dual wavelength probing, *Reviews of Scientific Instruments*, 81; 035111(5pp.); (2010).

30. Kersey, A., Davis, M., Patrick, H., LeBranc, M., Koo, K., Askins, C., Putnam, M., and Friebele, E., Fiber grating sensors, *Journal of Lightwave Technology*, 15(8); 1442–1463; 1997.

31. Othonos, A., Fiber Bragg gratings, *Review of Scientific Instruments*, 68(12); 4309–4341; 1997.

32. Ladicicco, A., Cusano, A., Campopiano, S., Cutolo, A., and Giordano, M., Thinned fiber Bragg gratings as refractive index sensors, *IEEE Sensors Journal*, 5(6); 1288–1295; 2005.

33. Erdogan, T. and Sipe, J., Tilted fiber phase gratings, *Journal of the Optical Society of America*, 13(2); 296–313; 1996.

34. Vengsarkar, A., Lemaire, P., Judkins, J., Bhatia, V., Erdogan, T., and Sipe, J., Long-period fiber gratings as band-rejection filters, *Journal of Lightwave Technology*, 14(1); 58–65; 1996.

35. James, S. and Tatam, R., Optical fibre long-period grating sensors: Characteristics and applications, *Measurement and Science Technology*, 14; R49–R61; 2003.

36. Alberto, N., Marques, C. A. F., Pinto, J. L., and Nogueira, R. N., Three-parameter optical fiber sensor based on a tilted fiber Bragg grating, *Applied Optics*, 49(31); 6085–6091; 2010.

37. Kang, J., Dong, X., Zhao, C., Qian, W., and Li, M., Simultaneous measurement of strain and temperature with a long-period fiber grating inscribed Sagnac interferometer, *Optics Communications*, 284(15); 2145–2148; 2011.

38. Xiong, Z., Peng, G., Wu, B., and Chu, P., Highly tunable Bragg gratings in single-mode polymer optical fibers, *IEEE Photonics Technology Letters*, 11(3); 352–354; 1999.

39. Dobb, H., Webb, D., Kalli, K., Argyros, A., Large, M., and Eijkelenborg, M., Continuous wave ultraviolet light-induced fiber Bragg gratings in few and single-mode microstructured polymer optical fibers, *Optics Letters*, 30(24); 3296–3298; 2005.

40. Chen, X., Zhang, C., Webb, D., Peng, G.-D., and Kalli, K., Bragg grating in a polymer optical fibre for strain, bend and temperature sensing, *Measurement and Science Technology*, 21; 094005(5pp.); 2010.

41. Xu, M., Archambault, J., Reekie, L., and Dakin, J., Discrimination between strain and temperature effects using dual-wavelength fibre grating sensors, *Electronics Letters*, 30(13); 1085–1087; 1994.

42. James, S., Dockney, M., and Tatam, R., Simultaneous independent temperature and strain measurement using in-fibre Bragg grating sensors, *Electronics Letters*, 32(12); 1133–1134; 1996.

43. Cavaleiro, P., Araújo, F., Ferreira, L., Santos, J., and Farahi, F., Simultaneous measurement of strain and temperature using Bragg gratings written in germanosilicate and boron-codoped germanosilicate fibers, *IEEE Photonics Technology Letters*, 11(12); 1635–1637; 1999.

44. Frazão, O. and Santos, J., Simultaneous measurement of strain and temperature using a Bragg grating structure written in germanosilicate fibre, *Journal of Optics A—Pure and Applied Optics*, 6(6); 553–556; 2004.

45. Shu, X., Liu, Y., Zhao, D., Gwandu, B., Floreani, F., Zhang, L., and Bennion, I., Dependence of temperature and strain coefficients on fiber grating type and its application to simultaneous temperature and strain measurement, *Optics Letters*, 27(9); 701–703; 2002.

46. Lima, H., Antunes, P., Nogueira, R., and Pinto, J., Simultaneous measurement of strain and temperature with a single fibre Bragg grating written in a tapered optical fibre, *IEEE Sensors Journal*, 10(2); 269–273; 2010.

47. Chehura, E., James, S., and Tatam, R., Temperature and strain discrimination using a single fibre Bragg grating, *Optics Communications*, 275(2); 344–347; 2007.

48. Du, W., Tao, X., and Tam, H., Fiber Bragg grating cavity sensor for simultaneous measurement of strain and temperature, *IEEE Photonics Technology Letters*, 11(1); 105–107; 1999.

49. Frazão, O., Romero, R., Rego, G., Marques, P., Salgado, H., and Santos, J., Sampled fibre Bragg grating sensors for simultaneous strain and temperature measurement, *Electronics Letters*, 38(14); 693–695; 2002.

50. Guan, B., Tam, H., Tao, X., and Dong, X., Simultaneous strain and temperature measurement using a superstructure fiber Bragg grating, *IEEE Photonics Technology Letters*, 12(6); 675–677; 2000.

51. Zhou, D., Wei, L., Liu, W., Liu, Y., and Lit, J., Simultaneous measurement for strain and temperature using fiber Bragg gratings and multimode fibers, *Applied Optics*, 47(10); 1668–1672; 2008.

52. Han, Y., Song, S., Kim, G., Lee, K., Lee, S., Lee, J., Jeong, C., Oh, C., and Kang, H., Simultaneous independent measurement of strain and temperature based on long-period fiber gratings inscribed in holey fibers depending on air-hole size, *Optics Letters*, 32(15); 2245–2247; 2007.

53. Ju, J. and Jin, W., Photonic crystal fiber sensors for strain and temperature measurement, *Journal of Sensors*, 2009; 476267(10pp.); 2009.
54. Dong, X., Liu, Y., and Liu, Z., Simultaneous displacement and temperature measurement with cantilever-based fiber Bragg grating sensor, *Optics Communications*, 192(3–6); 213–217; 2001.
55. Chen, G., Liu, L., Jia, H., Yu, J., Xu, L., and Wang, W., Simultaneous pressure and temperature measurement using Hi-Bi fiber Bragg gratings, *Optics Communications*, 228(1–3); 99–105; 2003.
56. Abe, I., Frazão, O., Kalinowski, H., Schiller, M., Nogueira, R., and Pinto, J., Characterization of Bragg gratings in normal and reduced diameter HiBi fibers, *International Microwave and Optoelectronics Conference*, 2; 887–891; 2003.
57. Watekar, P., Ju, S., and Han, W., A multi-parameter sensor system using concentric core optical fiber, *Optical and Quantum Electronics*, 40(7); 485–494; 2008.
58. Alberto, N., Marques, C., Pinto, J., and Nogueira, R., Simultaneous strain and refractive index sensor based on a TFBG, *Proceedings of SPIE*, 7653; 765324(1–4); 2010.

2 Microstructured and Solid Polymer Optical Fiber Sensors

Christian-Alexander Bunge and Hans Poisel

CONTENTS

2.1 OVERVIEW

Modern technical systems need a large number of sensors, which have to be versatile, simple to use, and immune to electromagnetic interferences. Optical sensors fulfill all these requirements and are being widely installed in increasing number. While many sensors rely on glass-optical fibers, which are mostly single mode and thus relatively delicate in terms of handling and operation, multimode fibers, especially robust, large-core polymer optical fibers (POF), are a cost-effective alternative. For quite a while, POF sensors have been about to enter the market, but only now they are considered seriously. Apparently, the first application of mass-produced POF sensors will be in automotive and industrial areas such as impact sensors or seat occupancy sensors.

In this chapter, we will present an overview on sensors based on POF. First, we will categorize the most common sensor types according to their fiber type and sensing effect. Here, we will differentiate between microstructured and solid POF sensors. Following this line of thought, we will then present several applications for microstructured POF (mPOF), their advantages, and application scenarios. We will then provide an overview on the principles for solid POF sensors and exemplary applications.

2.1.1 GENERAL REQUIREMENTS AND SENSING EFFECTS

There are many optical effects that can be used for sensing. Often those environmental influences, which are usually unwelcome side effects, are exploited for the generation of a sensing signal. In order to cover the magnitude of effects that can be used for sensing, we will organize them by an overview on the fundamental sensing principles.

Optical sensors may be classified in different ways:

- Intrinsic or extrinsic sensors
- Multimode or single-mode fiber sensors
- Parameters influenced, e.g., transmitted or reflected amplitude, phase, polarization, frequency, mode distribution, spectral distribution, etc.

Figure 2.1 provides a hierarchical structure of fiber optical sensors in general. They can be classified into glass and polymer fibers. Many glass-optical sensors are in use today. Most of them are based on single-mode fibers making use of interferometric or polarization effects in coherent light. These effects are well known and a lot of literature deals with this kind of sensors. We will therefore concentrate on polymer fiber sensors for the remainder of this chapter. Like in their glass counterparts, polymer fibers are available as solid fibers as well as microstructured with tiny holes in the cross section running along the whole length of the fiber. These holes alter the optical properties of the material they are embedded in and usually cause a decrease of the material's effective refractive index so that the material acts like being doped with air (see, e.g., [1]). Another approach is to form a crystal-like structure around the fiber core, in which energy bands develop like in semiconductors. These bands stand for allowed energy states for the photons so that a fiber can be constructed with energy states that are allowed in the core, but not in the cladding. Then, photons cannot help remaining in the core region. This effect is called band-gap guidance and can even be used for fibers

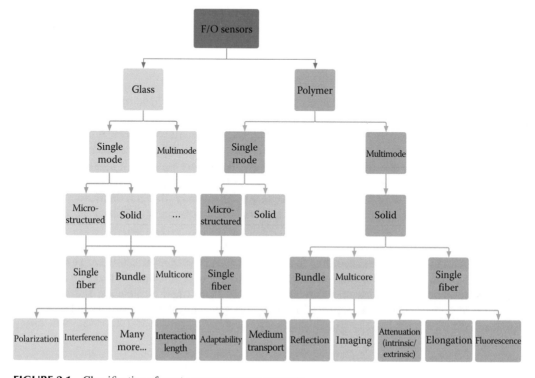

FIGURE 2.1   Classification of most common sensor concepts.

with air cores otherwise impossible to produce (see, e.g., [2]). This type of fibers offers the main advantage that the holes can be changed in terms of geometry (e.g., by pressure) or refractive index (e.g., by blowing in gases or liquids), or that they provide a means to bring the material to be studied closer to the light in order to interact.

Solid POFs are usually large in diameter and numerical aperture and thus massively multimode. There are also a couple of single-mode POF applications, but the main advantage of POF sensors is their robustness, ease of handling, and the easy reception of light (e.g., for reflectance measurements). One can also differentiate between single fibers, multicore fibers, and bundles. Most of the sensor effects themselves are based on single fibers, but these sensors can be easily extended or multiplexed by fiber bundles or in a more compact form as multicore structures. In this way, one can design distributed sensors that can sense at many different locations and bring their sensing signal to one centralized detector. Some effects, however, make direct use of fiber bundles or multicore fibers, e.g., one can use a bundle of fibers in order to receive the reflected light of a surface or use an additional fiber for calibration purposes. Solid POF sensors usually fall into several distinct categories. One can use their attenuation by a simple power measurement. Other sensors rely on refraction, which finally also leads to an increased attenuation. Since standard POF are multimode, one can also measure the modal power distribution and detect for instance bends that change the modal structure of the fiber. Although multimode, one can also use phase measurements for the detection of elongation or other mechanical changes of the fiber. In this case, not the optical phase can be measured, but a lower-frequency sine signal will be used and its phase can be compared with a reference signal.

In general, almost every physical parameter can be measured by an optical system, making use of the attractive properties inherent to these systems such as small dimensions, electromagnetic immunity, or high bandwidth. POF sensors offer cost-effective alternatives to otherwise ultraprecise, but expensive glass-optical sensors. In some cases, where a large diameter is needed, they provide the only viable solution for a fiber sensor.

2.2 SENSORS BASED ON mPOF

While standard polymer fibers provide many potential sensor applications that combine low cost and ruggedness, some well-known applications that require single-mode operation cannot be applied. While there is a solid single-mode POF available for a couple of years, they do not share POF main advantages. Especially, the widely used fiber gratings (both Bragg and Long Period [LPG]) provide simple high-resolution sensor applications. In addition, microstructured fibers open completely new possibilities for the detection of liquids or gases, which can be filled into the holes or have a much stronger interaction due to the largely increased surface area. While most these sensor principles can be realized with glass fibers as well, several applications forbid glass at all (all applications for which the fiber has to be put inside the body or several automotive applications due to vibrations). In the following, we will provide a short report on applications based on mPOF.

2.2.1 Mechanical Sensing with Fiber Using Gratings

Polymers have some advantageous mechanical properties that make mPOF better for mechanical sensing in some situations. Preliminary studies using mPOF sensors based on LPG [3] show that the use of polymer fiber increases the range of repeatable strain measurements by several times and the yield limit by an order of magnitude, compared to a silica-based sensor. Figure 2.2 shows an example for an LPG-based sensor. The viscoelastic properties of the polymer means there are time-dependent effects relating to strain rate and magnitude. These effects are small when the sensor is intermittently strained up to 2%, and are relatively small at strains of up to 4%–5%. Further testing is ongoing to characterize these effects at very high strains. The effect of stress relaxation has a small effect on the change in the wavelength of the loss features used in the measurement of strain.

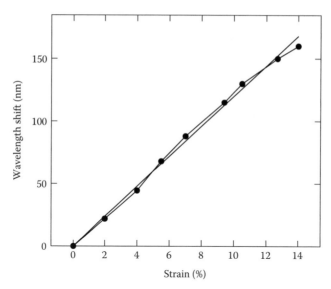

FIGURE 2.2 Strain results using an LPG in a mPOF, in which the strain removed rapidly after application. There is a small deviation from linearity at higher strains.

The use of high strains complicates the response of the sensor due to the viscoelastic properties that can experience a partly nonelastic deformation, requiring careful calibration. This technology is the only one that allows fiber strain sensors operating at strains of up to 30%–45%. These very high strain sensors are currently being developed.

2.2.2 SENSING OF FLUIDS

The fact that mPOFs have cladding holes, and may even have a hollow core (photonic band-gap fibers), opens up significant new opportunities for fluid sensing. Fluid sensing in conventional optical fibers is difficult because of their low refractive indices, which means guidance by total internal reflection (TIR) is impractical. The fluids have to be brought near the optical core of the fiber, which also provides some challenges [4].

Microstructured optical fibers offer two elegant solutions to this problem. In Figure 2.3, one can see a liquid-filled core and a photonic band-gap fiber, both of which have been experimentally demonstrated. A selectively filled core fiber can be used in conjunction with a high air fraction in the cladding

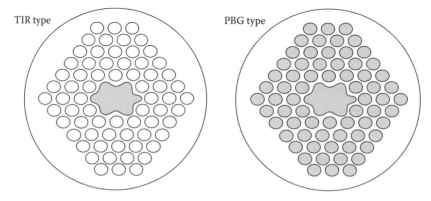

FIGURE 2.3 Liquid sensing is possible in liquid-filled mPOF: Open circles are air-filled; closed circles are liquid-filled. On the left a selectively filled core with total internal reflection with an air-filled cladding is used, while on the right a photonic band gap is used, in which both core and cladding holes are filled with liquid.

region, or fully filled band gap can be used [5]. In both cases of Figure 2.3, the microstructure confines both the light and the fluid, and critically allows almost a complete overlap of the optical field with the fluid, a dramatic improvement on evanescent field sensors that are the conventional fiber alternative. An alternative approach uses a microstructured core [6] though this relies on the evanescent field being maximized and produces a mode profile that makes efficient launching challenging. In both cases, the main benefit of the microstructure is the reduced interaction length that is needed because of the strong evanescent fields at the sensing region, which has been accomplished by the special field structure induced by the holes and the possibility to bring the fluid close to the waveguide.

2.2.3 Spectroscopy Using Evanescent-Field Interaction

Another use of the evanescent field is important for many surface-sensitive effects, like, e.g., gas sensing or the detection of antibodies. These are an area where microstructured fibers are likely to be extremely important because of their high surface-area/volume ratio. The selective detection of antibodies in mPOF has been shown in [7,8], and metal coating of the holes allows them to be used as a platform for surface-enhanced Raman spectroscopy (SERS), a technique that potentially combines very high sensitivity with detailed molecular spectra. SERS has already been demonstrated in mPOF using silver colloidal solution, easily showing detecting concentrations of 200 nM of Rhodamine 6G [9]. Aqueous sensing for medical or environmental applications makes this area a very active one, with a continuous online monitoring being one of the targets.

In order to bring the liquid to investigate near the sensitive area, a "side-hole" fiber [10] has been developed. It simplifies the filling of the fiber, and separates it from the coupling of light in and out of the ends. While the side-hole fiber allows the rapid ingress of fluids close to the sensing region, it also allows efficient surface treatment on the core. Most surface treatments on the interior surfaces of the holes in microstructured fibers have required the fluids to pass the full length of the fiber. This approach is problematic for some applications, both because it intrinsically treats a very long section of the fiber and because it does not allow the full range of processing techniques to be applied. In particular, surface plasmon resonance (SPR) requires the application of a very high-quality metal layer. Using the side-hole approach one can apply this layer externally using sputtering, an approach that is known for high-quality films.

With this geometry, the first experimental demonstration of a SPR sensor in a polymer microstructured fiber [11] was realized recently. The width and position of these shifts are consistent with theory, given the effect of measured surface roughness [12].

2.3 SOLID POF SENSORS

Despite the fact that mPOF allow for interesting sensing concepts, most POF sensors rely on solid fibers with large diameter and high numerical aperture. They often use so-called *standard* step-index POF according to IEC 60793-2 A4a.2 due to their robustness, easy handling, and preparation, and their possibility to receive light from a large range of angles and a big cross section. Since these fibers are highly multimode, they cannot make use of, e.g., interference or other very sensitive but single-mode effects, they usually rely on methods easier-to-implement like, e.g., attenuation measurements, where the sensing signal will be transformed into an additional attenuation, reflections that can be easily captured because of POF's good light-capture properties and so on. In the following sections, several sensor concepts will be presented that use different methods to generate the sensor signal.

2.3.1 Attenuation

Attenuation is a good effect for sensor applications since power meters can relatively easily measure it. Thus it is an ideal candidate for POF sensor concepts. Here, we will present two sensor applications that use the induced additional attenuation for sensing purposes: one relies on microbending losses of the fiber once it is subject to side pressure and deformation and the other one uses the refractive-index difference between core and cladding, which can lead to losses if the index difference becomes too small.

2.3.1.1 Attenuation Due to Microbending

Two attenuation-based sensors have been developed for the car industry. Due to European Union directive 2003/102/EC, pedestrian protection has to be provided for every new car since 2005. This can be fulfilled either by

- Passive means, i.e., structural measures such as "soft" front ends and sufficient deformation room between the hood and the engine
- Active means, i.e., sensors that identify a pedestrian's impact and then trigger protective means such as lifting the hood by means of actuators

Currently there are two fiber solutions in use by European car manufacturers: one is supplied by Leoni* and the other one by Magna.†

The Leoni system is based on attenuation of the evanescent electromagnetic field, which is restricted to a small range outside the core of a Leoni-proprietary POF core. When the vicinity of the core is changed, e.g., due to an absorbing material like a human finger coming close, the evanescent field will be absorbed leading to an absorption of the signal in the fiber itself without the need of a physical contact [13]. Practically, this can be realized through foam surrounding the bare core, making sure that in the most parts of the surface, air will be outside leading to TIR. When this foam will be compressed, the conditions for TIR are no longer fulfilled and absorption occurs (Figure 2.4).

In a different approach from Magna, the sensor is located in the front bumper of a car. A POF is fixed between two structures that enhance the bending the fiber, thus leading to increased attenuation (Figure 2.5). The temporal behavior of the signal is characteristic for each collision partner and thus allows for identifying if a pedestrian has been hit.

2.3.1.2 Attenuation Due to Change of Refractive Index

Attenuation-based sensors can also be used for liquid detection. A flexible quasi-distributed liquid level sensor based on the changes in the light transmittance in a POF cable has been proposed in [14] (Figure 2.6). The measurement points are constituted by small areas created by side polishing on a curved fiber and the removal of a portion of the core. These points are distributed on each full turn of a coil of fiber built on a cylindrical tube vertically positioned in a tank.

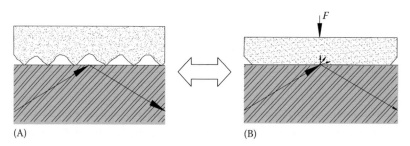

(A)   (B)

FIGURE 2.4   (A) Light rays totally reflected due to air outside the core. (B) TIR no longer possible due to the presence of absorbing material.

---

* http://www.leoni.com/fileadmin/leoni.com/downloads/pdf/produkte/en_pinchguard_flyer.pdf

† http://www.magnasteyr.com/xchg/en/43–66/electronic_systems/Products+%26+Services/Driver+Assistance+%26+
Safety+Systems

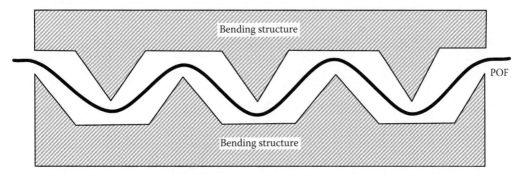

FIGURE 2.5 POF pedestrian impact sensor principle currently in use in several European cars.

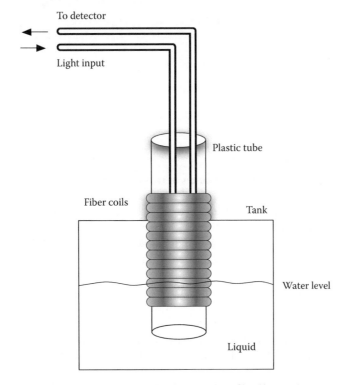

FIGURE 2.6 Schematic of the quasi distributed level sensor (one fiber/detector).

The changes between the refractive indices of air and liquid generate a signal power proportional to the position and level of the liquid. The sensor system has been successfully demonstrated in the laboratory, and experimental results of two prototypes with 15 and 18 measurement points and with bend radii of 5 and 8 mm have been achieved; the results of the 5 mm type are shown in Figure 2.7. The number of measurement points can be easily increased with the only limitation due to the dynamic range of the attenuation-measurement equipment.

2.3.2 REFLECTION

Another sensing approach uses reflections either outside or within the fiber. POF's large-core diameter and high numerical aperture make it an ideal fiber to capture reflected light and to guide it to a central location, where the evaluation of the sensor signal takes place. Reflections within the fiber, e.g., due to side pressure, localized elongation with consequent change of geometry, etc., can be

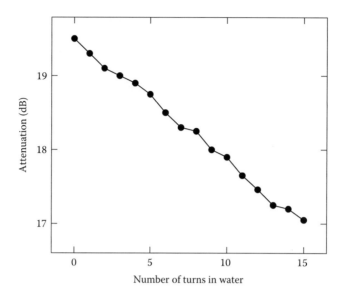

FIGURE 2.7   Experimental loss obtained as a function of turns immersed in water.

detected by optical time-domain reflectometry (OTDR), a measurement method widely used in fiber optics. With this approach, even distributed sensing is possible along the total length of the fiber.

2.3.2.1   Capture of Reflected Light

The changing intensity of reflected light from a mirror at a certain distance can be used for a displacement sensor. The power transmitted from the output end face of one fiber to the input end face of another depends on their separation distance, and closely follows an inverse square law. Utilizing this fact, for example, Ioannides et al. developed a linear displacement sensor usable between 15 and 80 mm [15–17]. In this case, two receiving, axially separated fibers are put in front of the surface whose displacement is to be measured, with a third fiber acting as the emitter (Figure 2.8). The accuracy, resolution, and stability of the sensor is better than 1%.

This sensor has also been shown to be suitable for the measurement of the frequency of large amplitude vibrating surfaces [18]. Main features of the POF that make this sensor possible are its high NA and its large-core size. Measurement of displacement can be made to relate to changes in liquid level [19,20]. Measurement of displacement has also been used to relate to water uptake by a plant with a resolution of 2.5×10^{-9} L. The same sensor was also used to monitor the reflectivity

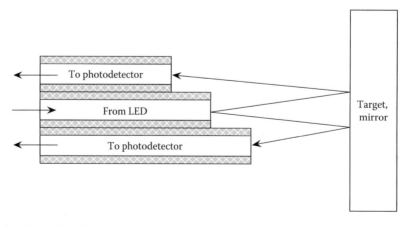

FIGURE 2.8   Schematics of the displacement sensor.

of a mirror surface as a function of condensation, as part of a condensation dew-point hygrometer [21]. There are also other cases where the POF displacement sensor can be used to indirectly probe on conditions of the environment, for example, detect the liquid nitrogen level present under quench conditions in a superconducting fault-current-limiter device using POF and not affected in performance up to 420 kV, as well as the sensor being inexpensive [22].

2.3.2.2 Mechanical Properties Using Optical Time-Domain Reflectometry

Mechanical stress can induce changes of the refractive index or the fiber geometry, which can lead to local reflections within the fiber. The evaluation of these reflections can be used for a strain sensor. Because of their high elasticity and high breakdown strain, POF are well suited for integration into technical textiles. Smart textiles with integrated POF sensors can be realized that are able to sense various mechanical, physical, and chemical quantities and can react and adapt themselves to environmental conditions.

Medical textiles with incorporated fiber-optic sensors that can measure vital physiological parameters like respiratory movement, cardiac activity, pulse oximetry, and temperature of the body are needed for wearable health monitoring, for patients requiring a continuous medical assistance and treatment.

The monitoring of anesthetized patients under medical resonance imaging (MRI) requires pure fiber optic solutions due to the immunity of fiber optics against electromagnetic radiation fields. POF sensors are advantageous because of their biocompatibility in case of fiber breakage. One can monitor the breathing movements of patients during a MRI examination by POF sensors embedded into a textile yarn and placed on an efficient area of the thorax or the abdomen [23,24]. Using an OTDR technique, the induced cyclic strain in the POF due to the respiratory movement can be measured distributed in the range between 0% and 3%. The distributed measurement provided by the OTDR technique allows to focus only on a special part of the fiber and so to differentiate between abdominal and thoracic respiration, and to neglect contributions from nonsensing parts.

POF can also be woven into geo textiles in order to add a sensing functionality to them. Geo-textiles are commonly used for reinforcement of geotechnical structures like dikes, dams, railway embankments, landfills, or slopes. The incorporation of optical sensor fibers into geo-textiles leads to additional functionalities of the textiles, e.g., monitoring of mechanical deformation, strain, temperature, humidity, pore pressure, detection of chemicals, measurement of the structural integrity, and the health of the geotechnical structure [25]. Especially, solutions for the distributed measurement of mechanical deformations over extended areas of some 100 m up to several kilometers are needed. Textile-integrated, distributed fiber-optic sensors can provide for any position of extended geotechnical structures information about critical soil displacement or slope slides via distributed strain measurement along the fiber with a high spatial resolution of less than 1 m. So the early detection of failure in geotechnical structures with high risk potential can be ensured.

The integration of POF as a sensor into geo textiles is very attractive because of their high elasticity and capability of measuring high strain values of more than 40% [26]. Especially, the monitoring of relatively small areas with expected high mechanical deformations such as endangered slopes takes advantage of the outstanding material properties of POF.

2.3.3 Elongation

As a low-cost alternative to FBG sensors targeting the lower sensitivity range, POF elongation sensors have been proposed (e.g., [27,28]). A recently recovered detection system known from laser distance meters turned out to be very sensitive while staying simple. The approach is based on measuring the phase shift of a sinusoidal light signal guided in a POF under different tensions resulting in different transit times. One of the setups under investigation is shown in Figure 2.9. The reference fiber is used for compensating temperature influence. Its output signal serves as reference in order to obtain a phase difference by comparison.

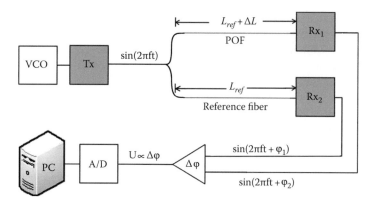

FIGURE 2.9 Schematic of the phase-measurement set up.

This sensor can be used to monitor the rotor blades of a windmill. Under hefty wind conditions, the blades can bend beyond an allowed degree and finally break. Thus they should be monitored and once wind conditions become too severe the blades will be turned out of wind direction. But also quite high-resolution measurements are feasible with this setup, which can be subject to further signal processing. Fourier transformation of the signals yields a very precise value of the oscillation frequency, thus offering the possibility to observe frequency shifts, e.g., due to layers of ice or snow that might cause dangerous situations in winter time. Oscillation frequencies above 100 Hz can be measured easily.

These sensors are supposed to be implemented in structures like rotor blades of a wind power generator, aircraft wings, or structures in general, which have to be monitored for their integrity (structural health monitoring). Transmitters and receivers have been directly adopted from POF transceivers developed for Gbit/s data transmission.

2.3.4 TEMPERATURE SENSORS BASED ON FLUORESCENCE

Temperature is a very important parameter for the electric power industry because insulators, copper conductors, core iron of transformers, insulator oil, and every equipment are very sensitive to the temperature, which has to be kept under strict control during all times. Nevertheless, when dealing with high voltage, sometimes one cannot use conventional electric sensors because they work in near ground level and therefore need to keep the creep distance from energized parts. Thus, some important parts simply cannot be monitored in a conventional way, such as copper conductors. For this reason, optical fiber sensors can offer many advantages over conventional sensors such as high immunity to electromagnetic interference, electrical isolation, no need of electric power to work and therefore can be placed at high electric potentials. Such fiber-optical temperature sensors can realized using the ruby fluorescence when pumped by a pulsed ultrabright green LED. In a field test, we used a 10 m single-fiber probe and could achieve temperature measurements in the range of 25°C–75°C with ±1°C accuracy with a time response of few seconds. The prototype with four probes was installed on a substation harmonic filter at 75 kV for temperature monitoring. The fluorescence-based optical sensors use a green LED to pump the ruby, which returns red light. Due to the distance between these two wavelengths, the sensor is potentially more sensitive and error immune than other fiber-optic temperature sensors that only rely on absorption difference when the temperature varies [29]. Previously, experiments with commercial polystyrene fluorescent fibers [30] and ruby [31] as temperature sensor were performed. Ruby has already been used for fluorescence thermometry [32]. It is low-cost, easily available, POF compatible, and can be driven with low-cost light sources (blue or green ultrabright LEDs). Si-based photodetection and simple electronics can be used because of strong intensity and long lifetime of fluorescence signal. The fluorescence peaking at 694 nm wavelength features a long decay time of 2–4 ms.

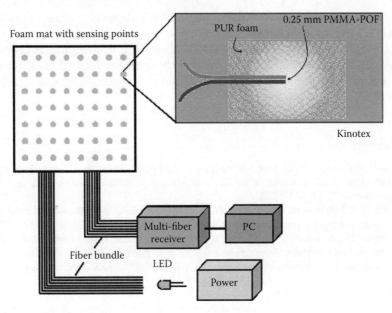

FIGURE 2.10 Application of the MFR in a Kinotex sensor mat.

2.4 SENSOR MULTIPLEXING BY FIBER BUNDLES

Fiber-optical sensors do not only provide interesting applications and sensor principles but can also easily be multiplexed and the sensor signals can be combined into compact fiber bundles or even more compact multicore fibers. Under certain conditions, it is even possible multiplex the sensor signals onto different frequencies or wavelengths so that only one single fiber may be used. We realized several multiple-fiber sensors with POF bundles. For instance in the liquid level sensor system we used a commercial USB camera at a central location, limited in temporal and spatial resolution and bad coupling efficiency that was still sufficiently good to receive the sensor signals accurately.

In a similar way, we developed a so-called "multi-fiber receiver (MFR)" [33]. A commercial CMOS sensor array was used with some modifications. In order to obtain efficient coupling of the fiber bundle to the sensitive chip, we had to remove the protective window. Thus the bundle could be put as close as possible to the array's surface (without breaking the bond wires). A sketch of a sensor application is shown in Figure 2.10.

Each sensing point consists of a pair of POF, one transporting the light of a single LED to the region of interest and the other one catching the scattered light and transporting it to the MFR. The intensity of the back-scattered light depends on the compression of the foam and serves as the sensor signal. For a first demonstration, we coupled 12 × 12 POF of 250 µm diameter each from a Kinotex™ sensor mat [34,35] to our MFR, thus resulting in 12 × 12 pressure-sensitive points distributed over an area of 50 cm by 50 cm. Using this mat, the shape of objects putting pressure on it can be recognized. By increasing the number of sensing points combined with a proper pressure calibration, one could further improve the resolution. Even dynamic processes can be detected through the fast response of the CMOS electronics, allowing at least 500 frames/s in our demo system.

2.5 CONCLUSION

Fiber sensors offer a great variety of applications. While glass-fiber sensors have a lot of applications, they are often single mode and sometimes critical in terms of handling and robustness. POF sensors, however, can be easily used in harsh environments. In addition to their mechanical and optical robustness, they provide cost-efficient alternatives to already existing sensor applications that often rely on simpler-to-implement methods. We have given a rough overview on available POF-sensor concepts with microstructured and solid fiber. While microstructured fiber sensors take the advantage of a stronger interaction with the media to sense and allow additional functionalities because of their capillaries and evanescent fields, solid POF sensors offer more robust and often simpler sensors for low-cost applications.

REFERENCES

1. M.A. van Eijkelenborg, M.C.J. Large, A. Argyros, J. Zagari, S. Manos, N.A. Issa, I. Bassett, S. Fleming, R.C. McPhedran, C.M. de Sterke, and N.A.P. Nicorovici, Microstructured polymer optical fibre, *OSA Optics Express*, 9(7), 319–327, 2001.
2. T.P. White, R.C. McPhedran, L.C. Botten, G.H. Smith, and C.M. de Sterke, Calculations of air-guided modes in photonic crystal fibers using the multipole method, *OSA Optics Express*, 9(11), 721–732, 2001.
3. M.C.J. Large, J.H. Moran, and L. Ye, The role of viscoelastic properties in strain testing using microstructured polymer optical fibres (mPOF), *Measurement Science and Technology*, 20(3), 034014, 2009.
4. M.C.J. Large and C.-A. Bunge, Microstructured polymer optical fibres compared to conventional POF: Novel properties and applications, *IEEE Sensors Journal*, Special Issue on *Photonic Crystal-Based Sensors*, 10(7), 1213–1217, 2010.
5. F.M. Cox, A. Argyros, and M.C.J. Large, Liquid-filled hollow core microstructured polymer optical fiber, *Optics Express*, 14(9), 4135–4140, 2006.
6. C.M.B. Cordeiro, M.A.R. Franco, G. Chesini, E.C.S. Barretto, R. Lwin, C.H. Brito Cruz, and M.C.J. Large, Microstructured-core optical fibre for evanescent sensing applications, *Optics Express*, 14, 13056–13066, 2006.
7. J.B. Jensen, J.P. Hoiby, G. Emiliyanov, O. Bang, L. Pedersen, and A. Bjarklev, Selective detection of antibodies in microstructured polymer optical fibers, *Optics Express*, 13(15), 5883–5889, 2005.
8. G. Emiliyanov, J.B. Jensen, O. Bang, P.E. Hoiby, L.H. Pedersen, E.M. Kjær, and L. Lindvold, Localized biosensing with Topas microstructured polymer optical fiber, *Optics Letters*, 32(5), 460–462 [Erratum: p. 1059, 2007].
9. F.M. Cox, A. Argyros, M.C.J. Large, and S. Kalluri, Surface enhanced Raman scattering in a hollow core microstructured optical fiber, *Optics Express*, 15(21), 13675–13681, 2007.
10. F.M. Cox, R. Lwin, M.C.J. Large, and C.M.B. Cordeiro, Opening up optical fibres, *Optics Express*, 15(19), 11843–11848, 2007.
11. A. Wang, A. Docherty, B.T. Kuhlmey, F. Cox, and M.C.J. Large, Surface plasmon resonance in slotted microstructured polymer optical fibres, *International Conference on Materials for Advanced Technologies, Symposium P: Optical Fiber Devices and their Applications*, Singapore, pp. 41–43, June 2003.
12. M. Kanso, S. Cuenot, and G. Louarn, Roughness effect on the SPR measurements for an optical fibre configuration: Experimental and numerical approaches, *Journal of Optics A: Pure and Applied Optics*, 9, 586–592, 2007.
13. G. Kodl, Large area optical pressure sensor based on evanescent field, *Proceedings of 12th POF Conference 2003*, Seattle, WA, p. 64, 2003.
14. M. Lomer, J. Arrue, C. Jaurequi et al., Lateral polishing of bends in plastic optical fibers applied to a multipoint liquid-level measurement sensor, *Sensors and Actuators A: Physical*, Volume 137, pp. 68–73, 2007.
15. T. Augousti, *Sensors VI, Technology Systems and Applications*, IOP Publishing Ltd., Bristol, U.K., pp. 291–293, 1994.
16. N. Ioannides, D. Kalymnios, and I. Rogers, A POF-based displacement sensor for use over long ranges, *Proceedings of 4th POF'95*, Boston, MA, pp. 157–161, 1995.
17. N. Ioannides, D. Kalymnios, and I.W. Rogers, An optimised plastic optical fiber (POF) displacement sensor, *Proceedings of 5th POF'96*, Paris, France, pp. 251–255, 1996.
18. N. Ioannides and D. Kalymnios, A plastic fiber (POF) vibration sensor, *Proceedings of Applied Optics Divisional Conference of the IOP*, Brighton, U.K., pp. 163–168, 1998.
19. J.D. Weiss, The pressure approach to fiber liquid level sensors, *Proceedings of 4th POF'95*, Boston, MA, pp. 167–170, 1995.
20. S. Vargas, C. Vázquez, A.B. Gonzalo, and J.M.S. Pena, A plastic fiber-optic liquid level sensor, *Proceedings of Second European Workshop on Optical Fiber Sensors*, Santander, España, SPIE Volume 5502, pp. 148–151, 2004.

21. S. Hadjiloucas, S. Karatzas, D.A. Keating, and M.J. Usher, Optical sensors for monitoring water uptake in plants, *Journal of Lightwave Technology*, 13(7), 1421–1428, 1995.
22. J. Niewisch, POF sensors for high temperature superconducting fault current limiters, *Proceedings of 6th POF'97*, Kauai, HI, pp. 130–131, 1997.
23. A. Grillet, D. Kinet, J. Witt, M. Schukar, K. Krebber, F. Pirotte, A. Depre, N.V. Barco, and N.V. Kortrijk, Optical fiber sensors embedded into medical textiles for healthcare monitoring, *IEEE Sensors Journal*, 8(7), 1215–1222, 2008.
24. J. Witt, C.-A. Bunge, M. Schukar, and K. Krebber, Real-time strain sensing based on POF OTDR, *Proceedings of 15th International POF Conference*, Paper SEN-II-4, Torino, Italy, pp. 210–213, September 2007.
25. N. Nöther, A. Wosniok, K. Krebber, and E. Thiele, Dike monitoring using fiber sensor-based geosynthetics, *Proceedings of the ECCOMAS Conference on Smart Structures and Materials*, Gdansk, Poland, July 2007.
26. P. Lenke, K. Krebber, M. Muthig, F. Weigand, and E. Thiele, Distributed strain measurement using polymer optical fiber integrated in technical textile to detect displacement of soil, *Proceedings of the ECCOMAS Conference on Smart Structures and Materials*, Gdansk, Poland, July 2007.
27. H. Doering, High resolution length sensing using PMMA optical fibers and DDS technology, *Proceedings of 15th POF Conference 2006*, Seoul, Korea, September 11–14, 2006.
28. H. Poisel et al., POF strain sensor using phase measurement techniques, *Proceedings of 16th POF Conference 2007*, Turin, Italy, September 10–13, 2007.
29. K. Asada and H. Yuuki, Fiber optic temperature sensor, *3rd POF Conference 1994*, Yokohama, Japan, pp. 49–51, 1994.
30. R.M. Ribeiro, L.A. Marques-Filho, and M.M. Werneck, Fluorescent plastic optical fibers for temperature monitoring, *12th POF Conference 2003*, Seattle, WA, pp. 282–285, 2003.
31. R.M. Ribeiro, L.A. Marques-Filho, and M.M. Werneck, Simple and low cost temperature sensor using the ruby fluorescence and plastic optical fibers, *14th POF Conference 2005*, Hong Kong, pp. 291–294, 2005.
32. K.T.V. Grattan and Z.Y. Zhang, *Optic Fluorescence Thermometry*, Chapman & Hall, London, 1995.
33. T. Hofmann, Multi-Faser-Receiver, Master thesis, GSO University of Applied Sciences, Nuernberg, Germany, 2009.
34. Canpolar East Inc. http://www.canpolar.com/principles.shtm (access date January, 2013)
35. H. Poisel, POF sensors—An update, *18th POF Conference 2009*, Sydney, Australia, September 2009.

3 Optical Fiber Sensors and Interrogation Systems for Interaction Force Measurements in Minimally Invasive Surgical Devices

Ginu Rajan, Dean Callaghan,
Yuliya Semenova, and Gerald Farrell

CONTENTS

3.1 INTRODUCTION

Minimally invasive surgical (MIS) procedures involving laparoscopic and endoscopic devices are often preferred over traditional open surgery due to a shorter postoperative recovery time and reduced intraoperative complications. Many ongoing research activities are focused on the use of strain/force sensors for the measurement of interaction forces occurring at the instrument–tissue interface of the surgical devices [1]. Electrical strain gauge technology has been either utilized in the form of a modular sensor [2] or attached onto the instrument trocar [3]. However, optical fiber sensors have many advantages over their electrical counterparts, for example, immunity to electromagnetic interference, which allows them to be used in magnetic resonance imaging (MRI) fields. Instruments in which the optical fiber sensor forms an integral part of the end effector are desirable to enable accurate measurement of complex interaction and cutting forces.

Different types of optical fiber sensors used for MIS devices [4] include fiber Bragg gratings (FBG), photonic crystal fiber (PCF) sensors, and intensity-modulated fiber sensors. FBGs have been a favorite option used in many telerobotic devices as well as in structural health monitoring systems [5–7]. For a telerobotic strain/force sensing system based on an FBG, the resolution of the system depends on the FBG interrogation system. Most commercial FBG interrogation systems can provide a wavelength resolution of approximately 5–10 $\mu\varepsilon$. Recent improvements in the fabrication of FBGs have reduced their cost, so that the interrogation unit, rather than the sensor, accounts for a large proportion of the cost of a complete sensing system. The advantages of choosing FBG sensors include compact dimensions and multiplexing capabilities. Additionally, a single fiber provides a low-loss, high-speed path for tactile force information to be transferred from the sensorized instrument to the remote robotic system [8]. However, the principal concern in this is the cross-sensitivity nature of the sensor whereby the accuracy of the force measurements is influenced by the effects of localized temperature variations. Therefore, error compensation is required through the use of additional FBG sensors for the purpose of temperature measurement only [9,10].

Recent advancements in the area of PCF sensors [11,12] open a new possibility for developing simple and miniature fiber sensors for surgical devices. PCF interferometers (PCFIs) can be fabricated in a number of ways and their properties and applications have been studied by many authors [13]. Different types of PCFIs include fiber loop mirror interferometers, microhole-collapsed interferometers, tapered interferometers, and hollow core Fabry–Perot interferometers [14,15]. These interferometers are used in many applications such as strain sensing, refractive index sensing, biosensing, gas sensing, and temperature sensing. The properties of a PCFI and its sensitivity to the measurands are determined by the type of interferometer, by its geometry, and by the type of the PCF used. Among the PCFI sensors, tapered PCFIs and modal interferometers are of particular interest due to their miniature size and ease in fabrication and are well suited for MIS applications.

The aim of this chapter is to provide an overview of FBG- and PCF-type fiber sensors and the interrogation system used for application in MIS. Issues relating to the integration of fiber sensors into surgical devices, static force calibration, sensor interrogation, and a comparison of both FBG- and PCF-sensorized surgical blades are discussed.

3.2 MINIMALLY INVASIVE SURGICAL DEVICES

The layout of a typical master–slave minimally invasive robotic surgical (MIRS) system is illustrated in Figure 3.1, highlighting the three primary communication modalities (positional, force, and visual information feedback) between the surgeon and the remote operating environment. Robotic assistance, as well as overcoming the disadvantages of MIS techniques, offers new advancements in areas such as provision of additional degrees of freedom, tremor filtering, and scaling of motions, particularly in the field of microsurgery [16–18].

MIRS systems are an accurate and reliable means of eliminating the shortcomings associated with traditional laparoscopic surgery. MIRS can overcome drawbacks of traditional laparoscopic surgery by using a teleoperated approach. In this scheme, the surgeon comfortably sits at a console controlling the surgical instruments guided by a patient sided surgical robot. Computational support is allowed for reestablishing hand–eye coordination, motion scaling, indexing (repositioning of the input devices to a comfortable working position while the instruments remain still), and even motion compensation.

The current commercially available MIRS systems greatly augment the surgeon's ability to carry out an operating procedure effectively but lack the ability to relay haptic feedback to the user [19]. Hence, there is a need for small, unobtrusive strain/force sensing transducers to facilitate the measurement

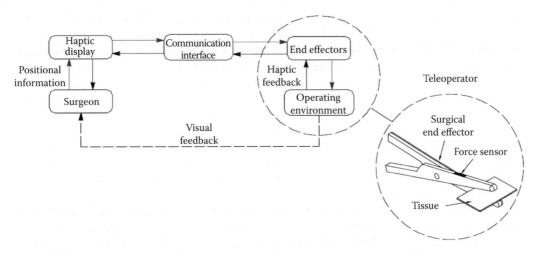

FIGURE 3.1 Minimally invasive robotic surgical system. (From Callaghan, D.J. et al., *Proceedings of the International Manufacturing Conference (IMc25)*, 2008, pp. 389–398. With permission.)

of forces at the instrument–tissue interface. In this chapter, as a proof of concept of optical fiber–sensorized surgical devices, a sensorized laparoscopic scissor blade and a standard scissor blade are presented. The different types of fiber sensors used with these surgical devices, the integration of fiber sensor with the devices, and force calibration of the surgical devices are presented in the following sections.

3.3 OPERATING PRINCIPLE OF FIBER SENSORS USED FOR MIS DEVICES

3.3.1 FIBER BRAGG GRATINGS

An FBG comprises of a short section of single-mode optical fiber in which the core refractive index is modulated periodically using an intense optical interference pattern [20], typically at UV wavelengths. The wavelength of light reflected by periodic variations of the refractive index of the Bragg grating, λ_G, is given by [7]

$$\lambda_G = 2n_{\text{eff}}\Lambda \tag{3.1}$$

where
  n_{eff} is the effective refractive index of the core
  Λ is the periodicity of the refractive index modulation

The basic principle of operation of any FBG-based sensor system is to monitor the shift in the reflected wavelength due to changes in measurands such as strain and temperature. The sensitivity of the Bragg wavelength to temperature arises from the change in period associated with the thermal expansion of the fiber, coupled with a change in the refractive index arising from the thermo-optic effect. The strain sensitivity of the Bragg wavelength arises from the change in period of the fiber coupled with a change in refractive index arising from the strain-optic effect.

3.3.2 PHOTONIC CRYSTAL FIBER SENSORS

The two types of PCF sensors that show potential for MIS applications include a hole-collapsed PCFI and a tapered PCFI, both operate in the single-ended reflection mode regime. A commercially available LMA-10 PCF can be used for the fabrication of the interferometers owing to its high insensitivity to temperature effects. To fabricate a hole-collapsed modal interferometer, a small section of the PCF is fusion spliced between two standard single-mode fibers (SMFs). In the vicinity of both splice regions, the holes of the PCF are collapsed in a microscopic region approximately 170–180 μm long. After splicing, one of the SMFs is cleaved using a standard cleaving machine ~2 cm from the PCF splicing point. The cleaved fiber end behaves as the mirror for the interferometer. The operating principle of the PCFI is based on the excitation and recombination of modes occurring in the region of the PCF where the voids of the PCF are collapsed. The fundamental SMF mode begins to diffract when it enters the collapsed section of the PCF. Because of diffraction, the mode broadens, allowing the excitation of multiple core and cladding modes in the stub of the PCF [21]. The modes propagate through the PCF until they reach the cleaved end from where they are reflected. The length of the PCF region is chosen to match the length of the scissor blade.

To fabricate the tapered interferometer initially, a small section of the PCF is spliced on to a standard SMF pigtail. A carefully controlled fusion splicing process ensured that no interference patterns arose due to hole collapse in the PCF during splicing [22]. The central region of the PCF was then collapsed and thinned down to a micron size to form the tapered region. The tapered region formed in the PCF is similar to an unclad multimode optical fiber. As a consequence of this, the fundamental mode in the PCF in the region where the holes are open is coupled to the modes of the solid fiber in the tapered region. As a result of this, mode coupling and beating between the multiple modes take place and the transmission spectrum shows an oscillatory behavior [23]. The normalized spectral responses of both the hole-collapsed and the tapered interferometers are shown in Figure 3.2. The visibility of the response can be further enhanced by applying a reflection coating to the free end of the PCF sensor.

It is necessary to know the strain and temperature sensitivity of the interferometric sensors before embedding the sensors into the devices to measure unknown strain/forces experienced by the surgical blades. The strain characterization responses of the PCFIs are shown in Figure 3.3a. The strain sensitivity for the tapered sensor with waist diameter 22 μm and length 0.3 mm is 1.9 με/pm and for a hole-collapsed PCF sensor with length 35 mm is 1.68 με/pm. It should be noted that this strain sensitivity is higher than other strain sensors such as FBGs. The temperature sensitivity of the PCF sensors must also be measured to ensure the low temperature dependency.

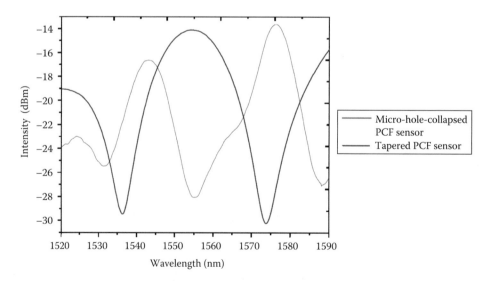

FIGURE 3.2 Spectral responses of the hole-collapsed and the tapered interferometers. (From Rajan, G. et al., *IEEE Trans. Biomed. Eng.*, 59(12), 332, 2012. With permission.)

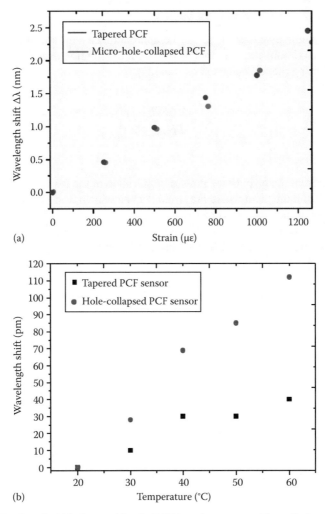

FIGURE 3.3 (a) Wavelength shift observed for the PCF interferometers with applied strain. (b) Plot showing the temperature dependence of the tapered and the hole-collapsed PCF interferometers. (From Rajan, G. et al., *IEEE Trans. Biomed. Eng.*, 59(12), 332, 2012. With permission.)

The observed wavelength shift for both the aforementioned tapered PCFI and the hole-collapsed PCFI with change in temperature is shown in Figure 3.3b. The observed temperature sensitivity of the tapered interferometer was 1 pm/°C, while that for the hole-collapsed interferometer was 3 pm/°C. This sensitivity is much less compared to that of the FBG and can be neglected especially in the case of the tapered PCFI.

3.4 SENSOR INTEGRATION WITH THE SURGICAL BLADES

Incorporating a fiber strain/force sensing into a surgical end effector necessitates consideration of many factors. Among them are (1) location of the proposed sensors so as not to interfere with the instrument functionality and (2) incorporating grooves or holes in the end-effector structure to facilitate routing of the optical fiber as well as protecting the sensing element. The following sections elaborate the importance of the optimization of the placement position of sensor on the blade and the effect of bonding length on the strain transfer from the blade to the optical fiber sensor.

3.4.1 Placement of the Optical Fiber Sensor

Assuming that the surgical scissor blade can be approximated as a uniformly tapered cantilever beam, the point of maximum strain as well as the strain distribution over the bonded region of the fiber is established using the following expression [24]:

$$\varepsilon_z = \frac{6F(L-z)}{Eb(mz+W)^2} \tag{3.2}$$

where

  L is the distance from the pivot to the point of application of the load F
  z is the distance from the pivot where the strain is to be known ($0 < z > L$)
  W is the thickness of the blade at its pivot
  b is the width of the blade
  E is the Young's modulus of the blade material (185 GN/m²)
  m is the uniform slope of the blade

The position on the blade where maximum strain occurs can be found by taking

$$\frac{d\varepsilon_z}{dz} = 0 \tag{3.3}$$

therefore,

$$\frac{d\varepsilon_z}{dz} = \frac{6F(2Lm - mz + W)}{Eb(mz+W)^3} = 0 \tag{3.4}$$

The position of maximum strain on the blade top surface is therefore

$$z_{\max} = 2L + \frac{W}{m} \tag{3.5}$$

From Equation 3.5, the maximum strain was found to occur at a location 14 mm from the blade pivot point for a standard surgical blade of 39 mm long [25]. For an FBG sensor, it is important that the strain distribution across the FBG is uniform in order to avoid distortion of the reflected spectrum. It was estimated that the strain variation across the 5 mm long FBG centered on the 14 mm point is only 0.003%. Such a small strain field variation will have negligible adverse effect on the reflected FBG signal.

3.4.2 Bonding Length and Strain Transfer

Strain transfer from the blade surface (host material) to the optical fiber sensor is influenced by the properties of the blade surface, adhesive layer thickness, and the protective coating, such as polyimide, on the fiber sensor. A 5 mm long FBG is written in the middle of an 11 mm long buffer-stripped portion of a standard SMF and is recoated with polyimide after inscription and is then attached to a steel substrate. The region of the fiber recoated with polyimide had a thickness of 4–4.5 μm. Since standard SMF with a buffer is 250 μm in diameter, the adhesive thickness between the metal and the FBG fiber will be approximately 58–60 μm. It is also important to estimate the

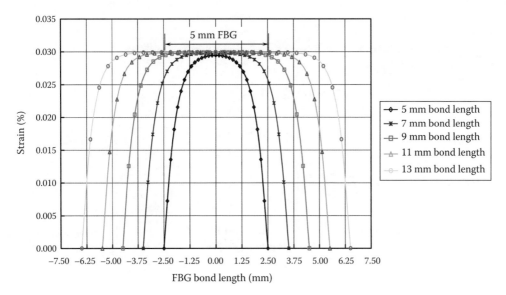

FIGURE 3.4 Strain distribution along the FBG for different bonding lengths. (From Rajan, G. et al., *IEEE Sensors J.*, 10(12), 1913, 2010. With permission.)

bonding length to ensure that the distribution of the strain over the FBG is uniform. In the earlier example where the adhesive thickness between the metal and the FBG is 60 μm and a polyimide layer thickness of 4.5 μm, a minimum bond length of 11 mm ensures uniform strain distribution along the 5 mm length of the FBG [26,27]. The strain measured by an FBG sensor is taken to be the average strain over the bonded portion of the fiber. The effectiveness of the strain transfer from the blade to the fiber core is influenced by the extent of shear concentrations through the adhesive layer thickness upon loading. Moreover, the stiffness of the adhesive layer also greatly influences the strain transfer effectiveness with a stiffer adhesive inducing great strain transfer to the fiber core. Figure 3.4 shows simulated strain distributions over the FBG sensor for different bonding lengths and it is clear from the figure that an 11 mm bonding length ensures a uniform strain distribution over the 5 mm long FBG sensor.

3.5 SENSOR INTERROGATION SYSTEM

3.5.1 MACROBEND FIBER–BASED FBG INTERROGATION SYSTEM

A macrobend fiber filter is an edge filter that gives a quasilinear transmission response for a range of the wavelength region [26]. Using this characteristic property, an edge filter can be used to measure optical wavelength when employed in a ratiometric scheme [27]. The ratiometric scheme improves the stability and accuracy of the system. Since FBGs are sensitive to both strain and temperature, temperature compensation FBGs should be used to eliminate the inaccuracy due to the ambient temperature variations especially when used with MIS devices as accuracy is very important in this application. An experimental arrangement for the interrogation system [24] for the FBG sensors using a macrobend fiber filter is shown in Figure 3.5. Two FBG sensors are used in the system: one attached to the top of the blade to measure the direct strain and the other for temperature compensation (only one end is fixed to the lateral side of the blade). In order to use a macrobend fiber edge filter ratiometric system to interrogate FBGs, the fiber edge filter system needs to be calibrated in the wavelength range of interest, normally from 1500 to 1600 nm since the FBGs peak reflected wavelength commonly lies in that wavelength region. In this arrangement, a broadband source is used as the input source. Since two FBG sensors are used, two identical characteristic macrobend fiber filter systems are used to interrogate the FBG sensors, which are used for direct strain

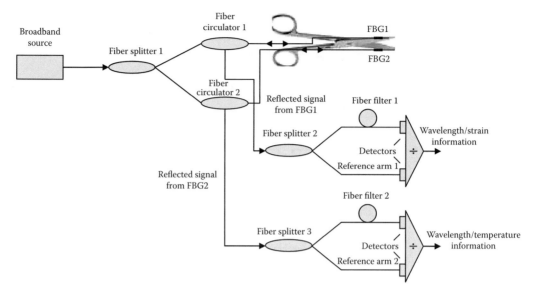

FIGURE 3.5 Schematic of the FBG interrogation system using macrobend fiber filter ratiometric systems. (From Rajan, G. et al., *IEEE Sensors J.*, 10(12), 1913, 2010. With permission.)

measurement and for temperature compensation. A 2×2 fiber coupler is used to split the signals between the two FBGs. The reflected signals from the FBGs are directed to the fiber edge filter ratiometric system by using fiber circulators as shown in Figure 3.5.

The slope of the fiber edge filter used was 0.15 dB/nm [28]. As the FBGs are sensitive to strain and temperature, the difference between the two ratios in the aforementioned configuration can provide temperature-independent strain information. This is attained because the slopes of the edge filters are set to the same and hence the wavelength change and power ratio due to change in the FBG temperature (for both FBG_1 and FBG_2) will be the same for both edge filter systems, and the difference between the ratios yields strain information from FBG_1 without the influence of temperature. To obtain a slope of 0.15 dB/nm, the fiber filters use 10 turns of standard SMF28 with a 10.5 mm bend radius [28]. The wavelength measurement resolution of such a ratiometric system was 20 pm, which gives a strain resolution of $\sim$16 $\mu\varepsilon$.

The resolution and accuracy of any ratiometric system are limited by signal-to-noise ratio of the optical source and also the noise in the receiver system [29]. In the earlier example, the strain inaccuracy due to this will be in the range of ±12.5 $\mu\varepsilon$. It is also known that the macrobend fibers are slightly temperature sensitive, and the temperature dependence of the fiber filter can lead to a measured strain error [30]. The variation in the output ratio of the system due to ambient temperature is oscillatory in nature, and hence the temperature compensation calibration is a complex task. However, in this configuration since two identical fiber edge filters are being used, the system has an advantage that the influence of ambient temperature on the fiber filters will be canceled out. The macrobend fiber filter interrogation system can therefore effectively measure strain with competitive accuracy and independent of the influence of ambient temperature on both the FBG sensors and also the interrogation system.

Other commercial interrogation systems can also be used to interrogate the FBG sensors. Examples of such interrogations include Wx-02 from Smart Fibres Ltd. [31]. Some experimental results obtained using Wx-02 are also presented in this chapter for demonstration.

3.5.2 INTERROGATION SYSTEM FOR PCF SENSORS

The PCF sensors operate in the wavelength domain with multiple wavelength peaks. Hence, for laboratory demonstration purpose, an optical spectrum analyzer (OSA) can be used to measure the wavelength shifts. A commercial FBG interrogation system is an alternative option, if the finesse of the interference pattern is very high. The experimental arrangement of the interrogation is similar to that of the FBG sensors, since the PCF sensors also operate in the reflection mode regime.

3.6 CALIBRATION OF SENSORIZED MIS DEVICES

Strain/force calibration of the sensorized surgical devices are necessary to calculate the sensitivity of the sensorized devices. Two types of surgical scissor blades presented here are a laparoscopic scissor blade and a standard surgical scissor blade. The laparoscopic scissor blade is 19 mm long and the surgical scissor blade is 39 mm long. The tapered interferometer and a 3 mm FBG sensor are attached to the laparoscopic blade, and the 35 mm long hole-collapsed sensor and a 5 mm FBG are attached to the surgical scissor blade for the strain/force measurements.

Surgical scissors are typically used over the first one-third of the blade length (measured from the tip) when making cuts as it allows for better control of the cutting process. The strain induced near the tip of the blade is substantially lower than that induced at the pivot during normal cutting procedures. Therefore, careful consideration must be given to the placement of the sensor so that the measurement sensitivity of the sensorized blade is not unduly compromised. For the surgical scissor blade as mentioned in Section 3.4.1, the 5 mm FBG can be placed at a distance of 14 mm from the pivot, while the hole-collapsed PCFI is attached to the blade from the pivotal point toward the tip of the blade. The sensor covers 35 mm of the blade from the pivot and measures the average strain along the length of the blade. In the case of the laparoscopic blade, the tapered strain sensor and the 3 mm FBG are placed proximal to the pivot point as far from the laparoscopic blade tip as possible. Placing the fiber sensor within this area means that the cutting range of the blades is maximized and the measurement sensitivity will also be maximized.

Test rigs for the calibration and sensitivity evaluation of both the sensorized surgical scissor blade and the laparoscopic blade are developed and the calibration procedure involves securing the scissor blades in a clamping fixture and applying a series of static loads at a number of locations along the prescribed cutting envelope of the blades [24]. The forces are manually applied to the blades via a low friction slider mechanism and monitored using a button load cell connected to a PC. The data from the button load cell is collected using a National Instruments load cell module SG-24, which is connected to a data acquisition board NI6221. To interrogate the PCF sensors, a broadband SLD source is used. A fiber circulator is used to direct the reflected interference spectrum from the PCF sensors to the OSA where the strain-induced wavelength shift is measured. For the interrogation of FBG sensors, the macrobend fiber interrogation system and the Wx-02 interrogation system are used.

3.6.1 FORCE SENSITIVITY OF THE FBG-SENSORIZED BLADES

3.6.1.1 FBG-Sensorized Standard Surgical Blade

Using the experimental setup described in Sections 3.5.1 and 3.6, direct strain on the sensorized surgical blade with the 5 mm FBG can be measured with a load applied at multiple points along the blade from its tip toward the pivot. The measured direct strains for different loads applied at different blade positions for the surgical blade using the macrobend FBG interrogation system are shown in Figure 3.6a. It can be observed that the maximum strain is registered by the FBGs when the load is applied to the tip of the blade and the strain response is linear with respect to the applied load. However, during a typical cutting cycle the forces on the blades vary along its length over a typical working envelope between 10° and 23°. This is equivalent to a linear range of between 0 and 26 mm from the blade tip. In practical cutting applications, the load position can be obtained if the blade opening angle is known, and hence the corresponding strain can be measured assuming the system is calibrated.

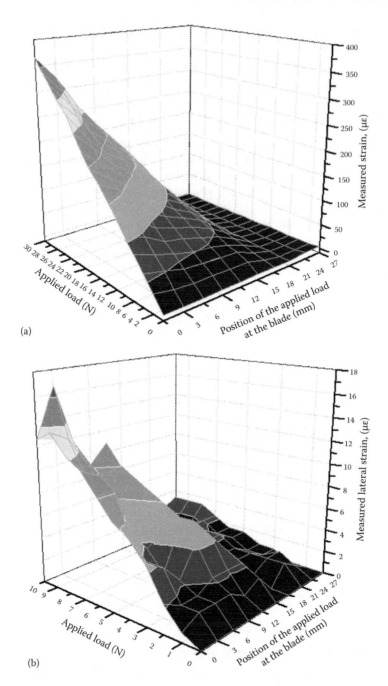

FIGURE 3.6 (a) Applied load vs. direct strain measurement at different locations in the blade. (b) Impact of lateral load on the direct strain measurements. (From Rajan, G. et al., *IEEE Sensors J.*, 10(12), 1913, 2010. With permission.)

In practice, during a typical cutting cycle, scissor blades experience laterally applied loading due to the curved nature of the blade. The strain resulting from the lateral load registered by the FBG attached to the top side of the blade is shown in the Figure 3.6b with an applied load in the range of 0–10 N to multiple points along the lateral side of the blade from the tip toward the pivotal region at 3 mm intervals. Although the FBG sensor is less sensitive to the lateral load, it is clear that the lateral loading of the blade impacts the direct strain output from the FBG sensor. A lateral load of 10 N applied to the tip of the blade introduces a maximum of 16 μɛ error in the measured direct strain and the value of error decreases when the applied load shifts toward the pivotal region. Thus the accuracy of the direct strain measurement is limited due to the inadvertent lateral loading, arising from the deflection of the blade during cutting. However, this can be minimized by characterizing the blade for a dry cut (without any tissue) and using the results a calibration correction factor can be made to eliminate the impact of the lateral force.

Strain measured using the FBG and the macrobend fiber filter interrogation system can be referenced against that of a standard electrical strain gauge for comparison, allowing the sensitivities of both measuring techniques to be directly compared under the same loading conditions. A comparison of the strain measured by the FBG with that of the strain gauge is shown in Figure 3.7a. It can be seen that both the strain gauge results and the strain measured from the macrobend fiber interrogation system agree. The FBG-measured strain, for a corresponding load, is unaffected by temperature variations, indicating effective compensation is being achieved with the compensation

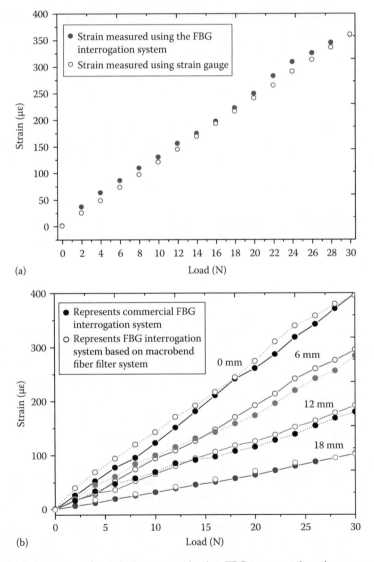

FIGURE 3.7   (a) Comparison of the strain measured using FBG sensor and strain gauge at the tip of the blade. (b) Comparison of measured strain using a macrobend fiber filter interrogation system and a commercial FBG interrogation system. (From Rajan, G. et al., *IEEE Sensors J.*, 10(12), 1913, 2010. With permission.)

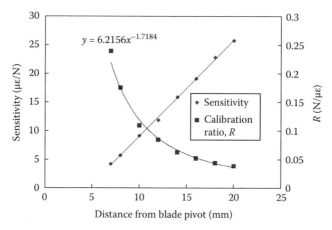

FIGURE 3.8 Force sensitivity values and calibration ratio for the FBG-sensorized scissor blade.

techniques described earlier. The strain measured by using the macrobend fiber interrogation system compared with that of the Wx-02 Smart Fibre interrogation system is shown in Figure 3.7b. The strain measured using both the interrogation systems agrees very well confirming that, to obtain the force information from surgical blades, the FBG-based sensing system together with a macrobend fiber filter–based interrogation system can be used effectively. An additional factor that introduces inaccuracy to the measured strain values is the inherent noise in the receiver system as explained in Section 3.5.1 [29]. However, the influence of receiver noise can be minimized by low-noise design of the receiver system and the strain resolution can be improved if higher resolution is required.

3.6.1.2 FBG-Sensorized Laparoscopic Scissor Blade

The calibration plot of a 19 mm long laparoscopic blade attached with a 3 mm FBG, interrogated with the Wx-02 FBG interrogation system is shown in Figure 3.8. It can be seen that the sensitivity of the sensorized blade is greatest when loaded at the tip and decreases as the position of the applied load moves toward the pivot point. This results in a sensitivity variation from 25.7 to 14 $\mu\varepsilon$/N over the first one-third of the blade length. Based on observations of the fluctuation in the measured strain caused by noise in the interrogation system, the force resolution was estimated to be 0.28 N from the unfiltered FBG signal; however, as the sensitivity increases toward the blade tip the estimated resolution improves to 0.15 N. System noise estimated in this case was equivalent to 4 $\mu\varepsilon$. Taking the sensitivity at the blade tip as 25.7 $\mu\varepsilon$/N, the force resolution is estimated to be 0.15 N. At 13 mm from the blade pivot, the sensitivity is 14 $\mu\varepsilon$/N; hence, the force resolution is 0.28 N. Obtaining force readings from the FBG sensor, representative of the friction and fracture forces acting on the blade during operation requires the use of the empirically derived calibration ratio R shown in Figure 3.8.

3.6.2 Force Sensitivity of the PCF Sensor–Attached Blades

3.6.2.1 PCF-Sensorized Standard Surgical Blade

The hole-collapsed PCFI attached to the standard surgical scissor blade from the pivot point toward the tip of the blade is considered. The wavelength shift observed for the microhole-collapsed PCF sensor attached to the surgical scissor blade for an applied load of 25 N is shown in Figure 3.9a. The average strain corresponding to the wavelength shifts for an applied load up to 27 N at the tip of the blade is shown in Figure 3.9b. The theoretical average strain for the surgical blade is also compared with that of the observed strain values and is also shown in Figure 3.9b. It can be seen that the measured and the calculated values agree well. The strain sensitivity range for the sensorized surgical scissor went from 13.3 $\mu\varepsilon$/N at the tip to 3.5 $\mu\varepsilon$/N at 18 mm from the pivot.

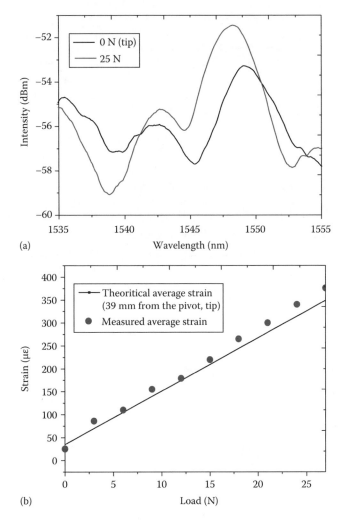

FIGURE 3.9   (a) Spectral shift observed with the hole-collapsed PCF interferometer attached to the surgical scissor blade with an applied load of 25 N. (b) Measured average strain in the scissor blade for different applied loads and its comparison with the calculated average strain. (From Rajan, G. et al., *IEEE Trans. Biomed. Eng.*, 59(12), 332, 2012. With permission.)

3.6.2.2   PCF-Sensorized Laparoscopic Blade

The wavelength shift for the tapered PCF sensor attached to the laparoscopic blade for a load of 14 N is shown in Figure 3.10a. With a strain sensitivity of 1.9 pm/$\mu\varepsilon$, the strain values corresponding to the wavelength shifts are calculated for the applied loads for a blade region of 10 mm from the tip of the blade. The strain/force sensitivity of the sensorized laparoscopic blade with the tapered sensor attached is shown in Figure 3.10b. It can be seen that the sensitivity of the sensorized blade is greatest when loaded at the tip and decreases as the position of the applied load moves toward the pivot point. This results in a sensitivity range from 26.7 to 12.5 $\mu\varepsilon$/N over the first one-third of the blade length.

The studies on the microhole-collapsed interferometer and the tapered PCFI show that both types of sensor can be used with MIS devices. With miniature MIS devices, tapered PCF sensors can be employed, while for standard devices such as surgical graspers, the microhole-collapsed sensor is also an option. However, the tapered PCFI is superior in practical applications due to its higher strain sensitivity and lower temperature sensitivity compared to the microhole-collapsed PCF sensors.

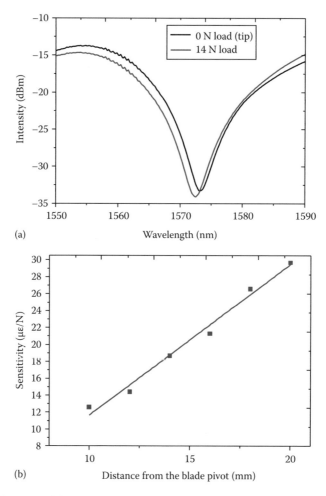

(a)

(b)

FIGURE 3.10 (a) Spectral shift observed with the tapered PCF interferometer attached to the laparoscopic blade with an applied load of 14 N. (b) Strain/force sensitivity of the sensorized laparoscopic blade for different locations along the length of the blade. (From Rajan, G. et al., *IEEE Trans. Biomed. Eng.*, 59(12), 332, 2012. With permission.)

3.6.3 COMPARISON BETWEEN FBG- AND PCF-SENSORIZED SURGICAL BLADES

A comparison of strain/force sensitivity of the PCF sensors to that of the FBG sensor is shown in Table 3.1. From the table, it can be seen that the sensitivity values obtained for the tapered PCF sensor and microhole-collapsed PCF sensor are comparatively close to those of the FBG sensors. It should also be noted that the wavelength/strain sensitivities of the PCF sensors are higher than the FBG sensors and hence the system can measure interaction forces with higher resolution than the FBG sensors and also with much lower level of errors due to the temperature fluctuation. The results therefore suggest the possibility of employing PCF optical force sensing in smart surgical devices where interaction force measurements are required without temperature compensation. However, since the interrogation system for FBGs is well established, it would be easier to use FBG-based systems in real-time applications. However, in the near future, more research and developments in low-cost interrogation systems for PCF sensors are expected, which will further enhance the use of such sensors in MIS and in other application areas.

TABLE 3.1

Comparison of the Performance of the Surgical Blades Sensorized with FBGs and PCF Sensors

Sensor/Blade Type	Laparoscopic Blade (με/N)	Standard Surgical Blade (με/N)	Strain/Wavelength Sensitivity (pm/με)	Temperature Sensitivity (pm/°C)
Tapered PCF sensor	31.1	—	1.9	1
Hole-collapsed PCF sensor	—	14.5	1.68	3
FBG	26.7	13.3	1.2	10

3.7 SUMMARY

An overview of optical fiber sensors for interaction force measurements in MIS devices such as laparoscopic blades is presented in this chapter. The operating principles of two types of optical fiber sensors, FBGs and PCF sensors, are presented. Placement of the fiber sensor and the strain transfer from the blade to the sensor must be considered in the design of the sensorized surgical devices. A macrobend fiber interrogation is amongst a low-cost interrogation that can be used for FBGs. The PCFs are attractive to MIS device due to their low temperature dependence and high resolution. In addition, PCF sensors have wavelength resolution higher than that of the FBG sensors. A comparison of the performance evaluation of both PCF sensor and FBG-sensorized blades is also presented in this chapter to provide an insight into the benefits of each sensor type. The research direction in this area suggests the possibility of employing PCF and FBG optical force sensing in smart surgical devices where interaction force measurements are required to enable accurate and reliable force feedback in future surgical robotic systems.

REFERENCES

1. Kuebler, B., Seibold, U., and Hirzinger, G., Development of actuated and sensor integrated forceps for minimally invasive robotic surgery, *Int. J. Med. Rob. Comp. Assist. Surg.*, 1, 96–107, 2005.
2. Berkelman, P. J., Whitcomb, L. L., Taylor, R. H., and Jensen, P., A miniature microsurgical instrument tip force sensor for enhanced force feedback during robot-assisted manipulation, *IEEE Trans. Rob. Automat.*, 19, 917–922, 2003.
3. Tavakoli, M., Patel, R. V., and Moallem, M., Haptic interaction in robot-assisted endoscopic surgery: A sensorized end-effector, *Int. J. Med. Rob. Comput. Assist. Surg.*, 1, 53–63, 2005.
4. Puangmali, P., Althoefer, K., Seneviratne, L. D., Murphy, D., and Dasgupta, P., State-of-the-art- in force and tactile sensing for minimally invasive surgery, *IEEE Sensors J.*, 8(4), 371–381, 2008.
5. Park, Y. L., Elayaperumal, S., Daniel, B. L., Kaye, E., Pauly, K. B., Black, R. J., and Cutkosky, M. R., MRI-compatible haptics: Feasibility of using optical fiber Bragg grating strain-sensors to detect deflection of needles in an MRI environment, *Int. Soc. Magn. Reson. Med.* (ISMRM), 2008.
6. Park, Y. L., Ryu, S. C., Black, R. J., Chau, K. K., Moslehi, B., and Cutkosky, M. R., Exoskeletal force sensing end-effectors with embedded optical fiber Bragg grating sensors, *IEEE Trans. Robot.*, 25(6), 1319–1331, 2009.
7. Rao, Y. J., In-fiber Bragg grating sensors, *Meas. Sci. Technol.*, 8, 355–375, 1997.
8. Ascari, L., Corradi, P., Beccai, L., and Laschi, C., A miniaturized and flexible optoelectronic sensing system for tactile skin, *J. Micromech. Microeng.*, 17, 2288, 2007.
9. Yong-Lae, P., Elayaperumal, S., Daniel, B., Seok Chang Ryu., Shin, M., Savall, J., Black, R. J., Moslehi, B., and Cutkosky, M. R., Real-time estimation of 3-D needle shape and deflection for MRI-guided interventions, *IEEE/ASME Trans. Mechatron.*, 15, 906–915, 2010.

10. Heijmans, J. A. C., Vleugels, M. P. H., Tabak, E., Dool, T. V. D., and Oderwald, M. P., The impact of electrosurgical heat on optical force feedback sensors, in *4th European Conference of the International Federation for Medical and Biological Engineering*, Vol. 22, Magjarevic, R. (Ed.), Springer, Berlin, Germany, 2009, pp. 914–917.
11. Frazao, O., Santos, J. L., Araujo, F. M., and Ferreira, L. A., Optical sensing with photonic crystal fiber, *Laser Photon. Rev.*, 2, 449–459, 2008.
12. Cerqueira, S. A., Jr., Recent progress and novel applications of photonics crystal fibers, *Rep. Prog. Phys.*, 73, 024401, 2010.
13. Villatoro, J., Finazzi, V., Badenes, G., and Pruneri, V., Highly sensitive sensors based on photonic crystal fiber modal interferometers, *J. Sensors*, 747803, 2009.
14. Frazao, O., Baptista, M., and Santos, J. L., Temperature-independent strain sensor based on a Hi-Bi photonic crystal fiber loop mirror, *IEEE Sensors J.*, 7(10), 1453–1455, 2007.
15. Shi, Q., Fuyun Lv, Wang, Z., Long, J., Hu, J. J., Liu, Z., Kai, G., and Dong, X., Environmentally stable Fabry–Perot type strain sensor based on hollow core photonic bandgap fiber, *IEEE Photon. Technol. Lett.*, 20, 237–239, 2008.
16. Preusche, C. and Hirzinger, G., Haptics in telerobotics: Current and future research and applications, *Visual Comput.*, 23, 273–284, 2007.
17. Callaghan, D. J., McGrath, M. M., and Coyle, E., Force measurement methods in telerobotic surgery: Implications for end-effector manufacture, in *Proceedings of the International Manufacturing Conference (IMc25)*, Dublin 2008, pp. 389–398.
18. Seibold, U., Kuebler, B., and Hirzinger, G., Prototypic force feedback instrument for minimally invasive robotic surgery, in *Medical Robotics*, Bozovic, V. (Ed.), InTech Publishers, Rijeka, Croatia, 2008, ISBN: 978-3-902613-18-9.
19. Reiley, C. E., Evaluation of augmented reality alternatives to direct force feedback in robot-assisted surgery: Visual force feedback and virtual fixtures, Department of Computer Science. Master of Science, The Johns Hopkins University, Baltimore, MD, 2007.
20. Hill, K. O. and Meltz, G., Fiber Bragg grating technology fundamentals and overview, *J. Lightwave Technol.*, 15, 1263–1276, 1997.
21. Villatoro, J., Finazzi, V., Minkovich, V. P., Pruneri, V., and Badenes, G., Temperature-insensitive photonic crystal fiber interferometer for absolute strain sensing, *Appl. Phys. Lett.*, 91, 091109(3), 2007.
22. Xiao, L., Demokan, M. S., Jin, W., Wang, Y., and Zhao, C. L., Fusion splicing photonic crystal fibers and conventional single-mode fibers: Microhole collapse effect, *J. Lightwave Tech.*, 25, 3563–3574, 2007.
23. Minkovich, V. P., Villatoro, J., Hernandez, D. M., Calixto, S., Sotsky, A. B., and Sotskaya, L. I., Holey fiber tapers with resonance transmission for high-resolution refractive index sensing, *Opt. Express*, 13(19), 7609–7614, 2005.
24. Rajan, G., Callaghan, D., Semenova, Y., McGrath, M., Coyle, E., and Farrell, G., A fiber Bragg grating based all-fiber sensing system for telerobotic cutting applications, *IEEE Sensors J.*, 10(12), 1913–1920, 2010.
25. Callaghan, D., McGrath, M. M., Rajan, G., Coyle, E., Semenova, Y., and Farrell, G. Analysis of strain transfer to FBG's for sensorized telerobotic end-effector applications, in *Advances in Robotic Research. Theory, Implementation, Application*, Torsten Kröger and Friedrich M. Wahl, Eds., Part 3, Springer-Verlag, New York, June 2009, pp. 65–75.
26. Wang, Q., Farrell, G., Freir, T., Rajan, G., and Wang, P., Low-cost wavelength measurement based on a macrobending single-mode fiber, *Opt. Lett.*, 31(12), 1785–1787, 2006.
27. Rajan, G., Semenova, Y., Hatta, A., and Farrell, G., Passive all-fiber wavelength measurement systems: Performance determination factors, in *Advances in Measurement Systems*, Milind Kr Sharma (Ed.), InTech Publishers, Rijeka, Croatia, 2010, ISBN 978-953-307-061-2.
28. Wang, Q., Rajan, G., Wang, P., and Farrell, G., Macrobending fiber loss filter, ratiometric wavelength measurement and application, *Meas. Sci. Technol.*, 18, 3082–3088, 2007.
29. Rajan, G., Semenova, Y., Freir, T., Wang, P., and Farrell, G., Modelling and analysis of the effect of noise on an edge filter based ratiometric wavelength measurement system, *J. Lightwave Technol.*, 26, 3434–3442, 2008.
30. Rajan, G., Semenova, Y., Wang, P., and Farrell, G., Temperature induced instabilities in macro-bend fiber based wavelength measurement systems, *J. Lightwave Technol.*, 27, 1355–1361, 2009.
31. Smart Fibres Ltd, Bracknell, United Kingdom, http://www.smartfibres.com/FBG-interrogators. Accesed on June 2010.

4 Recent Advances in Distributed Fiber-Optic Sensors Based on the Brillouin Scattering Effect

Alayn Loayssa, Mikel Sagues, and Ander Zornoza

CONTENTS

4.1 INTRODUCTION

Fiber-optic sensors based on the stimulated Brillouin-scattering effect can become a truly disruptive technology in the field of structural health monitoring. These sensors can be embedded within materials and structures and provide a fiber-optic "nerve system" that feels the "pain" that they are suffering in the form of strain, stress, deformation, delamination, cracks, temperature variations, and others [1]. Moreover, this type of sensor can be deployed in other monitoring applications such as in power lines, bridges, pipelines, fire detection, process control, etc. For instance, Brillouin distributed sensors (BDSs) can be used in electric power lines to measure temperature, which gives an indirect assessment of local current load and degradation of the line. Temperature change detection can also give an indication of leakage in oil or gas pipelines. In roads and railways, distributed strain measurements serve to detect ground displacements that may affect the stability of the infrastructure. Furthermore, BDSs can also be configured to measure displacement [2], birefringence [3], and chromatic dispersion [4] in addition to temperature and strain.

The main asset of BDSs in all these applications is their ability to perform distributed measurements. Conventional fiber-optic point sensors such as fiber Bragg gratings (FBGs) or interferometric sensors just measure at specific locations within a structure, whereas BDSs provide a quasi-continuous measurement of strain and temperature along an optical fiber. The sampling interval of this measurement can be made arbitrarily small and, for each sample, the BDS gives an average value of the measurands integrated over the spatial resolution of the sensor. Therefore, a single BDS can replace many point sensors. This translates to an important reduction in costs when monitoring large structures. Further reduction in costs comes from the fact that in BDS a standard single-mode fiber (SSMF) without any modifications is deployed as a transducer in contrast to point fiber-optic sensors that require an elaborate manufacturing process.

In this chapter, we focus on some of the main research trends currently unfolding in the field of BDS sensing. These can be broadly grouped in three lines: simplification and cost reduction of the experimental setups, enlargement of the measurement range, and enhancement of the spatial resolution of the sensors. The simplification and cost reduction of the implementation of BDSs are required in order to make feasible their widespread application. The other two main research lines are related to the enhancement of the performance of BDS in terms of range as well as resolution. These are complementary efforts. On the one hand, there is the need to monitor large structures that can extend over hundreds of kilometers such as railways, pipelines, power lines, etc. On the other hand, increasingly detailed information of the measurands' distribution is required in applications such as structural health monitoring. For instance, enhanced resolution can enable the detection of small cracks in reinforced concrete structures.

The organization of this chapter is as follows. We start by giving a basic introduction to BDS in Section 4.2. This includes a brief review of the Brillouin scattering effect and the description of the techniques that have been devised to harness this nonlinear effect in order to build viable sensor systems. Then, in Section 4.3, we describe the work that has been done lately in some of the research lines described earlier, with emphasis on the work performed within our group. Finally, conclusions drawn from the chapter material are presented.

4.2 FUNDAMENTALS OF BRILLOUIN DISTRIBUTED SENSORS

Brillouin scattering nonlinear effect results from the interaction in a material between optical photons and acoustic phonons [5]. In optical fibers, spontaneous Brillouin scattering takes place when narrow-band pump light interacts with thermally excited acoustic waves. Due to the acousto-optic effect, the acoustic wave generates a periodic perturbation of the refractive index that reflects part of the energy of the pump wave by Bragg diffraction. The reflected light (Stokes wave) experiences a shift in frequency due to the Doppler effect. This is the so-called Brillouin frequency shift (BFS), which is related to the velocity of the acoustic waves in the fiber. Spontaneous Brillouin scattering can become stimulated Brillouin scattering (SBS) if a Stokes wave whose optical frequency is downshifted from that of the pump by the BFS is injected in the fiber. In SBS, the counterpropagation of the pump and Stokes waves generates a moving interference pattern which induces an acoustic wave due to the electrostriction effect, reflecting part of the energy of the pump wave by Bragg diffraction. The frequency of the light reflected as a consequence of this process is downshifted by the BFS, which is given by

$$\mathrm{BFS} = \frac{2n\upsilon_\mathrm{A}}{\lambda_\mathrm{p}} \tag{4.1}$$

where
   υ_A is the velocity of the acoustic wave
   n is the refractive index
   λ_p is the wavelength of the pump wave

This shift is around 10.8 GHz at a wavelength of 1550 nm for silica fibers.

As a result of this interaction, there is a transfer of energy between the pump and Stokes waves that simultaneously enhances the amplitude of the acoustic waves leading to a stimulation of the process. Therefore, as the signals propagate along the fiber, the Stokes wave is amplified, while the pump wave losses energy and is attenuated. In terms of spectral response, this interaction can be described by the Brillouin gain coefficient given by

$$g_B(\Delta v) = g_B \frac{(\Delta v_B/2)^2}{\Delta v^2 + (\Delta v_B/2)^2} \quad (4.2)$$

with g_B being the peak gain coefficient, Δv the detuning from the center of the Brillouin resonance, and Δv_B the Brillouin linewidth, which is given by the inverse of the phonon lifetime and is usually on the order of a few tens of MHz for SSMF.

When using the Brillouin scattering effect for optical fiber sensing purposes, the dependence of Brillouin parameters on temperature and strain in the fiber is exploited. This linear dependence of the BFS with the applied strain $\delta\varepsilon$ or temperature change δT is given by [6]

$$BFS - BFS_0 = A \times \delta\varepsilon + B \times \delta T \quad (4.3)$$

where BFS_0 is the reference BFS value, measured at room temperature and in the loose state of the fiber, i.e., laid freely in order to avoid any artificial disturbances. A and B are the strain and temperature coefficients given in MHz/$\mu\varepsilon$ and MHz/°C units, respectively. Typical values for these strain and temperature coefficients are around 0.04 MHz/$\mu\varepsilon$ and 1 MHz/°C for most standard fibers at 1550 nm.

Therefore, the BFS in the fiber can be found by measuring the BGS, or Brillouin loss spectrum (BLS), and then it can be translated to a temperature or strain measurement. In order to perform this spectral measurement, a pump wave can be injected from one end of an optical fiber and a probe wave from the other, and the gain (or loss) experienced by the latter can be recorded as the wavelength separation between the two waves is scanned. Finally, the strain and temperature of the fiber can be obtained from the measured Brillouin spectrum peak (gain or loss) using the coefficients obtained from a previous calibration of the deployed sensing fiber.

Nevertheless, this procedure would measure the global gain spectrum of the total length of fiber, which is given by the integration of the local gain (or loss) at each position along the fiber. In order to implement a distributed sensor featuring spatially resolved measurements, the gain (or loss) spectra at individual locations in the fiber must be isolated. This can be done in the time, coherence, or frequency domains, giving rise to the three main analysis-type BDS families: Brillouin optical time-domain analysis (BOTDA), Brillouin coherence-domain analysis (BOCDA), and Brillouin optical frequency-domain analysis (BOFDA). This chapter is focused on BOTDA sensors because they are the BDS type with the greatest potential.

The fundamentals of BOTDA sensors are highlighted in Figure 4.1. A continuous wave (CW) probe wave is counterpropagated with a pump pulse. The pulse propagates along the fiber and at each location imparts Brillouin gain to the probe. Finally, the received probe signal is detected in the time domain, so that the position-dependent gain can be calculated by the classical time-of-flight method. If we consider t_0 the time when the trailing end of the pump pulse enters the fiber, then the probe signal received at $t = t_0 + 2z/c$ has interacted with the pulse between positions z and $z + u$, where c is the speed of light in the fiber and $u = \tau \times c/2$ is the interaction length, with τ the temporal duration of the pulse. Therefore, the spatial resolution of the measurement is given by u. For instance, in order to achieve a resolution of 1 m, a pulse duration of around 10 ns must be deployed. Notice that the two factor in the expression for u is due to the fact that the pump pulse and the CW probe are counterpropagating. From the previous discussion, it is clear that the time-dependent BOTDA signal can be translated to position-dependent gain by the simple relation $z = (t - t_0) \times c/2$. Therefore, as shown in the figure, it is possible to reconstruct the position-dependent Brillouin spectra distributed along the fiber by sweeping the wavelength separation between pump and probe and measuring multiple BOTDA time-domain traces. Finally, strain or temperature can be quantified from the measured BFS at each position.

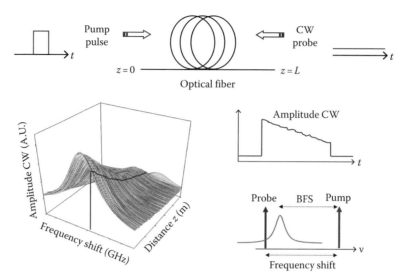

FIGURE 4.1   Fundamentals of BOTDA.

4.3   RESEARCH TRENDS IN BRILLOUIN DISTRIBUTED SENSORS

Current research trends in BDS are defined by the needs of the potential application fields of this sensing technology. These applications demand an ideal BDS that enables measurements in ranges larger than 100 km and with spatial resolutions under a meter, providing strain and temperature simultaneously, obtaining accurate results in fast quasi "real time" and all at a reasonable cost.

The effort to enlarge the measurement range of BDS is mainly driven by applications where there is the need to monitor large structures such as pipelines, railways, power lines, etc. As in fiber-optic communications, the longer the sensing fiber, the lower the SNR becomes in detection and the worse the performance of the sensor. Moreover, pump and probe powers cannot be increased indefinitely so as to compensate the SNR reduction, since other undesirable hindrances such as nonlinear and nonlocal effects will also deteriorate the sensor accuracy. The research to enlarge the measuring range in BOTDA sensors is thus focused on the development of techniques to compensate the SNR reduction in long fibers while mitigating undesirable effects.

Structural health monitoring and other applications require measurement of either temperature or strain or both simultaneously. However, Equation 4.3 highlights that basic BDSs, as other optical fiber sensor technologies, display a cross-sensitivity to temperature and strain that makes it impossible to isolate either of these measurands unless additional actions are taken. The simplest solution is to use two fibers in the measurements, with one of them in a loose state dedicated just to measure temperature. However, this is only possible in short fibers, where doubling the measuring range does not lead to a significant penalty in the sensor performance. Other approaches rely on using fibers with multiple acoustic modes [7] or using birefringent fibers [8] with the common goal of having two independent responses to strain and temperature.

Spatial resolution is another area of research that has concentrated great efforts recently. Many potential applications of BDS require the highest possible spatial resolution to obtain an accurate assessment of the distribution of a given measurand. However, in the standard configuration of a BOTDA sensor, the spatial resolution is limited to around 1 m. This limitation stems from the broadening of the BGS as the pump pulse duration is reduced. Indeed, the effective BGS is the convolution of the pulse spectrum and the Lorentzian Brillouin linewidth. The latter is of the order of a few tens of MHz as determined by the acoustic lifetime of phonons in the fiber ($\tau_A \approx 10$ ns). For long pulses (reduced spatial resolution), this convolution approximately equals the intrinsic Brillouin spectrum. On the contrary, as the pulse width approaches the duration of the acoustic lifetime, the pulse spectrum broadens over the natural Lorentzian linewidth and increasingly determines the final BGS linewidth. The BGS is measured in the presence of noise. Hence, the narrower the final spectra, the better the precision in the determination of the peak of the BGS and thus of the strain and temperature in the fiber. Therefore, for a given precision and measurement time, the pulse has to be longer than around 10 ns in most setups (≈1 m spatial resolution).

Solutions to enhance the resolution over the acoustic lifetime limit have been proposed. They are based on the observation that preexcitation of the acoustic wave in the Brillouin interaction can overcome the BGS broadening effect [9]. The basic idea is to have Brillouin interaction before the arrival of the pump pulse. This can be achieved by the presence of preexisting CW pump and Stokes signals. In this way, upon arrival of the pulse, it can reflect on the preexisting grating without the need to generate its own interaction. Therefore, the interaction spectrum experienced by the pump wave is defined by the natural Brillouin linewidth and not by the convolution of the pump spectrum and Lorentzian profile. This principle has been exploited in several configurations; for instance, using a pulse over a CW pump pedestal [10] or introducing a "dark pulse" [11], i.e., a temporal suppression of an otherwise continuous pump wave. More recently, an improved physical explanation for the acoustic wave preexcitation has been developed that has led to an optimized setup based on using optical phase-shift instead of optical intensity pulses [12].

The final major research trend is focused on cost reduction. In BDS, the transducer itself, i.e., SSMF, is very cheap. However, the interrogation setup is complex and expensive, using a number of broadband and microwave frequency components as well as sophisticated photonic devices.

Among the research trends just described, we will focus on the results of the research of our group in the cost reduction of BOTDA sensors and in the enlargement of their measuring range. We present two techniques with the aim of simplifying and reducing the cost of the setup: an enhanced setup that shapes the pump pulses in the radio frequency (RF) domain and a novel scanning method of the Brillouin spectra based on the dependence of Brillouin scattering with wavelength. Furthermore, in order to enlarge the measuring range we describe two additional techniques: one using distributed Raman amplification and another taking advantage of self-heterodyne detection.

4.3.1 SIMPLIFICATION OF EXPERIMENTAL SETUPS

When compared to other sensing technologies, BOTDA implementations are usually complex and use expensive components such as synthesized microwave generators, multiple electro-optic modulators, wideband detectors, or semiconductor optical amplifiers. For that reason, much effort has been recently devoted to simplify the experimental setups in order to achieve cost-effective commercial systems that can compete, for instance, with the simpler and less costly distributed Raman sensors for temperature measurements. Recent examples of simplified BOTDA schemes include the use of Brillouin generators or Brillouin fiber lasers to obtain the probe wave from the pump [13], the deployment of injection locking to generate the pump and probe waves using DFB lasers [14], or the use of offset locking [15]. In the following, we introduce with more detail two of the developments of our group in this area: first, a system to generate high extinction ratio (ER) pump pulses in a simplified manner and then a method to perform the scanning of the Brillouin spectrum without using a costly synthesized microwave generator.

4.3.1.1 RF Pulse Shaping

When optical devices such as optical modulators are used to shape the pump pulse, there is always a residual DC base or leakage. Thus, there is SBS interaction between the leakage and the probe in addition to that between the pulse and the probe. This leads to a distortion of the measured Brillouin spectra. Solutions to improve ER of pump pulses have been proposed using the injection locking method [14], using semiconductor optical amplifiers as optical switches [16], or using specialty ultrahigh ER electro-optic modulators [15].

The system that we propose provides an alternative solution in the form of a simplified BOTDA sensing scheme that shapes pump pulses in the RF domain, where it is easier and cheaper to obtain high ER pulses [17].

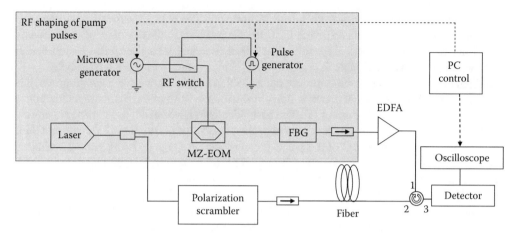

FIGURE 4.2 Experimental setup of a simplified BOTDA sensing scheme featuring high ER pulses. (From Zornoza, A., Olier, D., Sagues, M., and Loayssa, A., Brillouin distributed sensor using RF shaping of pump pulses, *Meas. Sci. Technol.*, 21, 094021, 2010. With permission from Institute of Physics.)

Figure 4.2 highlights the experimental setup of a simplified BOTDA system using the RF shaping method [17]. In this system, the output of a laser source is divided in two branches by an optical coupler. In this way, the same light source is used for pump and Stokes generation. The sensing fiber is located in the lower branch of the setup, where the output of the laser source is used as CW probe after being polarization-scrambled to compensate the polarization sensitivity of SBS. In the upper branch, the RF shaping of pump optical pulses is performed, so that the pulsing and frequency shifting are achieved in a single step. Figure 4.3 schematically describes this process.

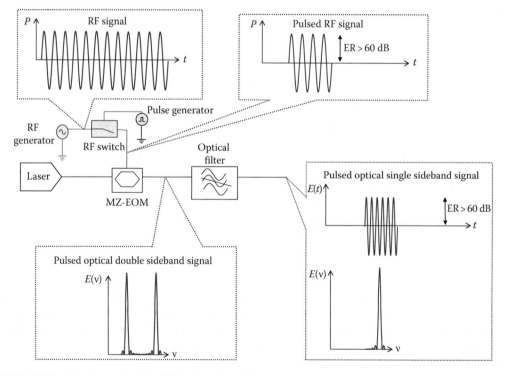

FIGURE 4.3 Fundamentals of the RF shaping of optical pump pulses. (From Zornoza, A., Olier, D., Sagues, M., and Loayssa, A., Brillouin distributed sensor using RF shaping of pump pulses, *Meas. Sci. Technol.*, 21, 094021, 2010. With permission from Institute of Physics.)

A microwave generator is deployed to generate RF signals at frequencies close to the BFS of the sensing fiber. This RF signal is injected into a single-pole single-throw microwave switch, driven by a baseband pulse generator, so that high ER RF pulses are generated. Low-cost microwave switches are commercially available, featuring extremely high isolation and very fast responses (up to 1 ns). Therefore, this is an inexpensive solution to obtain fast pulses of RF energy with extremely high ER. Then, a Mach–Zehnder electro-optic modulator (MZ-EOM) biased at minimum transmission and driven by these high ER RF pulses is deployed to modulate an optical beam. Therefore, the RF pulse shape is directly translated to the optical domain by the modulator, generating two pulsed sidebands with very high ER. Next, this pulsed optical double sideband suppressed carrier modulated signal is filtered with a FBG to select the upper sideband and an erbium-doped fiber amplifier (EDFA) is used to amplify the pulses before they are launched into the sensing fiber using a circulator. Finally, the resulting BOTDA signal is directed to a photodetector and then captured with a digital oscilloscope.

The BOTDA sensing scheme described earlier presents advantages over conventional BOTDA setups. First, a single MZ-EOM is required in the setup. Moreover, this can be a low-cost telecom-grade MZ-EOM with 20–30 dB ER even if higher ER pulses are required, instead of the much more expensive specialty high-ER MZ-EOMs that would be needed in conventional setups. Finally, as the CW probe comes directly from the unmodulated laser source, no additional deleterious spurious signals are present in the detected optical spectra, avoiding the need for additional optical filtering in the receiver.

Figure 4.4 depicts the results of a distributed temperature measurement in a 25 km length of SSMF with −10 dBm CW and 20 dBm pump pulses and where the RF was swept at 1 MHz steps. The resolution was set to 6 m by the 65 ns pulses provided by the specific microwave switch deployed, but latter refinements of this system have demonstrated resolutions as high as 1 m [18]. The last 200 m of the fiber was placed loose in a climatic chamber at 47°C, whereas the rest was held at room temperature in a spool. As shown in the figure, the heated section of the fiber is clearly detectable due to the shift in the BFS. The temperature accuracy of the measurement was estimated to be 0.44°C.

4.3.1.2 Spectral Scanning of the Brillouin Spectra Using Wavelength Tuning

Another approach to simplify the setup of BOTDA sensors is to deploy alternative Brillouin spectrum characterization methods. The majority of Brillouin distributed sensing techniques are based on the spectral characterization of this nonlinear effect along an optical fiber. As it was previously explained, this characterization is usually implemented using pump and probe schemes where the wavelength separation of the pump and the probe is swept to scan the spectral response of the interaction. In most proposals, this tuning of the wavelength separation is ultimately controlled by a synthesized microwave generator which can be a costly component. Recently, an alternative Brillouin spectral scan technique for BDSs has been proposed that is based on the wavelength dependence of the BFS [19]. Its main advantage is that it allows the substitution of the synthesized microwave generators or the complex wavelength sweeping schemes of other setups by the use of a low-cost coarse tunable laser.

The wavelength dependence of the BFS in single-mode fibers is given by Equation 4.1, which is repeated here for clarity:

$$\text{BFS} = \frac{2nv_a}{\lambda_p} = v_p \times \frac{2nv_a}{c} \tag{4.4}$$

Note that there is an explicit relationship with v_p, and also another implicit via the wavelength dependence of n. Nevertheless, the latter can be neglected for most fibers if a small wavelength range is considered. This wavelength dependence of BFS can be exploited to scan the Brillouin spectra following the technique depicted in Figure 4.5. This technique is based on tuning the wavelength of the pump

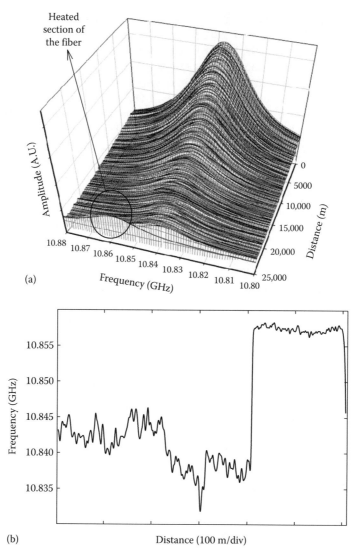

(a)

(b)   Distance (100 m/div)

FIGURE 4.4 Evolution of the (a) Brillouin spectra and the (b) measured Brillouin frequency shift in the fiber under test. The last 200 m corresponds to the heated section of the fiber. (From Zornoza, A., Olier, D., Sagues, M., and Loayssa, A., Brillouin distributed sensor using RF shaping of pump pulses, *Meas. Sci. Technol.*, 21, 094021, 2010. With permission from Institute of Physics.)

and probe waves simultaneously while their optical frequency difference is kept constant. According to Equation 4.4, BFS varies with pump wavelength; thus, a different detuning from the Brillouin resonance is experienced by the probe wave at each wavelength. This detuning is given by $\Delta v = v_S - v_p + BFS$, where v_S is the optical frequency of the Stokes wave. In conventional pump–probe Brillouin sensors, Δv is scanned by modifying $f_m = v_S - v_p$, while in the present method, it is via BFS. In addition, notice that this spectral scanning method can be applied to any of the three analysis-type BDS families.

Measured variations of BFS with pump wavelength in SSMF are typically of the order of 7 MHz/nm at 1550 nm [20]. Hence, in order to measure the complete Brillouin spectrum, a full C-band tunable laser is required. However, fine wavelength tuning is not needed since BDS requires spectrum scanning at around 1 MHz resolution, i.e., 0.15 nm steps. Therefore, this coarse tuning capability could be provided, for instance, by a low-cost wavelength-agile monolithic tunable laser as those deployed for wavelength division multiplexing networks.

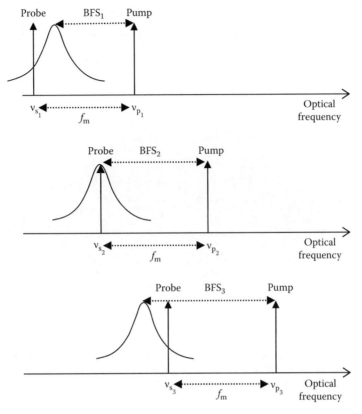

FIGURE 4.5 Fundamentals of the Brillouin spectral scanning method using wavelength tuning. (Zornoza, A., Olier, D., Sagues, M., and Loayssa, A., Brillouin spectral scanning using the wavelength dependence of the frequency shift, *IEEE Sensors J.*, 11(2), 382–383, 2011. With permission. Copyright 2011 IEEE.)

A possible setup implementing a BOTDA sensor using this technique was introduced in [19], which is a version of the one in Figure 4.2 but slightly modified to use a tunable wavelength laser. Figure 4.6 depicts measurements with this setup of the Brillouin spectra in a 10 km length of SSMF with 200 m of fiber placed in a climatic chamber at 42°C while the rest was at room temperature. The spatial resolution of this measurement was 24 m and the temperature was measured with a resolution of 0.3°C. The laser wavelength was tuned in 0.13 nm steps from 1540.3 to 1551.5 nm.

A factor to take into account in this system is the effect of chromatic dispersion in the optical fiber. This leads to a difference in propagation velocity in the fiber of the different wavelengths. In the measurements that have been presented, its effect is negligible due to the relatively long pulse temporal width and the short lengths of fiber involved. However, in long-range high-resolution measurements, the influence of chromatic dispersion should be calibrated out, which is not difficult as dispersion is a stable, deterministic effect.

4.3.2 Enlargement of the Sensor Measurement Range

The main practical problems in BOTDA setups when the length of the sensing fiber increases are SNR reduction, nonlinear effects, and nonlocal effects. The measuring range is limited because of the attenuation of the pump pulse and the probe wave while they propagate along the fiber. When these signals are small, the gain of the probe wave is buried in detection noise and more averaging is demanded to obtain significant measurements; thus, the measurement time is greatly increased.

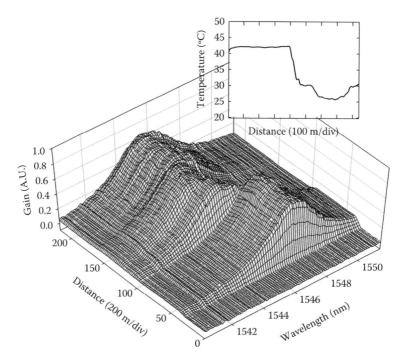

FIGURE 4.6 Distributed measurement of the Brillouin gain for every pump wavelength and (inset) distributed temperature. (Zornoza, A., Olier, D., Sagues, M., and Loayssa, A., Brillouin spectral scanning using the wavelength dependence of the frequency shift, *IEEE Sensors J.*, 11(2), 382–383, 2011. With permission. Copyright 2011 IEEE.)

A simple solution to this problem could be to increase the injected pump power, and consequently increase the Brillouin gain of the CW probe signal. However, the maximum pump power launched into the fiber is limited by the appearance of several nonlinear effects such as self-phase modulation, modulation instability, or spontaneous Raman scattering.

Self-phase modulation of pump pulses appears in long-range measurements (>10 km) when the shape of the pulses has slow rising and falling times [21]. In this case, the pump pulses suffer a spectral broadening that can have a significant effect on the measured Brillouin spectrum linewidth. Since the measured BFS accuracy is proportional to the Brillouin spectral width [22], it causes a reduction of the measurement resolution. However, self-phase modulation can be easily avoided by using perfectly shaped rectangular pump pulses with sharp rising and falling edges.

Modulation instability is harder to tackle. This effect is the result of the interplay between anomalous dispersion and the Kerr effect in the fiber [23]. It causes a spectral broadening of the pump pulse spectrum, and hence a broadening of the measured Brillouin spectrum, which, as with self-phase modulation, increases the measurement uncertainty. Moreover, the pump pulse suffers a strong depletion, which can completely suppress the Brillouin gain at some sections of the fiber. Modulation instability can be mitigated using normal dispersion fiber instead of SSMF [24]. However, there is a limit to the maximum pulse power that can be used despite using normal dispersion fiber. This is set by the depletion of the pulse suffered because of spontaneous Raman scattering [25].

In principle, apart from increasing the pump power, another possibility to mitigate the SNR reduction while enlarging the range of the sensor could be to increase the injected probe power. However, the limit in this case is set by the onset of nonlocal effects. Nonlocal effects refer to a situation in which the measurement of a given position along the fiber is dependent on the interaction at every preceding location of the fiber. Although ideally, as the pump pulse travels to a given measurement location it is just affected by the attenuation of the fiber, in reality when the probe wave is present there is Brillouin interaction at all the previous sections with energy been transferred from pump to probe. This means that the pump pulse suffers depletion caused by the CW as it propagates along the fiber. The frequency-dependent nature of the pump depletion introduces distortion in the measured gain spectra because of the dependence of the Brillouin gain with the pump. As a consequence, a systematic error in the measured BFS is introduced. This error increases with greater probe powers, and it reaches its maximum as the ending section of the fiber under test is approached. It should be noted that nonlocal effects are greatly accentuated by imperfections in the generated pump pulses, particularly, using pulses with deficient ER [17]. Therefore, in order to minimize nonlocal effects, it is of the essence to use high ER pump pulses. Recently, pulsing the CW to limit the interaction length between pulse and probe has been proven as a good solution to decrease nonlocal effects at the expense of increasing the measuring time [26]. Some other proposals are based on compensating the measured spectra by quantifying the pulse energy loss [27] or by numerically reconstructing the distorted BFS profile to fit the measurement data [28].

4.3.2.1 Use of Raman Amplification

As it has been discussed, the SNR of the detected BOTDA signal is limited by the attenuation of the fiber. As the pump pulses propagate along the fiber, they are attenuated; hence, the Brillouin gain imparted to the CW probe is very small at the end of long fibers. This hinders the proper measurement of the Brillouin spectra. One solution to this problem is to increase the pump pulses power injected in the fiber. However, as it was previously mentioned, the maximum power of the pump pulses injected in the fiber is limited by the onset of nonlinear effects. An alternative is to compensate the pump pulses attenuation as they propagate along the fiber by using distributed amplification based on stimulated Raman scattering (SRS). This has the main advantage that the gain is distributed along the fiber so that the pump pulse does not reach at any position the high power that causes nonlinear effects.

We have demonstrated the use of Raman amplification in a long-range hybrid sensor network comprising point and distributed sensors [29]. Previously, the use of Raman amplification had been demonstrated for the BOTDR-type sensor [30]. However, to our knowledge ours was the first experimental demonstration of its application to BOTDA sensors.

Figure 4.7 depicts the proof-of-concept network that was experimentally demonstrated integrating point vibration sensors based on FBGs and tapers with distributed temperature sensing along the network bus. In this network, the use of Raman amplification allows to compensate branching and fiber losses. In addition, Raman distributed amplification is deployed to simultaneously amplify both the Brillouin sensing and the interrogation signals of the point sensors. Therefore, the range of the BDS is increased and the optical SNR of the point sensor system is enhanced. The hybrid sensor network can be divided in two parts: the monitoring station equipment and the sensor network itself.

The monitoring station is where all the signal processing required for the interrogation of the point sensors and for the Brillouin distributed measurements takes place. The Brillouin distributed measurement system implemented in the network is based on a BOTDA setup optimized for long range, featuring the RF pulse-shaping technique described in Section 4.3.1.1. As it was previously explained, in long-range BOTDA sensors the problem of nonlocal effects is greatly enhanced by the use of pump pulses with insufficient extinction ratio. In this system, by deploying the RF pulse-shaping principle, we are able to generate pulses with extremely high extinction ratio.

The sensor network has a bus topology suitable for wavelength division multiplexed interrogation. Point sensors are introduced in the bus using fiber couplers and they are followed by a narrow-bandwidth FBG reflecting light at a unique wavelength. Finally, a Raman pump laser is injected at one end of the fiber bus using a fiber wavelength division multiplexer, so that distributed amplification is generated in the sensor bus, extending the fiber sensor network range. A second wavelength division multiplexer is used at the other end of the bus to prevent the residual Raman pump from reentering the monitoring station.

The specific network demonstrated had a 46 km bus length that was implemented using one 25.2 km and four 5.2 km fiber spools, all of them of SSMF. Four vibration sensors based on

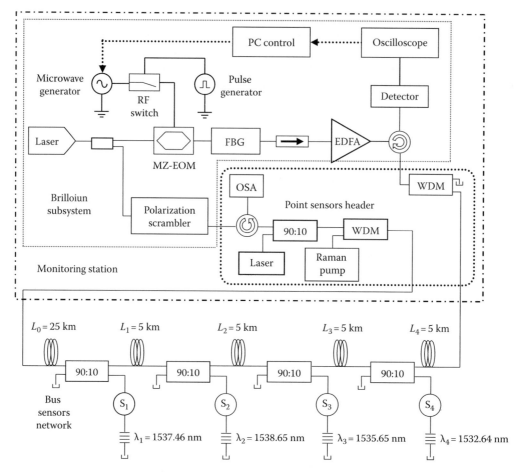

FIGURE 4.7 Experimental setup of the hybrid sensor network with point and distributed optical sensors. The monitoring station contains the Brillouin subsystem and the point sensors' header. (From Zornoza, A. et al. *Opt. Express*, 18, 9531, 2010. With permission. Copyright 2010 Optical Society of America.)

fiber-optic tapers (S_1, S_2, S_3, and S_4 in Figure 4.7) were located at 5.2 km intervals at the end of the fiber bus. Finally, a length of 200 m of fiber was inserted at the middle of the fiber bus in a temperature-controlled climatic chamber at 60°C.

Figure 4.8a depicts the complete characterization of the Brillouin spectra along the fiber with 13 m resolution. The 200 m of heated fiber in the middle of the bus can be clearly distinguished. The amplitude evolution of the measured Brillouin spectra along the fiber is highlighted by the figure. This is directly related to the amplitude of the pump pulse that the probe wave finds at each location. Initially, the amplitude of the pulses starts to decay due to fiber attenuation as they travel through the fiber. However, around the middle of the bus the trend is inverted due to Raman amplification and the amplitude progressively increases up to the fiber bus end. Several pumping configurations were analyzed to find an optimal backward Raman pumping of the pulses, so that the pump pulses get maximum amplification at the end of the fiber. The optimum Raman pump was found to be 1 W. The use of higher pump powers was prevented due to the undesirable increase in the probe wave amplitude, which leads to significant saturation of the Brillouin gain and depletion of the pump, introducing systematic errors in the measurement. If a network with more point sensors is desired, the Raman pump should be increased accordingly to compensate the additional coupler losses.

Figure 4.8b shows the measured BFS along the fiber, obtained by postprocessing the data in Figure 4.8a. The data reveals the four initial 5.2 km spools with slightly different BFS, then the

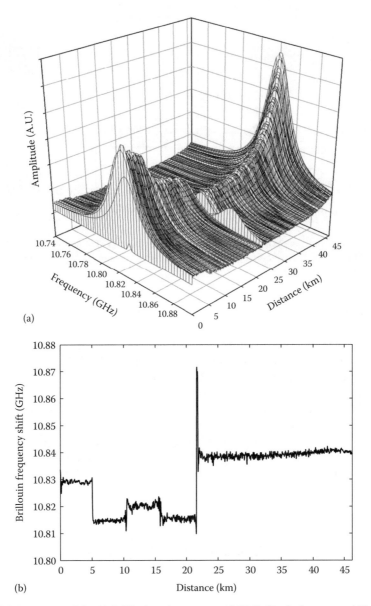

(a)

(b)

FIGURE 4.8 Measurement of the (a) Brillouin gain spectra and (b) Brillouin frequency shift along the fiber network. (From Zornoza, A. et al. *Opt. Express*, 18, 9531, 2010. With permission. Copyright 2010 Optical Society of America.)

temperature-controlled 200 m length and finally the 25.2 km spools. The BFS in the temperature-controlled area is found to be in good agreement with a previous calibration of the fiber that had given a temperature coefficient of 0.9 MHz/°C. The temperature resolution of the measurement was estimated to be 0.7°C.

4.3.2.2 Self-Heterodyne Detection to Increase Measurement SNR

As it has been already explained, another method to increase SNR in long-range BOTDA sensors is increasing the probe power. However, nonlocal effects set an upper limit to the maximum probe power that can be used in conventional BOTDA sensors. We have devised a method to counteract this limitation in which nonlocal effects are avoided by keeping the power of

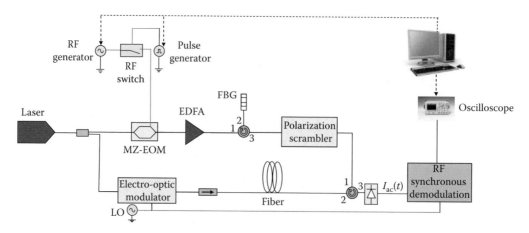

FIGURE 4.9 Experimental setup of the self-heterodyne detection BOTDA sensor. (From Zornoza, A. et al., *Proceedings of SPIE—The International Society for Optical Engineering 7753*, paper 77532F, 2011. With permission. Copyright 2011 SPIE.)

the probe wave low but nonetheless increasing the SNR at the detector. This BOTDA sensor scheme takes advantage of the enhanced characteristics obtained employing self-heterodyne optical detection combined with synchronous RF demodulation to increase the sensibility of the sensor [18]. Moreover, this technique allows measuring the full complex Brillouin spectrum instead of just its magnitude.

Figure 4.9 depicts a simplified coherent detection BOTDA setup [18]. Again, it is a modified version of the RF pulse-shaping setup described in Figure 4.2. The difference is mainly in the lower branch of the setup, which is used to generate the signals required to perform the self-heterodyne detection. An electro-optic modulator driven by an RF synthesizer is used to modulate the laser output. The rationale is to use one of the modulation sidebands as probe for the Brillouin interaction and the carrier as local oscillator (LO). Assuming an optical single-sideband modulation is deployed, the optical field at the output of the modulator is given by

$$E_T(t) = E_S \exp(j2\pi(\nu_0 + f_{RF})t) + E_0 \exp(j2\pi\nu_0 t) \qquad (4.5)$$

where
   E_0 and E_S are the complex amplitudes of the sideband and the optical carrier, respectively
   f_{RF} is the modulation frequency
   ν_0 is the optical frequency of the optical carrier

In the fiber, SBS interaction occurs between the pump pulse and the sideband of the modulation. Therefore, the optical field that reaches the photodetector after Brillouin interaction is given by

$$E_T(t) = E_S(1 + g_{SBS})\exp(j(2\pi(\nu_0 - f_{RF})t + \phi_{SBS})) + E_0 \exp(j2\pi\nu_0 t) \qquad (4.6)$$

where g_{SBS} and ϕ_{SBS} are the gain and phase shift generated on the probe wave due to Brillouin interaction.

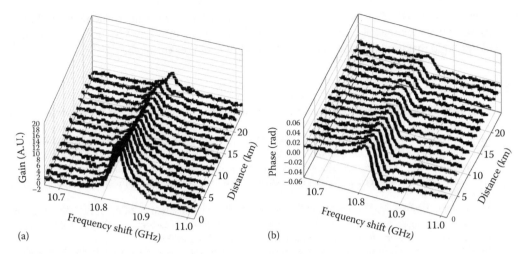

FIGURE 4.10 Distributed measurements of Brillouin (a) gain and (b) phase-shift spectra along a 25 km long sensing optical fiber. (From Zornoza, A. et al., *Proceedings of SPIE—The International Society for Optical Engineering 7753*, paper 77532F, 2011. With permission. Copyright 2011 SPIE.)

After quadrature detection in the photodetector, the detected RF current becomes

$$I_C(t) = 2R_C(1 + g_{SBS})\sqrt{P_S P_0}\, \cos(2\pi f_{RF} t - \phi_{SBS}) \tag{4.7}$$

where
  R_C is the responsivity of the detector
  P_0 and P_S are the optical powers of optical carrier and modulation sideband (probe), respectively

Notice that I_C contains not only the conventional Brillouin gain information but also information of the Brillouin phase shift along the fiber. Both can be recovered using RF synchronous demodulation. Finally, the detection SNR for the self-heterodyne BOTDA sensor is given by

$$SNR_C = \frac{4R_C{}^2 P_S g_{SBS}^2 P_0}{\sigma_n^2} \tag{4.8}$$

where σ_n^2 is the total noise of the system. This SNR is proportional to the optical carrier power. Therefore, by using self-heterodyne detection, we can maintain the probe wave power low to the limit in which nonlocal effects are negligible, and increase the SNR by using a stronger optical carrier. Moreover, as it was previously mentioned, the phase shift as well as amplitude of the Brillouin spectra can be measured. For example, Figure 4.10 depicts distributed measurements of Brillouin gain and phase shift along a 25 km fiber with its last 200 m heated in a thermal bath at 40°C. Note that in the heated section of the fiber, both gain and phase-shift spectra are shifted in frequency according to the temperature change.

4.4 CONCLUSIONS

In this chapter, we have presented an overview of the fundamentals and current research lines in BDSs. This is an exciting new technology that is bound to have a very significant impact in a number of application fields. As it has been shown, its fundamentals are now clearly understood and research is currently focused on improving the sensor performance and reducing its cost so as to make its application feasible and effective. This is an ongoing effort by several research groups around the world that has lately led to a number of significant breakthroughs.

REFERENCES

1. K. Hotate, Fiber optic nerve systems for smart materials and smart structures, *Optical Sensors, OSA Technical Digest (OSA)*, paper SWB2, 2010.
2. M. T. V. Wylie, A. W. Brown, and B. G. Colpitts, Fibre optic distributed differential displacement sensor, *Proceedings of SPIE—The International Society for Optical Engineering 7753*, paper 77532O, 2011.
3. T. Gogolla and K. Krebber, Distributed beat length measurement in single-mode optical fibers using stimulated Brillouin-scattering and frequency-domain analysis, *Journal of Lightwave Technology*, 18(3), 320–328, 2000.
4. H. González Herráez, L. Thevenaz, and P. Robert, Distributed measurement of chromatic dispersion by four-wave mixing and Brillouin optical time domain analysis, *Optics Letters*, 28(22), 2210–2212, 2003.
5. G. Agrawal, *Nonlinear Fiber Optics*, Academic Press, New York, 2006.
6. W. Zou, Z. He, and K. Hotate, Investigation of strain and temperature dependences of Brillouin frequency shifts in GeO_2-doped optical fibers, *Journal of Lightwave Technology*, 26, 1854–1861, 2008.
7. C. C. Lee, P. W. Chiang, and S. Chi, Utilization of a dispersion-shifted fiber for simultaneous measurement of distributed strain and temperature through Brillouin frequency shift, *IEEE Photonics Technology Letters*, 13, 1094–1096, 2001.
8. W. W. Zou, Z. He, and K. Hotate, Complete discrimination of strain and temperature using Brillouin frequency shift and birefringence in a polarization-maintaining fiber, *Optics Express*, 17, 1248–1255, 2009.
9. X. Bao, A. Brown, M. DeMerchant, and J. Smith, Characterization of the Brillouin-loss spectrum of single-mode fibers by use of very short (10-ns) pulses, *Optics Letters*, 24, 510–512, 1999.
10. V. P. Kalosha, E. A. Ponomarev, L. Chen, and X. Bao, How to obtain high spectral resolution of SBS-based distributed sensing by using nanosecond pulses, *Optics Express*, 14, 2071–2078, 2006.
11. A. W. Brown, B. G. Colpitts, and K. Brown, Dark-pulse Brillouin optical time-domain sensor with 20-mm spatial resolution, *Journal of Lightwave Technology*, 25, 381–386, 2007.
12. S. M. Foaleng, M. Tur, J. C. Beugnot, and L. Thévenaz, High spatial and spectral resolution long-range sensing using Brillouin echoes, *Journal of Lightwave Technology*, 28, 2993–3003, 2010.
13. V. Lecœuche, D. J. Webb, C. N. Pannell, and D. A. Jackson, Brillouin based distributed fibre sensor incorporating a mode-locked Brillouin fibre ring laser, *Optics Communications*, 152, 263–268, 1998.
14. L. Thévenaz, S. Le Floch, D. Alasia, and J. Troger, Novel schemes for optical signal generation using laser injection locking with application to Brillouin sensing, *Measurement Science and Technology*, 15, 1519–1524, 2004.
15. Y. Li, X. Bao, F. Ravet, and E. Ponomarev, Distributed Brillouin sensor system based on offset locking of two distributed feedback lasers, *Applied Optics*, 47, 99–102, 2008.
16. S. Diaz, S. F. Mafang, M. Lopez-Amo, and L. Thevenaz, A high-performance optical time-domain Brillouin distributed fiber sensor, *IEEE Sensors Journal*, 8, 1268–1272, 2008.
17. A. Zornoza, D. Olier, M. Sagues, and A. Loayssa, Brillouin distributed sensor using RF shaping of pump pulses, *Measurement Science and Technology*, 21, 094021, 2010.
18. A. Zornoza, D. Olier, and A. Loayssa, Self-heterodyne synchronous detection for SNR improvement and distributed Brillouin phase shift measurements in BOTDA sensors, *Proceedings of SPIE—The International Society for Optical Engineering 7753*, paper 77532F, 2011.
19. A. Zornoza, D. Olier, M. Sagues, and A. Loayssa, Brillouin spectral scanning using the wavelength dependence of the frequency shift, *IEEE Sensors Journal*, 11(2), 382–383, 2011.
20. M. Sagues, A. Loayssa, and J. Capmany, Multitap complex-coefficient incoherent microwave photonic filters based on stimulated Brillouin scattering, *IEEE Photonics Technology Letters*, 19, 1194–1196, 2007.
21. S. M. Foaleng, F. Rodríguez-Barrios, S. Martin-Lopez, M. González-Herráez, and L. Thévenaz, Detrimental effect of self-phase modulation on the performance of Brillouin distributed fiber sensors, *Optics Letters*, 36, 97–99, 2011.
22. T. Horiguchi, K. Shimizu, T. Kurashima, M. Tateda, and Y. Koyamada, Development of a distributed sensing technique using Brillouin scattering, *Journal of Lightwave Technology*, 13, 1296–1302, 1995.

23. D. Alasia, M. González Herráez, L. Abrardi, S. Martin-López, and L. Thévenaz, Detrimental effect of modulation instability on distributed optical fiber sensors using stimulated Brillouin scattering, *Proceedings of SPIE 5855*, pp. 587–590, 2005.

24. Y. Dong, L. Chen, and X. Bao, System optimization of a long-range Brillouin-loss-based distributed fiber sensor, *Applied Optics*, 49, 5020–5025, 2010.

25. S. M. Foaleng and L. Thévenaz, Impact of Raman scattering and modulation instability on the performances of Brillouin sensors, *Proceedings of SPIE 7753*, paper 77539V, 2011.

26. A. Zornoza, A. Minardo, R. Bernini, A. Loayssa, and L. Zeni, Pulsing the probe wave to reduce nonlocal effects in Brillouin optical time-domain analysis (BOTDA) sensors, *IEEE Sensors Journal*, 11, 1067–1068, 2011.

27. E. Geinitz, S. Jetschke, U. Ropke, S. Schroter, R. Willsch, and H. Bartelt, The influence of pulse amplification on distributed fibre-optic Brillouin sensing and a method to compensate for systematic errors, *Measurement Science and Technology*, 10, 112–116, 1999.

28. A. Minardo, R. Bernini, L. Zeni, L. Thevenaz, and F. Briffod, A reconstruction technique for long-range stimulated Brillouin scattering distributed fibre-optic sensors: Experimental results, *Measurement Science and Technology*, 16, 900–908, 2005.

29. A. Zornoza, R. A. Pérez-Herrera, C. Elosua, S. Diaz, C. Bariain, A. Loayssa, and M. Lopez-Amo, Long-range hybrid network with point and distributed Brillouin sensors using Raman amplification, *Optics Express*, 18, 9531–9541, 2010.

30. M. N. Alahbabi, Y. T. Cho, and T. P. Newson, 150-km-range distributed temperature sensor based on coherent detection of spontaneous Brillouin backscatter and in-line Raman amplification, *Journal of the Optical Society of America*, B22(6), 1321–1324, 2005.

5 Silicon Microring Sensors*

Zhiping Zhou and Huaxiang Yi

CONTENTS

A silicon microring resonator can be used to magnify the interaction between light and matter and its output is very sensitive to its index of refractive change caused by the interactions, which in turn will shift the resonating wavelength and the coupling coefficient of the resonator. These characters, plus the CMOS-compatible fabrication process, are making the silicon microring resonator a key building block of silicon-based smart-sensing platforms. The microring resonator has been proposed as a compact and highly sensitive optical sensor for the purposes of environment monitoring, health diagnosis, and drug development. This chapter describes the basic structure of the microring sensors, its working principle, and its unique properties. In particular, the sensitivity and the selectivity, the methods to improve dynamic range and thermal stability of the silicon microring sensors are discussed. Novel silicon microring sensors with better and more reliable sensing performance are introduced. Finally, specific applications are presented.

5.1   INTRODUCTION

Due to its closed-loop structure, the microring resonator can have a very high-quality factor value and its output intensity is very sensitive to different wavelengths. These characters make it a key building block for photonic integrated circuits due to their versatility in function and their capability of integration. Microring resonators have been utilized in various applications including filtering [1], switching [2], modulation [3], and wavelength conversion [4]. By implementing large index-contrast materials of silicon, the dimensions of microring resonators have been significantly reduced to 1 μm radius [5]. Their compactness and compatibility with mature semiconductor fabrication platforms facilitate integration with microelectronic devices, leading to a great opportunity

---

* Text pages 68–70 from Zhou, Z. and Yi, H., *Proc. SPIE*, 8236, 823617, 2012. With permission.

for mass production and commercialization [6]. The light field can be confined tightly within silicon waveguides, while the evanescent wave is surrounding it to provide the interaction between light and cladding materials.

Because of their small size and potential high sensitivity, sensors based on microring resonators have been proposed to detect biochemical analyte [7–13] or mechanical displacement [14]. In recent years, much progress has been reported in this field. The advantage of label-free sensing is demonstrated with antigen target based on the surface-modified waveguide [15,16]. And the specific sensing target is extended from gases to biomolecules and chemical materials [17–19]. These works are contributing to optical sensor development in the way of the biochemical area. Furthermore, the microring resonators arrays are investigated to provide more functions such as specific detection [20,21], thermal compensation [22], and even constructing a lab-on-chip system [23]. The fundamental sensing theory has also been developed for higher sensitivity of the device. One is to get the higher quality factor through folded microring resonators [24]. Other is to develop the sensing theory through cascade microring resonators [25,26].

5.2   MICRORING SENSOR BASICS

There are two configurations of the basic microring sensors. Both consist of single microring resonator and coupled waveguides; one has dual waveguide–coupled microring in Figure 5.1a and the other has single waveguide–coupled microring in Figure 5.1b. From the design point of view, the parameters for the configurations in Figure 5.1 can be divided into two categories. The first is to choose parameters based on the waveguide properties, which depend on the materials, thickness of the films, and the fabrication quality. These parameters, such as the indexes and loss coefficients, usually serve as the prerequisites for the design. On the contrary, the geometric parameters, such as the radius, waveguide width, and the gap between the bent and straight waveguides, are the parameters that can be modified in a relatively large range in the design.

In these sensors, the analyte causes a change in the refractive index of the cladding of the waveguide, which is probed by the evanescent tail of the modal field located in the wavelength, and in turn changes the transmission behavior of the light propagating in the ring. There are two basic sensing principles with the above microring sensor through monitoring different light parameters change, wavelength shift, and intensity change [27]. As shown in Figure 5.2, the spectrum shifts due to the analyte change. The resonance waveguide shifts $\Delta\lambda$, which can be observed to indicate the analyte change. Meanwhile, the output intensity also varies (ΔI). Both

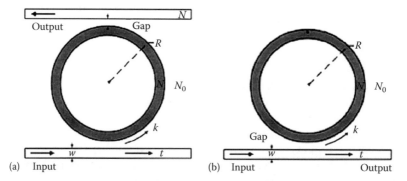

FIGURE 5.1   Basic microring sensor. (a) Schematic of the dual waveguide–coupled microring resonator. (b) Schematic of the single waveguide–coupled microring resonator. Note that the ring waveguide depicted by the gray region serves as the transducer for sensing. (From Xia, Z. et al., *IEEE J. Quantum Electron.*, 44(1), 100, 2008. With permission.)

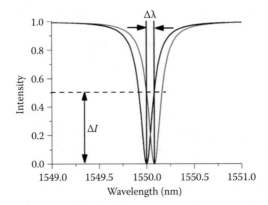

FIGURE 5.2 Spectrum shift due to the analyte change.

sensing principles are sensitive and depend on the spectrum shift. The waveguide shift way has larger dynamic range but complex monitoring of the spectrum scan. The intensity change way is easy but the dynamic range is limited.

As a good sensor, there are several key properties for the microring. At the first, the high-quality factor (Q) is required for a high-sensitive sensor, which is defined as

$$Q = \frac{\lambda_0}{\Delta\lambda} \tag{5.1}$$

where
  λ_0 is the resonance wavelength
  $\Delta\lambda$ is the full width at half maximum of the resonance peak

High Q means narrower resonance peak and sharp spectrum slope, which leads to smaller resolution in the wavelength shift way and larger intensity change in the intensity change way.

The FSR is another key property of the microring sensor:

$$FSR = \frac{\overline{\lambda_0}^2}{L \times n_{eff}} \tag{5.2}$$

where
  λ_0 is the average wavelength around the resonance
  L is the circumference of the microring
  n_{eff} is the effective index of the waveguide

The dynamic range of wavelength shift way is limited within one FSR. Thus the larger FSR leads to larger dynamic range of the microring sensor.

The sensitivity is the most important property of the sensor. According to the different monitored parameters, sensitivity is defined as

$$s = \frac{\Delta I}{\Delta n_c} \tag{5.3}$$

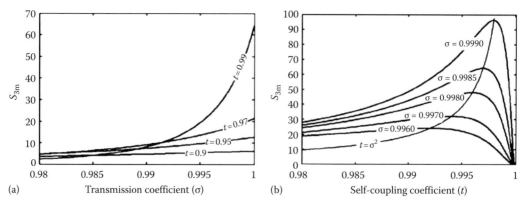

FIGURE 5.3 (a) Maximum sensitivity to the transmission coefficient and (b) sensitivity to the self-coupling coefficient. (From Xia, Z. et al., *IEEE J. Quantum Electron.*, 44(1), 100, 2008. With permission.)

for intensity change sensing, where ΔI is the intensity change and Δn_c is the index change of the analyte:

$$s = \frac{\Delta \lambda_0}{\Delta n_c} \tag{5.4}$$

for wavelength shift sensing, where $\Delta \lambda_0$ is the resonance wavelength change.

We have rigorously analyzed the device sensitivity for dual waveguide–coupled microring resonator sensors based on intensity detection. The analysis and simulation results indicate that the device sensitivity increases monotonically as the transmission coefficient approaches unity as shown in Figure 5.3a, while the optimum of self-coupling coefficient t equals σ^2, at which the maximal sensitivity is obtained (Figure 5.3b). Moreover, a wavelength offset is also required so as to operate at the best sensitivity, and this offset is determined by t and σ, together with the specific mode number.

5.3 MICRORING SENSOR PROPERTIES

Sensitivity, selectivity, dynamic range, and thermal stability are most important properties for microring optical sensor. Several works on these properties are presented here. Novel sensing mechanisms, such as Fano resonance, coupling-induced intensity sensing, are adopted to obtain higher sensitivity. The microring arrays with different probes and measurement strategies are used to increase the sensing selectivity. Multimicroring resonators, such as interferencing microrings and cascaded microrings, are applied to enlarge the dynamic range. To overcome one of the strongest influences from environment, temperature, the athermal microring sensor is also investigated.

5.3.1 SENSITIVITY

The conventional silicon microring sensor is based on single microring, where only the single symmetric Lorentzian resonance lineshape was used to respond to the index change. The fundamental sensing theory has several limitations for the higher sensitivity, such as the limited Q factor and limited waveguide index change to the light–matter interactions.

To overcome these limitations, a highly sensitive Fano resonance single microring sensor was proposed. The spectra of silicon microring resonators coupled with waveguides possessing an

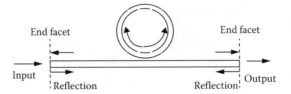

FIGURE 5.4 High-sensitivity Fano resonance single microring sensor. (From Yi, H. et al., *Opt. Express*, 18(3), 2967, 2010. With permission.)

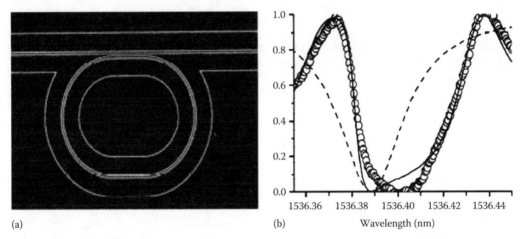

FIGURE 5.5 (a) SEM of single microring resonator and (b) asymmetric spectrum in experiment. (From Yi, H. et al., *Opt. Express*, 18(3), 2967, 2010. With permission.)

end-facet reflections are discussed theoretically and experimentally [28]. The end-facet reflection in a silicon waveguide forms a Fabry–Perot (FP) resonator that couples to the microring resonance, thus changing the symmetric Lorentzian resonance lineshape to a strongly asymmetric shape shown in Figure 5.4. The physics underlying this change in lineshape is closely related to the Fano lineshape [29], which results from a discrete resonance coupled to a continuum—here the quasicontinuum of densely spaced FP resonances of the waveguide.

The spectral slope is demonstrated to become steeper when the resonances of FP and microring have a π-phase difference as shown in Figure 5.5. This requirement is easy to meet because the long cavity of FP resonance leads to dense collection of resonances with small FSR. For applications that have much shorter length, an asymmetric resonance can also be obtained but by judicious choice of the phase difference. Therefore, our device can provide asymmetric resonance with easier design. Because a steeper slope is obtained in asymmetric resonance shape, Si microring resonators can provide enhanced sensitivity in chemical detection application. Thus, this feature can reduce the stringent requirement of a high-quality factor in microring resonators as in our demonstration experiment. With regard to device fabrication, this means that there can be greater tolerance to imperfections in the microring and in the waveguide end facets. This is an effective method to produce inexpensive and easily fabricated chemical sensors.

Other novel microring sensors were also proposed to permit highly sensitive intensity detection without necessitating monitoring of the entire resonance spectrum [30]. The physical mechanism underlying their superior performance is the significant coupling coefficient change that can change the critical coupling condition, resulting in a large change in the output intensity [31].

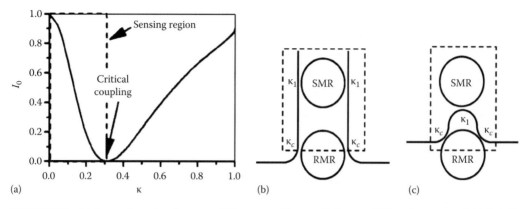

FIGURE 5.6 (a) Output intensity change to different coupling coefficient κ, (b) Type I sensor, and (c) Type II sensor. (From Yi, H. et al., *J. Soc. Am. B*, 28(7), 1611, 2011. With permission.)

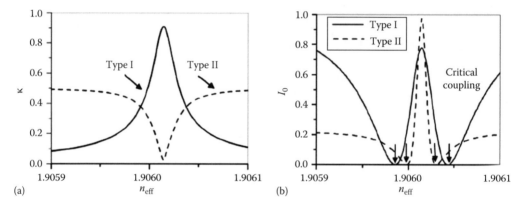

FIGURE 5.7 (a) Coupling coefficient change to different effective index and (b) the corresponding intensity to different effective index. (From Yi, H. et al., *J. Soc. Am. B*, 28(7), 1611, 2011. With permission.)

At the resonance wavelength λ_0, the output intensity I_0 is sensitive to κ, especially within the small κ region as in Figure 5.6a. Based on this property, two types of microring-assisted MZI coupler are investigated and applied in the resonator to function as an intensity sensor, shown in Figure 5.6b and c.

The sensitivity of the Type II sensor without phase bias is predicted to be $\delta_n \sim 4.9 \times 10^{-8}$ RIU from Figure 5.7, which is one order of magnitude smaller than that of conventional microring sensors. By comparing the intensity curves of the conventional microring sensor and the coupling-induced sensors, a higher sensitivity is observed in the latter due to the steeper slope of the curve for I_0 versus n_{eff}. Thus, the coupling-induced intensity sensor based on the microring is demonstrated to provide a different and superior approach to intensity sensing by providing enhanced sensitivity.

5.3.2 SELECTIVITY

Typically, there are multiple analytes at the same one measurement. One microring sensor with a specific probe is only able to sense one analyte. Multimicroring sensors are essential to distinguish different analytes from each others.

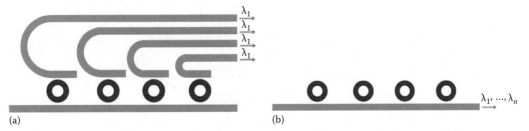

FIGURE 5.8 (a) Add-drop microring sensor array and (b) cascaded microring sensor array.

The sensor arrays were first choice to deal with the selectivity, or specificity. For example, multiplexed antibody was detected in complex samples with an array of three by four SOI microring resonators [21]. Four microring sensors shared one input light from the same input coupled waveguide in Figure 5.8a. Different bioprobes are bonded at each microring separately. By utilizing the add-drop microring, there are four output waveguides. With different antibody interactions with different sensors, the output ports indicated different sensing information. The selected sensing was realized through the interaction between the bioprobe and the analyte at different microrings.

Besides the add-drop microring arrays sensor, there is a cascaded microring array configuration in Figure 5.8b. Based on the WDM mechanism, different resonance wavelengths are determined for the cascaded sensors [32]. Different bioprobes are bonded on each microring separately. By monitoring the shift at each resonance, selective detection to multiple analytes can be realized.

Both microring arrays consist of multimicroring sensors, where specific probe was bonded. Only the unique analyte can interact with the probe and change the index of the microring, while others remain unaffected. By monitoring the output intensity or resonance shift, the selective sensing is demonstrated.

5.3.3 DYNAMIC RANGE

Dynamic range is defined as the ratio of the sensor's largest detectable wavelength shift to its smallest signal. In the conventional single resonance sensor, the largest detectable wavelength shift is same as the FSR. In order to enlarge the dynamic range, multimicrorings are suggested.

An optical vernier effect was adopted in sensing based on the dual-microring MZI sensor shown in Figure 5.9a [33]. The dramatic shift of the overlapping resonance within the multiresonance spectrum due to the presence of an analyte is found in Figure 5.9b, which can lead to high sensitivity of the device compared to an otherwise comparable single-microring sensor. The overlay resonance shift is larger than one FSR in the spectra, which enlarges the dynamic range significantly. To be specific, when the rings were chosen to have radii $R_1 = 505$ μm and $R_2 = 500$ μm, the device was predicted to have a detection limit 2×10^{-6} RIU with 0.31 nm overlap resonance shift. Moreover, the sensitivity is predicted to be 100 times that of the single-microring sensor. Since the overlay resonance shift is measured as the difference between λ_0 and λ_0' as shown in Figure 5.9b, there is no limitation to the dynamic range of this sensor configuration. Meanwhile, the detection limit and sensitivity of this configuration do not depend sensitively on the width of the resonance. That means this sensor can, up to a point, circumvent conventional limits due to Q. Finally, this dual-microring interference sensor is also suited for other sensing schemes, which rely on analyte-induced changes of cavity resonances. Using this configuration, both high sensitive and large dynamic range may be obtained simultaneously.

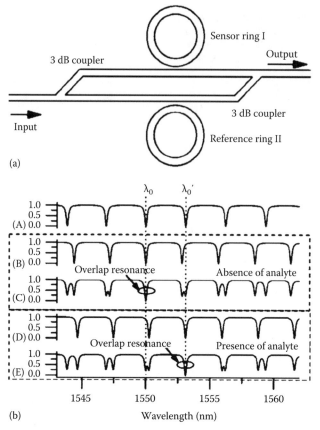

(a)

(b)   Wavelength (nm)

FIGURE 5.9 (a) Dual-microring MZI sensor and (b) overlapped spectrum in sensing. (From Yi, H. et al., *Appl. Phys. Lett.*, 95, 191112, 2009. With permission.)

Also, the cascade two microring resonators with different dimensions can be used to achieve a large dynamic range.

One microring has radius R_1 with higher sensitivity s_1 and smaller FSR_1, while the other has radius R_2 with lower sensitivity s_2 and larger FSR_2, where $s_1 > s_2$ and $FSR_1 < FSR_2$. Assuming the resonance shift induced by the analyte exceeds FSR_1, then

$$\lambda_{shift} = N \times FSR_1 + \Delta\lambda_1 \tag{5.5}$$

where
   N is the number of FSR
   $\Delta\lambda_1$ is the margin included in the wavelength shift

Therefore, the accuracy of the shift depends on the acquisition of N and $\Delta\lambda_1$.

Meanwhile, resonance shift of the other microring $\Delta\lambda_1$ is within the measurement range of FSR_2. In the presence of analyte change a, the resonance shifts for both microrings are

$$s_1 a = N_1 \times FSR_1 + \Delta\lambda_1$$
$$s_2 a = \Delta\lambda_2 \tag{5.6}$$

According to the earlier equations, we could have

$$N_1 = \frac{\Delta\lambda_2 s_1 - \Delta\lambda_1 s_2}{\text{FSR}_1 \times s_2} \qquad (5.7)$$

In this equation, the number of FSR_1, N_1, can be specified through measurement of the variable $\Delta\lambda_1$ and $\Delta\lambda_2$. Thus, the measurement scale is determined by the FSR_2, and the dynamic range can be enlarged without the limit of the FSR_1 of the R_1.

5.3.4 THERMAL STABILITY

The thermo-optical effect can make the microring resonance spectrum shifting, which typically changes the intensity output at operation wavelength to distort the sensing single. However, the flat-top resonance lineshape near the operation wavelength can provide the tolerance of shift for the output intensity. Based on this mechanism, an athermal intensity sensor based on coupled microring interference was presented in Figure 5.10a. The sensor provides a flat sensing zone (shown in Figure 5.10b) where no change with thermal fluctuation within ±5 K (this result can be optimized by designing a low Q factor of the microrings to broaden the sensing zone). Different from athermal processing application, the 10 K temperature independence is sufficient for sensing operation [34].

Moreover, the output intensity of the device is very sensitive to the difference between the sensing and reference microrings originating from the shift of the cladding index due to the presence of the analyte. The physical origin of this sensitivity is the interference between the outputs of the two individual microrings comprising the device. A detection limit of 6.8×10^{-7} RIU was predicted. The sensor was based on the low Q microrings and is robust with respect to variations in the coupling coefficients. Thus, a highly sensitive athermal optical intensity sensor may be realized by using microrings based on a silicon platform.

Alternatively, sensor arrays can also be used to overcome the thermal effect [20]. In this scheme, one idle microring was placed and covered with up-cladding, while the rest microrings were exposed to the analyte for sensing on the same chip. All microrings will suffer from the same thermal influence and have their spectrum shifted. By referring to the shift of the covered microring, the thermal-induced resonance shift can be subtracted from the shift of the sensing microrings. Thus, the athermal sensing is obtained.

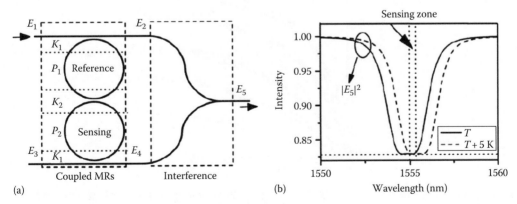

FIGURE 5.10 Athermal microring sensor. (a) Schematic of the coupled microring resonators and (b) spectrum change due to the temperature increased 5 K. (From Yi, H. et al., *IEEE J. Quantum Electron.*, 47(3), 354, 2011. With permission.)

5.4 APPLICATIONS

5.4.1 Biosensing

The label-free biosensing techniques have shown the potential to simplify clinical analyses. The silicon microring resonator was demonstrated for the robust and label-free detection. Based on the sensing theory, microring resonators for protein sensing have been demonstrated, such as measurements in buffer of avidin–biotin interactions [24] or detection of polyclonal IgG [35]. Furthermore, the ability to monitor in real time the steps involved in the chemical and biomolecular functionalization of the sensor surface was demonstrated utilizing an initial slope-based quantitation method to sensitively detect CEA at clinically relevant levels and to determine the CEA concentrations [15]. The experiments shows that the detection limit of our platform is comparable to that of a commercial enzyme-linked immunosorbent assay (ELISA) kit and is satisfactory for the quantitation of CEA over the clinically relevant range of 5–100 ng/mL as shown in Figure 5.11.

5.4.2 Seismic Sensing

In this study, a seismic sensor based on a stress-coupled racetrack resonator is proposed in Figure 5.12a [36]. The theoretical model is introduced to describe the interaction between the seismic acceleration and racetrack resonator. We find that the resonant wavelength shift is directly related to the length increment of the waveguides, which makes the sensing mechanism much simpler compared with other optical acceleration sensor. Moreover, to optimize the design, the resonant wavelength shifts under different configurations are analyzed in Figure 5.12b and c. Our results indicate that sensing performances of the device depend on both mechanical parameters and optical characteristics of the racetrack resonator. In the case of below the yield strength, the higher the mechanical sensitivity the larger the wavelength shift. The wavelength shift is 52 pm under 1 g acceleration with mechanical frequency up to 200 Hz as shown in Figure 5.12d. The proposed structure can be fabricated by mature silicon CMOS process and can be integrated with data processing circuit to form smart system, which can have prospective potential for seismic prospecting.

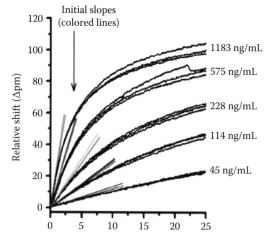

FIGURE 5.11 Real-time, label-free detection of CEA using microring resonators. Overlay of three time-resolved association curves for the same ring at each concentration of CEA. The colored traces are tangent lines to the association curve at $t = 0$ and are used to determine the initial slope of sensor response. (From Washburn, A.L. et al., *Anal. Chem.*, 81, 9499, 2009. With permission.)

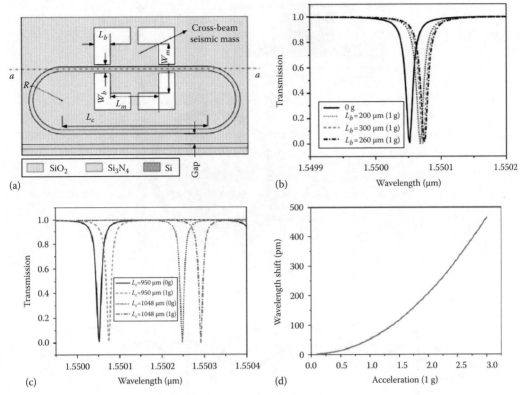

FIGURE 5.12 (a) Schematic of a racetrack resonator integrated with a crossbeam seismic mass, (b) transmission spectrums of the racetrack resonator with different beam lengths L_b, (c) transmission spectrum with different length L_c, and (d) wavelength shift with the creasing acceleration. (From Mo, W. et al., *IEEE Sensors J.*, 11(4), 1035, 2011. With permission.)

5.4.3 ELECTRICAL SENSING

The plasmonic dispersion effect in silicon material makes the free carries to affect the light field distribution. Based on this, silicon microring resonator was studied to detect the electrical single change and transform to the variation of light. The modulator is based on a microring resonant light-confining structure that enhances the sensitivity of light to small changes in refractive index of the silicon and also enables high-speed operation [3]. Injection electrical–optical modulator is realized based on the forward-bias PIN junction with a microring resonator as shown in Figure 5.13a. The corresponding optical spectrum was shifted to different electrical singles as shown in Figure 5.13b.

5.4.4 OPTICAL SENSING

The property light can inject the free carriers in the silicon waveguide. Based on this mechanism, the silicon microring can indicate the change of the light signal as a optical sensing. Light-induced microring switch is based on an estimated free-carrier concentration of 1×10^{19} cm^{-3} injected optically by a pump laser [37]. Light from an argon ion pump laser at 488 nm was used to inject free carriers into the silicon microring with an integrated PIN diode. The laser was focused onto a 2 mm radius spot over the coupling region between the third and fourth rings in the cascade, as shown in Figure 5.14. As the pump laser power was increased, the drop-port transmission at the center of the passband was significantly suppressed, and transmission at the through-port, which was initially less than −10 dB, was almost completely restored.

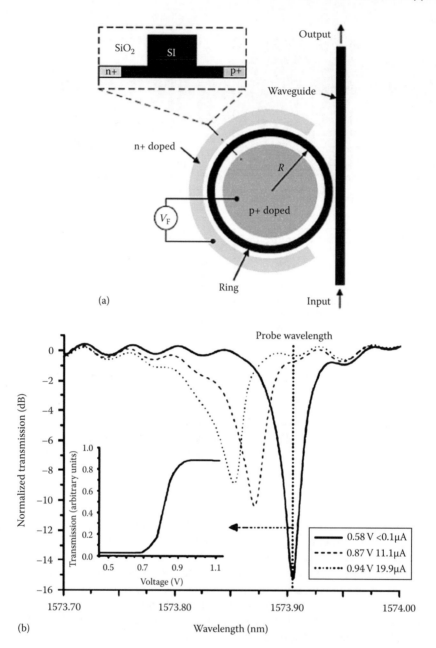

FIGURE 5.13 (a) Silicon electrical–optical modulator configuration and (b) spectral change versus different applied voltages. (From Xu, Q.-F. et al., *Nature*, 435, 325, 2005. With permission.)

5.4.5 ULTRASONIC SENSING

Acoustic waves irradiating the ring induce strain, deforming the waveguide dimensions and changing the refractive index of the waveguide via the elasto-optic effect [38]. The pulse stresses the structure, deforming the waveguide dimension and simultaneously modifying the refractive index of the core. Typically, the wavelength of the ultrasound wave is much larger than the waveguide dimension. The sharp wavelength dependence of the microring resonance

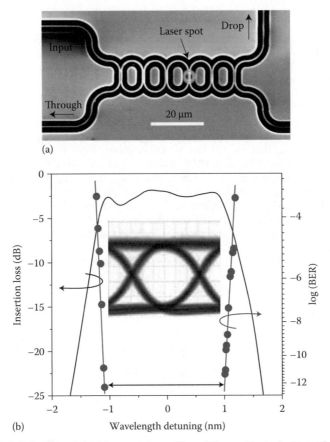

FIGURE 5.14 Cascaded silicon microring resonator. (From Vlasov, Y. et al., *Nat. Photon.*, 2, 242, 2008. With permission.)

can enhance the optical response to acoustic strain. Such polymer microring resonators are experimentally demonstrated in detecting broadband ultrasound pulses from a 50 MHz transducer. Measured frequency response shows that these devices have potential in high-frequency, ultrasound detection.

5.5 CONCLUSIONS

Silicon microring sensors can be highly sensitive with a very small footprint. The fabrication process for the sensors and the extended subsystems is CMOS compatible since they are building on a silicon material platform. Based on analyzing the basics of the single microring sensor using the resonance theory, the sensitivity, selectivity, dynamic range, and thermal stability of the sensors are discussed. In particular, the works about the microring sensor properties are introduced to elevate the sensing performance, such as the Fano resonance, multiresonance, coupling-induced resonance, microring arrays, and athermal resonance. The sensors based on these methods can be easily implemented on the silicon platform, which makes the truly low-cost, smart optical sensing technology possible. Through the different interactions to affect the index change, several microring sensor applications are presented.

REFERENCES

1. Absil, P. P., Hryniewicz, J. V., Little, B. E., Wilson, R. A., Joneckis, L. G., and Ho, P.-T., Compact microring notch filters, *IEEE Photon. Technol. Lett.*, 12(4), 398–400 (2000).
2. Almeida, V. R., Barrios, C. A., Panepucci, R. R., and Lipson, M., All-optical control of light on a silicon chip, *Nature*, 431, 1081–1083 (2004).
3. Xu, Q.-F., Schmidt, B., Pradhan, S., and Lipson, M., Micrometre-scale silicon electro-optic modulator, *Nature*, 435, 325–327 (2005).
4. Absil, P. P., Hryniewicz, J. V., Little, B. E., Cho, P. S., Wilson, R. A., Joneckis, L. G., and Ho, P.-T., Wavelength conversion in GaAs micro-ring resonators, *Opt. Lett.*, 25(8), 554–556 (2000).
5. Prabhu, A. M., Tsay, A., Han, Z., and Van, V., Extreme miniaturization of silicon adddrop microring filters for VLSI photonics applications, *IEEE Photon. J.*, 2(3), 436–444 (2010).
6. Xu, F. and Poon, A. W., Silicon cross-connect filters using microring resonator coupled multimode interference-based waveguide crossings, *Opt. Express*, 16(12), 8649–8657 (2008).
7. Krioukov, E., Klunder, D. J. W., Driessen, A., Greve, J., and Otto, C., Sensor based on an integrated optical microcavity, *Opt. Lett.*, 27(7), 512–514 (2002).
8. Ksendzov, A. and Lin, Y., Integrated optics ring-resonator sensors for protein detection, *Opt. Lett.*, 30(24), 3344–3346 (2005).
9. Guo, J.-P., Shaw, M. J., Vawter, G. A., Hadley, G. R., Esherick, P., and Sullivan, C. T., High-Q microring resonator for biochemical sensors, *Proc. SPIE*, 5728, 83–92 (2005).
10. Yalçin, A., Popat, K. C., Aldridge, J. C., Desai, T. A., Hryniewicz, J., Chbouki, N., Little, B. E., King, O., Van, V., Chu, S., Gill, D., Washburn, M. A., Ünlü, M. S., and Goldberg, B. B., Optical sensing of biomolecules using microring resonators, *IEEE J. Sel. Topics Quant. Electron.*, 12(1), 148–154 (2006).
11. Zhou, Z. and Yi, H., Silicon microring sensor, *Proc. SPIE*, 8236, 823617 (2012).
12. White, I. M., Oveys, H., and Fan, X.-D., Liquid-core optical ring-resonator sensors, *Opt. Lett.*, 31(9), 1319–1321 (2006).
13. Chao, C.-Y., Fung, W., and Guo, L. J., Polymer microring resonators for biochemical sensing applications, *IEEE J. Sel. Topics Quant. Electron.*, 12, 134–142 (2006).
14. Kiyat, I., Kocabas, C., and Aydinli, A., Integrated micro ring resonator displacement sensor for scanning probe microscopies, *J. Micromech. Microeng.*, 14, 374–381 (2004).
15. Washburn, A. L., Gunn, L. C., and Balley, R. C., Label-free quantitation of a cancer biomarker in complex media using silicon photonic microring resonators, *Anal. Chem.*, 81, 9499–9506 (2009).
16. De Vos, K., Girones, J., Popelka, S., Schacht, E., Baets, R., and Bienstman, P., SOI optical microring resonator with poly(ethylene glycol) polymer brush for label-free biosensor applications, *Biosens. Bioelectron.*, 24, 2528–2533 (2009).
17. Yebo, N. A., Taillaert, D., Roels, J., Lahem, D., Debliquy, M., Van Thourhout, D., and Baets, R., Silicon-on-insulator (SOI) ring resonator-based integrated optical hydrogen sensor, *IEEE Photon. Tech. Lett.*, 21(14), 960–962 (2009).
18. Luchansky, M. S., Washburn, A. L., Martin, T. A., Iqbal, M., Gunn, L. G., and Bailey, R. C., Characterization of the evanescent field profile and bound mass sensitivity of a label-free silicon photonic microring resonator biosensing platform, *Biosens. Bioelectron.*, 26, 1283–1291 (2010).
19. Orghici, R., Lützow, P., Burgmeier, J., Koch, J., Heidrich, H., Schade, W., Welschoff, N., and Waldvogel, S., A microring resonator sensor for sensitive detection of 1,3,5-trinitrotoluene (TNT), *Sensor*, 10, 6788–6795 (2010).
20. Carlborg, C. F., Gylfason, K. B., Kazmierczak, A., Dortu, F., Banuls Polo, M. J., Maquieira Catala, A., Kresbach, G. M., Sohlstrom, H., Moh, T., Vivien, L., Popplewell, J., Ronan, G., Barrios, C. A., Stemme, G., and van der Wijngaart, W., A packaged optical slot-waveguide ring resonator sensor array for multiplex label-free assays in labs-on-chips, *Lab on Chip*, 10, 281–290 (2010).
21. De Vos, K., Girones, J., Glaes, T., Popelka, S., Schacht, E., Baets, R., and Bienstman, P., Multiplexed antibody detection with an array of silicon-on-insulator microring resonators, *IEEE Photon. J.*, 1(4), 225–235 (2009).
22. Lee, H.-S., Kim, G.-D., and Lee, S.-S., Temperature compensated refractometric biosensor exploiting ring resonators, *IEEE Photon. Tech. Lett.*, 21(16), 1136–1138 (2009).
23. Gylfason, K. B., Carlborg, C. F., Kazmierczak, A., Dortu, F., Sohlstrom, H., Vivien, L., Barrios, C. A., van der Wijngaart, W., and Stemme, G., On-chip temperature compensation in an integrated slot-waveguide ring resonator refractive index sensor array, *Opt. Express*, 18(4), 3226–3237 (2010).
24. Xu, D.-X., Densmore, A., Delage, A., Waldron, P., McKinnon, R., Janz, S., Lapointe, J., Lopinski, G., Mischki, T., Post, E., Cheben, P., and Schmid, J. H., Folded cavity SOI microring sensors for high sensitivity and real time measurement of biomolecular binding, *Opt. Express*, 16(19), 15137–15148 (2008).

25. Dai, D., Highly sensitive digital optical sensor based on cascaded high-Q ring-resonators, *Opt. Express*, 17(26), 23817–23822 (2009).

26. Jin, L., Li, M., and He, J.-J., Optical waveguide double-ring sensor using intensity interrogation with a low-cost broadband source, *Opt. Lett.*, 36(7), 1128–1130 (2011).

27. Xia, Z., Chen, Y., and Zhou, Z., Dual waveguide coupled microring resonator sensor based on intensity detection, *IEEE J. Quant. Electron.*, 44(1), 100–107 (2008).

28. Yi, H., Citrin, D. S., and Zhou, Z., Highly sensitive silicon microring sensor with sharp asymmetrical resonance, *Opt. Express*, 18(3), 2967–2972 (2010).

29. Fano, U., Effects of configuration interaction on intensities and phase shifts, *Phys. Rev.*, 124(6), 1866–1878 (1961).

30. Yi, H., Citrin, D. S., and Zhou, Z., Coupling-induced high-sensitivity silicon microring intensity-based sensor, *J. Soc. Am. B*, 28(7), 1611–1615 (2011).

31. Zhou, L. and Poon, A. W., Electrically reconfigurable silicon microring resonator-based filter with waveguide-coupled feedback, *Opt. Express*, 15(15), 9194–9204 (2007).

32. Xu, D.-X., Vachon, M., Densmore, A., Ma, R., Delage, A., Janz, S., Lapointe, J., Li, Y., Lopinski, G., Zhang, D., Liu, Q. Y., Cheben, P., and Schmid, J. H., Label-free biosensor array based on silicon-on-insulator ring resonators addressed using a WDM approach, *Opt. Lett.*, 35(16), 2771–2773 (2010).

33. Yi, H., Citrin, D. S., Chen, Y., and Zhou, Z., Dual-microring-resonator interference sensor, *Appl. Phys. Lett.*, 95, 191112 (2009).

34. Yi, H., Citrin, D. S., and Zhou, Z., Highly sensitive athermal optical microring sensor based on intensity detection, *IEEE J. Quant. Electron.*, 47(3), 354–358 (2011).

35. Ramachandran, A., Wang, S., Clarke, J., Ja, S. J., Goad, D., Wald, L., Flood, E. M., Knobbe, E., Hryniewicz, J. V., Chu, S. T., Gill, D., Chen, W., King, O., and Little, B. E., A universal biosensing platform based on optical micro-ring resonators, *Biosens. Bioelectron.*, 23(7), 939–944 (2008).

36. Mo, W., Zhou, Z., Wu, H., and Gao, D., Silicon-based stress-coupled optical racetrack resonators for seismic prospecting, *IEEE Sensors J.*, 11(4), 1035–1039 (2011).

37. Vlasov, Y., Green, W. M. J., and Xia, F., High-throughput silicon nanophotonic wavelength insensitive switch for on-chip optical networks, *Nat. Photon.*, 2, 242–246 (2008).

38. Chao, C.-Y., Ashkenazi, S., Huang, S.-W., O'Donnell, M., and Guo, L.-J., High-frequency ultrasound sensors using polymer microring resonators, *IEEE Trans. Ultrason. Ferroelectr. Freq. Control*, 54(5), 957–965 (2007).

6 Laser Doppler Velocimetry Technology for Integration and Directional Discrimination

Koichi Maru and Yusaku Fujii

CONTENTS

6.1 INTRODUCTION

Laser Doppler velocimeters (LDVs) have been widely used to measure the velocity of a fluid flow or rigid object in various research and industries since the introduction of the concept in 1964 [1]. The measurement by using differential LDV has the advantage of contactless, small measuring volume, giving excellent spatial resolution and a linear response. However, conventional LDVs using bulk optical systems or fibers have large sizes and complex assembly and are often affected by environmental disturbances, such as vibrations, due to the large optical path length in the optical system. Hence, integrated LDVs with high precision and compact size have been highly demanded. Using a planar lightwave circuit (PLC) [2], in which several optical elements are arranged on a planar surface of a silica or silicon substrate, is a promising way to integrate an optical system in a small size. The PLC technology has been widely used for optical passive devices deployed for optical communication systems because of their reliability and ability to be manufactured in large volumes.

Meanwhile, multidimensional velocity needs to be measured in many applications of fluid experiments, whereas many techniques for LDVs for measuring one-dimensional velocity have been developed [3]. LDVs using a polarization method [4–6], different wavelengths [7,8], frequency shift discrimination [9–12], and self-mixing technique based on dynamical perturbation of a laser [13] have been developed for measuring two or three velocity components. In typical conventional LDVs for measuring multidimensional velocity, a complicated optical configuration or a three-dimensional arrangement of optical components is needed. When we use wavelength discrimination, lasers with different wavelengths and color filters or color splitters are needed. In some LDVs based on frequency shift discrimination, frequency shifters such as acousto-optic modulators (AOMs) are needed to discriminate the direction of velocity. Hence, the accuracy considerably depends on the performance of the frequency shifters to be used and its measurable velocity range is limited by preshifting frequencies.

In this chapter, several types of integrated LDVs using the PLC technology, especially, a wavelength-insensitive LDV [14–17] and a multipoint LDV [18] using arrayed waveguide gratings (AWGs), are described. AWGs have been widely used and deployed as key filtering devices in commercial wavelength-division multiplexing (WDM) optical communication systems. In addition, techniques for two-dimensional velocity measurement using a simple optical configuration without multiple colors or any optical modulator [19,20] are described.

6.2 INTEGRATED LDVs USING PLC TECHNOLOGY

6.2.1 ARRAYED WAVEGUIDE GRATING

The AWG is one of the planar devices. The concept of the AWG was first proposed in 1988 by Smit [21]. Also in 1988, Dragone [22] reported the star coupler configuration that has been mainly used as an element of the AWG. Since the early 1990s, much attention has been paid to the realization of devices with superior performance as well as their application to several functional optical circuits for WDM systems. Takahashi et al. [23,24] reported the first devices operating in the long wavelength window. Vellekoop and Smit [25] demonstrated the first devices operating in short wavelengths. Dragone extended the concept from $1 \times N$ to $N \times N$ that were so-called waveguide grating routers (WGRs) [26,27], which play an important role in wavelength routing network. Since 1993, system experiments involving AWGs have been reported [28,29].

Figure 6.1 illustrates the basic optical circuit of an AWG. The optical circuit consists of input waveguides, two slab waveguides (input and output slabs), waveguide array, and output waveguides. The waveguide array is designed with waveguides having a constant waveguide length difference to adjacent waveguides. The two slabs have a similar function to lenses and the waveguide array has a similar function to a grating. When the lightwave is launched into one of the input waveguides, it spreads out in the input slab and is coupled to the waveguides in the array. After passing through the array, each beam of lightwave interferes constructively or destructively according to the phase condition in the output slab. The constructively interfering lightwave focuses onto one of the output waveguides according to its wavelength.

Silica-based PLC technology [2] has been used to fabricate waveguide devices including AWGs. In a typical structure, the core buried by the cladding has a higher refractive index than that of the cladding by doping germanium, titanium, phosphorus, or boron into silica (SiO_2). The silica-based waveguides are formed on silica or silicon substrate by glass film deposition including flame hydrolysis deposition (FHD) [30], RF spattering [31], and chemical vapor deposition (CVD) [32], which

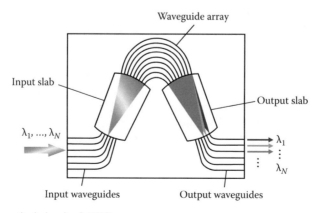

FIGURE 6.1 Basic optical circuit of AWG.

is followed by the etching process. Passive devices using silica-based PLC have been widely used in optical communication systems because of their reliability, low insertion loss, ease of coupling to optical fibers, integration capacity, and ability to produce optical filters with high accuracy [33].

6.2.2 Wavelength-Insensitive LDV

For an integrated LDV, using a semiconductor laser is desirable as a light source for integration. However, typical inexpensive semiconductor lasers suffer from the problem of instability in lasing wavelength due mainly to the dependence on temperature. In conventional differential LDVs, the Doppler frequency shift at a monitoring point depends on the signal wavelength to be used. Hence, wavelength instability in semiconductor lasers causes measurement errors in the Doppler frequency shift. To reduce the measurement error due to wavelength change, differential LDVs with diffractive gratings have been reported [34–36]. In these LDVs, the dependence of the diffraction angle on wavelength is utilized for the wavelength-insensitive operation. However, this type of LDVs needs assembly and alignment of a diffractive grating and other optical elements. To reduce the sizes and the cost of devices, wavelength-insensitive LDVs consisting of integrated optical waveguides are desirable.

In typical differential LDV, Doppler frequency shift is sensitive to the signal wavelength of the input laser beam. The Doppler frequency shift F_D is expressed as

$$F_D = \frac{2v_\perp \sin \psi}{\lambda} \tag{6.1}$$

where
  ψ is the incident angle of the beam to the object
  $v_\perp$ is the velocity of the object perpendicular to the bisector of the angles of the incident beams to the object
  λ is the wavelength

From this equation, if ψ appropriately changes depending on λ, the wavelength-insensitive operation can be expected. When F_D is to be wavelength-insensitive around $\lambda = \lambda_0$, the derivative of F_D with respect to λ should be zero at $\lambda = \lambda_0$.

There are several methods to change the incident angle of the beam to the object ψ according to the wavelength change. One method to obtain the appropriate change of ψ is the LDV using AWGs. Figure 6.2 illustrates the optical circuit of the wavelength-insensitive LDV using

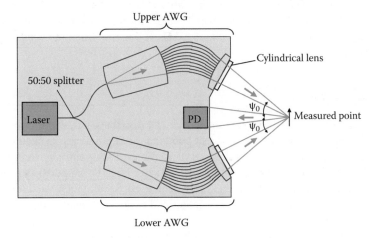

FIGURE 6.2 Optical circuit of wavelength-insensitive LDV using AWGs. (From Maru, K. and Fujii, Y., *Appl. Mechan. Mater.*, 103, 76, 2011. With permission.)

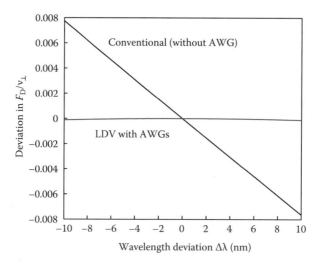

FIGURE 6.3 Deviation in $F_D/v_\perp$ for wavelength-insensitive LDV with AWGs as a function of wavelength deviation $\Delta\lambda = \lambda - \lambda_0$. (From Maru, K. and Fujii, Y., *Appl. Mechan. Mater.*, 103, 76, 2011. With permission.)

AWGs [15,17]. Input light is split with a 50:50 beam splitter. Each light passes through each AWG, output with diffraction, and incident on the object. The beams are scattered on the object and detected by a photodetector (PD). Each AWG is used to diffract the beam whose diffraction angle is changed depending on the wavelength. Each AWG is also used to focus the beam to the object. In other words, the AWGs can function as a lens system and dispersive elements having a planar structure.

In the proposed structure, ψ is changed with the system using the AWGs. Hence, wavelength-insensitive operation is expected if the design of these AWGs is optimized. Provided that the upper AWG and the lower AWG have the same design parameters, the condition of the design parameters for wavelength-insensitive operation is given by $\tan\psi_0 = m\lambda_0/(n_s d)$, where n_s is the effective refractive index for the slab waveguides in the AWGs, d is the interval of the waveguides in the array at the output side of the slab-to-array interface, and m is the grating order.

Figure 6.3 plots the deviation in $F_D/v_\perp$ determined as $[F_D/v_\perp - (F_D/v_\perp)|_{\lambda=\lambda_0}]/(F_D/v_\perp)|_{\lambda=\lambda_0}$ as a function of the wavelength deviation $\Delta\lambda = \lambda - \lambda_0$. For comparison, an LDV without wavelength-insensitive structure (i.e., without AWGs) is also shown in Figure 6.3. The absolute value of the deviation in $F_D/v_\perp$ can be significantly reduced to less than 1×10^{-4} by using AWGs whereas the maximum deviation for the conventional structure without AWGs is 7×10^{-3} when the wavelength deviation $\Delta\lambda$ is within 9.0 nm.

6.2.3 MULTIPOINT LDV

In some cases such as fluid flow in narrow pipes, velocity distribution in depth direction should be measured. In this case, the velocities of different points in the depth direction should be simultaneously measured. For this purpose, several types of LDVs for multipoint measurement have been proposed [3,37]. However, these multipoint LDVs consist of large bulk optical systems. Hence, an integrated type of multipoint LDVs has been highly demanded. Several integrated differential LDVs have been proposed [38,39] as applications of integrated optical sensors although these LDVs are used for single-point measurement.

Figure 6.4 illustrates the configuration of the proposed integrated multipoint differential LDV [18]. The LDV consists of laser diodes (LDs) with different lasing wavelengths as light

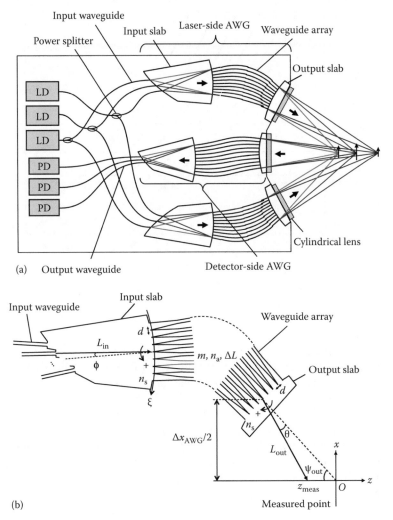

FIGURE 6.4 Integrated multipoint differential LDV. (a) Optical circuit and (b) schematic diagram of laser-side AWG. (From Maru, K. et al., *Opt. Express*, 18(1), 301, 2010. With permission.)

sources, power splitters, two laser-side AWGs with the same design, cylindrical lenses, a detector-side AWG, and PDs. The beam from each laser is split with a beam splitter. Each spilt beam is incident on the input slab of each laser-side AWG, phase-shifted through its waveguide array, output with diffraction, and incident on the measured point. The beam is scattered on the object at the measured point, input to the detector-side AWG, diffracted according to its wavelength, and detected by one of the PDs. The cylindrical lenses are used to collimate the beams in the vertical direction. The beams with different wavelengths are focused on different focusing points by the laser-side AWGs, and detected by different PDs by the detector-side AWG. Hence, the velocities at multiple measured points can be simultaneously measured. The laser-side AWGs are also used to diffract the beam whose diffraction angle is changed depending on the wavelength in order to reduce the dependence of the measured Doppler shift on the wavelength. This change in the diffraction angle contributes to reducing wavelength sensitivity, as described in Section 6.2.2.

In order to investigate the design and characteristics of the circuit, a model illustrated in Figure 6.4b is derived. The simulation was performed by assuming to use silica-based materials as the optical circuit [18]. Figure 6.5 shows the relation between the relative position of the

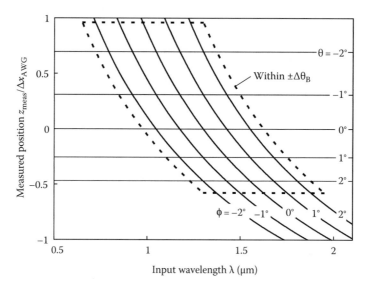

FIGURE 6.5 Relation between relative position of measured point $z_{meas}/\Delta x_{AWG}$ and input wavelength λ for various ϕ and θ. $m = 2$, $d = 10$ μm, and $\psi_{out} = 10.17°$. The condition of angles ϕ and θ within $\pm\Delta\theta_B$ is indicated as the area within the dotted line. (From Maru, K. et al., *Opt. Express*, 18(1), 301, 2010. With permission.)

measured point in the depth direction, $z_{meas}/\Delta x_{AWG}$, and the input wavelength λ for various input angles ϕ. The lines for the constant θ are also plotted in Figure 6.5. Obviously, when the diffraction angle θ is not changed, the measured position $z_{meas}/\Delta x_{AWG}$ becomes a constant. If the beam is diffracted to the direction out of the central Brillouin zone [40], its power is significantly decreased and the beam becomes no longer the main beam. The input beam to the waveguide array also suffers from considerable loss when the beam direction is out of the central Brillouin zone. Hence, the input angle ϕ and diffraction angle θ for each input wavelength must be within the central Brillouin zone defined as the angle within $\pm\Delta\theta_B$. The condition of angles ϕ and θ within $\pm\Delta\theta_B$ is indicated as the area within the dotted line. The positions of measured points can be derived from this figure once a set of input angles ϕ and the input wavelengths λ is determined. Table 6.1 shows an example of the parameters ϕ, λ, L_{in}, and z_{meas} for 5-point velocity measurement. Here, we assume $\Delta x_{AWG} = 30$ mm. In this case, the measured points are arranged over the range of 25.77 mm.

The absolute value of deviation in $F_D/v_\perp$ can be reduced for the multipoint LDV. We have simulated the absolute value of deviation in $F_D/v_\perp$ for the parameters shown in Table 6.1 as the nominal input wavelengths and the input angles. The simulation result indicates that the absolute value of

TABLE 6.1
Example of Parameters ϕ, λ, L_{in}, and z_{meas}

Input Wavelength, λ (μm)	Input Angle ϕ (Degree)	Input Array Distance L_{in} (mm)	Measured Position z_{meas} (mm)
1.20	0.7096	17.66	14.77
1.25	0.3548	17.93	6.80
1.30	0.0000	18.21	0.00
1.35	−0.3548	18.50	−5.87
1.40	−0.7096	18.80	−11.00

Source: Maru, K. et al., *Opt. Express*, 18(1), 301, 2010. With permission.

the deviation in $F_D/v_\perp$ for $\lambda = 1.3\ \mu m$ (i.e., the wavelength of the beam from the central input port) for the proposed structure can be significantly reduced to less than 10^{-4} of that for a conventional LDV and that the deviation for $\lambda = 1.2$ and $1.4\ \mu m$ (i.e., beams from marginal input ports) can also be reduced to less than 1/10 of that for a conventional LDV.

6.3 DIRECTIONAL DISCRIMINATION

Simultaneously measuring two components of velocity at a measured point has been required in many applications such as fluid experiments. Several methods for measuring multidimensional velocity components have been developed [4–13]. In typical conventional LDVs for measuring two-dimensional velocity, a complicated optical configuration or a three-dimensional arrangement of optical components is needed. In some LDVs using frequency shift, frequency shifters such as AOMs are needed to discriminate the direction of velocity. In these LDVs, measurable velocity range is limited by preshifting frequencies and the accuracy considerably depends on the performance of the optical modulators to be used. Several approaches have been reported for directional discrimination without optical modulators [39,41], although these approaches are for one-dimensional velocity measurement. In this section, two methods for two-dimensional velocity measurement using a simple optical configuration without any optical modulator are reviewed.

6.3.1 Two-Dimensional LDV Using Polarized Beams and 90° Phase Shift

Figure 6.6 illustrates the principle of the proposed LDV for two-dimensional velocity measurement using polarized beams and 90° phase shift [20]. The beam output from a laser is split into a signal beam to be incident on the measured point and a reference beam by a 1 × 2 splitter. The signal beam is split again into two orthogonally polarized beams by the polarization beam splitter 1 (PBS1). The two beams are incident on the measured point with different angles, scattered on the object at the measured point, and split again by PBS2. After passing through PBS2, the beam is split into the scattered beam from Port A and that from Port B. Each scattered beam is

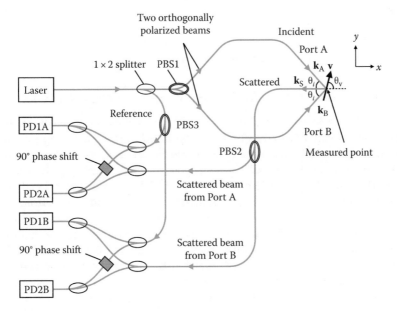

FIGURE 6.6 Principle of LDV for two-dimensional velocity measurement using polarized beams and 90° phase shift. (From Maru, K. et al., *Optik*, 122(11), 974, 2011. With permission.)

split into two beams, and each of the split beams is combined with one of the reference beams split by PBS3 and other 1×2 splitters. Here, before the reference beams are combined with each of the scattered beams, the phase of the reference beams is shifted so that the phase difference between the two reference beams becomes 90°. The power of each combined beams is detected by a PD as a beat signal produced by the interference of the scattered beam and the reference beam. For each scattered beams (from Port A and from Port B), two PDs and the 90° phase shift of the reference beam are used to detect the signs of the beat frequency and discriminate the direction of velocity.

The detected beat frequencies f_A at PD1A and PD2A and f_B at PD1B and PD2B are given by the wavenumber vectors of the incident beam from Port A and Port B, $\mathbf{k}_A$ and $\mathbf{k}_B$, the wavenumber vector of the scattered beam, $\mathbf{k}_S$, and the velocity vector at the measured point, $\mathbf{v}$ [20]. Here, $\mathbf{k}_A$, $\mathbf{k}_B$, and $\mathbf{k}_S$ are all placed in the x–y plane. The incident angle θ_i is defined by the angle between $\mathbf{k}_A$ ($\mathbf{k}_B$) and the x-axis, and the direction of the scattered beam is set to the direction opposite to the x-axis. The beat signal detected by each PD itself provides only the absolute value of its beat frequency. Only when the absolute values of the beat frequencies f_A and f_B are known, the direction of velocity cannot be discriminated but just the magnitudes of the components of the velocity. In order to discriminate the direction of velocity, the signs of the beat frequencies f_A and f_B also need to be discriminated. The sign can be discriminated by using the relation between the phases in the signals detected by the two PDs (PD1A and PD2A for the discrimination of the sign of f_A, and PD1B and PD2B for the discrimination of the sign of f_B).

To estimate the characteristics of the proposed LDV for two-dimensional velocity measurement, the relation among velocity vector, beat frequencies, and incident angle θ_i is simulated. Figure 6.7 shows the polar expression of the absolute value of the beat frequency normalized with $|\mathbf{v}/\lambda|$ and the direction of velocity θ_v for various θ_i. The distance from the origin O represents the absolute value of the normalized beat frequency and the angle from the x-axis represents θ_v as shown in Figure 6.7e. The sign of the beat frequency for each plot is also indicated in this figure. The proportion between f_A and f_B changes as changing the direction of velocity θ_v with one-to-one correspondence unless $\theta_i = 0°$. It indicates that the direction of velocity θ_v can be discriminated by the proposed LDV as well as the absolute values of the components of velocity. When the incident angle θ_i increases from 0° to 90°, the plots for f_A and f_B gradually separate each other. It implies that θ_i should be larger for good discrimination of the direction of velocity, although the optimal value of θ_i may also depend on allowable arrangement of the optical system and the measured point. In the plots for $\theta_i = 0°$ (Figure 6.7a), both the beat frequencies f_A and f_B are zero at $\theta_v = 90°$ and 270°. This indicates that there is no sensitivity to y-directed velocity.

6.3.2 Two-Dimensional LDV by Monitoring Beams in Different Directions

The method described in Section 6.3.1 uses the discrimination of two orthogonally polarized beams. In this method, polarization crosstalk between the polarized beams due mainly to imperfection of polarization-dependent optical components or depolarization on scattering easily results in a measurement error. Hence, a two-dimensional LDV without polarization discrimination is more desirable.

Figure 6.8 illustrates the principle of the proposed LDV for two-dimensional velocity measurement by monitoring beams in different directions [19]. The beam output from a laser is split into four beams, including two beams to be incident on the measured point and two reference beams with a 1×4 splitter. Two beams are incident on the measured point with different angles and scattered on the object at the measured point. The scattered beams in different directions are monitored with two detection blocks (Detection Block 1 and Detection Block 2). In each detection block, the scattered beam is split into three beams with two 1×2 splitters. Two of the three split scattered beams are combined with one of the reference beams split by another 1×2 splitter. Here, the phase of one of the split reference beams is shifted so that the phase difference between two split reference beams becomes 90°. The power of each combined beam is detected

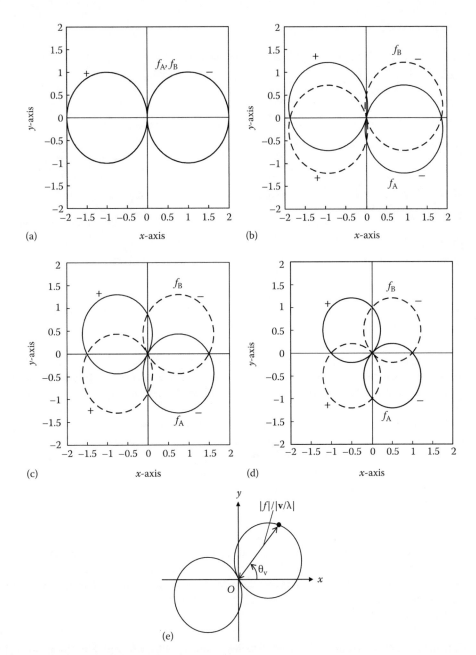

FIGURE 6.7 Polar expression of absolute value of beat frequency normalized with $|v/\lambda|$ and direction of velocity θ_v for various θ_i. (a) $\theta_i = 0°$, (b) $\theta_i = 30°$, (c) $\theta_i = 60°$, (d) $\theta_i = 90°$, and (e) explanation of plots. The distance from the origin O represents the absolute value of the normalized beat frequency and the angle from the x-axis represents θ_v. The sign of the beat frequency for each plot is also indicated. (From Maru, K. et al., *Optik*, 122(11), 974, 2011. With permission.)

by a PD as a beat signal with three beat frequencies produced by the interference among the scattered beam from Port A, the scattered beam from Port B, and the reference beam. Two PDs (PD1 and PD2) and the 90° phase shift of the reference beam are used to detect the signs of the beat frequencies. The power of the other one of the three split scattered beams is detected by PD3. The beat frequency of the beat signal produced by the interference between the scattered

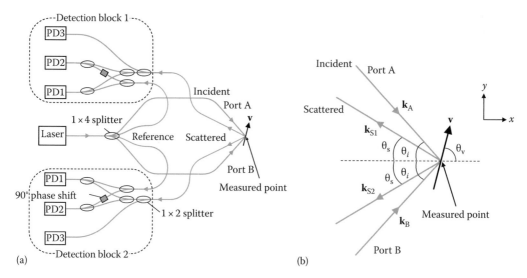

FIGURE 6.8 Principle of LDV for two-dimensional velocity measurement by monitoring beams in different directions. (a) Setting of optical system and (b) diagram of beams at measured points. (From Maru, K. and Fujii, Y., *IEEE Sens. J.*, 11(2), 312, 2011. With permission.)

beams from Port A and Port B can be discriminated by PD3 in each detection block because of the differential operation of the LDV. In each detection block, the beat frequency produced by the interference between the scattered beam from Port A and the reference beam cannot be discriminated from that produced by the interference between the scattered beam from Port B and the reference beam. Hence, the direction of velocity cannot be detected only by using one detection block. In order to discriminate the direction of velocity, scattered beams in different directions are monitored with two detection blocks.

The beat frequencies detected with Detection Block n (n = 1, 2), f_{ASn} and f_{BSn}, are derived as a similar procedure to the derivation in Section 6.3.1. The principle of the discrimination of velocity direction in the proposed LDV is as follows. In each detection block, the beat frequencies f_{ASn} and f_{BSn} cannot be discriminated from each other whereas the signs of f_{ASn} and f_{BSn}, as well as their absolute values, can be detected because of the 90° phase shift of the reference beam as described earlier. Hence, the direction of velocity cannot be discriminated in case that only one detection block is used. In the proposed velocimetry, the direction of velocity is discriminated by using two detection blocks as the following principle. The actual velocity vector at the measurement point $\mathbf{v}_r$ can be discriminated by the following procedure even if the two beat frequencies cannot be discriminated with each detection block [19]:

1. Calculate $\mathbf{v}_1$ and $\mathbf{v}_1'$ from the beat frequencies f_1 and f_1' obtained from Detection Block 1 as the following equations:

$$\mathbf{v}_1 = 2\pi \begin{pmatrix} (\mathbf{k}_{S1}-\mathbf{k}_A)^T \\ (\mathbf{k}_{S1}-\mathbf{k}_B)^T \end{pmatrix}^{-1} \begin{pmatrix} f_1 \\ f_1' \end{pmatrix} \tag{6.2}$$

$$\mathbf{v}_1' = 2\pi \begin{pmatrix} (\mathbf{k}_{S1}-\mathbf{k}_A)^T \\ (\mathbf{k}_{S1}-\mathbf{k}_B)^T \end{pmatrix}^{-1} \begin{pmatrix} f_1' \\ f_1 \end{pmatrix} \tag{6.3}$$

2. Calculate $\mathbf{v}_2$ and $\mathbf{v}_2'$ from the beat frequencies f_2 and f_2' obtained from Detection Block 2 as the following equations:

$$\mathbf{v}_2 = 2\pi \left(\begin{matrix} (\mathbf{k}_{S2} - \mathbf{k}_A)^T \\ (\mathbf{k}_{S2} - \mathbf{k}_B)^T \end{matrix} \right)^{-1} \begin{pmatrix} f_2 \\ f_2' \end{pmatrix} \tag{6.4}$$

$$\mathbf{v}_2' = 2\pi \left(\begin{matrix} (\mathbf{k}_{S2} - \mathbf{k}_A)^T \\ (\mathbf{k}_{S2} - \mathbf{k}_B)^T \end{matrix} \right)^{-1} \begin{pmatrix} f_2' \\ f_2 \end{pmatrix} \tag{6.5}$$

3. When $\mathbf{v}_1$ differs from $\mathbf{v}_1'$, $\mathbf{v}_2$ also differs from $\mathbf{v}_2'$. When $\mathbf{v}_1$ corresponds to $\mathbf{v}_2$ or $\mathbf{v}_2'$, $\mathbf{v}_1$ is the same as $\mathbf{v}_r$. When $\mathbf{v}_1'$ corresponds to $\mathbf{v}_2$ or $\mathbf{v}_2'$, $\mathbf{v}_1'$ is the same as $\mathbf{v}_r$.
4. When $\mathbf{v}_1$ corresponds to $\mathbf{v}_1'$, $\mathbf{v}_2$ also corresponds to $\mathbf{v}_2'$ and all the four vectors correspond to $\mathbf{v}_r$.

Here, $\mathbf{k}_A$ and $\mathbf{k}_B$ are the wavenumber vectors of the incident beam from Port A and Port B to each block, respectively, $\mathbf{k}_{Sn}$ is the wavenumber vector of the scattered beam detected with Detection Block n. The incident angle θ_i is defined by the angle between $\mathbf{k}_A$ ($\mathbf{k}_B$) and the x-axis, and the angle θ_s is defined by the angle between $\mathbf{k}_{S1}$ ($\mathbf{k}_{S2}$) and the direction opposite to the x-axis as shown in Figure 6.8b.

Let $\mathbf{v}_{i1}$ be defined as either of $\mathbf{v}_1$ or $\mathbf{v}_1'$, which does not correspond to $\mathbf{v}_r$, and $\mathbf{v}_{i2}$ be defined as either of $\mathbf{v}_2$ or $\mathbf{v}_2'$, which does not correspond to $\mathbf{v}_r$. Provided that we consider the transformation from $\mathbf{v}_r$ to $\mathbf{v}_{i1}$, there are two eigenvectors $\mathbf{x}_{S1,1}$ and $\mathbf{x}_{S1,-1}$ with the eigenvalues of 1 and −1, respectively. Similarly, there are two eigenvectors $\mathbf{x}_{S2,1}$ and $\mathbf{x}_{S2,-1}$ with the eigenvalues of 1 and −1 on the transformation from $\mathbf{v}_r$ to $\mathbf{v}_{i2}$. This implies that when the actual velocity vector $\mathbf{v}_r$ is directed to $\mathbf{x}_{S1,1}$ (or $\mathbf{x}_{S2,1}$), both $\mathbf{v}_{i1}$ and $\mathbf{v}_{i2}$ become the same as $\mathbf{v}_r$. On the other hand, when $\mathbf{v}_r$ is directed to a different direction from $\mathbf{x}_{S1,1}$ (or $\mathbf{x}_{S2,1}$), $\mathbf{v}_r$ has the component of $\mathbf{x}_{S1,-1}$ and the component of $\mathbf{x}_{S2,-1}$. Here, the eigenvectors $\mathbf{x}_{S1,-1}$ and $\mathbf{x}_{S2,-1}$ are different from each other because the wavenumber vectors $\mathbf{k}_{S1}$ and $\mathbf{k}_{S2}$ are different from each other, whereas their eigenvalues are the same. It means that the resultant vectors of these components are different from each other even after the transformations.

Figure 6.9 shows the examples of the directional relation among $\mathbf{v}_r$, $\mathbf{v}_{i1}$, and $\mathbf{v}_{i2}$ for various directions of the velocity, θ_{vr}, defined as the angle between the direction of $\mathbf{v}_r$ and the x-axis. The magnitude of $\mathbf{v}_r$ is normalized to unity, and it is assumed that $\theta_i = 60°$ and $\theta_s = 50°$. In this case, the eigenvectors $\mathbf{x}_{S1,1}$ and $\mathbf{x}_{S2,1}$ are directed to the parallel direction to the x-axis. Consequently, both $\mathbf{v}_{i1}$ and $\mathbf{v}_{i2}$ correspond to $\mathbf{v}_r$ when θ_{vr} is 0° or 180°. When the θ_{vr} is 45°, 90°, or 135°, $\mathbf{v}_{i1}$ and $\mathbf{v}_{i2}$ are different from each other, and both $\mathbf{v}_{i1}$ and $\mathbf{v}_{i2}$ are also different from $\mathbf{v}_r$. It is because the eigenvectors $\mathbf{x}_{S1,-1}$ and $\mathbf{x}_{S2,-1}$ are directed to different directions (56.2° and 123.8° from the x-axis, respectively) and both the $\mathbf{x}_{S1,-1}$ component and the $\mathbf{x}_{S2,-1}$ component of $\mathbf{v}_r$ are oppositely directed after the transformation.

Figure 6.10 shows the magnitudes and directions of the velocities $\mathbf{v}_r$, $\mathbf{v}_{i1}$, and $\mathbf{v}_{i2}$ as a function of the direction of the velocity θ_{vr}. Also in Figure 6.10, the magnitude of $\mathbf{v}_r$ is normalized to unity and it is assumed that $\theta_i = 60°$ and $\theta_s = 50°$. Both $\mathbf{v}_{i1}$ and $\mathbf{v}_{i2}$ correspond to $\mathbf{v}_r$ only at $\theta_{vr} = 0°$ or 180°. When the θ_{vr} is different from 0° or 180°, $\mathbf{v}_{i1}$ and $\mathbf{v}_{i2}$ are different from each other, and both $\mathbf{v}_{i1}$ and $\mathbf{v}_{i2}$ are also different from $\mathbf{v}_r$. In the proposed structure, $\mathbf{v}_{i1}$ is obtained from Detection Block 1, $\mathbf{v}_{i2}$ is obtained from Detection Block 2, and $\mathbf{v}_r$ is obtained from both of the detection blocks. Consequently, for all the directions of $\mathbf{v}_r$, the velocity vector $\mathbf{v}_r$ can be discriminated as one of two velocity vectors obtained from Detection Block 1, which corresponds to one of two velocity vectors obtained from Detection Block 2.

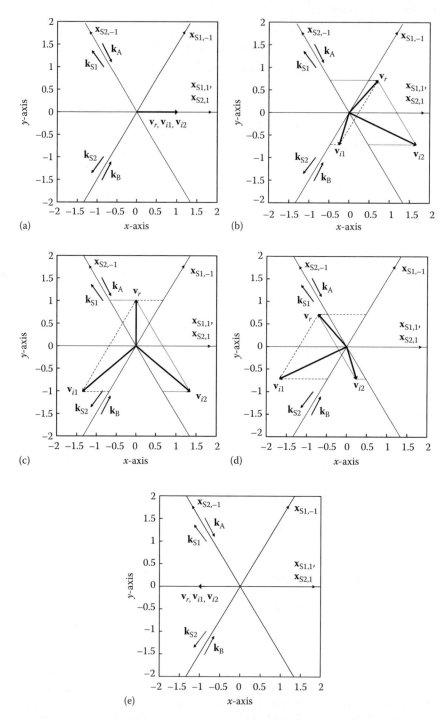

FIGURE 6.9 Directional relation among $\mathbf{v}_r$, $\mathbf{v}_{i1}$, and $\mathbf{v}_{i2}$ at $\theta_i = 60°$ and $\theta_s = 50°$ for (a) $\theta_{vr} = 0°$, (b) $\theta_{vr} = 45°$, (c) $\theta_{vr} = 90°$, (d) $\theta_{vr} = 135°$, and (e) $\theta_{vr} = 180°$. (From Maru, K. and Fujii, Y., *IEEE Sens. J.*, 11(2), 312, 2011. With permission.)

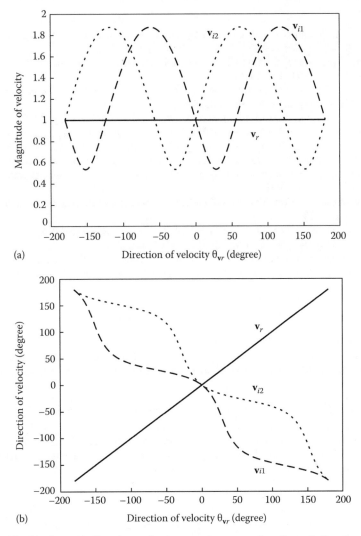

(a)

(b)

FIGURE 6.10 Magnitudes and directions of $\mathbf{v}_r$, $\mathbf{v}_{i1}$, and $\mathbf{v}_{i2}$ as a function of direction of velocity θ_{vr}. (a) Magnitudes and (b) directions. $\theta_i = 60°$ and $\theta_s = 50°$. The magnitude of $\mathbf{v}_r$ is normalized to unity. (From Maru, K. and Fujii, Y., *IEEE Sens. J.*, 11(2), 312, 2011. With permission.)

6.4   SUMMARY

Several types of integrated LDVs using the PLC technology, especially wavelength-insensitive LDVs and a multipoint LDV using AWGs, are described. The PLC technology has a potential for drastically reducing the sizes of LDVs. By optimizing the design parameters, a wavelength-insensitive operation can be obtained by using AWGs without assembly of a bulk diffractive grating and other optical elements. The use of AWGs also has a potential to simultaneously measure velocities at different points in compact optical systems.

In addition, techniques for two-dimensional velocity measurement using a simple optical configuration without multiple colors or any optical modulator are described. In the proposed methods, the combination of polarized beams and the 90° phase shift or the monitoring of the scattered beams in different directions with two detection blocks is used to discriminate the direction of velocity. In each method, the simulation result indicates that the two-dimensional velocity vector can be discriminated. These methods would be applicable to various velocity measurement fields.

REFERENCES

1. Y. Yeh and H. Z. Cummins, Localized fluid flow measurements with an He–Ne laser spectrometer, *Appl. Phys. Lett.*, 4(10), 176–178, 1964.
2. M. Kawachi, Silica waveguides on silicon and their application to integrated-optic components, *Opt. Quant. Electron.*, 22, 391–416, 1990.
3. H.-E. Albrecht, M. Borys, N. Damaschke, and C. Tropea, *Laser Doppler and Phase Doppler Measurement Techniques*, Chap. 7, Springer-Verlag, Berlin Germany, 2003.
4. K. A. Blake, Simple two-dimensional laser velocimeter optics, *J. Phys. E*, 5, 623–624, 1972.
5. N. Nakatani, M. Tokita, and T. Yamada, LDV using polarization-preserving optical fibers for simultaneous measurement of two velocity components, *Appl. Opt.*, 23(11), 1686–1687, 1984.
6. N. Nakatani, M. Tokita, T. Izumi, and T. Yamada, Laser Doppler velocimetry using polarization-preserving optical fibers for simultaneous measurement of multidimensional velocity components, *Rev. Sci. Instrum.*, 56, 2025–2029, 1985.
7. G. T. Grant and K. T. Orloff, Two-color dual beam backscatter laser Doppler velocimeter, *Appl. Opt.*, 12(12), 2913–2916, 1973.
8. H.-E. Albrecht, M. Borys, N. Damaschke, and C. Tropea, *Laser Doppler and Phase Doppler Measurement Techniques*, Section 7.4.2, Springer-Verlag, Berlin Germany, 2003.
9. R. J. Adrian, A bipolar, two component laser-Doppler velocimeter, *J. Phys. E*, 8, 723–726, 1975.
10. X. Shen, J. Zhang, Z. Wang, and H. Yu, Two component LDV system with dual-differential acousto-optical frequency shift and its applications, *Acta Mechan. Sin.*, 2, 81–92, 1986.
11. S. Kato and K. Hasegawa, Novel techniques of laser Doppler velocimetry using optical integrated circuit, *R&D Rev. Toyota Cent. R&D Labs.*, 34, 35–42, 1999 (in Japanese).
12. K. Hasegawa, S. Kato, and H. Itoh, Two-dimensional fiber laser Doppler velocimeter by integrated optical frequency shifter, *Proc. SPIE*, 3740, 294–297, 1999.
13. L. Kervevan, H. Gilles, S. Girard, and M. Laroche, Two-dimensional velocity measurements with self-mixing technique in diode-pumped Yb:Er glass laser, *IEEE Photon. Technol. Lett.*, 16(7), 1709–1711, 2004.
14. K. Maru and Y. Fujii, Wavelength-insensitive laser Doppler velocimeter using beam position shift induced by Mach–Zehnder interferometers, *Opt. Express*, 17(20), 17441–17449, 2009.
15. K. Maru and Y. Fujii, Integrated wavelength-insensitive differential laser Doppler velocimeter using planar lightwave circuit, *J. Lightwave Technol.*, 27(22), 5078–5083, 2009.
16. K. Maru and Y. Fujii, Differential laser Doppler velocimeter with enhanced range for small wavelength sensitivity by using cascaded Mach–Zehnder interferometers, *J. Lightwave Technol.*, 28(11), 1631–1637, 2010.
17. K. Maru and Y. Fujii, Laser Doppler velocimetry with small wavelength sensitivity using planar lightwave circuit, *Appl. Mechan. Mater.*, 103, 76–81, 2011.
18. K. Maru, K. Kobayashi, and Y. Fujii, Multi-point differential laser Doppler velocimeter using arrayed waveguide gratings with small wavelength sensitivity, *Opt. Express*, 18(1), 301–308, 2010.
19. K. Maru and Y. Fujii, Laser Doppler velocimetry for two-dimensional directional discrimination by monitoring scattered beams in different directions, *IEEE Sens. J.*, 11(2), 312–318, 2011.
20. K. Maru, L. Y. Hu, R. S. Lu, Y. Fujii, and P. P. Yupapin, Two-dimensional laser Doppler velocimeter using polarized beams and 90° phase shift for discrimination of velocity direction, *Optik*, 122(11), 974–977, 2011.
21. M. K. Smit, New focusing and dispersive planar component based on an optical phased array, *Electron. Lett.*, 24(7), 385–386, 1988.
22. C. Dragone, Efficient $N \times N$ star coupler based on Fourier optics, *Electron. Lett.*, 24(15), 942–944, 1988.
23. H. Takahashi, S. Suzuki, K. Kato, and I. Nishi, Arrayed-waveguide grating for wavelength division multi/demultiplexer with nanometer resolution, *Electron. Lett.*, 26(2), 87–88, 1990.
24. H. Takahashi, I. Nishi, and Y. Hibino, 10 GHz spacing optical frequency division multiplexer based on arrayed-waveguide grating, *Elecron. Lett.*, 28(4), 380–382, 1992.
25. A. R. Vellekoop and M. K. Smit, Four-channel integrated-optic wavelength demultiplexer with weak polarization dependence, *J. Lightwave Technol.*, 9(3), 310–314, 1991.
26. C. Dragone, An $N \times N$ optical multiplexer using a planar arrangement of two star couplers, *IEEE Photon. Technol. Lett.*, 3, 812–815, 1991.
27. C. Dragone, C. A. Edwards, and R. C. Kistler, Integrated optics $N \times N$ multiplexer on silicon, *IEEE Photon. Technol. Lett.*, 3, 896–899, 1991.
28. Y. Tachikawa, Y. Inoue, M. Kawachi, H. Takahashi, and K. Inoue, Arrayed-waveguide grating add-drop multiplexer with loop-back optical paths, *Electron. Lett.*, 29(24), 2133–2134, 1993.

29. O. Ishida, H. Takahashi, S. Suzuki, and Y. Inoue, Multichannel frequency-selective switch employing an arrayed-waveguide grating multiplexer with fold-back optical paths, *IEEE Photon. Technol. Lett.*, 6(10), 1219–1221, 1994.
30. M. Kawachi, M. Yasu, and M. Kobayashi, Flame hydrolysis deposition of SiO_2–TiO_2 glass planar optical waveguides on silicon, *Jpn. J. Appl. Phys.*, 22(12), 1932, 1983.
31. S. Kashimura, M. Takeuchi, K. Maru, and H. Okano, Loss reduction of GeO_2-doped silica waveguide with high refractive index difference by high-temperature annealing, *Jpn. J. Appl. Phys.*, 39(Pt 2) (6A), L521–L523, 2000.
32. C. H. Henry, G. E. Blonder, and R. F. Kazarinov, Glass waveguides on silicon for hybrid optical packaging, *J. Lightwave Technol.*, 7(10), 1530–1539, 1989.
33. C. R. Doerr and K. Okamoto, Advances in silica planar lightwave circuits, *J. Lightwave Technol.*, 24(12), 4763–4789, 2006.
34. J. Schmidt, R. Volkel, W. Stork, J. T. Sheridan, J. Schwider, and N. Steibl, Diffractive beam splitter for laser Doppler velocimetry, *Opt. Lett.*, 17(17), 1240–1242, 1992.
35. R. Sawada, K. Hane, and E. Higurashi, *Optical Micro Electro Mechanical Systems*, Section 5.2, Ohmsha, Tokyo, 2002 (in Japanese).
36. H.-E. Albrecht, M. Borys, N. Damaschke, and C. Tropea, *Laser Doppler and Phase Doppler Measurement Techniques*, Section 7.2.2, Springer-Verlag, Berlin Germany, 2003.
37. T. Hachiga, N. Furuichi, J. Mimatsu, K. Hishida, and M. Kumada, Development of a multi-point LDV by using semiconductor laser with FFT-based multi-channel signal processing, *Exp. Fluids*, 24, 70–76, 1998.
38. M. Haruna, K. Kasazumi, and H. Nishihara, Integrated-optic differential laser Doppler velocimeter with a micro Fresnel lens array, in *Proceedings of Conference on Integrated & Guided-Wave Optics (IGWO '89)*, MBB6. Houston, TX.
39. T. Ito, R. Sawada, and E. Higurashi, Integrated microlaser Doppler velocimeter, *J. Lightwave Technol.*, 17(1), 30–34, 1999.
40. C. R. Doerr, M. Cappuzzo, E. Laskowski, A. Paunescu, L. Gomez, L. W. Stulz, and J. Gates, Dynamic wavelength equalizer in silica using the single-filtered-arm interferometer, *IEEE Photon. Technol. Lett.*, 11(5), 581–583, 1999.
41. K. Plamann, H. Zellmer, J. Czarske, and A. Tünnermann, Directional discrimination in laser Doppler anemometry (LDA) without frequency shifting using twinned optical fibres in the receiving optics, *Meas. Sci. Technol.*, 9, 1840–1846, 1998.

7 Vision-Aided Automated Vibrometry for Remote Audio–Visual Range Sensing

Tao Wang and Zhigang Zhu

CONTENTS

7.1 INTRODUCTION

Remote object signature detection is becoming increasingly important in noncooperative and hostile environments for many applications (Dedeoglu et al., 2008; Li et al., 2008). These include (1) remote and large area surveillance in frontier defense, maritime affairs, law enforcement, and so on; (2) perimeter protection for important locations and facilities such as forest, oil fields, railways, and high voltage towers; and (3) search and rescue in natural and man-made disasters such as earthquakes, floods, hurricanes, and terrorism attacks. In these situations, target signature detection, particularly signatures of humans, vehicles, and other targets or events, at a large distance, is critical in order to watch out for the trespassers or events before taking appropriate actions, or make quick decisions to rescue the victims, with minimum risks.

Although imaging and video technologies (including visible and IR) have had great advancement in object signature detection at a large distance, there are still many limitations in non-cooperative and hostile environments because of intentional camouflage and natural occlusions. Audio information, another important data source for target detection, still cannot match the range and signal qualities provided by video technologies for long-range sensing, particularly under a variety of large background noises. For obtaining better performance of human tracking in a near to mediate range, Beal et al. (2003) and Zou and Bhanu (2005) have reported the integrations of visual and acoustic sensors. By integration, each modality may compensate for the weaknesses of the other one. But in these systems, the acoustic sensors (microphones) need to be placed near the subjects in monitoring, therefore cannot be used for long-range surveillance. A parabolic microphone, which can capture voice signals at a fairly large distance in the direction pointed by the microphone, could be used for remote hearing and surveillance. But it is very sensitive to noise caused by the surroundings (i.e., wind) or the sensor motion, and all the signals on the way are captured. Therefore, there is a great necessity to find a new type of acoustic sensor for long-range voice detection.

Laser Doppler vibrometers (LDV) such as those manufactured by Polytec (2009) and Ometron (2009) can effectively detect vibration within 200 m with sensitivity in the order of 1 µm/s. Larger distances could be achieved with the improvements of sensor technologies and the increase of the laser power while using a different wavelength (e.g., infrared instead of visible). In our previous work (Li et al., 2006; Zhu et al., 2007), we have presented very promising results in detecting and enhancing voice signals of people from large distances using a Polytec LDV. However, the user had to manually adjust the LDV sensor head in order to aim the laser beam at a surface that well reflects the laser beam, which was a tedious and difficult task. In addition, it was very hard for the user to see the laser spot at a distance above 20 m, and so it was extremely difficult for the human operator to aim the laser beam of the LDV on a target in a distance larger than 100 m. Of course, human eyes cannot see infrared laser beams so it would be a serious problem if the LDV uses infrared. Also, it takes quite some time to focus the laser beam even if the laser beam is pointed to the surface. Therefore, reflection surface selection and automatic laser aiming and focusing are greatly needed in order to improve the performance and the efficiency of the LDV for long-range hearing.

Here, we present a novel multimodal sensing system, which integrates the LDV with a pair of pan–tilt–zoom (PTZ) cameras to aid the LDV in finding a reflective surface and focusing its laser beam automatically, and consequently the system captures both video and audio signals synchronously for target detection using multimodal information: in addition to video and audio, this sensing system can also obtain range information using the LDV–PTZ triangulation as well as stereo vision using the two cameras. The range information will further add values to object signature detection in addition to the audio and video information, and improve the robustness and the detection rate of the sensor. The main contribution of this work is the collaborative operation of a dual-PTZ camera system and a laser pointing system for long-range acoustic detection. To our knowledge, this is the first work that uses a PTZ stereo for automating the long-range laser-based voice detection. Meanwhile, the combination is a natural extension of the already widely used PTZ camera–based video surveillance system toward multimodal surveillance with audio, video, and range information.

The rest of this chapter is organized as follows. Section 7.2 presents some background and related work. Section 7.3 describes an overview of our vision-aided automated vibrometry system. Section 7.4 discusses the calibration issues among the multimodal sensory components. Section 7.5 shows the algorithms for feature matching and distance measuring using the system. Section 7.6 describes the adaptive and collaborative sensing approach. Section 7.7 provides some experimental results. Finally, we conclude our work in Section 7.8.

7.2 BACKGROUND AND RELATED WORK

7.2.1 PRINCIPLE OF THE LASER DOPPLER VIBROMETER

The LDV works according to the principle of laser interferometry. Measurement is made at the point where the laser beam strikes the structure under vibration. In the heterodyning interferometer (Figure 7.1), a coherent laser beam is divided into object and reference beams by a beam splitter BS1. The object beam strikes a point on the moving (vibrating) object and light reflected from that point travels back to beam splitter BS2 and mixes (interferes) with the reference beam at beam splitter BS3. If the object is moving (vibrating), this mixing process produces an intensity fluctuation in the light as

$$I_1 = \frac{1}{2} A^2 \left\{ 1 - \cos \left[2\pi \left(f_B + \frac{2v}{\lambda} \right) t \right] \right\} \tag{7.1}$$

where
I_1 is light intensity
A is the amplitude of the emitted wave
f_B is modulation frequency of the reference beams
λ is the wavelength of the emitted wave
v is the object's velocity
t is observation time

A detector converts this signal to a voltage fluctuation. And from the fluctuating of light patterns, the velocity of object can be decoded by a digital quadrate demodulation method (Scruby and Drain, 1990). An interesting finding of our study is that most objects vibrate while wave energy (including that of voice waves) is applied on them. Although the vibration caused by the voice energy is very small compared with other vibration, it can be detected by the LDV, and be extracted with advanced signal filtering. The relation of voice frequency f, velocity v, and magnitude m of the vibration is

$$v = 2\pi f m \tag{7.2}$$

As seen from the earlier principle of the LDV, there are three requirements to be considered in order to use the LDV to measure the vibration of a target caused by sounds:

1. An appropriate surface close to the sounding target with detectable vibration and good reflection index
2. The focus of the LDV laser beam on the refection surface, otherwise very weak reflection signals are obtained due to the scattering of coherent light and path length differences
3. A necessary signal enhancement process to filter out the background noise and the inherent noise of the LDV

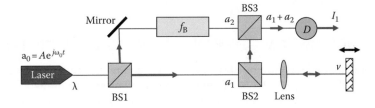

FIGURE 7.1 Principle of the laser Doppler vibrometer (LDV).

In close-range and lab environments, it is not a serious problem for a human operator to find an appropriate reflective surface, focus the laser beam, and acquire the vibration signals. But at a large distance (from 20 to 100 m), the manual process becomes extremely difficult because it is very hard for a human operator to aim the laser beam to a good reflective surface. Also, it takes quite some time to focus the laser beam even if the laser beam is pointed to the surface. Therefore, there are great unmet needs in facilitating the process of surface detection, laser aiming, laser focusing, and signal acquisition of the emerging LDV sensor, preferably through system automation.

7.2.2 RELATED WORK

Acoustic sensing and event detection can be used for audio-based surveillance, including intrusion detection (Zieger et al., 2009), abnormal situations detection in public areas such as banks, subways, airports, and elevators (Clavel et al., 2005; Radhakrishnan et al., 2005). It can also be used as a complementary source of information for video surveillance and tracking (Cristani et al., 2007; Dedeoglu et al., 2008). In addition to microphones, an LDV, another type of acoustic sensors, is a novel type of measurement device to detect a target's vibration in a noncontact way, in applications such as bridge inspection (Khan et al., 1999), biometrics (Lai et al., 2008), and underwater communication (Blackmon and Antonelli, 2006). It has also been used to obtain the acoustic signals of a target (e.g., a human or a vehicle) in a large distance by detecting the vibration of a reflecting surface caused by the sound of the target next to it (Zhu et al., 2005, 2007; Li et al., 2006; Wang et al., 2011a). The LDVs have been used in the inspection industry and other important applications concerning environment, safety, and preparedness that meet basic human needs. In bridge and building inspection, the noncontact vibration measurements for monitoring structural defects eliminate the need to install sensors as a part of the infrastructure (e.g., Khan et al., 1999). In security and perimeter applications, an LDV can be used for voice detection without having the intruders in the line of the sight (Zhu et al., 2005). In medical applications, an LDV can be used for noncontact pulse and respiration measurements (Lai et al., 2008). In search and rescue scenarios where reaching humans can be very dangerous, an LDV can be applied to detect survivors which are even out of visual sight. Blackmon and Antonelli (2006) have tested and shown a sensing system to detect and receive underwater communication signals by probing the water surface from the air, using an LDV and a surface normal tracking device.

However, in most of the current applications, such systems are manually operated. In close-range and lab environments, this is not a very serious problem. But in field applications, such as bridge/building inspection, area protection, or search and rescue application, the manual process takes a very long time to find an appropriate reflective surface, focus the laser beam, and get a vibration signal; more so, if the surface is at a distance of 100 m or more. A vision-aided LDV system can improve the performance and the efficiency of the LDV for automatic remote hearing. In this work, we improved the flexibility and usability of the vision-aided automated vibrometry system from our previous design with a single PTZ camera (Qu et al., 2010) to the current design with a pair of PTZ cameras (Section 7.3) and by providing adaptive and collaborative sensing (Section 7.6).

7.3 VISION-AIDED AUTOMATED VIBROMETRY: SYSTEM OVERVIEW

The system consists of a single-point LDV sensor system, a mirror mounted on a pan–tilt unit (PTU), and a pair of PTZ cameras, one of which is mounted on the top of the PTU (Figure 7.2). The sensor head of the LDV uses a helium–neon laser with a wavelength of 632.8 nm and is equipped with a super long-range lens. It converts velocity of the target into interferometry signals and magnitude signals, and sends them to the controller of the LDV that are controlled by the computer via an RS-232 port. The controller processes signals received from the sensor

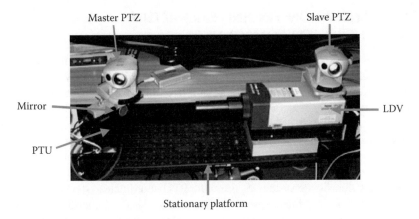

FIGURE 7.2 The multimodal sensory platform.

head of the LDV, and then outputs either voltage or magnitude signals to the computer using an S/P-DIF output. The Polytec LDV sensor OFV-505 and the controller OFV-5000 that we use in our experiments can be configured to detect vibrations under several different velocity ranges: 1, 2, 10, and 50 mm/s/V, where V stands for velocity. For voice vibration of a basic frequency range from 300 to 3000 Hz, we usually use the 1 mm/s/V range. The best resolution is 0.02 μm/s under the range of 1 mm/s/V according to the manufacturer's specification with retro-reflective tape treatment. Without the retro-reflective treatment, the LDV still has sensitivity on the order of 1.0 μm/s. This indicates that the LDV can detect vibration (due to voice waves) at a magnitude in nanometers without retro-reflective treatment; this can even get down to picometer with retro-reflective treatment.

The LDV sensor head weighs about 3.4 kg; this is the major reason that a mirror mounted on the PTU is used in our system to reflect the laser beam to freely and quickly point it to various directions in a large field of view. The laser beam points to the mirror at the center of the panning tilting of the PTU. The vision component consists of a pair of Canon VC-C50i (26×) PTZ cameras with one mounted on the top of the PTU, which is called the *master PTZ* since it is the main camera to track the laser beam, and another one mounted on the top of the LDV, which is called the *slave PTZ*. Each PTZ camera (Canon VC-C50i) has a 720 × 480 focal plane array and an auto-iris zoom lens that can change from 3.5 to 91 mm (26× optical power zoom). The pan angle of the PTZ is ±100° with rotation speed 1° to 90° per second and the tilt angle of it is from −30° to +90° with rotation speed 1° to 70° per second. The PTU is the model PTU-D46-70 of Directed Perception, Inc. It has a pan range from −159° to +159° and a tilt range from −47° to +31°. Its rotation resolution is 0.013° and maximum rotation speed is 300° per second. The reason to use zoom cameras is to detect targets and to assist the laser pointing and focusing at various distances. However, at a long distance, the laser spot is usually hard to be seen by the cameras, either zoomed or with wide views, if the laser is unfocused or not pointed on the right surface. Therefore, the master PTZ camera is used to rotate synchronously with the reflected laser beam from the mirror in order to track the laser spot. Although the laser point may not be observed from the master PTZ, we always control the pan and tilt angles of the master PTZ camera so that its optical axis is in parallel to the reflected laser beam, and therefore the laser spot is always close to the center of the image. Then, the master PTZ camera and the slave PTZ form a stereo vision system to obtain the distance to focus the laser spot as well as guide the laser to the right surface for acoustic signal collection. The baseline between of the two PTZ cameras is about 0.6 m for enabling long-range distance measurements. In order to obtain the distance from the target surface to the LDV, the calibration among the two PTZ cameras and the LDV is the first important step, which will be elaborated in the next section before the discussion of our method for distance measurement.

7.4 SYSTEM CALIBRATION: FINDING PARAMETERS AMONG THE SENSOR COMPONENTS

There are two stereo vision components in our system: stereo vision between the two PTZ cameras, and stereo triangulation between the slave PTZ camera and the mirrored LDV laser projection. The first component is used to obtain the range of a point in a reflective surface by matching its image projection (x, y) in the master camera to the corresponding image point (x', y') in the salve camera. The second component is mainly used to obtain the pan (α) and tilt (β) rotation angles of the PTU so that the LDV points to the image point (x, y) in the master image. Before determining the distance, several coordinate systems corresponding to the multisensory platform (Figure 7.2) are illustrated in Figure 7.3a: the master PTZ camera coordinate system (S_c), the slave PTZ camera coordinate system $(S_{c'})$, the LDV coordinate system (S_L), and the PTU coordinate system (S_u).

We assume the mirror coordinate system is the same as the PTU coordinate system since the laser will point to the mirror at the origin of the PTU system. The mirror normal direction is along the Z_u axis and initially points to the outgoing laser beam along Z_l. In order to always actually track the reflected laser beam visible or invisible (by having the optical axis of the master PTZ parallel to the reflected laser beam), the master PTZ not only rotates the same base angles with the PTU, α and β, which are the pan and tilt angles of the PTU around the X_u and Z_u axes, but also undergoes additional pan and tilt rotations (α' and β') around the Y_c and X_c axes. We will explain in detail how to determine these angles in Section 7.6.

The stereo matching is performed after the full calibration of the stereo component of the two PTZ cameras, and that between the slave PTZ camera and the "mirrored" LDV. Given a selected point on a reflective surface in the image of the master camera, we first find its corresponding point in the image of the slave camera, meanwhile calculating the pan and tilt angles of the PTU and the master and slave PTZ camera so that the laser spot is right under the center of the image of the master PTZ camera; the offset to the center is a function of the distance of the surface to the sensor system. The farther the surface is, the closer is the laser spot to the center. The distance from the target point to the optical center of the LDV is estimated via the stereo PTZ and then used to focus the laser beam to the surface.

7.4.1 CALIBRATION OF THE TWO PTZ CAMERAS

The calibration between the two PTZ cameras is carried out by estimating both the intrinsic and extrinsic parameters of each camera on every possible zoom factor when the camera is in focus, using the same world reference system. We use the calibration toolbox by Bouguet (2008) to find a camera's parameters under different zoom factors. We have found that the estimated extrinsic

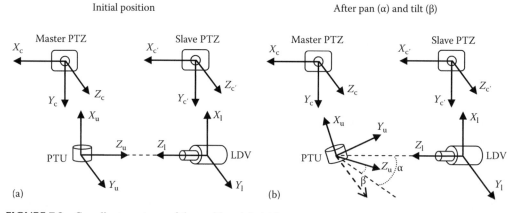

FIGURE 7.3 Coordinate systems of the multimodal platform.

parameters do not change much with the changes of zooms. However, the focal lengths of the cameras increase nonlinearly with the changes of different zooms; therefore, we have calibrated the camera under every possible zoom. Also note that the focal lengths of two PTZ cameras may not be same under the same zoom factor. In order to achieve similar fields of view (FOVs) and to ease the stereo matching between two images, the correct zoom of the slave PTZ camera corresponding to the actual focal length of the master PTZ camera should be selected. After the calibration, we obtain the effective focal lengths and image centers of the two cameras under every zoom factor k, and the transformation between the two cameras, represented by R and T:

$$P_{C'} = RP_{C} + T \tag{7.3}$$

where P_{C} and $P_{C'}$ are the representations of a 3D point in the master and slave PTZ coordinate systems (S_{C} and $S_{C'}$), respectively.

7.4.2 Calibration of the Slave Camera and the LDV

Since the intrinsic parameters of the slave PTZ camera have been obtained previously, we only need to estimate the extrinsic parameters characterizing the relation between the LDV coordinate system (S_{L}) and the slave PTZ camera coordinate system ($S_{C'}$), defined as

$$P_{L} = R_{C'}P_{C'} + T_{C'} \tag{7.4}$$

where P_{L} and $P_{C'}$ represent the coordinates of a 3D point in S_{L} and $S_{C'}$, respectively. The $R_{C'}$ and $T_{C'}$ are the rotation matrix and translation vector between S_{L} and $S_{C'}$. Next, the relation between the LDV coordinate system (S_{L}) and the mirrored LDV coordinate system (S_{ML}, not shown in Figure 7.3) is defined as

$$P_{L} = R_{U}R_{LR}R_{U}^{T}(P_{ML} - T_{U}) + T_{U} \tag{7.5}$$

where
  P_{L} and P_{ML} are the 3D point representations in the S_{L} and the S_{ML}, respectively
  R_{U} and T_{U} are the rotation matrix and translation vector between S_{L} and the PTU coordinate system
  R_{LR} is the rotation matrix that converts a right-hand coordinate system to a left-hand coordinate system

Then the extrinsic parameters are estimated by combining Equations 7.4 and 7.5 as

$$R_{C'}P_{C'} = R_{U}R_{LR}R_{U}^{T}(P_{ML} - T_{U}) + (T_{U} - T_{C'}) \tag{7.6}$$

For the calibration between the LDV and the slave PTZ, the LDV laser beam is projected at preselected points in a checkerboard placed at various locations/orientations. Because both the variables $P_{ML} - T_{U}$ and $T_{U} - T_{C'}$ are not independent in Equation 7.6, the distance between the fore lens of the LDV and the laser point on the mirror is estimated initially. Also, to avoid the complexity of the nonlinear equation we assume that the initial rotation matrix is the identified matrix, which can be manually adjusted by pointing both cameras parallel to the same direction. Then this initial distance and initial rotation matrix can be refined iteratively. Giving n 3D points, $3n$ linear equations that include $n + 14$ unknowns are constructed using Equation 7.6. Therefore, at least 7 points are needed.

7.5 STEREO VISION: FEATURE MATCHING AND DISTANCE MEASURING

7.5.1 STEREO MATCHING

After calibration, distance of a point can be estimated when the corresponding point in the slave image of a selected point in the master image is obtained. In the master camera, a target point can be selected either manually or automatically. We assume that both left and right images can be rectified given the intrinsic matrices for both cameras and the rotation matrix and translation vector. So, given any point $(x, y, 1)^T$ in original (right) image, the new pixel location $(x', y', 1)^T$ in rectified right image is $R'_r(x, y, 1)^T$. To simplify, the task radial distortion parameters are ignored. The rectified matrices for both cameras (virtually) make both camera images plane the same plane. Thus, the stereo matching problem turns into a simple horizontal searching problem since all epipolar lines are parallel. For example, in Figure 7.4, the right image is captured by the master PTZ camera and the left image by the slave PTZ camera. The same target points are shown in white circles. The epipolar line is shown in green line cross both images. Note that due to the calibration error, the corresponding point may not be exactly on the epipolar line. To solve the problem, a small search window is used to match the region around the selected point with a small range in the vertical direction as well. Since we are only interested in the selected point on a particular reflective surface, this step is very fast.

7.5.2 DISTANCE MEASURING

Once two corresponding points lying on the same horizontal epipolar line are identified, the distance can be calculated based on the triangulation using the baseline (B) of two rectified cameras. The relation between B and the range (D) of the target surface represented in the master camera system is defined as

$$\frac{B}{D} = \left[x_r - x_l \right] \left[\frac{1}{F_{x_r}} \quad \frac{1}{F_{x_l}} \right]^T \tag{7.7}$$

where
   x_r and x_l are the x coordinates of the selected point in the right and left images
   F_{x_r} and F_{x_l} are the focal lengths of the two PTZ cameras

Ideally, both PTZ cameras should have the same focal length after adjusting their zoom factors.

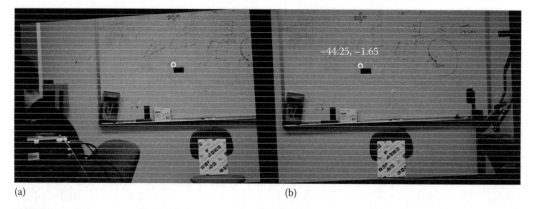

(a)   (b)

FIGURE 7.4   Stereo matching of the corresponding target points, on the images of the (a) master camera and (b) slave camera. The numbers on the right image show the pan and tilt angles of the PTU in order to point the laser beam to the target point.

The calibration result of the slave camera and the LDV is mainly used to determine the pan (α) and tilt (β) angles of the PTU in order to direct the laser beam to the selected point. The conventional triangulation method [16] is used to match the ray from the optical center of the PTZ to the ray of the reflected laser beam. Then the corresponding pan and tilt angles are estimated. Figure 7.4 shows an example of the calculated pan and tilt angles (on the right image) corresponding to the point (in white circle) in the left image. Then, giving the pan and tilt rotations of the PTU and knowing the corresponding 3D point in the slave camera system as

$$P_{C'} = R'[P_{C'X}, P_{C'Y}, D]^T \tag{7.8}$$

where R' is the pan and tilt rotation of the slave PTZ. Initially, it equals identity matrix if the slave PTZ camera is in its initial pose when it was calibrated. The estimated LDV distance $D_L = \|P_{ML}\|$ can be then defined based on Equation 7.6 that will be used for focusing the laser beam to the target point.

7.6 ADAPTIVE AND COLLABORATIVE SENSING

The overall goal of this system is to acquire meaningful audio signatures with the assistance of video cameras by pointing and focusing the laser beam to a good surface. However, a target location either manually or automatically selected may not return signals with a sufficient signal-to-noise ratio (SNR). Then a reselection of new target points is required. Figure 7.5 shows the basic idea of adaptive sensing to adaptively adjust the laser beam based on the feedback of its returned signal levels.

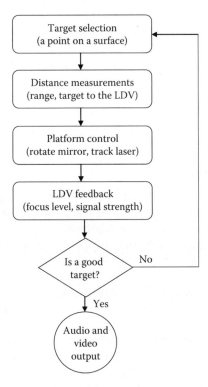

FIGURE 7.5 Flowchart of adaptive sensing for laser pointing and tracking for audio and video signature acquisition.

The stereo matching here is used to obtain the target distance to the system platform, and then we can automatically focus the laser point to the selected target. This involves the following procedures:

1. A point on a surface close to a designated target is selected either manually or automatically.
2. The target range and the distance from the point to the optical center of the LDV are measured.
3. The laser spot is moved to the new location, and the master PTZ camera is rotated synchronously to put the laser spot in the center of images.
4. The laser beam of the LDV is automatically and rapidly focused based on estimated distance and the signal levels, as we did (Qu et al., 2010).

If the selected target point does not have sufficient good returning signals for voice detection, we need to reselect new target points. If the target point is good enough, we can use it to record the audio signature as well as video signatures. In this procedure, there are two key issues need to be emphasized. First, what is a good surface and how to select a surface? Second, how to align the laser beam with the optical center of the camera accurately?

7.6.1 SURFACE SELECTION

The selection of reflection surfaces for LDV signals is important since it is a major factor that determines the quality of acquired vibration signals. There are two basic requirements for a good surface: *vibration to the voice energy* and *reflectivity to the helium–neon laser*. We have found that almost all natural objects vibrate more or less with normal sound waves. Therefore, the key technique in finding a good reflection surface is to measure its reflectivity. Based on the principle of the LDV sensor, the relatively poor performance of the LDV on a rough surface at a large distance is mainly due to the fact that only a small fraction of the scattered light (approximately one speckle) can be used because of the coherence consideration. A stationary, highly reflective surface usually reflects the laser beam of the LDV very well. Unfortunately, the body of a human subject does not have such good reflectivity to obtain LDV signals unless (1) it is treated with retro-reflective materials and (2) it can keep still relative to the LDV. Also, it is hard to have a robust signal acquisition on a moving object. Therefore, background objects nearby to the interested target are selected and compared in order to detect useful acoustic signals. Typically, a large and smooth background region that has a color most close to red is selected for the LDV pointing location.

7.6.2 LASER–CAMERA ALIGNMENT

The next issue is how to automatically aim and track the laser spot, especially for long-range detection. The laser spot may not be observable at a long range particularly if it is not focused or it does not point on the surface accurately. We solve this problem by keeping the reflected laser beam always in parallel to the optical axis of the maser PTZ camera so that the laser spot is right under and very close to the center of the master image. Figure 7.6 shows a typical example of a laser spot (in red spot) that is right below the image center (in yellow circle) in few pixels. We make the ray from the optical center of the master PTZ camera parallel to the reflected laser beam by rotating the PTZ camera with the PTU synchronously (since the PTZ is mounted on the PTU), then with additional PTZ camera rotations.

The main issue now is how to obtain the additional pan (α') and tilt (β') angles of the PTZ camera given the pan (α) and tilt (β) angles of the PTU. Figure 7.7 shows the relationship between the outgoing laser beam from the LDV ($\overrightarrow{BA}$) and the reflected laser ray ($\overrightarrow{AD}$), with the mirror normal ($\overrightarrow{AC}$). By projecting the reflected ray and the mirror normal on the YZ plane (in both the PTU and the LDV coordinate

FIGURE 7.6 Two examples of laser point tracking. Both images show the same cropped size (240 × 160) around the image center (a filled circle on the left image and an unfilled circle on the right) with focused laser spot close to it (sparkling white dots in both images). The laser point on a white board (indoor, about 6 m) in the left image is closer to the LDV; therefore, it is further to the image's center than the laser point on a metal pole (outdoor, about 100 m) in the right image.

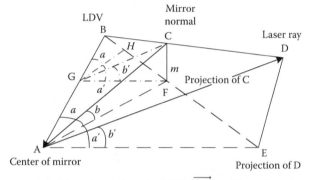

FIGURE 7.7 Geometry model of laser beam from the LDV ($\overrightarrow{BA}$) and its reflected laser ray ($\overrightarrow{AD}$) after the pan (α) and tilt (β).

systems in Figure 7.3), as $\overrightarrow{AE}$ and $\overrightarrow{AF}$, respectively, we see that the angle $\angle BAF$ is α and the angle $\angle FAC$ is β. Now the camera optical axis is parallel to the mirror normal AC. Two additional angles are defined in the figure, the pan angle α' as the angle $\angle FAE$, and the tilt angle β' as the angle $\angle EAD$. If the master PTZ camera (mounted on top of the PTU) is tilted back by β, then its optical axis will be parallel to AF. Therefore, by further panning the PTZ by the angle α' and tilting the PTZ by the angle β', its optical axis will be in parallel with the reflected laser ray AD.

Defining a helping line GC parallel to AD, we can easily solve the additional pan (α') and tilt (β') based on triangulation. The detailed derivation is straightforward and is neglected here. As a result, the pan angle (α') is

$$\alpha' = \tan^{-1}\left(\frac{\tan\alpha}{\cos 2\beta}\right) \tag{7.9}$$

and the tilt angle (β')

$$\beta' = \sin^{-1}(\sin 2\beta * \cos\alpha) \tag{7.10}$$

7.7 EXPERIMENTAL RESULTS

In this section, we provided some results on distance measuring, surface selection, auto-aiming using laser–camera alignment, and surface focusing and listening using our multimodal sensory system.

7.7.1 DISTANCE MEASURING VALIDATION

This experiment is used to verify the accuracy of the calibration among sensory components. Therefore, the test is performed under controlled environments, inside a lab room (with distances up to 10 m), and in the corridor of a building (with distances up to 35 m). Note that the camera's focal lengths do not increase linearly with the change of the zoom levels. In order to perform accurate distance measurement on a large distance, we calibrated the focal lengths under different zoom factors. Figure 7.8 shows the focal lengths (in both x- and y-directions) of the main PTZ camera and the slave PTZ camera.

Next, we verified the correctness of calibration parameters, especially with changes of the focal lengths of each camera. We used the same feature point on a check board pattern at various distances. At each zoom level, the distance from the check board to the platform was manually obtained as the ground truth, and then we used the calibrated parameters to estimate the distance at that zoom level. Figure 7.9 shows the comparison of the true and estimated distances under various zoom factors, which has an average relative error of 6%. The accuracy is sufficient for performing the adaptive focus of the LDV sensor.

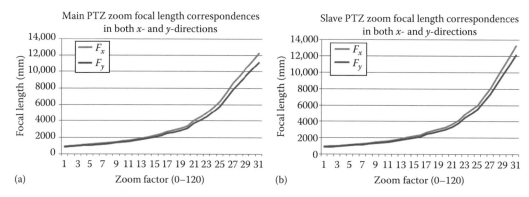

FIGURE 7.8   (a) Calibrated focal lengths of the master PTZ camera and (b) the slave PTZ camera under different zooms.

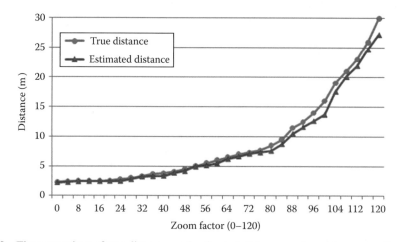

FIGURE 7.9   The comparison of true distances and estimated distances under various zoom factors.

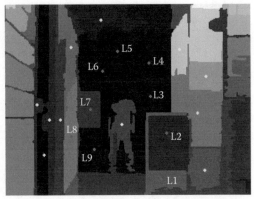

FIGURE 7.10 The cropped (320 × 240) original image (under zoom factor 48) with a target (inside a rectangular bounding box) is shown on left. On right, interested target points close to the human target in the segmented regions are selected and labeled (as L1–L9). The distance of the target to camera is about 31 m.

7.7.2 Surface Selection

In this experiment (Figure 7.10), several interested target points are automatically selected in the segmented regions close the human target in an image of the master PTZ. Note that a static target object such as human or vehicle can be easily detected using histograms of oriented gradients (HOG) (Dalal and Triggs, 2005) in the image. If a target is moving, then frame difference can be used to separate the target from the background surfaces. Then conventional color segmentation can be performed. The region centroid points close to the center of the target can be selected to point the laser.

7.7.3 Auto-Aiming Using Laser–Camera Alignment

When an interested surface point is selected, the master camera is centered to that point; then the laser–camera alignment technique automatically aims the laser spot close to or right below the image center in focus using the calculated distance via stereo vision of the two PTZ cameras. Here, we test our system in two environments: one is indoor (Figure 7.11) and another is outdoor (Figure 7.12).

The indoor experiment is performed at the corridor about 30 m. A metal box on a chair is placed on a fixed location at about 9 m. We manually selected three surfaces points: the points on the metal box (9 m), metal door handler (11 m), and extinguisher metal box (28 m). The laser spots can be clearly observed in the images that are close to the image centers with pixel errors of 2.3, 5.2, and 5.5 (from left to right in Figure 7.11).

FIGURE 7.11 The cropped images (with gray circles show the image center of the original image) include surfaces of metal cake box, door handler, and extinguisher box (from left to right). The calculated distances are 8.9, 10.6, and 26.6 m with corresponding true distances at 9.0, 11.0, and 28 m.

FIGURE 7.12 Target surfaces are selected at (a) the metal box under a tree, (b) the tape on a poster board, and (c) the right turn sign. Their distances are 45.4, 45.5, and 53.3 m, respectively.

The outdoor experiment is performed near a highway when the sensor platform has a standoff distance of about 60 m from the highway. Three sample surface targets to the side of the highway close to the sensor platform are selected: a metal box under a tree (45.4 m), a poster with a tape (45.5 m), and a right turn sign (53.3 m). All images are zoomed so that both the image centers (in yellow circles) and the laser spots (in red) right below are visible. The average pixel difference between the laser spot to the image center for the three examples is 6 pixels.

7.7.4 SURFACE FOCUSING AND LISTENING

The experimental results related to the distance measuring, surface selection, and laser pointing for those labeled points (Figure 7.10) are presented in Table 7.1. The estimated camera distance (D) are listed in column 3 with the "ground truth" data (D^*) at column 6. The LDV distance (D_L) in column 4, the distance from the target point to the optical center of the LDV, is calculated based on the pan and tilt angles of the PTU. Based on that, the focus step (in the range of 0–3300) in column 5 is determined and the laser beam is focused in about 1 s for each point. For comparison, the focus step using the full-range searching takes 15 s, and is presented in column 8. The signal returning levels (0–512) in column 6 can be used to determine what the best point is among the candidates for audio acquisition. As a result, the metal box (L7) has the strongest signal return level so that it is selected for the voice detection. Note that all selected surfaces do not have retroreflective tape treatment.

The experiment results of the focus positions and signal levels of the outdoor surface targets (Figure 7.12) are shown in Table 7.2. According to the signal return levels at the last column, the surface of the metal box under a tree should be selected as the best listening target. In addition, the

TABLE 7.1

Surface Selection, Laser Pointing, and Focusing

L#	Surface	Measurements				Ground Truth		
		D (m)	D_L (m)	Step	Level	D* (m)	Step	Level
L1	Floor	26.56	27.14	2642	10	27.74	2581	11
L2	Chalkboard	28.30	28.88	2750	31	27.74	2764	22
L3	Wall	27.67	28.25	2732	12	30.63	2734	12
L4	Wall	28.62	29.20	2734	11	30.63	2734	12
L5	Wall	29.67	30.26	2786	12	30.63	2845	14
L6	Mirror	27.90	28.56	2758	12	30.63	2757	12
L7	Metal box	29.67	30.26	2745	118	30.32	2839	121
L8	Side wall	21.10	21.60	2391	9	23.16	2410	10
L9	Wall	30.85	31.43	2410	11	30.63	2757	11

TABLE 7.2

Focus Positions and Signal Levels of Three Outdoor Surfaces

No.	Target	Distance (m)	Focus Position	Signal Level
001	Box under a tree	45.4	2890	285
002	Poster with tape	45.5	2890	116
003	Right turn sign	53.3	2904	14

poster with tape is also a good listening surface with moderate signal level. Therefore, it can be used as a substitute for the first one with some signal enhancing treatments, such as amplifying, noise removal, and filtering. Unfortunately, the right turn sign does not provide sufficient signal returns.

7.8   CONCLUSIONS

In this chapter, we presented a dual-PTZ camera–based stereo vision system for improving the automation and time efficiency of LDV long-range remote hearing. The close-loop adaptive sensing using the multimodal platform allowed us to determine good surface points and to quickly focus the laser beam based on target detection, surface point selection, distance measurements, and LDV signal return feedback. The integrated system greatly increased the performance of the LDV remote hearing and therefore its feasibility for audio–visual surveillance and long-range other inspection and detection applications. Experimental results showed the capability and feasibility of our sensing system for long-range audio–video-range data acquisition.

ACKNOWLEDGMENTS

This work has been supported by the US Air Force Office of Scientific Research (AFOSR) under Award #FA9550-08-1-0199 and the 2011 Air Force Summer Faculty Fellow Program (SFFP), by Army Research Office (ARO) under DURIP Award #W911NF-08-1-0531, by National Science Foundation (NSF) under grant No. CNS-0551598, by the National Collegiate Inventors and Innovators Alliance (NCIIA) under an E-TEAM grant (No. 6629-09), and by a PSC-CUNY Research Award. The work is also partially supported by NSF under award #EFRI-1137172. We thank Dr. Jizhong

Xiao at City College of New York, and Dr. Yufu Qu at Beihang University, China for their assistance in the prototyping of the hardware platform, and Dr. Rui Li for his discussions in the mathematical model for sensor alignment. An early and short version of this work was presented at the 2011 IEEE Workshop on Applications of Computer Vision (Wang et al., 2011b).

REFERENCES

Beal, M. J., Jojic, N., and Attias, H. 2003. A graphical model for audiovisual object tracking. *IEEE Transactions on Pattern Analysis and Machine Intelligence*, 25, 828–836.

Blackmon, F. A. and Antonelli, L. T. 2006. Experimental detection and reception performance for uplink underwater acoustic communication using a remote, in-air, acousto-optic sensor. *IEEE Journal of Oceanic Engineering*, 31(1), 179–187.

Bouguet, J. Y. 2008. Camera calibration toolbox for Matlab. Available at: http://www.vision.caltech.edu/bouguetj/calib_doc/index.html.

Clavel, C., Ehrette, T., and Richard, G. 2005. Events detection for an audio-based surveillance system. In *IEEE ICME'05*, Amsterdam, the Netherlands, 2005, pp. 1306–1309.

Cristani, M., Bicego, M., and Murino, V. 2007. Audio–visual event recognition in surveillance video sequences. *IEEE Transactions on Multimedia*, 9(2), 257–267.

Dalal, N. and Triggs, B. 2005. Histogram of oriented gradient for human detection. In *Proceedings of IEEE Conference on Computer Vision and Pattern Recognition*, San Diego, CA, 2005, pp. 886–893.

Dedeoglu, Y., Toreyin, B. U., Gudukbay U., and Cetin, A. E. 2008. Surveillance using both video and audio. In *Multimodal Processing and Interaction: Audio, Video, Text*, P. Maragos, A. Potamianos, and P. Gros, Eds., Springer, LLC, 2008, pp. 143–156.

Khan, A. Z., Stanbridge, A. B., and Ewins, D. J. 1999. Detecting damage in vibrating structures with a scanning LDV. *Optics and Lasers in Engineering*, 32(6), 583–592.

Lai, P. et al. 2008. A robust feature selection method for noncontact biometrics based on laser Doppler vibrometry. In *IEEE Biometrics Symposium*, Tampa, FL, 2008, pp. 65–70.

Li, X., Chen, G., Ji, Q., and Blasch, E. 2008. A non-cooperative long-range biometric system for maritime surveillance. In *Proceedings of IEEE Conference on Pattern Recognition*, Tampa, FL, 2008, pp. 1–4.

Li, W., Liu, M., Zhu, Z., and Huang, T. S. 2006. LDV remote voice acquisition and enhancement. In *Proceedings of IEEE Conference on Pattern Recognition*, Hong Kong, China, 2006, Vol. 4, pp. 262–265.

Ometron. 2009. Ometron systems. http://www.imageautomation.com/ (last visited December 2010).

Polytec. 2009. Polytec laser vibrometer. http://www.polytec.com/ (last visited December 2010).

Qu, Y., Wang, T., and Zhu, Z. 2010. An active multimodal sensing platform for remote voice detection. In *IEEE/ASME International Conference on Advanced Intelligent Mechatronics (AIM 2010)*, Montreal, Quebec, Canada, July 6–9, 2010, pp. 627–632.

Radhakrishnan, R., Divakaran, A., and Smaragdis, A. 2005. Audio analysis for surveillance applications. In *IEEE WASPAA'05*, Mohonk, NY, 2005, pp. 158–161.

Scruby, C. B. and Drain, L. E. 1990. *Laser Ultrasonics Technologies and Applications*. New York: Taylor & Francis.

Wang, T., Li, R., Zhu, Z., and Qu, Y. 2011b. Active stereo vision for improving long range hearing using a laser Doppler vibrometer. In *IEEE Computer Society's Workshop on Applications of Computer Vision (WACV)*, Kona, HI, January 5–6, 2011, pp. 564–569.

Wang, T., Zhu, Z., and Taylor, C. 2011a. Multimodal temporal panorama for moving vehicle detection and reconstruction. In *IEEE ISM International Workshop on Video Panorama (IWVP)*, Dana Point, CA, December 5–7, 2011.

Zhu, Z., Li, W., Molina, E., and Wolberg, G. 2007. LDV sensing and processing for remote hearing in a multimodal surveillance system. In *Multimodal Surveillance: Sensors, Algorithms and Systems*, Zhu and Huang, Eds. Frederick, MD: Artech House Publisher, pp. 59–88.

Zhu, Z., Li, W., and Wolberg, G. 2005. Integrating LDV audio and IR video for remote multimodal surveillance. In *IEEE Workshop on Object Tracking and Classification In and Beyond the Visible* Spectrum, San Diego, CA, 2005.

Zieger, C., Brutti, A., and Svaizer, P. 2009. Acoustic based surveillance system for intrusion detection. In *Sixth IEEE International Conference on Advanced Video and Signal Based Surveillance*, Genoa, Italy, 2009, pp. 314–319.

Zou, X. and Bhanu, B. 2005. Tracking humans using multimodal fusion. In *Proceedings of IEEE Conference on Computer Vision and Pattern Recognition*, San Diego, CA, Vol. 3, pp. 4–12.

8 Analytical Use of Easily Accessible Optoelectronic Devices

Colorimetric Approaches Focused on Oxygen Quantification

Jinseok Heo and Chang-Soo Kim

CONTENTS

8.1 INTRODUCTION

Rapid technological progress in digital color imaging devices, such as charge-coupled devices (CCD), complementary metal–oxide–semiconductor (CMOS) cameras, liquid crystal display (LCD), and digital light projection (DLP) devices, makes these optoelectronic products nearly ubiquitous in our daily lives. For example, relatively high-quality color images can be obtained anytime with mobile phone cameras and wireless webcams (some products as low as several tens of US dollars) and then be transmitted wirelessly. These color imaging devices were originally designed as "sensory" or "perceptual" devices that mimic the human eye when responding to visible wavelengths. This chapter introduces the potential use of these color imaging devices as economic analytical instruments. After we briefly review prospective luminophores amenable for colorimetric chemical quantification using color imaging devices, we will describe oxygen quantification as exemplary applications of this approach.

8.1.1 Color Optoelectronic Devices

The most widely used instruments for optical interrogation of biochemical agents are spectrometric systems that detect emission, absorption, or reflection signals from samples; however, they are generally expensive and bulky to use for cost-effective, portable sensors. Therefore, many lab-on-a-chip or batch-fabricated spectrometers have been actively developed recently for their common use as economic analytical instruments (Babin et al., 2009; Syms, 2009).

In parallel with this approach, the color image sensor is considered as an alternative device to the spectrometry because of its photometric detection capability (Yotter and Wilson, 2003). Over the last decade, the use of color image sensors for chemical quantification has been increasing very rapidly. These include color flatbed scanners (Lavigne et al., 1998; Rakow and Suslick, 2000; Taton et al., 2000) and digital color cameras (Jenison et al., 2001; Filippini et al., 2003; Abe et al., 2008; Martinez et al., 2008; Stich et al., 2009). Figure 8.1 shows the conceptual diagrams of the color sensing and color emission devices utilizing thin film color filter arrays composed of three primary colors (red, green, and blue). The photodetector array in the color image sensor records the incident photon intensity spectrally separated according to the three different colors of on-chip Bayer filters as in Figure 8.2a. Since a wide variety of luminophores emit lights in the visible range, this device can serve as the analytical instrument for colorimetric chemical quantification.

Another innovation in the recent photonics development is the advent of optoelectronic display devices including LCD and DLP devices. Similarly, the ability of emitting light in a selected range with the built-in color filters, as shown in Figure 8.2b, makes it very attractive as a ubiquitous light source for chemical quantification. This type of emission devices can be utilized as the illumination (for transmittance and/or reflection measurements) and excitation light sources (for fluorescence emission) required for optical interrogation (Batchelor and Jones, 1998; Filippini and Lundstrom, 2002; Filippini et al., 2003). We envision that the color imaging devices will be key components to develop low cost, portable sensors for environmental and industrial monitoring, and point-of-care testers in resource-poor settings.

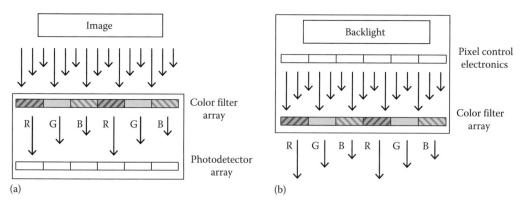

(a)   (b)

FIGURE 8.1 Simplified cross-sectional views of (a) color image sensor and (b) color display devices. Arrays of thin film color filters (red, green, and blue), originally intended for sensory purpose, provide spectral sensitivities for analytical applications.

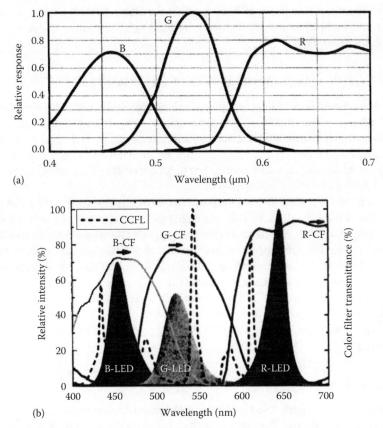

FIGURE 8.2 (a) Spectral ranges of three primary color filters of typical image sensors. (After Holst, G.C. and Lomheim, T.S., *CMOS/CCD Sensors and Camera Systems*, 2nd edn., SPIE Press, Bellingham, WA, 2011, Fig. 5-17 from p. 143. With permission from SPIE Press.) (b) Emission spectra of backlights (cold cathode fluorescent lamp and light-emitting diode) and transmission ranges of three color filters from typical liquid crystal display screens. (After Lee, J.-H. et al., *Introduction to Flat Panel Displays*, John Wiley & Sons, New York, 2008, Fig. 6.31 from p. 170. With permission from Wiley.)

8.1.2 COLOR SPACE

There are several color space models to quantitatively describe the color intensity with respect to the color matching function established by the International Commission on Illumination (CIE). The representative color systems are RGB (red, green, blue), CMYK (cyan, magenta, yellow, black), CIELAB/CIELUV (lightness, color differences), and HSL (hue, saturation, lightness) systems. Each system has its own characteristic advantages in describing the colors quantitatively and thus is complimentary to other ones. Both the RGB and CMKY systems are rather device-dependent as compared to the HSL and CIE systems that are designed to be perceptually uniform to approximate human vision. Some recent reports on colorimetric chemical quantification show that the CIELAB or HSL system is useful to express a variation in brightness of one color (i.e., a single emission peak), while the CMKY or RGB system appears to be more suitable in the case of active color changes (i.e., ratiometric change) (Abe et al., 2008; Martinez et al., 2008; Stich et al., 2009).

Despite a lossy compression format, the Joint Photographic Experts Group (JPEG) file format with the Exchangeable Image File Format (EXIF) algorithm can be commonly used for storing the color information because it is supported by most of imaging devices and image software.

Other lossless formats such as RAW or BMP are also available, but with larger file sizes. All data can be processed by one of these formats with 8-bit per single color information (i.e., 24-bit RGB with 256-level per each color).

8.2 PROSPECTIVE LUMINOPHORES FOR COLORIMETRIC DETERMINATION

Luminescence is an emission of light resulting from electronically excited molecules via chemical reaction, light, or other stimuli including electricity, temperature, sound, and pressure. The molecule showing the luminescence is termed as a luminophore. In this chapter, we will focus on the sensing based on photoluminescence, i.e., luminescence induced by light. Furthermore, photoluminescence can be divided into fluorescence and phosphorescence depending on the electron spin states involved during the photon emission process.

Luminescence can be used for sensing purposes in two different ways. It may directly respond to the concentration of target analytes by showing a change in the emission intensity or in the Stokes shift. Alternatively, luminophore can be attached to a reporter molecule that can specifically interact with target analyte. Luminophores can originate from various sources, such as organic dyes, polymer conjugates, semiconductor quantum dots (QDs), Au and Ag nanoparticles, carbon nanomaterials, fluorescent proteins, and metal complexes. The peak absorption and emission wavelengths of these materials vary and even can be tuned chemically or physically. Since the wavelength of the light source and detector in our colorimetric sensor platform is currently limited to the visible range, this chapter will focus on luminescent materials showing peak excitation and emission wavelengths in the visible range.

8.2.1 ORGANIC DYES

Organic dyes are the most widely used reporters in fluorescence sensing because of their economical cost and easy modification of their structures that can diversify spectral properties. In addition, the derivative forms of a fluorescence dye can be used to label proteins and other biological samples. Organic fluorescent dyes that can be used in our sensor platform are shown in Table 8.1. One of the important selection criteria is that the dye should show a large Stokes shift (>50 nm) to reduce the interference of excitation light in sensing fluorescence emission. However, fluorescence dyes showing a small Stokes shift (<50 nm) may be also used in our sensor. For example, the peak excitation and emission wavelengths of fluorescein are 488 and 514 nm, respectively. The blue-filtered light from LCD source can excite the molecule but the selection of green color in CCD sensor cannot completely eliminate the excitation light, thus resulting in poor signal-to-background ratio. In this case, a proper cutoff filter can be introduced in front of the image sensor.

The applications of organic dye for sensing are diverse. Some fluorescence dyes respond to a change in the environment, such as solvent polarity, pH, intermolecular interaction, or electric field. Generally, organic dye molecules are easily photobleached. Their photostability can be improved by incorporating them into polymer (Landfester, 2006) or silica particles (Gerion et al., 2001). The organic dyes encapsulated in these particles exhibit higher fluorescence quantum yield and are less prone to the photobleaching than free organic dyes (Yao et al., 2006).

8.2.2 NANOPARTICLES

QDs are typically 2–60 nm diameters of nanocrystals made from Group II/VI and III/V semiconductors. The examples are CdSe, CdTe, InP, and InGaP, which are all commercially available now. In general, the QDs can absorb light ranging from UV to near IR but their absorption decreases as the incident light wavelength increases. The emission bands of the QDs are very narrow and symmetric. This makes the QDs more attractive than the organic dye because of easy separation of the emission photons from the excitation photons. QDs can be efficiently excited with the blue-filtered

TABLE 8.1

Peak Excitation and Emission Wavelengths of Various Organic Luminophores Potentially Suitable for Colorimetric Determination

Luminophore	Excitation Wavelength (nm)	Emission Wavelength (nm)
Acridine Yellow	470	550
Acriflavin	436	520
7-Aminoactinomycin D (AAD)	546	647
Astrazon Orange R	470	540
Auramine	460	550
Aurophosphine G	450	580
Berberine Sulphate	430	550
Brilliant Sulphoflavin FF	430	520
Coriphosphine O	460	575
DiA	456	590
Fura Red	472 (low [Ca^{2+}]), 436 (high [Ca^{2+}])	657 (low [Ca^{2+}]), 637 (high [Ca^{2+}])
Genacryl Pink 3G	470	583
Mithramycin	450	570
NBD	465	535
NBD Amine	450	530
Phosphine 3R	465	565
Pontochrome Blue Black	535–553	605
Procion Yellow	470	600
Rhodamine 5 GLD	470	565
Rhodamine B	540	625
Rhodamine B 200	523–557	595
Rhodamine B Extra	550	605
Sevron Orange	440	530
Sulpho Rhodamine G Extra	470	570

light from LCD light source. The emission light can be observed using green or red color selection in CCD camera. Particularly, red color QDs will show better signal-to-background ratio than green or yellow–orange QDs because of less overlap between the excitation and emission bands. Since QDs for bioconjugation are commercially available, constructing quantum-dot biosensor is possible using our sensor platform. QDs show strong brightness because of their high molar absorptivity ($\sim 10^6$ M^{-1} cm^{-1}) and quantum yield, which cannot be easily achieved with organic dyes. In addition, unlike organic dyes the QDs have shown excellent chemical and photostability. While the QDs have gained popularity for labeling or tagging biological samples, the toxicity of Cd-based QDs has been a major concern.

Recently, porous silicon and silicon nanoparticles are receiving attention because they may replace Cd-containing QDs. Silicon nanoparticles show a broad luminescence spectrum in the visible range. The peak emission wavelength shifts depending on the excitation wavelength, because the silicon particles have a size distribution (Veinot, 2006). Therefore, the silicon nanoparticles can be excited with blue-filtered light and detected using red color option in CCD detector.

Nanoparticles consisting of noble metal atoms, such as Au and Ag, absorb light in a visible range in a size-dependent manner. Unlike semiconductor QDs, the light absorption and emission of noble metal nanoparticles are related with plasmons, a collective oscillation of free electrons upon the incidence of light. Au nanoparticle solution scatters and absorbs light efficiently but exhibits very weak fluorescence; therefore, the absorption property of Au nanoparticles is mainly used for sensing purpose. In addition, since Au acts as an efficient fluorescence quencher, the Au–fluorescence dye conjugate can be used for various sensing applications using dequenching effect.

8.2.3 Luminescent Metal Complexes

The metal complexes of lanthanides and noble metal ions exhibit long-lifetime luminescence. Among the lanthanide ions, Tb^{3+} and Eu^{3+} exhibit relatively high emission intensity in the visible range. Lanthanide ions themselves do not absorb photons very well and so they are not good photon emitters. But the formation of complex with heterocyclic chelating agent enhances the luminescence. The aromatic heterocycle chelating groups, called antenna, primarily absorb photon energy and then relay the excitation energy to the lanthanide. This energy transfer can increase the molar absorptivity to above 10,000 M^{-1} cm^{-1}, which is sufficient for luminescence sensing. Unfortunately, these metal complexes cannot be employed in our sensor because of their weak absorption in 400–600 nm range.

Ru^{2+}, Os^{2+}, and Re^{2+} ions can form different types of metal–ligand complexes showing long-lifetime emission. A representative example is the Ru(bpy)$_3^{2+}$ complex formed between Ru^{2+} and tris-(2,2′-bipyridine) ligand (Balzani et al., 2001). Its absorption spectrum is very broad showing the peak absorbance around 470 nm (molar absorptivity: 10,000–30,000 M^{-1} cm^{-1}). A broad emission band is observed ranging from 600 to 670 nm. Thus, these compounds have the spectral characteristic required for our sensor platform. Its origin of photon emission is different from the lanthanide complexes. It is phosphorescence resulting from the metal–ligand charge transfer (MLCT). The lifetime of this photon emission is much longer than fluorescence emission. The emission is highly sensitive to oxygen concentration and has been used in our sensors that will be discussed in the next section.

A porphyrin complex containing Pt or Pd is another example of a phosphorescent metal complex. A porphyrin structure can be easily found in nature from hemoglobin or chlorophyll. The porphyrin complex absorbs light at narrow band regions at 360–400 nm and 500–550 nm and emits red phosphorescence having long lifetime (0.01 to 1 ms) (Papkovsky et al., 2000). These porphyrin complexes can be applied in our sensor. It has been used for sensing oxygen, glucose, and lactate concentrations.

8.2.4 Conjugate Polymers

Conjugated polymers are the compounds with alternating single and double bonds (or aromatic units) along the polymer chain. The polarizable π-electrons extended along the conjugated backbone will determine the optical properties of conjugate polymers. They have high molar absorptivity reaching 10^6 M^{-1} cm^{-1} and display strong fluorescence. The conjugate polymer can be prepared as a thin film and used as a sensor relying on superquenching or superenhancement effect (Thomas et al., 2007). The quenching in conjugate polymer arises from the termination of exciton migration and occurs in a collective manner. One quencher molecule can simultaneously quench a large number of fluorescent monomeric units, and thus it is called "superquenching." On the other hand, superenhancement effect occurs when the quencher is removed from the conjugate polymer. The change in chain conformation of conjugate polymer modifies the spectral properties of the polymer, i.e., absorption and emission properties. Examples of conjugate polymers are polyacetylene, polythiophene, polypyrrole, polyaniline, and poly-*para*-phenylene vinylene. A rational design of the conjugate polymer can provide fine-tune for its excitation and emission wavelengths that fit for the colorimetric sensing modality.

8.2.5 Fluorescent Proteins

The green fluorescent protein (GFP) is a naturally fluorescent protein that was first extracted from a jellyfish species called *Aequorea victoria*. It is an exceptional protein that can be stably expressed as a fusion protein in a species other than the jelly fish. The wild-type GFP has two excitation peaks at 395 and 470 nm and the emission peak is at 509 nm. The GFP variants have revealed a shift in excitation peaks while maintaining the similar emission peak position to the wild-type GFP (Tsien, 1998). So the overall excitation and emission spectra of the GFP variant are similar to those of fluorescein dye. This made the GFP variant useful for fluorescence imaging. Furthermore, different colors of fluorescent protein were derived from the GFP-mutant variants. Blue, red, cyan, yellow fluorescent proteins are possible (Shaner et al., 2004). Recently, other GFP-like fluorescent proteins were derived from sources other than the jelly fish, which will allow diverse spectral characteristics for fluorescent proteins. GFP and its variants have served as unique tools in live cell imaging for protein tagging, monitoring gene expression, and drug and genetic screens.

8.3 COLOR CAMERA AS PHOTODETECTOR FOR DISSOLVED OXYGEN QUANTIFICATION

8.3.1 OPTICAL OXYGEN SENSING

Optical oxygen sensors are becoming dominant over electrochemical types because they are nondestructive sensors that do not consume oxygen, thus not perturbing oxygen environment during the measurement. Other advantages include (1) easy and simple device miniaturization; (2) capability of remote sensing in a noninvasive and noncontact mode; and (3) capability of two-dimensional imaging of oxygen distribution. Optical oxygen sensors use the mechanism of oxygen quenching as shown in Figure 8.3a. Quenching is a phenomenon of emission intensity decrease of a luminophore in presence of a quencher. In this case, oxygen molecule acts as a powerful quencher of the electronically excited state of the luminophore. Oxygen-sensitive luminophores include metal–complex organic dyes, dual emitters, and fullerenes (Amao, 2003; Wang et al., 2010a). Recent studies have also shown the use of ruthenium and porphyrin complexes as a luminophore for oxygen sensor. These agents are typically immobilized within oxygen-permeable matrices for implementing a sensor device and their spectral properties are not significantly altered by the immobilization. Oxygen concentration can be determined from Stern–Volmer relationship by measuring intensity, lifetime, or phase shift as follows (Mills, 1997):

$$\frac{I_0}{I}\left(=\frac{\tau_0}{\tau}=\frac{\Phi_0}{\Phi}\right)=1+K_{SV}[O_2] \tag{8.1}$$

where I_0 and I (τ, Φ) represent the steady-state luminescence intensities without and with the quencher (here oxygen). Alternately, τ and Φ represent the lifetime (decay time) of luminescence and the phase angle shift of luminescence (time delay) from a sinusoidal excitation

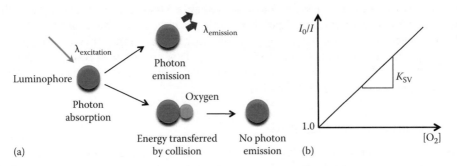

FIGURE 8.3 (a) Principle of luminescence quenching by molecular oxygen depicting the luminescence process in the absence of oxygen and the deactivation of luminophore by oxygen. (b) Stern–Volmer plot based on equation (8.1). (From Park, J. et al., *IEEE Sensors J.*, 10(12), 1855, 2010. With permission.)

light, respectively. K_{SV} is the Stern–Volmer quenching constant and [O_2] is the oxygen concentration, respectively. Therefore, for an ideal condition, the ratio of luminescence intensities without and with oxygen (I_0/I) becomes linearly proportional to the oxygen concentration as plotted in Figure 8.3b.

Several colorimetric oxygen-sensing methods based on absorption or emission measurements have been reported (Eaton, 2002; Evans et al., 2006; Evans and Douglas, 2006). Although the results were semiquantitative, these approaches pioneered the concept of colorimetric oxygen quantification. Recently, several research groups including us have reported the use of color cameras for quantitatively determining the oxygen concentration. This intensity-based oxygen sensing has demonstrated the analytical capability of color imaging devices (Thomas et al., 2009; Park et al., 2010, 2011; Wang et al., 2010b; Park and Kim, 2011; Shen et al., 2011). We used a ruthenium complex as the oxygen-sensing luminophore. This shows a great Stokes shift by emitting orange light (about 590 nm peak) when excited with blue light (about 470 nm peak). The complex is ideal for colorimetric oxygen quantification. This section summarizes how a color CCD camera and simple color analysis can serve as a reliable analytical instrument.

8.3.2 COMMERCIAL SOL–GEL SENSOR

An oxygen sensor patch (RedEye™, RE-FOX-8, 8 mm diameter, OceanOptics), which is commercially available, was used for a proof-of-concept demonstration. This oxygen sensor patch contained ruthenium complex immobilized within a sol–gel matrix. A color camera and a spectrophotometer were used to measure the oxygen concentration and the two results were compared under the same condition. A color camera (DS-5M, Nikon) that employs a CCD device (ICX282AQ, 5-megapixel, Bayer-masked, Sony) was used for image capturing. A spectrofluorometer was used as the detector. A blue light-emitting diode (LED) was used as the excitation source (peak wavelength 470 nm) for both methods. A long-wave pass filter (cut-on wavelength 500 nm) was used to minimize the intense blue excitation wavelengths that can also interfere with the sensitivity of red pixel photodetectors.

Figure 8.4a shows the emission spectra of the ruthenium complex between 550 and 700 nm range in response to the oxygen concentration. The data clearly demonstrates that the peak emission intensity at 595 nm decreases as the oxygen concentration increases. This oxygen-responsive emission range is almost identical with the spectral range of the red-responsive pixels of color imager as in Figure 8.2a. An exemplary image analysis result is shown in Figure 8.4b that includes a red-extracted image of the sensor patch at 0% oxygen solution. The Stern–Volmer plots (Equation 8.1) in Figure 8.5 include results of various image analysis and spectrometric data. The spectrometric data represent the intensities at the peak wavelength (595 nm) of Figure 8.4a. The image analysis data represent the mean values of digital color intensities (between 0 and 255) from the histograms. The red color intensity data obtained by the CCD camera shows a broader linearity and a better sensitivity than the integrated fluorescence intensity data collected by the conventional spectrometer (Figure 8.5). Only the red color component from the original RGB color image was extracted by ImageJ, a free image processing software developed by the National Institutes of Health (Abramoff et al., 2004). This procedure serves as a virtual bandpass filter by removing unnecessary "noise" colors (i.e., green–blue color range) to effectively "mine" only the oxygen-related information.

8.3.3 OPTOFLUIDIC HYDROGEL SENSOR

We have extended our previous study by using lab-made sensor assemblies. Ruthenium complex, dichlorotris(1,10-phenanthroline)ruthenium (II) hydrate, was embedded in photo-patterned poly(ethylene glycol) (PEG) hydrogels to use as the sensing element. The fabrication steps for the array are illustrated in Figure 8.6a. The hydrogel precursor solution was injected into an assembly which

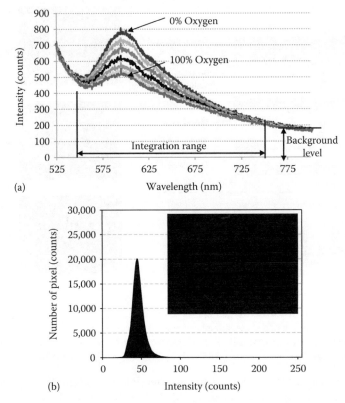

(a)

(b)

FIGURE 8.4 (a) The emission spectra of a commercial oxygen-sensitive patch in various dissolved oxygen concentrations. (b) Red-extracted images of the RedEye patch (8 mm diameter) and its histogram of red color intensity. (After Park, J., Hong, W., Kim, C.-S., Color intensity method for hydrogel optical sensor array, *IEEE Sensors J.*, 10(12), 1855–1861, 2010. Copyright 2010 IEEE.)

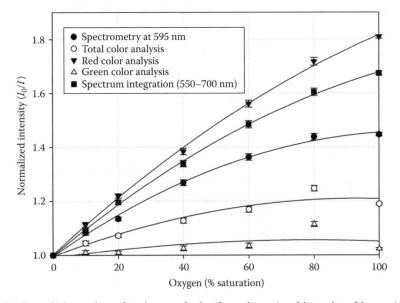

FIGURE 8.5 Stern–Volmer plots of various methods: (from the top) red intensity of image (▼), spectrum areal integration of spectrometry (■), emission peak intensity of spectrometry (●), total color intensity of image (○), green intensity of image (△). (After Park, J., Hong, W., Kim, C.-S., Color intensity method for hydrogel optical sensor array, *IEEE Sensors J.*, 10(12), 1855–1861, 2010. Copyright 2010 IEEE.)

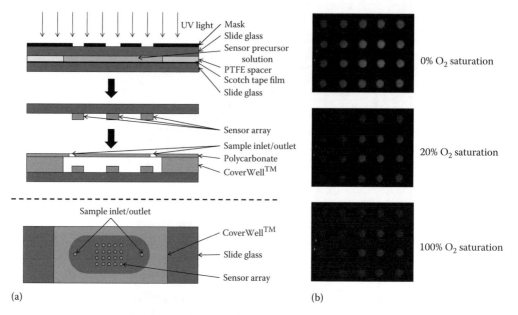

FIGURE 8.6 Opto-fluidic dissolved oxygen sensor assembly. (a) Photo-patterning of polyethylene glycol (PEG) hydrogel array and layout of the assembled system. (b) Red-extracted images of PEG array in 0%, 20%, and 100% oxygen-saturated water. (After Park, J., Hong, W., Kim, C.-S., Color intensity method for hydrogel optical sensor array, *IEEE Sensors J.*, 10(12), 1855–1861, 2010. Copyright 2010 IEEE.)

consisted of two glass slides (1″ × 3″) and a spacer. One slide was surface-treated with a silanization agent to promote the adhesion between the glass and PEG. A scotch tape film was adhered on the other slide to prevent the PEG layer from being attached to the slide. After the PEG layer was photo-patterned by illuminating UV light through a photomask and rinsing unreacted residue, a silicone CoverWell™ sheet (32 × 19 × 0.5 mm) was placed to cover the hydrogels and form a sealed optofluidic assembly. Twenty circular sensors (0.5 mm diameter, 120 μm thick) were patterned within a 1 cm² area.

Figure 8.6b is the red-extracted images of two-dimensional oxygen sensors in 0%, 20%, and 100% oxygen-saturated water. It can be recognized that the red intensity decreases with

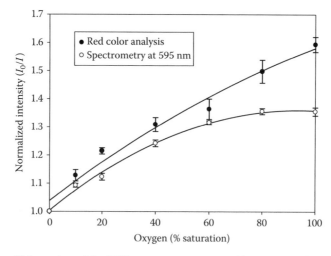

FIGURE 8.7 Stern–Volmer plots of the PEG oxygen sensor array with respect to dissolved oxygen based on spectrum and red color analysis. (After Park, J., Hong, W., Kim, C.-S., Color intensity method for hydrogel optical sensor array, *IEEE Sensors J.*, 10(12), 1855–1861, 2010. Copyright 2010 IEEE.)

increasing dissolved oxygen concentration due to the quenching reaction. The absolute red intensity of each sensing spot showed variations because of three reasons: inhomogeneity in excitation, illumination intensity and variations of the sensor thickness, and the amount of ruthenium complex in the sensors. However, the calibration curve for each sensor can be constructed by comparing the intensity of each sensor in the image, which is the inherent advantage of the two-dimensional image analysis. Figure 8.7 shows the Stern–Volmer plot of the average of the red intensities of 20 sensors and that obtained with the spectrometer. The red intensity of each sensor was obtained by averaging the red intensity over the sensor area (0.5 mm diameter). Similarly to the earlier, the red color analysis data showed better sensitivity and broader linear range than spectrometric analysis data.

8.4 NONTRADITIONAL EMISSION DEVICES AS LIGHT SOURCE FOR OXYGEN QUANTIFICATION

LED or traditional bulky, broadband lamps with filters can be commonly used to excite luminophores. A white LED or a LCD screen has been employed for determining gaseous oxygen both quantitatively and qualitatively. The same color camera setup was successfully used to characterize red emission from sensor films based on colorimetric intensity measurements.

8.4.1 White LED as Excitation Source

We further explored the possibility of using a filter-free white light source for quantitative imaging of gaseous oxygen. The principal method of implementing the white LED is to add a phosphor material in the blue LED. This implies that the blue wavelength range is a major emission peak of the white LED that can excite the oxygen-sensitive ruthenium complex. As shown in Figure 8.8a, a broadband white LED (LS-450 with LED-WHITE, OceanOptics) was used for quantifying gaseous oxygen. The same commercial oxygen-sensitive patch as earlier was used to examine this new approach. Gas samples of various oxygen percentages were prepared with mass flow controllers. The sample gas was uniformly delivered over the sensor surface through a glass tube (1 mm inner diameter) at a moderate flow rate of 2 L/min.

The spectral output of the white LED has a peak emission around 475 nm that matches well with the excitation wavelength of the ruthenium complex. A simple red color analysis of CCD images showed that the white LED without any filter performed similarly to the blue LED with a filter.

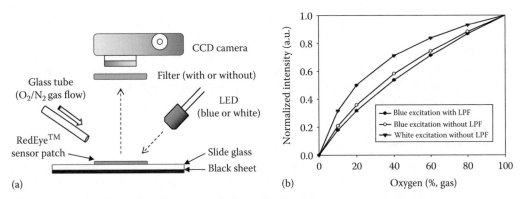

(a)   (b)

FIGURE 8.8 (a) The sensor imaging setup with a color CCD camera for gaseous oxygen quantification. A long-wave pass filter, blue LED, and white LED were selectively used for a series of measurements. The microscope setup is not shown. (b) Normalized Stern–Volmer plots to compare the performance of the three different imaging configurations. (After Park, J. and Kim, C.-S., *Sensor Lett.*, 9(1), 118, 2011. With permission from American Scientific Publishers.)

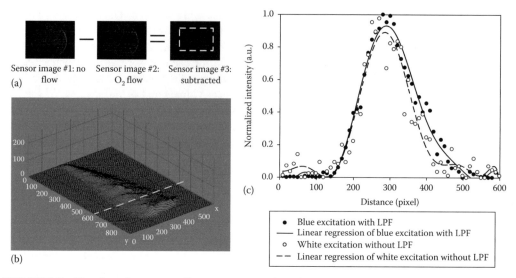

FIGURE 8.9 Mapping of oxygen gradient on sensor surface (8 mm diameter) created with a capillary tube. (a) Subtraction of two images (with oxygen flow and without oxygen flow) to eliminate background. (b) Plot of the differential intensity profile of the ROI (region of interest) in the subtracted image in (a). (c) Normalized intensity profiles along the dashed line in (b) (-•- blue LED excitation with filter, -o- white LED excitation without filter). (After Park, J. and Kim, C.-S., *Sensor Lett.*, 9(1), 118, 2011. With permission from American Scientific Publishers.)

Figure 8.8b shows a graph of comparing the normalized red color sensitivities of three different optical configurations. Although the white LED configuration without filtering shows the least linearity over the entire 0%–100% oxygen range, it has the highest Stern–Volmer (I_0/I) sensitivity in the range of 0%–20% oxygen. This concentration range is used for various medical and biochemical process monitoring.

For an oxygen distribution mapping, only a part of the sensor patch was directly exposed to the oxygen flow to generate an oxygen gradient over the patch surface. The right-side image in Figure 8.9a was the result of subtracting a red fluorescence intensity image in the presence of the oxygen flow (middle) from the reference image obtained with no oxygen flow (left side). Figure 8.9b shows the two-dimensional intensity gradient in the selected region of the subtracted image in Figure 8.9a (dashed line). We compared two measurement configurations: standard blue excitation with filter and broadband white illumination without filter. Normalized intensity profiles of the two configurations along the dashed line showed a good agreement between the two results (Figure 8.9c). We anticipate that the broadband white excitation configuration can be used for simultaneous quantification of multiple target analytes. Several luminophores that have different excitation and emission wavelengths can be analyzed by using the white light source, the digital color imagers, and color analysis methods.

8.4.2 LCD Monitor as Excitation Source

The traditional excitation light sources such as diodes and lamps are rather limited to one-dimensional illumination over sample surfaces. A uniform illumination of excitation light over a target area is important for analyzing the spatial distribution of chemicals. Two-dimensional display screens can be utilized for this application. Their emission colors (i.e., wavelength ranges) and intensity can be easily controlled by a computer. Commercial LCD monitors mix three primary colors (red, green, and blue) to display true color (i.e., 16,777,216 different colors with 24-bit RGB color space). In principle, three major wavelength ranges from its backlight (usually the broadband cold cathode

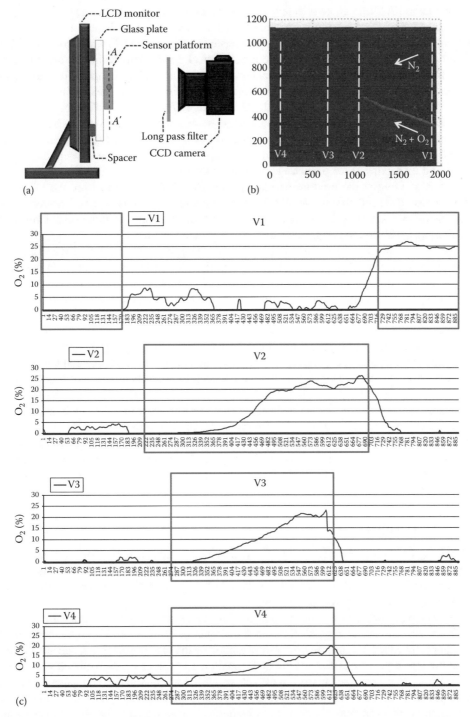

FIGURE 8.10 (a) Measurement setup with an LCD monitor as excitation light source and a color camera as photodetector. A color camera takes pictures of the fluidic sensor platform (8 × 8 cm^2) installed in close proximity to an LCD screen that provides an excitation blue light (470 nm) with uniform intensity over the sensor coating. (After Park, S. et al., *Progress in Biomedical Optics and Imaging, Proceedings of SPIE, 8025*, Paper ID: 802509, Orlando, FL, April 25–29, 2011. With permission from SPIE.) (b) Stern–Volmer image of oxygen distribution (equivalent to I_0/I). (c) Oxygen profiles at various locations defined in (b) (V1, V2, V3, and V4), showing a nitrogen and 20% oxygen fluxes at upper and lower branches, respectively.

fluorescent lamp [CCFL]) can be selected with controlled intensities. Several new techniques have been reported where these monitors are applied for analytical purposes. Especially, Filippini et al. implemented the idea of computer screen photo-assisted technique (CSPT) where commercial monitors were used as illumination light source for analytical applications (Filippini and Lundstrom, 2002; Filippini et al., 2003). These examples include illuminating samples for transmission/absorption measurements and exciting luminescent samples for emission measurements.

For oxygen quantification, we used an LCD monitor (HP 2159 m, 21.5″ color LCD monitor, Hewlett–Packard) as the excitation light source. Meso-scale test platforms (8×8 cm^2) incorporating a fluidic channel and a sensor coating were prepared and attached on an LCD screen as shown in Figure 8.10a. Planar sensor films formulated in our laboratory were used for oxygen imaging. This was prepared by mixing silica particles containing ruthenium complex with silicone prepolymer, spin-coating, and curing the mixture on glass plates. Blue light (470 nm) from the LCD screen was used to excite the ruthenium-complex films that respond to gaseous oxygen. The color camera provided significant advantages to store colorimetric digital data with two-dimensional information and analyze its color change and gradient over a captured image area.

Figure 8.10b shows a reconstructed image showing the spatial distribution of relative intensity (I_0/I) of Stern–Volmer relation. Nitrogen gas was initially injected into two channel branches to obtain a reference image (equivalent to I_0), followed by switching the gas for the lower branch to air (21% oxygen) for a sample image (equivalent to I). The reference image was divided by the second image on a pixel-by-pixel basis to obtain the final Stern–Volmer image. Figure 8.10c shows the oxygen concentration profiles obtained from four different locations of the fluidic channel as indicated by dotted lines in Figure 8.10b. Each profile represents the oxygen concentration gradient along the dotted lines. The difference of oxygen content between the upper and lower branches was clearly observed (V1 location). The oxygen concentration gradient becomes less steep along the downstream (from the right to the left in Figure 8.10b), because the gases introduced from the two channels mix together in the main channel. These results show that the combination of LCD and color camera enables a uniform illumination over a large area to image spatial distribution of chemicals. It is anticipated that the quantification of multiple target analytes is possible with variable wavelength ranges emitted from the LCD monitor. Furthermore, time-resolved imaging will also be possible with an application-specific camera paired with a synchronized display device with sinusoidal or pulsatile emission.

8.5   CONCLUSION

A wide variety of luminophores from organic dyes, inorganic nanoparticles, polymers, and proteins are available that exhibit large Stokes shifts within visible ranges. Therefore, these agents are potentially amenable for colorimetric determination based on this approach of adopting easily accessible colorimetric devices as analytical instruments.

As an exemplary proof-of-concept demonstration, rather nontraditional optoelectronic devices in analytical sciences were successfully used for oxygen determination. The ruthenium-complex luminophore embedded in sensor matrices were excited by the blue wavelength range (about 470 nm peak) emitted from the white LED and the LCD monitor. The color camera exhibited good sensitivity and linearity to gaseous and dissolved oxygen samples. Both qualitative and quantitative analyses were possible with relatively simple colorimetric image analysis. Especially, the combination of LCD and color camera enables a uniform illumination over a large area to image spatial distribution of chemicals and to analyze multiple target analytes simultaneously.

In general, the CCD cameras, especially the cooled monochromatic ones, are highly sensitive and low-noise imaging devices compared to their CMOS counterpart, enabling them more suitable for high-quality imaging in scientific research. The advent of submicron-scale memory chip technology, however, allowed the performance of CMOS camera chips to become more competitive to that of CCD. The CMOS chips are fabricated by economical standard IC processes without tailoring its process sequences. More importantly, they consume less power, which is essential for portable applications. In fact, the majority of portable imaging devices in mobile cellular phones and webcams in the market are economical CMOS devices. Furthermore, novel emission devices based on new materials are emerging to provide new functionalities. One representative example is the organic LEDs (OLEDs) that can be seamlessly integrated in flexible platforms. Many preceding researches relied on commercially available products. However, customized optoelectronic components can be developed to have suitable emission intensity, photonic sensitivity, spectral responsivity, dynamic range, etc. These innovations in optoelectronics areas will largely expedite the progress of economic analytical instruments based on colorimetric interrogation.

REFERENCES

Abe, K., K. Suzuki, and D. Citterio, Inkjet-printed microfluidic multianalyte chemical sensing paper, *Analytical Chemistry*, 80, 6928–6934, 2008.

Abramoff, M. D., P. J. Magelhaes, and S. J. Ram, Image processing with ImageJ, *Biophotonics International*, 11, 36–42, 2004.

Amao, Y., Probes and polymers for optical sensing of oxygen, *Microchimica Acta*, 143(1), 1–12, 2003.

Babin, S., A. Bugrov, S. Cabrini, S. Dhuey, A. Goltsov, I. Ivonin, E.-B. Kley, C. Peroz, H. Schmidt, and V. Yankov, Digital optical spectrometer-on-chip, *Applied Physics Letters*, 95(4), 041105, 2009.

Balzani, V., P. Ceroni, A. Juris, M. Venturi, S. Campagna, F. Puntoriero, and S. Serroni, Dendrimers based on photoactive metal complexes. Recent advances, *Coordination Chemistry Reviews*, 219–221, 545–572, 2001.

Batchelor, J. D. and B. T. Jones, Development of a digital micromirror spectrometer for analytical atomic spectroscopy, *Analytical Chemistry*, 70, 4907–4914, 1998.

Eaton, K., A novel colorimetric oxygen sensor: Dye redox chemistry in a thin polymer film, *Sensors and Actuators B: Chemical*, 85, 42–51, 2002.

Evans, R. C. and P. Douglas, Controlling the color space response of colorimetric luminescence oxygen sensors, *Analytical Chemistry*, 78, 5645–5652, 2006.

Evans, R. C., P. Douglas, J. A. G. Williams, and D. L. Rochester, A novel luminescence-based colorimetric oxygen sensor with a "traffic light" response, *Journal of Fluorescence*, 15, 201–206, 2006.

Filippini, D. and I. Lundstrom, Chemical imaging by a computer screen aided scanning light pulse technique, *Applied Physics Letter*, 81(20), 3891–3893, 2002.

Filippini, D., S. P. S. Svensson, and I. Lundstrom, Computer screen as a programmable light source for visible absorption characterization of (bio)chemical assays, *Chemical Communication*, 9, 240–241, 2003.

Gerion, D., F. Pinaud, S. C. Williams, W. J. Parak, D. Zanchet, S. Weiss, and A. P. Alivisatos, Synthesis and properties of biocompatible water-soluble silica-coated CdSe/ZnS semiconductor quantum dots, *Journal of Physical Chemistry B*, 105(37), 8861–8871, 2001.

Holst, G. C. and T. S. Lomheim, *CMOS/CCD Sensors and Camera Systems*, 2nd edn., p. 143, SPIE Press, Bellingham, WA, 2011.

Jenison, R., S. Yang, A. Haeberli, and B. Polisky, Interference-based detection of nucleic acid targets on optically coated silicon, *Nature Biotechnology*, 19, 62–65, 2001.

Landfester, K., Synthesis of colloidal particles in miniemulsions, *Annual Review of Materials Research*, 36, 231–279, 2006.

Lavigne, J. J., S. Savoy, M. B. Clevenger, J. E. Ritchie, B. McDoniel, S.-J. Yoo, E. V. Anslyn, J. T. McDevitt, J. B. Shear, and D. Neikirtk, Solution-based analysis of multiple analytes by a sensor array: Toward the development of an "electronic tongue," *Journal of the American Chemical Society*, 120, 6429–6430, 1998.

Lee, J.-H., D. N. Liu, and S.-T. Wu, *Introduction to Flat Panel Displays*, p. 170, John Wiley & Sons, New York, 2008.

Martinez, A. M., S. T. Philips, E. Carrilho, S. W. Thomas, H. Sindi, and G. M. Whitesides, Simple telemedicine for developing regions: Camera phones and paper-based microfluidic devices for real-time, off-site diagnosis, *Analytical Chemistry*, 80, 3699–3707, 2008.

Mills, A., Optical oxygen sensors, *Platinum Metals Review*, 41(3), 115–127, 1997.

Papkovsky, D. B., T. O'Riordan, and A. Soini, Phosphorescent porphyrin probes in biosensors and sensitive bioassays, *Biochemical Society Transactions*, 28, 74–77, 2000.

Park, S., S. G. Achanta, and C.-S. Kim, Fluorescence intensity measurements with display screen as excitation source, *Progress in Biomedical Optics and Imaging, Proceedings of SPIE, 8025*, Paper ID: 802509, Orlando, FL, April 25–29, 2011.

Park, J., W. Hong, and C.-S. Kim, Color intensity method for hydrogel optical sensor array, *IEEE Sensors Journal*, 10(12), 1855–1861, 2010.

Park, J. and C.-S. Kim, A simple oxygen sensor imaging method with white light-emitting diode and color charge-coupled device camera, *Sensor Letters*, 9(1), 118–123, 2011.

Rakow, N. A. and K. S. Suslick, A colorimetric sensor array for odour visualization, *Nature*, 406, 710–713, 2000.

Shaner, N. C., R. E. Campbell, P. A. Steinbach, B. N. G. Giepmans, A. E. Palmer, and R. Y. Tsien, Improved monomeric red, orange and yellow fluorescent proteins derived from *Discosoma* sp. red fluorescent protein, *Nature Biotechnology*, 22, 1567–1572, 2004.

Shen, L., M. Ratterman, D. Klotzkin, and I. Papautsky, Use of a low-cost CMOS detector and cross-polarization signal isolation for oxygen sensing, *IEEE Sensors Journal*, 11(6), 1359–1360, 2011.

Stich, M. I. J., S. M. Borisov, U. Henne, and M. Schaferling, Read-out of multiple optical chemical sensors by means of digital color cameras, *Sensors and Actuators B*, 139, 204–207, 2009.

Syms, R. R. A., Advances in microfabricated mass spectrometers, *Analytical and Bioanalytical Chemistry*, 393(2), 427–429, 2009.

Taton, T. A., C. A. Mirkin, and R. L. Letsinger, Scanometric DNA array detection with nanoparticle probes, *Science*, 289, 1757–1760, 2000.

Thomas, P. C., M. Halter, A. Tona, S. R. Raghavan, A. L. Plant, and S. P. Forry, A noninvasive thin film sensor for monitoring oxygen tension during in vitro cell culture, *Analytical Chemistry*, 81, 9239–9246, 2009.

Thomas, S. W. III, G. D. Joly, and T. M. Swager, Chemical sensors based on amplifying fluorescent conjugated polymers, *Chemical Reviews*, 107, 1339–1386, 2007.

Tsien, R. Y., The green fluorescent protein, *Annual Review of Biochemistry*, 67, 509–544, 1998.

Veinot, J. G. C., Synthesis, surface functionalization, and properties of freestanding silicon nanocrystals, *Chemical Communications*, 40, 4160–4168, 2006.

Wang, X.-D., H.-X. Chen, Y. Zhao, X. Chen, and X.-R. Wang, Optical oxygen sensors move towards colorimetric determination, *Trends in Analytical Chemistry*, 29(4), 319–338, 2010a.

Wang, X.-D., R. J. Meier, M. Link, and O. S. Wolfbeis, Photographing oxygen distribution, *Angewandte Chemie—International Edition*, 49(29), 4907–4909, 2010b.

Yao, G., L. Wang, Y. R. Wu, J. Smith, J. S. Xu, W. J. Zhao, E. J. Lee, and W. H. Tan, FloDots: Luminescent nanoparticles, *Analytical and Bioanalytical Chemistry*, 385, 518–524, 2006.

Yotter, R. A. and D. M. Wilson, A review of photodetectors for sensing light-emitting reporters in biological systems, *IEEE Sensors Journal*, 3, 288–303, 2003.

9 Optical Oxygen Sensors for Micro- and Nanofluidic Devices

*Volker Nock, Richard J. Blaikie,
and Maan M. Alkaisi*

CONTENTS

9.1   INTRODUCTION

Ever since the emergence of photosynthesis and subsequent appearance of eukaryotic organisms, oxygen and life have been connected through a complicated interdependence. The use of oxygen as a substrate for energy production, although very efficient, is not without risk [1]. A few electrons combining prematurely with an O_2 molecule can lead to the formation of reactive oxygen species (ROS), such as H_2O_2. The presence of these ROS in turn can lead to the oxidation of lipids, nucleic acids, and proteins and hence result in cellular dysfunction or even cell death, something exacerbated even further by the fact that acute increases or decreases in cellular oxygen (hyperoxia/hypoxia) can induce further the generation of excess ROS. One of the most important lessons to be drawn from this intricate relationship is that efficient cellular function only occurs within a narrow range of oxygen concentrations.

This indicates that efficient control of the cellular microenvironment is of utmost importance in in vitro cell culture. One successful approach to increased environmental control is to reduce the dimensions and thus sample size in a system. This reduces intermixing and increases the influence of diffusion-limited processes predominant at the micron scale. The characteristics of laminar flow, as observed in microfluidic devices, allow one to, for example, generate parallel multistream flows with stable interstream interfaces in a single channel. Material transport across these interfaces is by diffusion only and can be controlled using the flow speed of the individual streams. In cell biology, this phenomenon can be applied to produce controlled chemical microenvironments down to subcellular dimensions [2,3]. To this day, the use of multiple parallel flow streams has been explored mostly for the delivery of biochemical reagents to cells [2–4]. Recently though, their potential for the generation of cellular microenvironments with controlled oxygen concentrations has begun to attract increasing interest.

In cell-based microfluidic applications in particular, the oxygen concentration of a sample stream itself represents a parameter with significant effect on cellular development and function. For example, the concentration of dissolved oxygen (DO) a cell is exposed to has been found to be intimately linked to cell survival, metabolism, and function [5]. Devices capable of exposing defined areas of a cell culture, individual cells, and regions on the surface of an individual cell to controlled DO levels therefore have the potential to yield novel insights into cell biology and tissue formation [6–9]. Furthermore, measuring and controlling the DO concentrations of sample streams will also increase the relevance of existing small-molecule delivery applications, which to this day have been performed with media equilibrated under atmospheric oxygen conditions only [2–4,10,11]. Common to all these examples is the need for a simple and robust means of measuring DO in the respective microfluidic devices. In the following, we aspire to demonstrate that thin-film optical oxygen sensors provide such a means and discuss their fabrication, calibration, and use for spatially resolved measurement of DO_2 in microfluidic devices related to the control of cellular microenvironments.

9.2 DISSOLVED OXYGEN IN BIOLOGICAL APPLICATIONS

In a natural environment, such as in mammalian organs, cellular oxygen concentration is maintained to normoxic (12% to <0.5% O_2) conditions [12]. Regulation to within this relatively narrow range of normoxia is necessary in vivo to prevent oxidative damage to the cell from excess oxygen (hyperoxia) and metabolic demise from insufficient oxygen (hypoxia) [13]. Absolute normoxic values for a specific cell are furthermore dependent on the cell localization within a particular organ, such as the liver for example. In the liver, localization is exhibited in form of oxygen-modulated zonation. Gradients along the length of a sinusoid result in regionally dominant metabolic and detoxification functions [14]. This natural sensitivity to local oxygen concentration can be replicated in vitro to selectively increase specific liver cell function like urea synthesis [15].

Mounting evidence indicates that *in vitro* cell-culture experiments performed in air (≈21% O_2) may introduce excessive stress on cells due to exposure to unnaturally high oxygen concentrations [12]. Fibroblasts exposed to different oxygen concentrations were found to adjust to high oxygen levels by reversible growth inhibition and differentiation. For neural and other stem cells in comparison, hypoxic conditions were observed to promote growth and influence differentiation [16]. Measurement of oxygen concentration is therefore of special importance when in vitro results are to be compared to cell behavior observed *in vivo*. In summary, it can be stated that oxygen uptake of cells is a powerful marker for metabolic status, health, and response to exogenous and endogenous stimuli [17].

Beyond simple cellular constructs, microfluidic devices are also increasingly used to handle higher organisms such as nematodes and mammalian embryos. The health and behavior of these organisms has again been found to be strongly linked to the oxygen microenvironment. For preimplantation mammalian embryos, a direct link exists between oxygen consumption and their development status [17], while *Caenorhabditis elegans* nematodes show strong behavioral preference for 5%–12% O_2 [18]. All these examples indicate that biomedical and agricultural applications would profit extensively from a reliable, nonintrusive tool to investigate in vitro oxygen concentrations.

9.2.1 PRINCIPLES OF OXYGEN SENSING

Traditional laboratory procedures for the measurement of oxygen in solution require the extraction of a sample volume for external analysis. This sampling approach is limited by the difficulty and time needed for extraction and analysis. With fluid volumes in the range of microliters and below, as found frequently in current microfluidic Lab-on-a-Chip (LOC) devices, analyte sampling constitutes a major disturbance of the system to be measured. This is a particular problem for sensing of less stable solutes such as oxygen, where retrieval is likely to significantly alter the sample characteristics [19].

9.2.2 AMPEROMETRIC VS. OPTICAL OXYGEN SENSING

Two main technologies, amperometric electrochemical and optical sensing, currently constitute the bulk of integrated sensors for the measurement of DO_2 in biological applications. Both are tolerant to liquid exposure and exhibit the high sensitivity needed to detect the small changes of oxygen encountered in the cellular environment. The first principle, amperometric electrochemical sensing, has been applied in a variety of biomedical LOC devices [20–22]. Despite continuous interest, several significant limitations have been identified concerning the use of amperometric sensors. When exposed to organic matter, sensor lifetime has been found to decrease through membrane fouling. Oxidation of the electrode surface area after prolonged usage and electrolyte depletion can be responsible for inconsistency between measurement data. A second, more important concern relates to the fundamental method of operation of the amperometric electrode [23]. The Clarke sensor operates by reducing oxygen and thus is prone to show signal dependence on flow rate in a low flow environment like a bioreactor device. Additional problems are posed by miniaturization itself, mainly due to the need to integrate a reference electrode [19].

9.2.3 FLUORESCENCE-BASED OPTICAL OXYGEN SENSING

In contrast, fluorescent dye–based optical sensing does not exhibit analyte depletion, and has therefore emerged as a promising alternative in biomedical applications. The principle of measurement with these types of sensors relies on the quenching of either the intensity or lifetime of light emitted by a dye in the presence of molecular oxygen. Fluorescence occurs upon excitation of the dye molecule (fluorophore*) to an exited state S^* and subsequent relaxation back to the ground state S_0 via the emission of light [24]:

$$S_0 + h \times \upsilon_{ex} \rightarrow S^* \tag{9.1}$$

$$S^* \rightarrow S_0 + h \times \upsilon_{em} \tag{9.2}$$

$$S^* + O_2 \rightarrow S_0 \big|_K \tag{9.3}$$

where
  h is Planck's constant
  $\upsilon_{ex} > \upsilon_{em}$ are the frequencies of the excitation and emission light, respectively
  K is the molecular rate constant for oxygen quenching

---

* Although also often referred to as a phosphor, the terms fluorophore and fluorescence are used in context with PtOEPK here to emphasize fluorescence microscopy as the tool for sensor readout.

One pathway for relaxation from the excited state of the fluorophore is via interaction with a secondary molecule. In case of oxygen, fluorescence quenching has a pronounced effect on the quantum yield as a result of the triplet ground state of the O_2 molecule. The fluorescence intensity, I, and lifetime, τ, can be obtained for dissolved or gaseous oxygen from the Stern–Volmer equation for intensity [25]:

$$\frac{I}{I_0} = \frac{\tau}{\tau_0} = 1 + K_{SV}^S \times [O_2] = 1 + K_{SV}^G \times pO_2 \qquad (9.4)$$

where
K_{SV}^S and K_{SV}^G are the Stern–Volmer constants for solution and gas, respectively
I_0 and τ_0 are the reference values in absence of oxygen
$[O_2]$ is the oxygen concentration in solution
pO_2 is the gaseous partial pressure of oxygen

Figure 9.1 illustrates the experimental setup and measurement principle of oxygen sensing based on fluorescence intensity quenching. As illustrated in Figure 9.1a, the intensity signal of a sensor pattern when no oxygen is present (0% O_2) decreases upon exposure to gaseous or DO molecules (100% O_2). This decrease in intensity is recorded using a standard fluorescence microscope (see Figure 9.1b) and translated into oxygen concentration by comparison to a prerecorded reference value. Since the fluorescence quenching is reversible, the intensity reverses to its initial value once the oxygen is removed again (0% O_2).

As indicated in Equation 9.4, the change in fluorescence lifetime can also be used to measure oxygen. However, if spatial information is to be extracted this significantly complicates the experimental setup [26]. The following sections will thus focus on intensity-based oxygen measurement.

9.2.4 OPTICAL SENSOR MATERIALS

While oxygen-dependent quenching is exhibited by the majority of fluorescent dyes, a small subgroup has been found to be especially suited due to high sensitivity and long fluorescence lifetimes. These dyes are commonly categorized into organic fluorescent or organometallic compound probes [25] and used in solution [17,26,27] or immobilized on a support matrix [19,28–31] for the detection of DO_2. An excellent overview of the different forms of sensor application can be found in [32]. For use with low-cost LOCs, dye immobilization has the advantages of increasing the ease of handling and reducing the amount of fluorescent dye required. Table 9.1 shows a list of organometallic complexes used as oxygen sensors together with absorption/emission wavelengths, suitable solvents for immobilizing polymer films, and relative intensity change depending on the polymer matrix.

Out of these, platinum(II) octaethylporphyrin ketone (PtOEPK) suspended in a polystyrene (PS) matrix has been identified as well suited for use in LOC devices due to its desirable optical properties, compatibility with standard optical components, and being readily available commercially. The PtOEPK dye molecule shows strong phosphorescence with a high quantum yield and long lifetime at room temperature. Sensors immobilized in PS have been shown to diminish only 12% under continuous illumination for 18 h, detected by absorbance measurement [25]. It further exhibits both a long-wave shift and an extended long-term photostability compared to other fluorescent dyes [33]. This allows for the use of standard optical filters and makes sample handling less critical. PS as the polymer matrix complements the advantages of PtOEPK by providing good oxygen permeability, biocompatibility, and low autofluorescence.

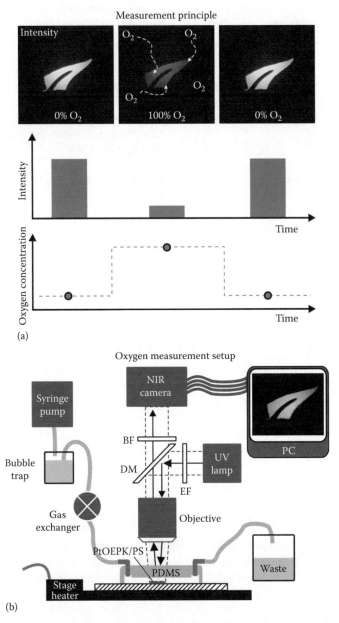

FIGURE 9.1 Principle of optical oxygen sensing: (a) the fluorescent signal intensity of a sensor film pattern (0% O_2) is reversibly quenched with exposure to oxygen molecules (100% O_2). Using a prerecorded calibration curve the change in intensity can be related to oxygen concentration. (b) Experimental setup for spatially resolved measurement of O_2 in microfluidic devices. A standard fluorescence microscope with a custom filter set and a digital camera sensitive in the NIR is used for data acquisition. (Adapted from Nock, V. and Blaikie, R.J., *IEEE Sensors J.*, 10, 1813, 2010.)

TABLE 9.1

List of Organometallic Luminescence Probes for Optical Oxygen Sensing

Probe	λ_{max} (abs) [nm]	λ_{max} (em) [nm]	Suitable Solvents for Immobilizing Polymer Films	I_0/I_{100} Depending on Polymer Matrix
$Ru(dpp)_3^{2+}$	337, 457	610	Dichloromethane	4.4 in silicone film
				1.1 in PS film
				3.5 in PVC film
$Os(dpp)_3^{2+}$	454, 500, 580, 650	729	Dichloromethane	4.5 in poly(DMS) film
$Ir(ppy)_3$	376	512	THF, dichloromethane	1.2 in PS film
PtOEP	381, 535	646	THF, dichloromethane, toluene	4.5 in PS film
PtTFPP	395, 541	648	THF, dichloromethane, toluene	3.0 in PS film
PtOEPK	398, 592	758	Toluene, chloroform	2.0 in PVC film
				20 in PS film
PdOEP	393, 512, 546	663	THF, dichloromethane, toluene	11.5 in PS film
PdOEPK	410, 603	790	Toluene, chloroform	8.0 in PVC film
				28 in PS film

Source: Adapted from Amao, Y., *Microchim. Acta*, 143, 1, 2003.
Note: Both palladium OEPK and platinum OEPK exhibit very high intensity ratios I_0/I_{100} immobilized in PS films.

9.3 INTEGRATION OF OPTICAL SENSORS INTO MICROFLUIDIC DEVICES

For any sensor system to find widespread use in microfluidic devices, a straightforward integration process compatible with common low-cost fabrication techniques is essential. In this section, the PtOEPK/PS optical sensor system is used as an example to discuss the limitations of existing fabrication methods and two alternative processes for the patterning of polymer-encapsulated oxygen sensors are introduced. The process flow and patterning results for each are described in detail, as well as methods for the integration of the patterns into devices. While developed for PtOEPK/PS, the fundamental fabrication principles can be applied to other oxygen sensors and sensing applications.

9.3.1 DEPOSITION OF SENSOR FILMS

Beyond the advantages mentioned earlier, PtOEPK/PS shares one major limitation with most other material systems, namely the challenge to fabricate micron-scale patterns of the material and integrate them with commonly used microfluidic devices. Organic solvents such as acetone readily dissolve PS, making it impossible to subsequently apply photoresist on PtOEPK/PS films for standard lithographic patterning [34]. As an alternative to direct lithography, Vollmer et al. [29] proposed photoresist liftoff in combination with pipetting of individual PtOEPK/PS patches to overcome this problem. They demonstrated an integrated oxygenator device with two large manually deposited sensor patches and successively calibrated these for DO_2 measurement. Recently, Sinkala and Eddington [35] proposed imprinting into cast PtOEPK/PS films with polydimethylsiloxane (PDMS) stencils as a method to fabricate oxygen-sensitive microwells. Due to limited film homogeneity or not fully removing the sensor material, the sensor patterns fabricated with both these methods have rather limited applicability. To improve the PtOEPK/PS integration into microfluidic devices, we have thus developed two novel fabrication methods combining deposition by spin coating with either patterning by soft lithography [36] or a sacrificial metal mask [37], which will be introduced in the following.

9.3.2 SOFT-LITHOGRAPHY PROCESS

Figure 9.2 shows a schematic of the sensor fabrication process using spin coating and soft lithography [36]. The sensor film is deposited with high film thickness uniformity and repeatability on a standard spin coater. When used as an oxygen sensor, these film properties lead to increased homogeneity and stability of the signal intensity. Thus improved films, in turn, allow one to implement fluorescent intensity-based, laterally resolved measurement of oxygen concentration. While spin-coated films can be applied to various substrates and directly used as sensors, their integration into microfluidic and PDMS-based devices in particular requires a means to produce design-specific sensor patterns. Due to differences in surface chemistry, a substrate fully covered in a thin film of PS makes it significantly more difficult to achieve a permanent bond to PDMS devices using the established process of surface-activated bonding. This problem can be avoided by limiting the sensor film to areas within a particular fluidic feature, such as a channel or reactor chamber. The PDMS device can thus be bonded to the exposed glass substrate surrounding the patterned sensor patches.

Oxygen-permeable films of the PtOEPK/PS sensor material are prepared by pipetting PtOEPK/PS dissolved in toluene onto clean glass microscope slides, as indicated in Figure 9.2a. The final film thickness is controlled through the spin speed and depends on the concentration of the solution and the solvent used. Coated PtOEPK/PS forms a solid film by solvent evaporation when left overnight at room temperature. Being a common prototyping material in microfluidics, PDMS is used to form the microfluidic channels and as a masking material to pattern the sensor films. The latter takes advantage of the flexibility of cured PDMS and poor adhesion of it to PS. This allows one to use prepatterned PDMS stamps to temporarily protect certain areas of the sensor film. Both positive and negative elastomer stamps can be fabricated. The negative stamps (stencil masks) contain vertical vias for the reactive-ion etch (RIE) patterning, which

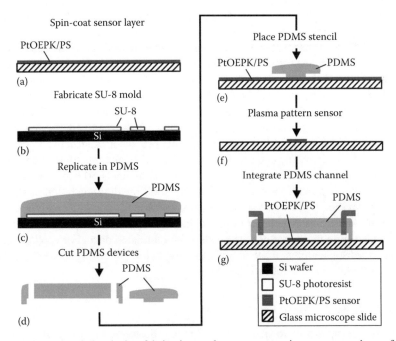

FIGURE 9.2 Schematic of the device fabrication and sensor patterning process using soft-lithography. (a) Application of the sensor layer to substrate by spin-coating. (b) Fabrication of photoresist mold master on separate substrate. (c) Replica-molding of master into PDMS. (d) Removal of PDMS microfluidic devices and stencils and cutting to size. (e) Placement of stencils on sensor layer. (f) Reactive ion etching of sensor layer. (g) Integration of sensor substrate with PDMS microfluidic devices. (Adapted from Nock, V. and Blaikie, R.J., *IEEE Sensors J.*, 10, 1813, 2010. With permission.)

can be problematic for isolated features. Positive stamps are typically made up of a 100–200 μm thick pattern detail part and a thicker backing that laterally overhangs the pattern by several millimeters. The extent of this overhang is determined by cutting away excess PDMS around the pattern areas and simplifies handling during stamp placement.

As depicted in Figure 9.2b, both microchannels and stamps are produced by fabricating a master with the inverse of the desired design in SU-8 negative tone photoresist on a silicon wafer. To facilitate later removal of the cured replica, a fluoropolymer layer is deposited onto the resist mould. Meanwhile, liquid PDMS prepolymer is prepared and degassed in vacuum to remove any trapped air bubbles. The prepolymer is poured onto the resist master and cured on a hotplate. Once cured, the PDMS replicas are peeled off, the backing cut to size using a scalpel and, in case of microchannels, access holes are cut using a hole punch. To produce negative stamps, the process is modified by using the process of exclusion molding [38]. For patterning, the stamps are brought into conformal contact with the PtOEPK/PS layer (Figure 9.2e) and placed in a RIE tool. Dry-etching conditions are based on data published for PS films [39] and adapted for use with the RIE setup by increasing the etch pressure and power. After the etching step, the stamps are peeled off and can be reused for further patterning.

In a final step, illustrated in Figure 9.2g, the patterned sensor films are integrated with a microfluidic network. Depending on the intended application, integration can be performed by either irreversible O_2 plasma bonding or pressure clamping. For the former, both the surface of the PDMS channel network and the glass substrate with the sensor pattern are activated in a RIE using oxygen plasma, manually aligned and brought into conformal contact [38]. Whilst short, this surface activation step will lead to some isotropic etching of the sensor layer by the O_2 plasma. Under certain circumstances, such as for characterization of different substrates or sensor systems, it can be advantageous to be able to remove the channel structure from the sample after use. Reversible clamping can be achieved using a custom-machined plate of transparent polymer, such as polymethylmethacrylate (PMMA). The seal is formed by sandwiching the glass substrate with oxygen sensor and microfluidic channels between two polymer plates or the metal plate of a microscope stage warmer. The stack is then secured using screws, which can be individually adjusted to provide the best possible seal. By applying this method, different sensor samples can be calibrated using a single channel structure.

A range of PtOEPK/PS test patterns fabricated using positive and negative stamps are shown in Figure 9.3. The minimum feature size achievable is limited to around 25 μm, mainly by the high aspect ratio SU-8 lithography required for the stamp mold master. Due to the thick PDMS backing supporting the positive stamps enclosed features cannot be fabricated easily. However, arrays of lines and dots, as well as other complex shapes such as text, can be readily replicated. Negative stencils made of 100 μm thick PDMS films (see Figure 9.3e) provide an alternative to positive stamps, but require more delicate handling and also have problems with the replication of enclosed features. In spite of these limitations and with typical microfluidic channels 50 μm and wider, the use of soft-lithographic stamps provides a straightforward method for the integration of oxygen sensing into microfluidic devices. Once molded off the primary master, individual stamps can be reused to pattern multiple substrates. The latter in particular makes the process well suited for biological laboratories as the stamp mold master only needs to be fabricated once. For applications requiring higher-resolution sensor features, a second, photolithographic patterning method has been developed, which is introduced in the following section.

9.3.3 Sacrificial Metal Mask Process

Extending the sensor patterning to smaller dimensions is desirable in particular for cell biological applications. The introduction of LOC devices in the field of cell biology has revolutionized assay techniques and made it possible to perform experiments on individual cells as opposed to Petri dish–sized population studies [2,40–43]. Assuming attached cells with typical diameters in the 10–50 μm range, a decrease in the pattern size to below 5 μm would allow one to integrate a single discreet sensor patch or whole arrays thereof subjacent to each individual cell. To facilitate such

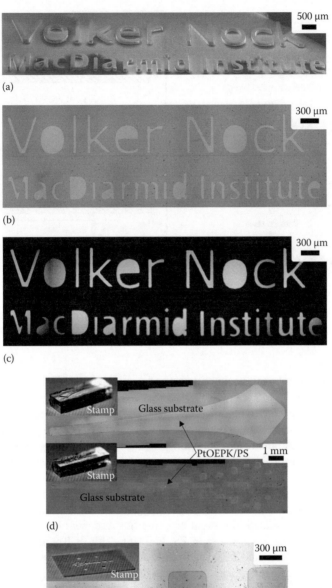

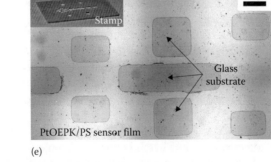

FIGURE 9.3 Results of the sensor film patterning using soft lithography. (a) SEM micrograph of a PDMS stamp (positive). (b) Optical micrograph of the corresponding PtOEPK/PS pattern on glass after transfer by reactive ion etching. Enclosed features such as in the letters *a*, *e*, *o*, and *d* cannot be fabricated using positive stamps. (c) Fluorescence intensity micrograph of the sensor film response for exposure to air. (d) and (e) Both positive and negative PDMS stamps can be used for patterning. (Adapted from Nock, V., Control and measurement of oxygen in microfluidic bioreactors, PhD thesis, Department of Electrical and Computer Engineering, University of Canterbury, Christchurch, New Zealand, 2009. With permission.)

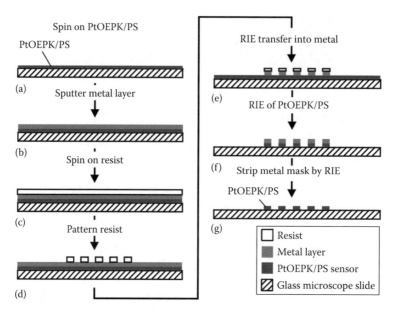

FIGURE 9.4 Schematic of the oxygen sensor patterning process using optical or electron beam lithography (EBL). (Adapted from Nock, V. et al., *Microelectron. Eng.*, 87, 814, 2009. With permission.)

single-cell oxygen measurements, we have developed a modified process based on the previous one by replacing the limiting PDMS stamps with sacrificial metal masks [37]. The process retains the homogeneous film characteristics obtained through spin coating the sensor material, while eliminating the need to fabricate separate polymer stamps. This simplifies the process significantly and makes the achievable minimum resolution mainly dependent on the lithography step used. A schematic of the modified sensor fabrication process using spin coating and sacrificial metal masks is shown in Figure 9.4.

The process begins with the application of an initial PtOEPK/PS sensor film to a substrate as outlined in the description of the soft lithographical process. Once the solvent has evaporated, the sensor film is thermally annealed to prevent cracking of subsequent layers when the PtOEPK/PS film is heated over the glass temperature of PS. Following this, a 100 nm thick metal layer (i.e., tungsten) is sputter-deposited onto the sensor, thus protecting the encapsulating PS matrix from solvents used during lithographic processing. Due to this sacrificial cover, a variety of conventional resists can now be applied and patterned using conventional photo- [37] or electron beam lithography (EBL) [44]. After exposure and development of the chosen resist, RIE is used to transfer the pattern into the subjacent tungsten layer. The etch gas is then switched to O_2 to transfer the pattern in the metal layer into the PtOEPK/PS film (see Figure 9.4f) and simultaneously strip the remainder of the resist. A final switch back to the metal etch gas removes the masking layer and thus concludes the patterning process. At this stage, the substrate and sensor patterns are ready to be integrated with prefabricated microfluidic devices using the techniques described earlier.

By using this method in combination with photolithographic chrome-on-glass masks and a standard projection exposure tool, we were able to fabricate large-scale arbitrary shapes and regular arrays of oxygen sensor patches with minimum feature sizes down to 3 μm [37]. Smaller patterns can be resolved by simply changing to a higher-resolution lithography technique such as interference lithography or EBL and the corresponding resist material. Figure 9.5 shows an example of a test pattern fabricated using EBL and the sacrificial metal mask process. Multiple arrays of oxygen sensor patches with edge lengths down to 500 nm were successfully replicated into a 600 nm thick PtOEPK/PS film. As opposed to the soft lithographic process, enclosed features can be readily

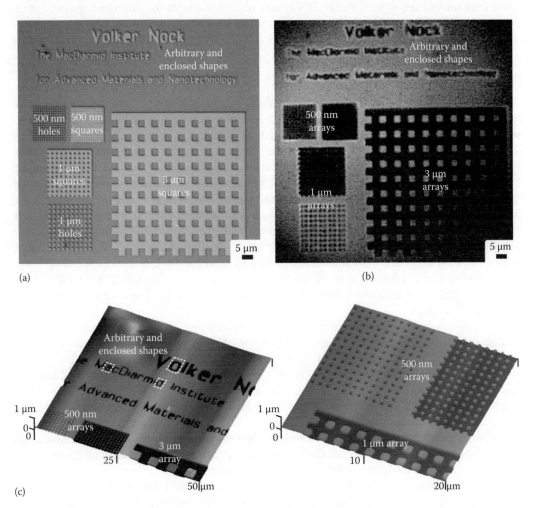

FIGURE 9.5 Results of the sensor film patterning using electron beam lithography. (a) Optical micrograph of the resist after e-beam exposure and development showing the replicated high-resolution patterns. (b) Fluorescence intensity micrograph of the patterned sensor film response for exposure to air. (c) AFM micrographs showing the replication of arbitrary enclosed and regular patterns in PtOEPK/PS. (Adapted from Nock, V. et al., Patterning of polymer-encapsulated optical oxygen sensors by electron beam lithography, in *Third International Conference on Nanoscience and Nanotechnology (ICONN 2010)*, Sydney, New South Wales, Australia, 2011, pp. 237–240. With permission.)

fabricated using this process and, in case of EBL, the amplification available for sensor readout becomes the limiting factor for the minimum practical pattern size.

9.4 LAB-ON-A-CHIP APPLICATIONS

With the full range of sensor patterning now available, integrated oxygen sensing can be realized in a variety of microfluidic devices and novel applications. In the following, three specific examples of the use of PtOEPK/PS sensor films in LOC-type devices are described to illustrate the versatility of the technology. Common to all potential application scenarios is the need to characterize and calibrate the response of the sensor for exposure to different oxygen concentrations. This is illustrated in the first part through an overview of the calibration process and a discussion of the intrinsic characteristics of the PtOEPK/PS system observed. Use of the calibrated sensors is then demonstrated for visualization and measurement of DO in multistream laminar flow and in contact with attached cancer cells.

9.4.1 CHARACTERIZATION AND CALIBRATION OF SENSOR FILMS

Prior to use, the integrated sensors have to be calibrated for measurement of gaseous or DO concentration. This is achieved by comparing the change in fluorescent sensor signal intensity for different solutions of oxygen concentration to a reference value provided by an external sensor [34]. After the calibration curve is recorded, oxygen concentration is determined by comparing the change of sensor intensity for an unknown concentration to the corresponding value determined during calibration. Due to the noncontact nature of the optical measurement, sensor readout can be performed on any fluorescence microscope fitted with the appropriate filter combination and camera (see Figure 9.2b). For fluid actuation, the device under test is interfaced with a syringe or peristaltic pump and an external or integrated gas exchanger providing custom DO concentrations. A bubble trap placed before the gas exchanger and bioreactor chip removes potential bubbles in the fluidic circuit. The inlet oxygen concentration is determined using a flow-through Clarke-type oxygen sensor, which provides a reference value for calibration of the integrated oxygen sensor. LOC devices undergoing calibration are further mounted on a temperature-controlled microscope stage to provide stable conditions.

Typically, the sensor signal intensity of a freshly prepared PtOEPK/PS films is first measured for exposure to different gaseous oxygen concentrations directly after spin coating and patterning. For this, partial oxygen pressures of 0% (I_0) and 100% (I_{100}) are produced by blowing oxygen-free nitrogen and industrial grade oxygen, respectively, through a microfluidic PDMS channel and thus onto the integrated sensor patterns. During this characterization, several parameters were found to affect the signal ratio [36]. These parameters include, amongst others, temperature, molecular weight (M_w) of the PS used, and the PS concentration (% w/w) of the initial toluene solution. By far, the strongest influence on the intensity ratio I_0/I_{100} was found to be related to the thickness of the sensor films. From an engineering point of view this is very advantageous, as the final thickness can be easily adjusted via the spin speed during coating of the sensor layer.

While the thickness of the PtOEPK/PS films decreases linearly with increasing spin speed, the sensor signal ratio actually increases with decreasing film thickness. Figure 9.6a summarizes results for two PtOEPK/PS mixtures with 5% and 7% w/w PS at various thicknesses. For the 7% PS solution, I_0/I_{100} increases almost twofold from 3.6 at 1.3 μm film thickness to 6.8 at 0.6 μm. One possible explanation for this increase in intensity ratio is the corresponding increase in the surface area-to-volume ratio. With decreasing film thickness, the overall intensity contribution of dye close to the surface will increase and, simultaneously, the permeability-limited contribution from dye molecules in the PS bulk is reduced. In addition, a thinner film also requires less time for oxygen to migrate inside the film so that equilibrium with the environmental oxygen pressure is reached faster [45]. The second observation from Figure 9.6a is the dependence of the increase in intensity ratio on PS solution. A 0.7 μm thick film in 7% PS solution ($M_w = 280k$) shows a threefold increase in I_0/I_{100} compared to a 5% PS solution ($M_w = 200k$) of equal thickness. This significant difference is mainly attributed to the higher-molecular-weight PS used in the 7% solution. It is thought that during solvent evaporation the system of PS chains strives to attain a state of minimal energy by contraction. Molecules above the entanglement molecular weight of PS (Me, PS = 18k) contract more slowly with increasing molecular weight and freeze in place before complete solvent evaporation [46]. This formation of disordered molecular chains can lead to an increase in oxygen permeability in PS films of higher M_w.

In general, these results indicate further possibilities for optimization of the sensor sensitivity through, for example, the use of thinner films and PS of different molecular weights. A fourth possible parameter not investigated here is the influence of the substrate material on sensor

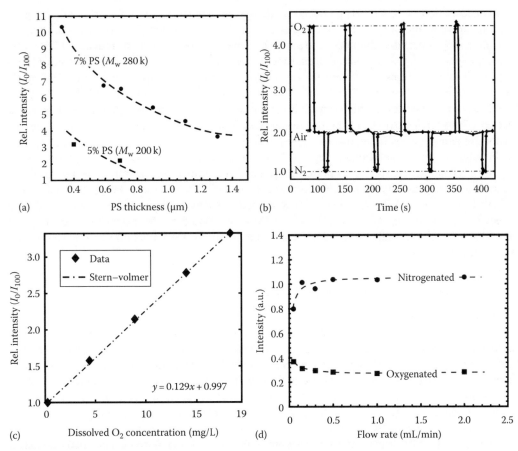

FIGURE 9.6 Sensor film characterization and calibration plots. (a) Relative intensity ratio of the PtOEPK/PS sensor films as a function of film thickness for two different PS concentrations and exposure to gaseous oxygen. (b) Dynamic sensor response of a 1.1 μm thick film for exposure to alternating gaseous oxygen concentrations. (c) Stern–Volmer curve for detection of dissolved oxygen in DI water with a 600 nm thick 7% w/w PtOEPK/PS sensor at a flow rate of $Q = 1$ mL/min ($T = 37.4°C$). (d) Plot of the sensor intensity data corresponding to flows of nitrogenated and oxygenated water across the tested dynamic operating range. At low flow rates, the operating range decreases slightly due to parasitic convective losses to the surrounding PDMS matrix. (Adapted from Nock, V. et al., *Lab on a Chip*, 8, 1300, 2008.)

signal intensity. The maximum intensity ratio of 11 for a 350 nm thick 7% PtOEPK/PS film is the highest reported on a LOC-compatible glass substrate. Even higher values have been reported for films on a Mylar substrate [47]. This indicates that the choice of substrate material can provide an additional parameter in optimizing the signal ratio. However, for PDMS-based devices this has to be balanced with the processing compatibility of the substrate material, especially if surface-activated plasma bonding is to be used. Notwithstanding the potential for further optimization, it should be noted that all current sensor films described here exceed the minimum intensity ratio of 3 necessary to be considered suitable for oxygen sensing [25].

Aside from a high sensitivity, a second major advantage of luminescent probes is the excellent signal reversibility and a nearly instantaneous sensor response. This is a direct consequence of the electronic nature of the probe–analyte interaction and illustrated in Figure 9.6b for exposure to different gaseous oxygen concentrations. For measurement of DO, the sensor films follow the linear Stern–Volmer calibration model given in Equation 9.4. An example of a calibration at a constant flow rate of 1 mL/min and five different oxygen concentrations is shown in Figure 9.6c. The maximum

DO concentration of 18.4 mg/L produced by the gas exchanger caused a factor of 3.4 change in total fluorescent intensity for this specific film. For flow rates below 0.5 mL/min, the dynamic operating range decreases slightly due to parasitic convective losses to the surrounding PDMS [29]. Depending on the initial concentration entering the device, oxygen is either added or removed from the liquid by mass transfer through the device walls, which is illustrated in Figure 9.6d by the decrease in intensity below 0.5 mL/min. This effect is due to the high permeability of oxygen in PDMS and demonstrates the high sensitivity of the sensor films. The obvious differences in signal decrease between hypoxic and hyperoxic water stem from the permeation coefficient of N_2 being a factor 2 smaller in PDMS than that of O_2 [48].

Finally, the overall stability and repeatability of the sensor response has also been found to be excellent. Dynamic measurements performed under 1 h illumination typically show no significant change in sensor intensity. Continuous illumination of PtOEPK/PS over 3 weeks with only minor signal degradation and the retaining of spectral and quenching characteristics for up to 2 years under storage have been reported [33]. These durations should be more than sufficient for most extended cell-culture studies where data acquisition usually is distributed over several shorter sessions per day.

9.4.2 VISUALIZATION OF DO CONCENTRATIONS

Essential to all experiments concerned with measuring and controlling oxygen concentrations is a reliable means of generating oxygenated flow. In microfluidic devices in particular biological cells need to be protected from excessive shear caused by gas bubbles [49], thus excluding the use of non-diffusive gas exchangers. PDMS-based versions on the other hand prevent the formation of inflow bubbles by limiting gas transfer to diffusion through a thin membrane [29]. This and their integral design make them ideally suited for integration with complex microfluidic LOCs. Photographs of such a PDMS gas exchanger and corresponding hydrodynamic focusing and multistream devices used in the following for DO visualization experiments are shown in Figure 9.7.

The hydrodynamic focusing device uses a buffer flow inlet, which is divided into two channels to provide the side sheath flows for focusing and a central sample stream. Buffer and sample streams are then focused in a 200 μm wide rectangular microchannel with a total length of 100 mm and a common outlet. The multistream device on the other hand combines two streams from external versions of the PDMS gas exchanger with a third stream from an on-chip exchanger in a 1.6 mm wide rectangular parallel-plate microchannel with integrated oxygen sensor film. The output flows with varying oxygen concentration from the three gas exchangers are combined in this channel to yield the parallel laminar flow streams.

In the following, the use of these two devices with integrated sensors for oxygen control and visualization is demonstrated. The inlet flows provided by the PDMS gas exchangers can be varied from hypoxic (0 mg/L O_2) and aerated (8.6 mg/L O_2) to fully saturated hyperoxic water (34 mg/L O_2). Depending on the application, additional concentrations within this range can be produced by simply adjusting the gas mixture in the gas exchangers. The local oxygen concentration in the devices is visualized using fluorescence microscopy on the subjacent PtOEPK/PS layer.

9.4.3 MULTISTREAM FLOW

Due to the absence of convective mixing at low Reynolds numbers diffusion is the predominant transport mechanism between two or more input streams. Several schemes for solute sorting and detection, such as the H-Filter and T-Sensor [50,51], operate using this principle. The T-Sensor in particular has been used extensively to measure the diffusion coefficients of solutes between adjacent flow streams [52] and this is demonstrated for oxygen in this section. A second application of multistream flow can be found in form of chip-based multistream assays for cell–cell networks [4]. The capability to selectively expose certain regions of an interconnected 2D cell culture to different

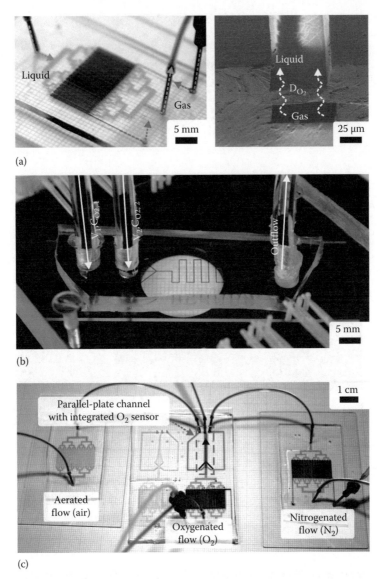

FIGURE 9.7 Photographs showing the microfluidic devices used to demonstrate oxygen measurement. (a) Close-ups of the integrated membrane-based PDMS gas exchanger used to generate flow with custom-dissolved oxygen concentrations. (b) Device used to generate hydrodynamically focused flow. Gas exchangers are not shown. (c) Device with three attached gas exchangers used to generate parallel flow streams. (Adapted from Nock, V. and Blaikie, R.J., *IEEE Sensors J.*, 10, 1813, 2010.)

environmental conditions and stimuli such as reduced or increased oxygen has the potential to yield novel insight into cell–cell interactions and signaling [53]. As opposed to well-based assays, colonies of cells in a multistream flow device remain in direct physical contact while experiencing different oxygen concentrations.

An example of the generation and spatially resolved visualization of multiple equal-width parallel laminar flow streams with individually controllable DO concentrations is shown in Figure 9.8. To demonstrate the oxygen sensor applicability, the resulting flow streams were imaged and analyzed using an integrated sensor film on the bottom of the channel [54]. Upon entering the central parallel-plate chamber, the three parallel flow streams remain separated over the total length of 18 mm

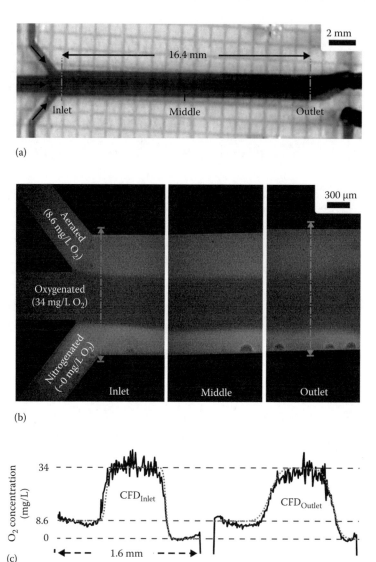

FIGURE 9.8 Demonstration of oxygen visualization and measurement in hydrodynamically focused flow. (a) Optical micrograph of the channel showing the central sample stream (middle) and buffer streams (top and bottom). Dashed rectangles indicate the imaging locations at the inlet and 6 cm downstream. (b) Oxygen-dependent optical intensity images of the sensor film at the locations indicated in (a). Regions of higher and lower intensity indicate lower and higher oxygen concentrations, respectively. In the two leftmost images, the central sample stream corresponds to a DO concentration of 34 mg/L (hyperoxia) and the buffer streams to 0 mg/L (hypoxia), whereas in the image on the right, the concentrations are inverted to yield a hypoxic sample stream. (c) Calibrated plots of the oxygen concentration across the channel at the locations indicated by the arrows in (b). The oxygen-rich sample stream remains focused to a width of 20 µm over the full length of 64 mm, while the generation of the oxygen-depleted stream indicates device versatility. (Adapted from Nock, V. and Blaikie, R.J., *IEEE Sensors J.*, 10, 1813, 2010.)

due to the predominantly laminar flow regime. This results in three distinct oxygen concentration levels being generated across the width of the microchamber. Figure 9.8b shows the corresponding intensity images recorded via the integrated sensor film for the reactor inlet, midpoint, and outlet. Aerated water enters the chamber through the top inlet, hyperoxic through the central, and hypoxic through the bottom inlet. The difference in oxygen concentration between the individual streams is easily discernible and remains stable over the full length of the device. Using the recorded intensity images, the local oxygen concentration and profiles across the reactor width can now be quantified.

Cross-width plots of the DO concentration at the inlet and outlet obtained from the intensity images are shown in Figure 9.8c. The oxygen concentration levels of hyperoxic, aerated, and hypoxic water are determined by precalibration of the sensors as described earlier. As can be observed, the width of the individual streams or oxygen levels varies over the reactor length. These differences in level width from inlet to outlet of the individual streams are indicative of lateral diffusion of oxygen from the central oxygenated stream to regions of lower oxygen concentration (aerated and hypoxic streams). Since the device dimensions, initial oxygen concentrations of the three streams and the flow conditions are known; this phenomenon can be used to deduce the diffusion coefficient of oxygen in the fluid using Fick's law [6]. By solving the diffusion equation, the coefficient of diffusion is found as a function of the characteristic length:

$$D_{O_2} = \frac{x^2}{4 \times t} \tag{9.5}$$

where
  D_{O_2} is the coefficient of diffusion of oxygen in the medium
  t is the residence time in the channel
  x is the diffusion length perpendicular to the flow direction

For a total flow rate of 0.3 mL/min in this example, the residence time $t = 1.05$ s. The diffusion length x can then be deduced from the difference of the slopes of the profiles at the inlet and outlet. By combining these values and solving Equation 9.5, the coefficient of diffusion for oxygen in water at a temperature of 37°C can be determined to $D_{O_2} = 2.57 \times 10^5$ cm^2/s. This value compares well with oxygen-diffusion coefficients in water of 2.52×10^5 cm^2/s at 35.1°C and 2.78×10^{-5} cm^2/s at 40.1°C published in literature [55], demonstrating the applicability of the integrated oxygen sensor films.

To further validate measurement result, computational fluid dynamic simulations of a microchannel with dimensions corresponding to the fabricated geometry can be performed. The simulation uses the experimental flow conditions and the measured coefficient of diffusion as model parameters [34]. Flow conditions are modeled using a Navier–Stokes application mode with three parallel inlet streams of equal flow rate. Cross-stream species transport is modeled using convection and diffusion. The measured D_{O_2} of 2.57×10^{-9} m^2/s is used as the isotropic diffusion coefficient of the liquid. Each inlet is assigned a constant species concentration and the boundary condition of the outlet is set to convective flux. To solve the model system, flow and species transport are coupled via the fluid velocity u parallel to the long axis of the reactor chamber.

For the multistream flow device in this example, simulated cross-chamber oxygen concentration profiles were evaluated at two points. The simulation results are plotted as dotted lines superimposed onto the measured profiles in Figure 9.8c. As can be seen, good agreement exists between the shape of the measured and the simulated concentration profiles at both the inlet and outlet and considering the noise on the measured data. In addition, the slopes of the profiles coincide, indicating the validity of the diffusion model for analysis, as well as the value of the diffusion coefficient measured using the integrated optical sensor film. This demonstration illustrates the potential of the integrated sensors to be used to catalog oxygen-diffusion coefficients in a variety of biologically relevant liquids either as a stand-alone device or inline on a LOC prior to on-chip microfluidic cell-culture bioreactors.

9.4.4 HYDRODYNAMICALLY FOCUSED FLOW

In two-dimensional symmetric hydrodynamic flow focusing, nonmixing streams can be further constrained laterally within the center of a microchannel by neighboring sheath flows from side channels [56]. The generation and spatially resolved detection of oxygen concentrations in such a setup are shown in Figure 9.9. With a single pump, typical buffer/sample flow rate combinations from 0.1 to 0.5 mL/min can be produced and oxygen-dependent fluorescence intensity images recorded at different positions along the meandering microchannel using the integrated oxygen sensor film [57]. Intensity images of the sensor film inside the microchannel at the inlet and 64 mm downstream are shown on the left and in the middle in Figure 9.9b. In this example, the oxygen concentration of the buffer streams was set to hypoxia and that of the sample stream to hyperoxia. The outline of the microchannel in the images indicates the capability of the sensor to resolve the different DO levels inside the channel boundaries with a lateral resolution of <1 μm.

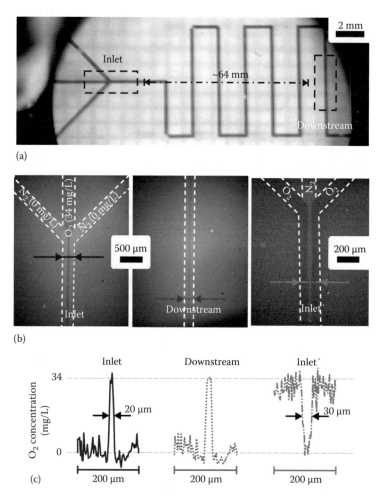

FIGURE 9.9 Visualization and measurement of oxygen in multistream laminar flow. (a) Optical micrograph of the parallel-plate rectangular microchannel with integrated optical oxygen sensor film indicating device dimensions, measurement locations, and flow layout. (b) Oxygen-dependent fluorescent intensity images recorded at the inlet, middle, and outlet. The intensity response for the top stream corresponds to aerated (8.6 mg/L O_2), the middle stream to hyperoxic (34 mg/L O_2), and the bottom stream to hypoxic (0 mg/L O_2) conditions. Interstream boundaries remain stable over the full channel length of 18 mm. (c) Oxygen concentration plots across the channel width at positions indicated by the dashed rectangles in (b). Concentration levels (dashed lines) were obtained through calibration of the sensor prior to use and coincide well with the individual streams. By applying diffusion theory, a coefficient of diffusion of oxygen in water of $D_{O_2} = 2.57 \times 10^{-5}$ cm^2/s can be calculated from the decrease in slope from the inlet to outlet profile. This compares well with CFD simulation results obtained using the measured coefficient as parameter and overlaid as dotted lines in the plot. (Adapted from Nock, V. and Blaikie, R.J., *IEEE Sensors J.*, 10, 1813, 2010.)

With the sample streams in this configuration, one would for example be able to study spatially resolved the reversible growth inhibition and differentiation exhibited by individual fibroblasts when exposed to high oxygen levels [12]. By limiting the hyperoxic region to the narrow sample stream, the behavior of individual cells in a culture can be studied. For other types of cells, such as neural and some stem cells, hypoxic conditions promote growth and directly influence cell differentiation [16]. Switching the setup to these conditions is easily realized in this device by simply changing the gases used in the gas exchangers to yield a hypoxic sample stream and oxygen-saturated buffer streams. The intensity image corresponding to this particular configuration is shown on the right in Figure 9.9b. In both cases the oxygen concentration of the buffer and sample streams can be finely tuned to the desired experimental conditions within the full range produced by the gas exchangers.

In addition to visualizing the DO content of the flow streams, the intensity response of the sensor film can also be used to measure the absolute local O_2 concentration. This is achieved by precalibration of the intensity change as described earlier. The calibration curve is then used to convert the change in intensity for an unknown concentration to the corresponding oxygen value in mg/L. Due to the homogeneous films obtained using spin coating and by using fluorescent intensity quenching for detection, the sensor can thus be used for spatially resolved oxygen sensing [36]. Precalibrated plots of the DO concentration profile across the channel width close to the focus point and 64 mm downstream are shown in Figure 9.9c. As can be observed from the measured profiles, hydrodynamic focusing produces a dimensionally well-defined stream with stable oxygen concentration over long stream lengths. In addition, the excellent lateral resolution of the sensor film means that the submicron wide interstream boundary between the hypoxic and hyperoxic streams is well defined in the intensity images and the oxygen concentration profiles. Only minor broadening of the hyperoxic sample stream occurs, which is due to lateral oxygen diffusion over the extended channel length. This can be controlled via the overall flow rate and thus the residence time of the liquid in the channel. The cross-channel plot for a hypoxic sample stream is shown on the right of Figure 9.9c and demonstrates the range of conditions the device can generate and detect.

Due to the intended use of the devices with hepatocyte liver cells (Ø ~ 20 μm) [58] or endometrial cancer cells (Ø ~ 35 μm) [59], the flow conditions shown are optimized to yield a sample stream width of around 20 μm. However, if needed, this can be further reduced significantly by adjusting the buffer/sample stream flow ratio. Stable sample streams of widths as small as 50 nm have been reported for mixing applications [56]. While sample streams of these dimensions enable high-resolution stimulation of certain areas on the surface of a single cell, the resulting fluid shear forces will have to be closely monitored so as not to influence the cell physiology and thereby reduce the relevance of the delivered stimuli.

9.4.5 OXYGEN SENSING IN CELL CULTURE

Both examples of sensor use introduced earlier localize the solid-state optical oxygen sensor films on the microchannel base either by simply covering the whole substrate [6] or at certain positions in the device by prepatterning the sensor [36]. The latter in particular allows one to physically separate oxygen sensors and cell-culture areas inside a single LOC [9,28]. However, with biological applications trending toward higher resolution, this separation of sensor and cells and thus averaging nature of the oxygen measurement will no longer be sufficient to study effects on a single-cell level. One possible solution to overcome this limitation is to directly culture cells on the sensor film and this final section will summarize recent progress made toward this goal.

In general, the PtOEPK/PS material system is well suited for this since the PS matrix used to encapsulate the probe molecules is a common cell-culture substrate and shows good general biocompatibility [60]. Initial tests with cell adhesion enhancing extracellular matrix proteins indicated that the long-term oxygen sensor capability of a PtOEPK/PS film remains unaffected by a covering layer of type I collagen [61]. In total the signal ratio decreases by only 8% over a measurement time of 14 days during which the sensor film was illuminated for an accumulated duration of 5 h

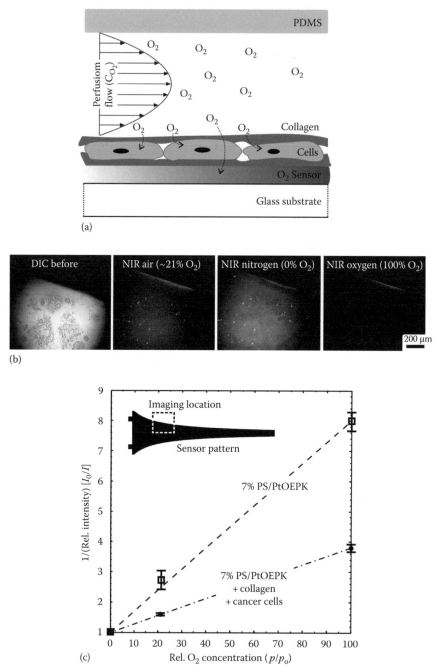

(a)

(b)

(c)

FIGURE 9.10 Demonstration of oxygen visualization and measurement in cell culture. (a) Schematic of the measurement system indicating the perfusion flow, sensor location, and oxygen transport. (b) Fluorescent intensity micrographs showing a region of the bioreactor indicated by the outline on the top right. From left to right: DIC microscopy of Ishikawa cancer cells growing on the sensor film prior to application of oxygen. Consecutive images of fluorescent sensor response of the same area in air ($\approx$21% O_2), pure nitrogen (0% O_2), and pure oxygen (100% O_2). (c) Graph showing a plot of the relative sensor intensity as a function of oxygen concentration with and without cell culture ($T = 37.4°C$). (Adapted from Nock, V. et al., Oxygen control for bioreactors and in-vitro cell assays, in *Fourth International Conference on Advanced Materials and Nanotechnology*, Dunedin, New Zealand, 2009, pp. 67–70.)

or 20 min per day. The result indicates that an overlying collagen film does not significantly affect the sensor film. Furthermore, taking into account photo-bleaching and measurement accuracy, it can be said that the reduction of intensity is small enough to make the sensor suitable for long-term observation of oxygen concentrations.

The application of the sensor in this form is however only an intermediate step on the way to its use in conjunction with live cells. Ideally, cells are either cultured in an alternating pattern with interspersed sensor stripes or directly on top of the collagen-covered PtOEPK/PS layer, as shown schematically in Figure 9.10a. In both cases, chemicals secreted by the cells as part of metabolism and signaling have the potential to cause biofouling of the sensor [62]. In addition, diffusion of oxygen through the cell/collagen sandwich to the sensor layer underneath may be reduced due to oxygen uptake of the superimposed cells. To demonstrate the compatibility of the PtOEPK/PS oxygen sensor with direct cell culture, oxygen measurements were performed with cancer cells seeded onto a sensor layer [63].

Oxygen concentration–dependent fluorescence micrographs of a PtOEPK/PS sensor patch seeded with *Ishikawa* human endometrial cancer cells [64] after 3 days in culture are shown in Figure 9.10b. Prior to testing of the sensor function, all remaining culture media was rinsed off the cells and the sample was exposed alternatively to air, nitrogen, and oxygen. A graph of the relative intensity vs. the relative gaseous oxygen concentration for the sensor sample is shown in Figure 9.10c. The measured intensity ratio I_0/I_{100} exhibited by the sensor when covered with a film of type I collagen and cancer cells was measured to be 3.8. The second line in the graph indicates the sensor calibration result prior to deposition of collagen and culturing of the cells. In this case, the intensity ratio $I_0/I_{100} = 8.2$, approximately double the value recorded with superimposed cells and collagen.

Despite the significant decrease in intensity, these results provide a successful initial demonstration of the applicability of the PtOEPK/PS sensor films to in situ measurement of oxygen in active biological environments. The observed reduction in measured intensity ratio is likely due to reduced diffusion of oxygen through the cell/collagen sandwich. Further studies comparing areas of plain sensor with such covered in cells should be able to help shed light on this and simultaneously provide a tool to model the oxygen permeability of cell layers.

In addition to live cell assays in oxygen-sensitive microwells [35] and microchannel-covering sensor films [63], the newly available sensor patterning techniques discussed earlier further allow one to integrate complete arrays of micro- or nanoscale sensor patches subjacent to individual cells [37,44]. At sensor patch edge lengths of 1 µm and below, several independent patches can be localized underneath a single cell and thus provide local oxygen concentration information at a subcellular level. The number of patches per cell and hence the spatial resolution of the array can be increased by simply decreasing the size of the patches. Initial tests with *Ishikawa* endometrial cancer cells cultured on a 1 µm PtOEPK/PS sensor array demonstrate the feasibility of unobstructed sensor readout even through the overlying layer of biological material [44].

Further work is clearly required to fully classify sensor interaction with cells and other biological material. While changes in biologically relevant oxygen levels have been measured on a several hundred to a few tens of cells basis using PtOEPK/PS sensors [35], oxygen tension measurements at a single-cell level have yet to be demonstrated. Caution should also be applied to the interpretation of such localized measurements and the extrapolation of results obtained at a two-dimensional interface into the three-dimensional space earlier, as this will depend on a variety of factors such as the flow characteristics, boundary layer effects, and the cell type used.

9.5   OUTLOOK AND CONCLUSIONS

The deposition of solid-state polymer-encapsulated optical oxygen sensors has been demonstrated using simple casting and spin-coating techniques. Homogeneous spin-coated films allow for better thickness control and thus can be used for spatially resolved measurement of oxygen. Deposited films can be patterned using soft-lithographic stamps or photo- and EBL combined with reactive ion etching. Sensing characteristics of the films can be optimized by adjusting the composition and thickness of the encapsulating polymer matrix. Substrates with patterned sensor films can be integrated into common PDMS microfluidic devices by plasma bonding and the sensor signal can be recorded using a fluorescence microscope. Application of the integrated optical oxygen sensors to microfluidic LOC devices capable of generating microenvironments with controlled oxygen concentrations has been demonstrated. Spatially resolved in situ measurements of DO have been shown for parallel laminar and hydrodynamically focused streams and for oxygen concentrations ranging from 0 to 34 mg/L using low-cost reusable sensor films. Water-based sample streams with controlled oxygen concentration have been focused to widths of 20 µm and could be maintained over lengths exceeding 6 cm in these devices. Visualization of DO in multistream laminar flow with stream-independent control of oxygen concentration has also been demonstrated and devices based on this have been used to determine the coefficient of diffusion of oxygen in water by measuring diffusive stream broadening.

While optical oxygen sensor are increasingly being commercialized as water-soluble bioassay kits such as MitoXpress by Luxcel Biosciences [65], the current work in solid-state optical oxygen sensors has largely been of an academic nature. Apart from the use of sensor films as active layers on handheld laboratory optrodes, thin-film oxygen sensor have not yet found their way into commercially available standardized cell-culture wells or plates. However, the advantages of these thin-film sensors combined with their good biocompatibility should see increasing interest in the future. Compared to DO probes, polymer-encapsulated dye sensor films such as PtOEPK/PS are less expensive due to them being reusable and, through encapsulation in the PS matrix, less likely to influence the measurement by interacting with cells and microorganisms cultured in close proximity. Sensor films have been shown to remain fully functional under cell-culture conditions and in direct contact with live cells. Their solid-state nature combined with the noncontact optical readout further allows for oxygen measurements to be performed inside fully enclosed devices. This in particular will enable the study of effects of oxygen concentration on cell development and function in general. In enclosed microfluidic LOC devices with reduced sample volumes, solutes like oxygen are less prone to interaction with the surrounding atmosphere and thus these devices provide higher sensitivity, which should enable oxygen tension assays on a single-cell level. The sensors will further allow one to provide better environmental control for molecular delivery experiments, in particular by adding the ability to visualize and control the oxygen concentration of the perfusion media used to deliver other chemical stimuli.

REFERENCES

1. G. L. Semenza, Life with oxygen, *Science*, 318, 62–64, 2007.
2. S. Takayama, E. Ostuni, P. LeDuc, K. Naruse, D. E. Ingber, and G. M. Whitesides, Laminar flows: Subcellular positioning of small molecules, *Nature*, 411, 1016–1016, 2001.
3. S. Takayama, E. Ostuni, P. LeDuc, K. Naruse, D. E. Ingber, and G. M. Whitesides, Selective chemical treatment of cellular microdomains using multiple laminar streams, *Chemistry & Biology*, 10, 123–130, 2003.
4. H. Kaji, M. Nishizawa, and T. Matsue, Localized chemical stimulation to micropatterned cells using multiple laminar fluid flows, *Lab on a Chip*, 3, 208–211, 2003.
5. H. Zhang and G. Semenza, The expanding universe of hypoxia, *Journal of Molecular Medicine*, 86, 739–746, 2008.
6. V. Nock and R. J. Blaikie, Spatially resolved measurement of dissolved oxygen in multistream microfluidic devices, *IEEE Sensors Journal*, 10, 1813–1819, 2010.
7. P. Abbyad, P.-L. Tharaux, J.-L. Martin, C. N. Baroud, and A. Alexandrou, Sickling of red blood cells through rapid oxygen exchange in microfluidic drops, *Lab on a Chip*, 10, 2505–2512, 2010.
8. M. Polinkovsky, E. Gutierrez, A. Levchenko, and A. Groisman, Fine temporal control of the medium gas content and acidity and on-chip generation of series of oxygen concentrations for cell cultures, *Lab on a Chip*, 9, 1073–1084, 2009.
9. R. H. W. Lam, M.-C. Kim, and T. Thorsen, Culturing aerobic and anaerobic bacteria and mammalian cells with a microfluidic differential oxygenator, *Analytical Chemistry*, 81, 5918–5924, 2009.

10. B. Kuczenski, W. C. Ruder, W. C. Messner, and P. R. LeDuc, Probing cellular dynamics with a chemical signal generator, *PLoS ONE*, 4, e4847, 2009.

11. F. Wang, H. Wang, J. Wang, H.-Y. Wang, P. L. Rummel, S. V. Garimella, and C. Lu, Microfluidic delivery of small molecules into mammalian cells based on hydrodynamic focusing, *Biotechnology and Bioengineering*, 100, 150–158, 2008.

12. S. Roy, S. Khanna, A. A. Bickerstaff, S. V. Subramanian, M. Atalay, M. Bierl, S. Pendyala, D. Levy, N. Sharma, M. Venojarvi, A. Strauch, C. G. Orosz, and C. K. Sen, Oxygen sensing by primary cardiac fibroblasts: A key role of p21Waf1/Cip1/Sdi1, *Circulation Research*, 92, 264–271, 2003.

13. G. L. Semenza, HIF-1, O_2, and the 3 PHDs: How animal cells signal hypoxia to the nucleus, *Cell*, 107, 1–3, 2001.

14. K. Jungermann and T. Kietzmann, Oxygen: Modulator of metabolic zonation and disease of the liver, *Hepatology*, 31, 255–260, 2000.

15. J. W. Allen and S. N. Bhatia, Formation of steady-state oxygen gradients in vitro—Application to liver zonation, *Biotechnology and Bioengineering*, 82, 253–262, 2003.

16. L.-L. Zhu, L.-Y. Wu, D. Yew, and M. Fan, Effects of hypoxia on the proliferation and differentiation of NSCs, *Molecular Neurobiology*, 31, 231–242, 2005.

17. F. C. O'Mahony, C. O'Donovan, J. Hynes, T. Moore, J. Davenport, and D. B. Papkovsky, Optical oxygen microrespirometry as a platform for environmental toxicology and animal model studies, *Environmental Science and Technology*, 39, 5010–5014, 2005.

18. J. M. Gray, D. S. Karow, H. Lu, A. J. Chang, J. S. Chang, R. E. Ellis, M. A. Marletta, and C. I. Bargmann, Oxygen sensation and social feeding mediated by a *C. elegans* guanylate cyclase homologue, *Nature*, 430, 317–322, 2004.

19. D. A. Chang-Yen and B. K. Gale, An integrated optical oxygen sensor fabricated using rapid-prototyping techniques, *Lab on a Chip*, 3, 297–301, 2003.

20. C.-C. Wu, T. Saito, T. Yasukawa, H. Shiku, H. Abe, H. Hoshi, and T. Matsue, Microfluidic chip integrated with amperometric detector array for in situ estimating oxygen consumption characteristics of single bovine embryos, *Sensors and Actuators, B: Chemical Sensors and Materials*, 125, 680–687, 2007.

21. J. Karasinski, L. White, Y. C. Zhang, E. Wang, S. Andreescu, O. A. Sadik, B. K. Lavine, and M. Vora, Detection and identification of bacteria using antibiotic susceptibility and a multi-array electrochemical sensor with pattern recognition, *Biosensors and Bioelectronics*, 22, 2643–2649, 2007.

22. E. Akyilmaz, A. Erdogan, R. Ozturk, and I. Yasa, Sensitive determination of L-lysine with a new amperometric microbial biosensor based on *Saccharomyces cerevisiae* yeast cells, *Biosensors and Bioelectronics*, 22, 1055–1060, 2007.

23. Y.-J. Chuang, F.-G. Tseng, J.-H. Cheng, and W.-K. Lin, A novel fabrication method of embedded microchannels by using SU-8 thick-film photoresists, *Sensors and Actuators A: Physical*, 103, 64–69, 2003.

24. S. Fischkoff and J. M. Vanderkooi, Oxygen diffusion in biological and artificial membranes determined by the fluorochrome pyrene, *Journal of General Physiology*, 65, 663–676, 1975.

25. Y. Amao, Probes and polymers for optical sensing of oxygen, *Microchimica Acta*, 143, 1–12, 2003.

26. D. Sud, G. Mehta, K. Mehta, J. Linderman, S. Takayama, and M.-A. Mycek, Optical imaging in microfluidic bioreactors enables oxygen monitoring for continuous cell culture, *Journal of Biomedical Optics*, 11, 050504-3, 2006.

27. J. Alderman, J. Hynes, S. M. Floyd, J. Kruger, R. O'Connor, and D. B. Papkovsky, A low-volume platform for cell-respirometric screening based on quenched-luminescence oxygen sensing, *Biosensors and Bioelectronics*, 19, 1529–1535, 2004.

28. P. Roy, H. Baskaran, A. W. Tilles, M. L. Yarmush, and M. Toner, Analysis of oxygen transport to hepatocytes in a flat-plate microchannel bioreactor, *Annals of Biomedical Engineering*, 29, 947–955, 2001.

29. A. P. Vollmer, R. F. Probstein, R. Gilbert, and T. Thorsen, Development of an integrated microfluidic platform for dynamic oxygen sensing and delivery in a flowing medium, *Lab on a Chip*, 5, 1059–1066, 2005.

30. X. Xiong, D. Xiao, and M. M. F. Choi, Dissolved oxygen sensor based on fluorescence quenching of oxygen-sensitive ruthenium complex immobilized on silica–Ni–P composite coating, *Sensors and Actuators B: Chemical*, 117, 172–176, 2006.

31. S. Lee, B. L. Ibey, G. L. Cote, and M. V. Pishko, Measurement of pH and dissolved oxygen within cell culture media using a hydrogel microarray sensor, *Sensors and Actuators B: Chemical*, 128, 388–398, 2008.

32. S. M. Grist, L. Chrostowski, and K. C. Cheung, Optical oxygen sensors for applications in microfluidic cell culture, *Sensors*, 10, 9286–9316, 2010.

33. D. B. Papkovsky, G. V. Ponomarev, W. Trettnak, and P. O'Leary, Phosphorescent complexes of porphyrin ketones: Optical properties and application to oxygen sensing, *Analytical Chemistry*, 67, 4112–4117, 1995.
34. V. Nock, Control and measurement of oxygen in microfluidic bioreactors, PhD thesis, Department of Electrical and Computer Engineering, University of Canterbury, Christchurch, New Zealand, 2009.
35. E. Sinkala and D. T. Eddington, Oxygen sensitive microwells, *Lab on a Chip*, 10, 3291–3295, 2010.
36. V. Nock, R. J. Blaikie, and T. David, Patterning, integration and characterisation of polymer optical oxygen sensors for microfluidic devices, *Lab on a Chip*, 8, 1300–1307, 2008.
37. V. Nock, M. Alkaisi, and R. J. Blaikie, Photolithographic patterning of polymer-encapsulated optical oxygen sensors, *Microelectronic Engineering*, 87, 814–816, 2009.
38. B. H. Jo, L. M. Van Lerberghe, K. M. Motsegood, and D. J. Beebe, Three-dimensional micro-channel fabrication in polydimethylsiloxane (PDMS) elastomer, *Journal of Microelectromechanical Systems*, 9, 76–81, 2000.
39. G. N. Taylor, T. M. Wolf, and J. M. Moran, Organosilicon monomers for plasma-developed x-ray resists, *Journal of Vacuum Science and Technology*, 19, 872–880, 1981.
40. S. Kobel, A. Valero, J. Latt, P. Renaud, and M. Lutolf, Optimization of microfluidic single cell trapping for long-term on-chip culture, *Lab on a Chip*, 10, 857–863, 2010.
41. A. Salehi-Reyhani, J. Kaplinsky, E. Burgin, M. Novakova, A. J. deMello, R. H. Templer, P. Parker, M. A. A. Neil, O. Ces, P. French, K. R. Willison, and D. Klug, A first step towards practical single cell proteomics: A microfluidic antibody capture chip with TIRF detection, *Lab on a Chip*, 11, 1256–1261, 2011.
42. M. C. Park, J. Y. Hur, H. S. Cho, S.-H. Park, and K. Y. Suh, High-throughput single-cell quantification using simple microwell-based cell docking and programmable time-course live-cell imaging, *Lab on a Chip*, 11, 79–86, 2011.
43. X. Li, Y. Chen, and P. C. H. Li, A simple and fast microfluidic approach of same-single-cell analysis (SASCA) for the study of multidrug resistance modulation in cancer cells, *Lab on a Chip*, 11, 1378–1384, 2011.
44. V. Nock, L. Murray, M. M. Alkaisi, and R. J. Blaikie, Patterning of polymer-encapsulated optical oxygen sensors by electron beam lithography, in *Third International Conference on Nanoscience and Nanotechnology* (*ICONN 2010*), Sydney, New South Wales, Australia, 2011, pp. 237–240.
45. Y. Amao, K. Asai, T. Miyashita, and I. Okura, Novel optical oxygen pressure sensing materials: Platinum porphyrin–styrene–trifluoroethylmethacrylate copolymer film, *Chemistry Letters*, 28, 1031–1032, 1999.
46. J. Zhao, S. Jiang, Q. Wang, X. Liu, X. Ji, and B. Jiang, Effects of molecular weight, solvent and substrate on the dewetting morphology of polystyrene films, *Applied Surface Science*, 236, 131–140, 2004.
47. P. Hartmann and W. Trettnak, Effects of polymer matrices on calibration functions of luminescent oxygen sensors based on porphyrin ketone complexes, *Analytical Chemistry*, 68, 2615–2620, 1996.
48. M. Ohyanagi, H. Nishide, K. Suenaga, and E. Tsuchida, Oxygen-permselectivity in new type polyorganosiloxanes with carboxyl group on the side chain, *Polymer Bulletin*, 23, 637–642, 1990.
49. J. Sung and M. Shuler, Prevention of air bubble formation in a microfluidic perfusion cell culture system using a microscale bubble trap, *Biomedical Microdevices*, 11, 731–738, 2009.
50. A. E. Kamholz, E. A. Schilling, and P. Yager, Optical measurement of transverse molecular diffusion in a microchannel, *Biophysical Journal*, 80, 1967–1972, 2001.
51. A. E. Kamholz and P. Yager, Theoretical analysis of molecular diffusion in pressure-driven laminar flow in microfluidic channels, *Biophysical Journal*, 80, 155–160, 2001.
52. M. S. Munson, K. R. Hawkins, M. S. Hasenbank, and P. Yager, Diffusion based analysis in a sheath flow microchannel: The sheath flow T-sensor, *Lab on a Chip*, 5, 856–862, 2005.
53. C. Michiels, T. Arnould, and J. Remacle, Endothelial cell responses to hypoxia: Initiation of a cascade of cellular interactions, *Biochimica et Biophysica Acta (BBA)—Molecular Cell Research*, 1497, 1–10, 2000.
54. V. Nock, R. J. Blaikie, and T. David, Generation and detection of laminar flow with laterally-varying oxygen concentration levels, in *12th International Conference on Miniaturized Systems for Chemistry and Life Sciences*, San Diego, CA, 2008, pp. 299–301.
55. P. Han and D. M. Bartels, Temperature dependence of oxygen diffusion in H_2O and D_2O, *Journal of Physical Chemistry*, 100, 5597–5602, 1996.
56. J. B. Knight, A. Vishwanath, J. P. Brody, and R. H. Austin, Hydrodynamic focusing on a silicon chip: Mixing nanoliters in microseconds, *Physical Review Letters*, 80, 3863, 1998.
57. V. Nock and R. J. Blaikie, Visualization and measurement of dissolved oxygen concentrations in hydrodynamic flow focusing, in *Sensors, 2009 IEEE*, Christchurch, New Zealand, 2009, pp. 1248–1251.

58. V. Nock, R. J. Blaikie, and T. David, Microfluidics for bioartificial livers, *New Zealand Medical Journal*, 120, 2–3, 2007.
59. V. Nock, L. Murray, F. Samsuri, M. M. Alkaisi, and J. J. Evans, Microfluidics-assisted photo nanoimprint lithography for the formation of cellular bioimprints, *Journal of Vacuum Science & Technology B*, 28, C6K17–C6K22, 2010.
60. K. E. Geckeler, R. Wacker, and W. K. Aicher, Biocompatibility correlation of polymeric materials using human osteosarcoma cells, *Naturwissenschaften*, 87, 351–354, 2000.
61. V. Nock, R. J. Blaikie, and T. David, In-situ optical oxygen sensing for bio-artificial liver bioreactors, in *13th International Conference on Biomedical Engineering*, Singapore, 2009, pp. 778–781.
62. B. Starly and A. Choubey, Enabling sensor technologies for the quantitative evaluation of engineered tissue, *Annals of Biomedical Engineering*, 36, 30–40, 2008.
63. V. Nock, R. J. Blaikie, and T. David, Oxygen control for bioreactors and in-vitro cell assays, in *Fourth International Conference on Advanced Materials and Nanotechnology*, Dunedin, New Zealand, 2009, pp. 67–70.
64. M. Nishida, The Ishikawa cells from birth to the present, *Human Cell*, 15, 104–117, 2002.
65. C. Diepart, J. Verrax, P. B. Calderon, O. Feron, B. F. Jordan, and B. Gallez, Comparison of methods for measuring oxygen consumption in tumor cells in vitro, *Analytical Biochemistry*, 396, 250–256, 2010.

10 Multidirectional Optical Sensing Using Differential Triangulation

Xian Jin and Jonathan F. Holzman

CONTENTS

10.1 INTRODUCTION AND BACKGROUND

Remote sensing is the basis for probing external environments, and improvements in this sensing have followed the development of various signal emission and detection technologies. Remote sensor technologies gather information from distributed targets by way of two general formats: passive remote sensing and active remote sensing.

The first format, being passive remote sensing [1–3], is unidirectional in nature, as information is simply gathered from the external environment. Natural or artificial targets emit visible, infrared, or radio frequency (RF) signals, and these signals are detected by the passive sensor network. Charge-coupled devices (CCD) [4] and complementary metal oxide semiconductor (CMOS) sensors [5] are classic examples of such a passive system for visible/IR spectra. Global positioning systems (GPS) are an example of passive reception in the RF regime [6,7]. Ultimately, the selected technology must be appropriate for the scale of the desired detection environment, with long-wavelength RF systems being well suited to global scales and short-wavelength optical systems being better suited to applications requiring high sensitivities and small spatial resolutions.

The second format, being active remote sensing [8–10], differs from its passive counterpart through its increased level of control and bidirectional implementation. Active remote sensing can effectively probe its external environment by emitting signals and sampling their reflected/backscattered light level, as the source characteristics can be controlled (through modulation, polarization, wavelength, etc.) to improve the signal sensitivity. This principle has been successfully applied in long-wavelength remote sensing applications such as millimeter-wave radar [11], short range radar [12], and even mid-IR terahertz (THz) imaging [13] and radar ranging [14]. Technologies based upon active remote sensing can, therefore, be adapted for use in multidirectional sensor links.

An intriguing application for active remote sensing exists in modern Smart Dust technologies [15]. These technologies establish bidirectional links between multiple sensor nodes. Two-way communication is established between each sensor node through the realization of three distinct tasks: detection of the incident light, retroreflection of the incident light, and modulation of the retroreflected light. Incident light detection is typically accomplished with a photodetector, while retroreflection is provided by a separate corner-cube retroreflector (CCR). Modulation of the returning signal, allowing communication in the bidirectional link, is typically accomplished by mechanical beam deflectors on the incorporated corner-cube mirrors [16], liquid crystal (LC) shutters [17], or multiple quantum well (MQW) modulations [18].

Given the motivations for emerging multidirectional technologies, we present an architecture that merges the detection and retroreflection capabilities of active remote sensing. An integrated retrodetector is introduced in a discrete package. The device is comprised of three mutually orthogonal photodiodes (PDs) in a corner-cube architecture. Unlike standard PD [19] and photoconductive [20] switching technologies, which record a single magnitude for the incident optical power, such a structure allows for simultaneous retroreflection of incident signals back to their source (through the retroreflecting nature of the corner-cube geometry) and local detection of the incident power level (through the parallel sum of the individual photocurrents). At the same time, the three-channel PD structure provides real-time control and optimization of the optical alignment. Indeed, differential combinations of PD photocurrents are used to triangulate the direction of the optical source and align the structure along the optimal orientation.

In this chapter, the novel active remote sensing structure is introduced in Section 10.2 associated with its fabrication and operational principle. In Section 10.3, the techniques for optical differential triangulation are presented with theory and methodology. The experimental free-space optical (FSO) sensing and communication applications are given in Section 10.4. Section 10.5 draws the final conclusions.

10.2 SENSING ARCHITECTURE

10.2.1 SYSTEM DESIGN AND FABRICATION

The proposed integrated retrodetector of interest is assembled as the corner cube–based structure shown in Figure 10.1a. As metal–semiconductor–metal PDs lead to diffraction, scattering, and losses on the reflections, the multilayered (i.e., vertical electrode–semiconductor–electrode) silicon PDs with the uniform and smooth surfaces are selected to provide enhanced reflectivity with minimal surface scattering/diffraction. Each of them has a 9.7×9.7 mm^2 electrically isolated active area with negligible inactive exposed edges around the periphery. The silicon PD retrodetector is then optically bonded into its mutually orthogonal interior corner-cube form during an alignment/ calibration process, with PD$_1$ lying in the yz-plane, PD$_2$ lying in the xz-plane, and PD$_3$ lying in the xy-plane. To ensure the required retroreflection accuracy and signal levels, the retroreflection is monitored over the 5 m length scale of interest while the unit is UV-cured. Moreover, a high-speed Pi-cell LC optical modulator is bonded to the entrance interface of this retrodetector to modulate both incident and retroreflected optical beams.

10.2.2 SYSTEM OPERATION AND PRINCIPLE

The orthogonal nature of the integrated retrodetector provides a distinct retroreflecting characteristic which allows incident light to be reflected directly back to its source. This is accomplished through three internal reflections of the three constituent PDs. Interestingly, the employment of PDs offers a unique opportunity to probe the power levels of incident light by sampling the nonreflected

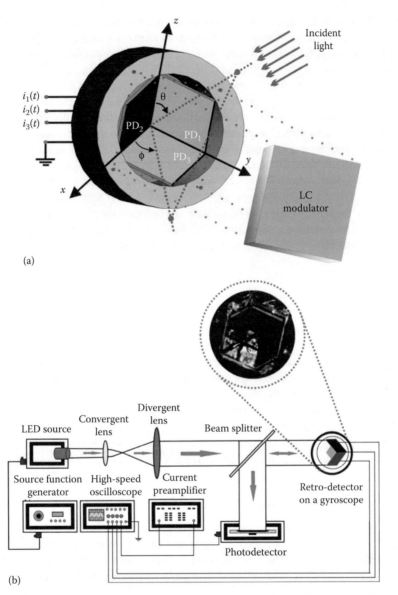

FIGURE 10.1 (a) Solid Works schematic of the integrated silicon PD retrodetector and (b) a typical FSO experimental setup are shown with a light-emitting device (LED) as the light source and the retrodetector shown in the inset photograph. (From Jin, X., *IEEE Sensors J.*, 2010. With permission.)

fraction of the beam power. The three resulting photocurrents are recorded as $i_1(t)$, $i_2(t)$, and $i_3(t)$, as shown in Figure 10.1a, with each corresponding to PD_1, PD_2, and PD_3, respectively. The sum of these photocurrents provides a measure of the total incoming optical signal power level. The constituent photocurrents of the integrated retrodetector play an important role in triangulating the incoming optical signal orientations, and the differential combinations of these photocurrents can ultimately offer an effective mechanism for this differential optimization and triangulation of the incident light directions by way of the azimuthal ϕ and polar θ angles shown in Figure 10.1a.

The differential photocurrents are defined here as $i_{1-2}(t) = i_1(t) - i_2(t)$, $i_{1-3}(t) = i_1(t) - i_3(t)$, and $i_{2-3}(t) = i_2(t) - i_3(t)$. As these differential photocurrents are in a function of the azimuthal ϕ and polar θ angles, they thus can be used to optimize the optical alignment between the light source and the retrodetector itself. Typically, ϕ is ranging from 0° to 90°, and it can be chosen as the first orientation–optimization control phase. Over this orientation range, the PD_1 photocurrent falls from its maximum (full-illumination) signal level down to its minimum (negligible/parallel-illumination) signal level. In contrast to this fact on PD_1, the photocurrent on PD_2 rises from its minimum (negligible/parallel-illumination) signal level up to its maximum (full-illumination) signal level. In this way that the system could be rotated and realigned in the xy-plane by a careful tracking of the differential photocurrent between PD_1 and PD_2, and the $\phi = 45°$ midpoint gives a balanced differential photocurrent with $i_{1-2}(t) = 0$. It is apparent that this position offers the equal signal detection of PD_1 and PD_2, which indicates an optimal optical alignment for ϕ orientation. In the second orientation–optimization control phase, the same optimization process is employed to balance θ while monitoring both $i_{1-3}(t)$ and $i_{2-3}(t)$. Once all three differential photocurrents are zero, i.e., $i_{1-2}(t) = i_{2-3}(t) = i_{1-3}(t) = 0$, this perfect optical alignment orients the system with the incident light rays entering the corner along the azimuthal angle $\phi = 45°$ and the polar angle $\theta = \cos^{-1}(1/\sqrt{3}) \approx 54.7°$, where this retrodetector is now oriented with its $(x, y, z) = (1, 1, 1)$ coordinate aligned directly toward the illumination source. Thus, it is possible to monitor the system for conditions in which one or more differential photocurrents are not equal to zero, $i_{1-2}(t)$ or $i_{2-3}(t)$ or $i_{1-3}(t) \neq 0$, where the retrodetector is exhibiting optical misalignment.

10.3   NUMEROUS THEORETICAL ANALYSES

In the previous section, the proposed integrated retrodetector shown in Figure 10.1a was introduced. The geometrical nature of this architecture allows it to act as a retroreflector by reflecting a fraction of the incident light back to its source, while local sensing is facilitated by the use of PDs (as the nonreflected fraction of incident light can be sampled and summed to probe the incident light levels and orientations). Ultimately, the geometrical and active PD construction together provides a method for differentiating between orthogonal components of the incident light intensity levels, and this offers an effective mechanism for differential optimization, realignment, and control. The theory and methodology for the angular response of these optical retroreflection, photodetection, and active control characteristics are presented in details within the following subsections.

10.3.1   OPTICAL RETROREFLECTION CHARACTERISTIC ANALYSES

Active remote sensing targets must effectively redirect large fractions of incident light back to their respective sources. A suitable structure for accomplishing this retroreflection is the standard CCR. Such an architecture consists of three reflective surfaces being mutually perpendicular and oriented as an interior corner as shown in Figure 10.1a. When incident light enters this corner along the unit-normal vector

$$\hat{r} = -n_1\hat{x} - n_2\hat{y} - n_3\hat{z}, \tag{10.1}$$

it undergoes a series of reflections, reversing the n_1, n_2, and n_3 incident ray directional cosines in the x, y, and z directions, respectively. The uniform incident intensity

$$I_0\hat{r_1} = -I_0n_1\hat{x} - I_0n_2\hat{y} - I_0n_3\hat{z}, \tag{10.2}$$

for a uniform intensity magnitude of I_0 leads to intensities of

$$\vec{I}_1 = +RI_0n_1\hat{x} - RI_0n_2\hat{y} - RI_0n_3\hat{z}$$
$$\vec{I}_2 = -RI_0n_1\hat{x} + RI_0n_2\hat{y} - RI_0n_3\hat{z} \tag{10.3}$$
$$\vec{I}_3 = -RI_0n_1\hat{x} - RI_0n_2\hat{y} + RI_0n_3\hat{z},$$

after the light undergoes reflections from PD_1, PD_2, and PD_3, respectively. The angular-independent surface reflectivity here is denoted by R. After three successful internal reflections, light exits the corner-cube antiparallel to the incident ray with an intensity defined by

$$-I_0 R^3 \hat{r} = +I_0 R^3 n_1 \hat{x} + I_0 R^3 n_2 \hat{y} + I_0 R^3 n_3 \hat{z}. \tag{10.4}$$

The degree to which this structure acts as a retroreflector can be quantified by the reflection solid angle subtended by the azimuthal ϕ and polar θ angles. The definitions for these angles have been shown in Figure 10.1 along with the directional cosine components:

$$n_1 = \cos\phi \cdot \sin\theta$$
$$n_2 = \sin\phi \cdot \sin\theta \tag{10.5}$$
$$n_3 = \cos\theta.$$

The retroreflective response of this corner-cube structure has been successfully applied to short [21] and long [22] range bidirectional applications, and the angular analyses for the retroreflection directionality is shown elsewhere [23].

10.3.2 OPTICAL PHOTODETECTION CHARACTERISTIC ANALYSES

The desired optical detection process for the proposed retrodetector is fundamentally based on the use of three orthogonal silicon PDs. The incident light with a uniform intensity illuminates the element shown in Figure 10.1a and (typically) leads to three PD photocurrents with differing amplitudes. The disparity between these amplitudes is due to imbalances between the incident light ray directional cosine components. Each PD will have a differing cross-sectional illumination area as viewed from the source: incident light with a large n_1 component along the x-axis will preferentially illuminate PD_1; incident light with a large n_2 component along the y-axis will preferentially illuminate PD_2; and incident light with a large n_3 component along the z-axis will preferentially illuminate PD_3. Quantifying this relationship becomes further complicated by the fact that the recorded PD photocurrents will also have contributions from light rays after primary and secondary internal reflections. The complete theoretical model for the PD photocurrents, including these internal reflections, is developed in this detection characteristics subsection. Power levels incident upon PD_1, PD_2, and PD_3 are investigated as three individual responses with results defined in terms of six illumination condition cases: (1) $n_1 < n_2 < n_3$, (2) $n_1 < n_3 < n_2$, (3) $n_2 < n_1 < n_3$, (4) $n_2 < n_3 < n_1$, (5) $n_3 < n_1 < n_2$, and (6) $n_3 < n_2 < n_1$. Each of these cases is, itself, characterized by six individual subcases, where the photocurrent levels will depend upon the ordering of the primary and secondary reflections. Fortunately, arguments based on rotational and mirror symmetries can be employed, and the required analysis is demonstrated here specifically for the $n_1 < n_2 < n_3$ case.

The first analysis case corresponds to incident illumination with the x-, y-, and z-directional cosine components in increasing order of magnitudes. This $n_1 < n_2 < n_3$ situation is analyzed for the first subcase, in which the incident light strikes PD_1, reflects onto PD_2, and reflects onto PD_3 (and subsequently exits the retrodetector to return to its source). In this case, the light ray unit-normal vectors that are incident on PD_1, PD_2, and PD_3 are

$$\hat{r}_1 = -n_1 \hat{x} - n_2 \hat{y} - n_3 \hat{z}$$
$$\hat{r}_{12} = +n_1 \hat{x} - n_2 \hat{y} - n_3 \hat{z} \tag{10.6}$$
$$\hat{r}_{123} = +n_1 \hat{x} + n_2 \hat{y} - n_3 \hat{z},$$

respectively, where the subscripts indicate the successive illumination and reflection sequence. Figure 10.2a shows the resulting illumination areas for this subcase. Here, PD_1 is fully illuminated and shaded, and its four corner points in the yz-plane are projected along $\hat{r}_{12}$ onto the xz-plane. The resulting shaded PD_2 illumination area is then defined as the overlap between this projected area and the PD_2 surface. Similarly, the PD_2 illumination area is projected along $\hat{r}_{123}$

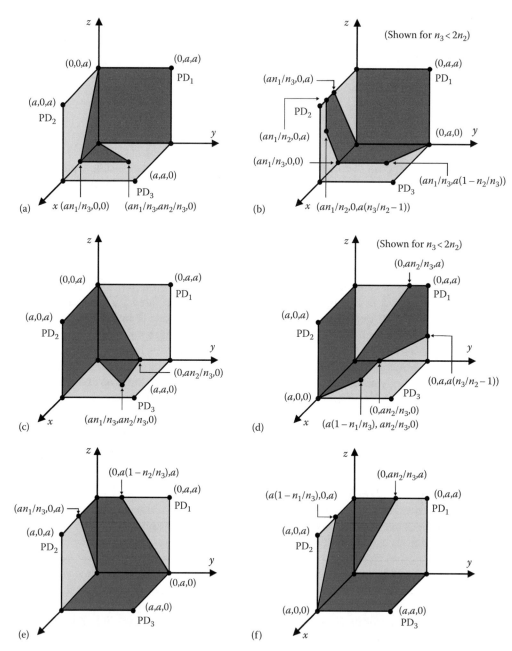

FIGURE 10.2 Schematics are shown for the internal reflection processes occurring in the retrodetector for directional cosine conditions $n_1 < n_2 < n_3$. Unilluminated areas are shown with light shading. Illuminated areas are shown with dark shading. The order of internal reflections proceeds as (a) PD_1 onto PD_2 onto PD_3, (b) PD_1 onto PD_3 onto PD_2, (c) PD_2 onto PD_1 onto PD_3, (d) PD_2 onto PD_3 onto PD_1, (e) PD_3 onto PD_1 onto PD_2, and (f) PD_3 onto PD_2 onto PD_1. (From Jin, X., *IEEE Sensors J.*, 2010. With permission.)

onto the xy-plane, and the shaded PD_3 illumination is defined as the overlap between this projected area and the PD_3 surface. The resulting illumination area normal vectors can now be defined for this subcase as

$$\vec{A}_1 = a^2 \hat{x}$$

$$\vec{A}_{12} = a^2 \frac{n_1}{2n_3} \hat{y} \tag{10.7}$$

$$\vec{A}_{123} = a^2 \frac{n_1 n_2}{2n_3^2} \hat{z}.$$

The corresponding incident power levels associated with PD_1, PD_2, and PD_3 are then found by taking the component of these illumination area normal vectors in (10.7) along the respective reflected light ray unit-normal vectors in (10.6). The result is

$$P_1 = -I_0 \hat{r}_1 \cdot \vec{A}_1 = I_0 a^2 n_1$$

$$P_{12} = -RI_0 \hat{r}_{12} \cdot \vec{A}_{12} = RI_0 a^2 \frac{n_1 n_2}{2n_3} \tag{10.8}$$

$$P_{123} = -R^2 I_0 \hat{r}_{123} \cdot \vec{A}_{123} = R^2 I_0 a^2 \frac{n_1 n_2}{2n_3}.$$

The second subcase for $n_1 < n_2 < n_3$ corresponds to the situation with the incident light illuminating PD_1, PD_3, and PD_2, in order. In this case, the respective light ray unit-normal vectors incident on the surfaces of PD_1, PD_3 and PD_2 are

$$\hat{r}_1 = -n_1 \hat{x} - n_2 \hat{y} - n_3 \hat{z}$$

$$\hat{r}_{13} = +n_1 \hat{x} - n_2 \hat{y} - n_3 \hat{z} \tag{10.9}$$

$$\hat{r}_{132} = +n_1 \hat{x} - n_2 \hat{y} + n_3 \hat{z}.$$

Again, the successive illumination areas are referred onto their adjoining neighbors, and the illumination areas are displayed in Figure 10.2b (shown for $n_3 < 2n_2$). Using the shaded areas on this figure, the illumination area normal vectors can be defined as

$$\vec{A}_1 = a^2 \hat{x}$$

$$\vec{A}_{13} = a^2 \left(\frac{n_1}{n_3} - \frac{n_1 n_2}{2n_3^2} \right) \hat{z} \tag{10.10}$$

$$\vec{A}_{132} = a^2 \begin{cases} \left(2\dfrac{n_1}{n_2} - \dfrac{n_1}{n_3} - \dfrac{n_1 n_3}{2n_2^2} \right) \hat{y}, & n_3 < 2n_2 \\[2ex] \dfrac{n_1}{n_3} \hat{y}, & n_3 > 2n_2 \end{cases}.$$

The corresponding incident power levels associated with PD_1, PD_3, and PD_2 are then found by taking the component of these illumination area normal vectors from (10.10) along the respective reflected light ray unit-normal vectors from (10.9). The result is

$$P_1 = -I_0 \hat{r}_1 \cdot \vec{A}_1 = I_0 a^2 n_1$$

$$P_{13} = -R I_0 \hat{r}_{13} \cdot \vec{A}_{13} = R I_0 a^2 \left(n_1 - \frac{n_1 n_2}{2 n_3} \right)$$

$$P_{132} = -R^2 I_0 \hat{r}_{132} \cdot \vec{A}_{132} \tag{10.11}$$

$$= R^2 I_0 a^2 \begin{cases} \left(2 n_1 - \dfrac{n_1 n_2}{n_3} - \dfrac{n_1 n_3}{2 n_2} \right), & n_3 < 2 n_2 \\[2ex] \dfrac{n_1 n_2}{n_3}, & n_3 > 2 n_2 \end{cases}.$$

The third subcase for $n_1 < n_2 < n_3$ is shown in Figure 10.2c, and it is characterized by illumination of PD_2, followed by subsequent reflections of PD_1 and PD_3. The light ray unit-normal vectors incident on the surfaces of PD_2, PD_1, and PD_3, respectively, are

$$\hat{r}_2 = -n_1 \hat{x} - n_2 \hat{y} - n_3 \hat{z}$$

$$\hat{r}_{21} = -n_1 \hat{x} + n_2 \hat{y} - n_3 \hat{z} \tag{10.12}$$

$$\hat{r}_{213} = +n_1 \hat{x} + n_2 \hat{y} - n_3 \hat{z}.$$

Using these light ray components and the resulting illumination area projections, the respective illumination area normal vectors can be written for PD_2, PD_1, and PD_3 as

$$\vec{A}_2 = a^2 \hat{y}$$

$$\vec{A}_{21} = a^2 \frac{n_2}{2 n_3} \hat{x} \tag{10.13}$$

$$\vec{A}_{213} = a^2 \frac{n_1 n_2}{2 n_3^2} \hat{z}.$$

The resulting incident power levels associated with PD_2, PD_1, and PD_3 are then found by taking the component of the illumination area normal vectors from (10.13) along the respective reflected light ray unit-normal vectors from (10.12), such that

$$P_2 = -I_0 \hat{r}_2 \cdot \vec{A}_2 = I_0 a^2 n_2$$

$$P_{21} = -R I_0 \hat{r}_{21} \cdot \vec{A}_{21} = R I_0 a^2 \frac{n_1 n_2}{2 n_3} \tag{10.14}$$

$$P_{213} = -R^2 I_0 \hat{r}_{213} \cdot \vec{A}_{213} = R^2 I_0 a^2 \frac{n_1 n_2}{2 n_3}.$$

The fourth subcase for $n_1 < n_2 < n_3$ is shown in Figure 10.2d, and it corresponds to the case for which PD_2, PD_3, and PD_1 are successively illuminated. The respective light ray unit-normal vectors incident on PD_2, PD_3, and PD_1 can be expressed as

$$\hat{r}_2 = -n_1 \hat{x} - n_2 \hat{y} - n_3 \hat{z}$$

$$\hat{r}_{23} = -n_1 \hat{x} + n_2 \hat{y} - n_3 \hat{z} \tag{10.15}$$

$$\hat{r}_{231} = -n_1 \hat{x} + n_2 \hat{y} + n_3 \hat{z},$$

and the PD_2, PD_3, and PD_1 illumination area normal vectors are

$$\vec{A}_2 = a^2\hat{y}$$

$$\vec{A}_{23} = a^2\left(\frac{n_2}{n_3} - \frac{n_1 n_2}{2n_3^2}\right)\hat{z} \qquad (10.16)$$

$$\vec{A}_{231} = a^2 \begin{cases} \left(2 - \dfrac{n_2}{n_3} - \dfrac{n_3}{2n_2}\right)\hat{x}, & n_3 < 2n_2 \\[2ex] \dfrac{n_2}{n_3}\hat{x}, & n_3 > 2n_2 \end{cases}.$$

The PD_2, PD_3, and PD_1 incident powers are then found by taking the illumination area normal vector components in (10.16) along the reflected light ray unit-normal vectors in (10.15), giving

$$P_2 = -I_0\hat{r}_2 \cdot \vec{A}_2 = I_0 a^2 n_2$$

$$P_{23} = -RI_0\hat{r}_{23} \cdot \vec{A}_{23} = RI_0 a^2\left(n_2 - \frac{n_1 n_2}{2n_3}\right)$$

$$P_{231} = -R^2 I_0\hat{r}_{231} \cdot \vec{A}_{231} \qquad (10.17)$$

$$= R^2 I_0 a^2 \begin{cases} \left(2n_1 - \dfrac{n_1 n_2}{n_3} - \dfrac{n_1 n_3}{2n_2}\right), & n_3 < 2n_2 \\[2ex] \dfrac{n_1 n_2}{n_3}, & n_3 > 2n_2 \end{cases}.$$

The fifth subcase for $n_1 < n_2 < n_3$ has illumination and reflection conditions shown in Figure 10.2e. Here, light rays are reflected in the order of PD_3, PD_1, and PD_2. The light ray unit-normal vectors incident on the surfaces of PD_3, PD_1, and PD_2 become

$$\hat{r}_3 = -n_1\hat{x} - n_2\hat{y} - n_3\hat{z}$$

$$\hat{r}_{31} = -n_1\hat{x} - n_2\hat{y} + n_3\hat{z} \qquad (10.18)$$

$$\hat{r}_{312} = +n_1\hat{x} - n_2\hat{y} + n_3\hat{z},$$

and the resulting illumination area normal vectors for illumination of PD_3, PD_1, and PD_2 are

$$\vec{A}_3 = a^2\hat{z}$$

$$\vec{A}_{31} = a^2\left(1 - \frac{n_2}{2n_3}\right)\hat{x} \qquad (10.19)$$

$$\vec{A}_{312} = a^2\frac{n_1}{2n_3}\hat{y}.$$

The resulting incident power levels recorded by PD_3, PD_1, and PD_2 for this subcase are then found from (10.18) and (10.19) to be

$$P_3 = -I_0\hat{r}_3 \cdot \vec{A}_3 = I_0 a^2 n_3$$

$$P_{31} = -RI_0\hat{r}_{31} \cdot \vec{A}_{31} = RI_0 a^2 \left(n_1 - \frac{n_1 n_2}{2n_3} \right) \tag{10.20}$$

$$P_{312} = -R^2 I_0\hat{r}_{312} \cdot \vec{A}_{312} = R^2 I_0 a^2 \frac{n_1 n_2}{2n_3}.$$

The sixth and final subcase for $n_1 < n_2 < n_3$ is displayed in Figure 10.2f. The surface of PD_3 is illuminated first, followed by successive reflections of PD_2 and PD_1. The light ray unit-normal vectors for this succession are

$$\hat{r}_3 = -n_1\hat{x} - n_2\hat{y} - n_3\hat{z}$$

$$\hat{r}_{32} = -n_1\hat{x} - n_2\hat{y} + n_3\hat{z} \tag{10.21}$$

$$\hat{r}_{321} = -n_1\hat{x} + n_2\hat{y} + n_3\hat{z}.$$

The resulting illumination area normal vectors and detected power levels for PD_3, PD_2, and PD_1 are

$$\vec{A}_3 = a^2\hat{z}$$

$$\vec{A}_{32} = a^2 \left(1 - \frac{n_1}{2n_3} \right)\hat{y} \tag{10.22}$$

$$\vec{A}_{321} = a^2 \frac{n_2}{2n_3}\hat{x}$$

and

$$P_3 = -I_0\hat{r}_3 \cdot \vec{A}_3 = I_0 a^2 n_3$$

$$P_{32} = -RI_0\hat{r}_{32} \cdot \vec{A}_{32} = RI_0 a^2 \left(n_2 - \frac{n_1 n_2}{2n_3} \right) \tag{10.23}$$

$$P_{321} = -R^2 I_0\hat{r}_{321} \cdot \vec{A}_{321} = R^2 I_0 a^2 \frac{n_1 n_2}{2n_3}.$$

The six subcases described earlier are all defined for the directional cosine case of $n_1 < n_2 < n_3$, though each differs in its successive ordering of illumination and reflection. The ultimate photocurrents observed for each of the three PDs will be a result of all the derived possible reflection permutations that culminate in their respective PD area being illuminated. For example, the total incident power level detected on PD_1 will include five components: (i) the direct incident power onto PD_1, (ii) the reflected power due to a primary reflection of PD_2, (iii) the reflected power due to a primary reflection of PD_3, (iv) the reflected power due to a primary reflection of PD_2 followed by a secondary reflection of PD_3, and (v) the reflected power due to a primary reflection of PD_3 followed by a secondary reflection of PD_2. The resulting total power permutations are

$$P_{1-\text{total}} = P_1 + P_{21} + P_{31} + P_{231} + P_{321}$$

$$P_{2-\text{total}} = P_2 + P_{12} + P_{32} + P_{312} + P_{132} \tag{10.24}$$

$$P_{3-\text{total}} = P_3 + P_{13} + P_{23} + P_{123} + P_{213},$$

respectively, where the subscripts indicate the successive illumination and reflection sequence. The corresponding photocurrents, $i_1(t)$, $i_2(t)$, and $i_3(t)$, can then be found by summing the respective subcase power levels and applying a responsivity constant of proportionality, R, that describes both the internal PD quantum efficiency and optical transmissivity at the interface. For comparison, the PD photocurrents can be normalized with respect to RI_0a^2 (representing photocurrents with normal optical incidence). For the first case of $n_1 < n_2 < n_3$, the photocurrents can be written as

$$
i_1\left(n_1 < n_2 < n_3\right) = n_1 + Rn_1 + R^2 \begin{cases} \left(2n_1 - \dfrac{n_1 n_2}{2n_3} - \dfrac{n_1 n_3}{2n_2}\right), & n_3 < 2n_2 \\[2ex] \dfrac{3n_1 n_2}{2n_3}, & n_3 > 2n_2 \end{cases}
$$

$$
i_2\left(n_1 < n_2 < n_3\right) = n_2 + Rn_2 + R^2 \begin{cases} \left(2n_1 - \dfrac{n_1 n_2}{2n_3} - \dfrac{n_1 n_3}{2n_2}\right), & n_3 < 2n_2 \\[2ex] \dfrac{3n_1 n_2}{2n_3}, & n_3 > 2n_2 \end{cases} \tag{10.25}
$$

$$
i_3\left(n_1 < n_2 < n_3\right) = n_3 + R\left(n_1 + n_2 - \frac{n_1 n_2}{n_3}\right) + R^2 \frac{n_1 n_2}{n_3}.
$$

While the earlier PD$_1$, PD$_2$, and PD$_3$ photocurrents have been defined explicitly for the directional cosine case of $n_1 < n_2 < n_3$, the process employed in deriving (10.6) through (10.24) can be modified for the remaining five directional cosine cases. Rotational and mirror symmetries can be employed in this derivation, and the resulting three photocurrent expressions can be explicitly stated for each permutation. For the second incident light ray case, the x-, z-, and y-components of the incident light ray are in ascending order of magnitudes, and the inequality $n_1 < n_3 < n_2$ leads to three normalized photocurrents with

$$
i_1\left(n_1 < n_3 < n_2\right) = n_1 + Rn_1 + R^2 \begin{cases} \left(2n_1 - \dfrac{n_1 n_3}{2n_2} - \dfrac{n_1 n_2}{2n_3}\right), & n_2 < 2n_3 \\[2ex] \dfrac{3n_1 n_3}{2n_2}, & n_2 > 2n_3 \end{cases}
$$

$$
i_2\left(n_1 < n_3 < n_2\right) = n_2 + R(n_1 + n_3 - \frac{n_1 n_3}{n_2}) + R^2 \frac{n_1 n_3}{n_2} \tag{10.26}
$$

$$
i_3\left(n_1 < n_3 < n_2\right) = n_3 + Rn_3 + R^2 \begin{cases} \left(2n_1 - \dfrac{n_1 n_3}{2n_2} - \dfrac{n_1 n_2}{2n_3}\right), & n_2 < 2n_3 \\[2ex] \dfrac{3n_1 n_3}{2n_2}, & n_2 > 2n_3 \end{cases}.
$$

For the third case, with $n_2 < n_1 < n_3$, the y-, x-, and z-components of the incident light ray are in an ascending order of magnitudes, and the three normalized photocurrents become

$$i_1\left(n_2 < n_1 < n_3\right) = n_1 + Rn_1 + R^2 \begin{cases} \left(2n_2 - \dfrac{n_1 n_2}{2n_3} - \dfrac{n_2 n_3}{2n_1}\right), & n_3 < 2n_1 \\[3mm] \dfrac{3n_1 n_2}{2n_3}, & n_3 > 2n_1 \end{cases}$$

$$i_2\left(n_2 < n_1 < n_3\right) = n_2 + Rn_2 + R^2 \begin{cases} \left(2n_2 - \dfrac{n_1 n_2}{2n_3} - \dfrac{n_2 n_3}{2n_1}\right), & n_3 < 2n_1 \\[3mm] \dfrac{3n_1 n_2}{2n_3}, & n_3 > 2n_1 \end{cases} \qquad (10.27)$$

$$i_3\left(n_2 < n_1 < n_3\right) = n_3 + R\left(n_1 + n_2 - \dfrac{n_1 n_2}{n_3}\right) + R^2 \dfrac{n_1 n_2}{n_3}.$$

For the fourth case, with $n_2 < n_3 < n_1$, the y-, z-, and x-components of the incident light ray are in an ascending order of magnitudes. The resulting three normalized photocurrents for PD_1, PD_2, and PD_3 can then be written as

$$i_1\left(n_2 < n_3 < n_1\right) = n_1 + R\left(n_2 + n_3 - \dfrac{n_2 n_3}{n_1}\right) + R^2 \dfrac{n_2 n_3}{n_1}$$

$$i_2\left(n_2 < n_3 < n_1\right) = n_2 + Rn_2 + R^2 \begin{cases} \left(2n_2 - \dfrac{n_2 n_3}{2n_1} - \dfrac{n_1 n_2}{2n_3}\right), & n_1 < 2n_3 \\[3mm] \dfrac{3n_2 n_3}{2n_1}, & n_1 > 2n_3 \end{cases} \qquad (10.28)$$

$$i_3\left(n_2 < n_3 < n_1\right) = n_3 + Rn_3 + R^2 \begin{cases} \left(2n_2 - \dfrac{n_2 n_3}{2n_1} - \dfrac{n_1 n_2}{2n_3}\right), & n_1 < 2n_3 \\[3mm] \dfrac{3n_2 n_3}{2n_1}, & n_1 > 2n_3 \end{cases} .$$

For the fifth case, with $n_3 < n_1 < n_2$, the z-, x-, and y-components of the incident light ray are in an ascending order of magnitudes, and the normalized PD_1, PD_2, and PD_3 photocurrents become

$$i_1\left(n_3 < n_1 < n_2\right) = n_1 + Rn_1 + R^2 \begin{cases} \left(2n_3 - \dfrac{n_1 n_3}{2n_2} - \dfrac{n_2 n_3}{2n_1}\right), & n_2 < 2n_1 \\[3mm] \dfrac{3n_1 n_3}{2n_2}, & n_2 > 2n_1 \end{cases}$$

$$i_2\left(n_3 < n_1 < n_2\right) = n_2 + R\left(n_1 + n_3 - \dfrac{n_1 n_3}{n_2}\right) + R^2 \dfrac{n_1 n_3}{n_2} \qquad (10.29)$$

$$i_3\left(n_3 < n_1 < n_2\right) = n_3 + Rn_3 + R^2 \begin{cases} \left(2n_3 - \dfrac{n_1 n_3}{2n_2} - \dfrac{n_2 n_3}{2n_1}\right), & n_2 < 2n_1 \\[3mm] \dfrac{3n_1 n_3}{2n_2}, & n_2 > 2n_1 \end{cases} .$$

For the sixth and final case, with $n_3 < n_2 < n_1$, the z-, y-, and x-components of the incident light ray are in an ascending order of magnitudes. The resulting three normalized photocurrents are

$$i_1\left(n_3 < n_2 < n_1\right) = n_1 + R\left(n_2 + n_3 - \frac{n_2 n_3}{n_1}\right) + R^2 \frac{n_2 n_3}{n_1}$$

$$i_2\left(n_3 < n_2 < n_1\right) = n_2 + Rn_2 + R^2 \begin{cases} \left(2n_3 - \dfrac{n_2 n_3}{2n_1} - \dfrac{n_1 n_3}{2n_2}\right), & n_1 < 2n_2 \\[2ex] \dfrac{3n_2 n_3}{2n_1}, & n_1 > 2n_2 \end{cases}$$ (10.30)

$$i_3\left(n_3 < n_2 < n_1\right) = n_3 + Rn_3 + R^2 \begin{cases} \left(2n_3 - \dfrac{n_2 n_3}{2n_1} - \dfrac{n_1 n_3}{2n_2}\right), & n_1 < 2n_2 \\[2ex] \dfrac{3n_2 n_3}{2n_1}, & n_1 > 2n_2 \end{cases}.$$

The PD_1, PD_2, and PD_3 photocurrents earlier are summarized in Table 10.1 for the six light ray directional cosine component inequalities. These analytic expressions have been confirmed with a brute-force MATLAB® ray-tracing program (data not shown), in which the structure is illuminated by a fine mesh of light rays, and the three photocurrents are recorded for all possible incident angle combinations.

10.3.3 OPTICAL ACTIVE CONTROL CHARACTERISTIC ANALYSES

In the previous two subsections, the reflection and detection characteristics of the corner-cube geometry were explored for the orthogonal PD structure. It is shown in this section that the complete retrodetector construction can be used as an active control device for optimizing the incident optical signal alignment, as a strong dependence exists among the individual PD photocurrent levels and the incident light ray components. This dependence is readily apparent in Figure 10.3a–c, where the three normalized theoretical photocurrents from (10.25) through (10.30) are displayed. The photocurrents here are plotted as a function of the azimuthal angle ϕ and polar angle θ. Notice from this figure that the photocurrent maxima for each of PD_1, PD_2, and PD_3 correspond to situations in which the respective PD surface is orthogonal to the incident light rays and the greatest amount of light is absorbed. Likewise, negligible photocurrents are observed when the respective PD surface is parallel to the incident light rays and the least amount of light is absorbed.

The directionality of the three PD photocurrents can be used for active control and optimization through a combination of differential sums. If, for example, PD_1 and PD_2 have the same photocurrent level, it can be expected that the incident light rays are balanced between the xz- and yz-planes. This alignment corresponds to the situation for which the optical source lies along the $\phi = 45°$ plane. The same directionality arguments can be made by quantifying and balancing the differential signals between PD_2 and PD_3 and then PD_1 and PD_3.

The complete procedure for balancing and optimizing the photocurrent signals can be observed most easily by summing the three differential sum magnitudes. The result is shown as a surface in Figure 10.3d, where this photocurrent differential sum is defined as

$$i_{\text{diff}}(\phi,\theta) = \left|i_3(\phi,\theta) - i_2(\phi,\theta)\right| + \left|i_3(\phi,\theta) - i_1(\phi,\theta)\right| + \left|i_2(\phi,\theta) - i_1(\phi,\theta)\right|.$$ (10.31)

Minimizing this photocurrent differential sum corresponds to an optimization of the angular alignment with the balanced condition achieved when the absolute minimum of the displayed surface

TABLE 10.1

Theoretical Normalized Differential Photocurrents for PD$_1$, PD$_2$, and PD$_3$, Given Various Directional Cosine Component Inequality Permutations

Cases	PD$_1$ Normalized Photocurrents	PD$_2$ Normalized Photocurrents	PD$_3$ Normalized Photocurrents
$n_1 < n_2 < n_3$	$n_1 + Rn_1 + R^2 \begin{cases} \left(2n_1 - \dfrac{n_1 n_2}{2n_3} - \dfrac{n_1 n_3}{2n_2}\right), & n_3 < 2n_2 \\ \dfrac{3n_1 n_2}{2n_3}, & n_3 > 2n_2 \end{cases}$	$n_2 + Rn_2 + R^2 \begin{cases} \left(2n_1 - \dfrac{n_1 n_2}{2n_3} - \dfrac{n_1 n_3}{2n_2}\right), & n_3 < 2n_2 \\ \dfrac{3n_1 n_2}{2n_3}, & n_3 > 2n_2 \end{cases}$	$n_3 + R\left(n_1 + n_2 - \dfrac{n_1 n_2}{n_3}\right) + R^2 \dfrac{n_1 n_2}{n_3}$
$n_1 < n_3 < n_2$	$n_1 + Rn_1 + R^2 \begin{cases} \left(2n_1 - \dfrac{n_1 n_3}{2n_2} - \dfrac{n_1 n_2}{2n_3}\right), & n_2 < 2n_3 \\ \dfrac{3n_1 n_3}{2n_2}, & n_2 > 2n_3 \end{cases}$	$n_2 + R\left(n_1 + n_3 - \dfrac{n_1 n_3}{n_2}\right) + R^2 \dfrac{n_1 n_3}{n_2}$	$n_3 + Rn_3 + R^2 \begin{cases} \left(2n_1 - \dfrac{n_1 n_3}{2n_2} - \dfrac{n_1 n_2}{2n_3}\right), & n_2 < 2n_3 \\ \dfrac{3n_1 n_3}{2n_2}, & n_2 > 2n_3 \end{cases}$
$n_2 < n_1 < n_3$	$n_1 + Rn_1 + R^2 \begin{cases} \left(2n_2 - \dfrac{n_1 n_2}{2n_3} - \dfrac{n_2 n_3}{2n_1}\right), & n_3 < 2n_1 \\ \dfrac{3n_1 n_2}{2n_3}, & n_3 > 2n_1 \end{cases}$	$n_2 + Rn_2 + R^2 \begin{cases} \left(2n_2 - \dfrac{n_1 n_2}{2n_3} - \dfrac{n_2 n_3}{2n_1}\right), & n_1 < 2n_3 \\ \dfrac{3n_1 n_2}{2n_3}, & n_1 > 2n_3 \end{cases}$	$n_3 + R\left(n_1 + n_2 - \dfrac{n_1 n_2}{n_3}\right) + R^2 \dfrac{n_1 n_2}{n_3}$
$n_2 < n_3 < n_1$	$n_1 + R\left(n_2 + n_3 - \dfrac{n_2 n_3}{n_1}\right) + R^2 \dfrac{n_2 n_3}{n_1}$	$n_2 + Rn_2 + R^2 \begin{cases} \left(2n_2 - \dfrac{n_2 n_3}{2n_1} - \dfrac{n_1 n_2}{2n_3}\right), & n_1 < 2n_3 \\ \dfrac{3n_2 n_3}{2n_1}, & n_1 > 2n_3 \end{cases}$	$n_3 + Rn_3 + R^2 \begin{cases} \left(2n_2 - \dfrac{n_2 n_3}{2n_1} - \dfrac{n_1 n_2}{2n_3}\right), & n_1 < 2n_3 \\ \dfrac{3n_2 n_3}{2n_1}, & n_1 > 2n_3 \end{cases}$
$n_3 < n_1 < n_2$	$n_1 + Rn_1 + R^2 \begin{cases} \left(2n_3 - \dfrac{n_1 n_3}{2n_2} - \dfrac{n_2 n_3}{2n_1}\right), & n_2 < 2n_1 \\ \dfrac{3n_1 n_3}{2n_2}, & n_2 > 2n_1 \end{cases}$	$n_2 + R\left(n_1 + n_3 - \dfrac{n_1 n_3}{n_2}\right) + R^2 \dfrac{n_1 n_3}{n_2}$	$n_3 + Rn_3 + R^2 \begin{cases} \left(2n_3 - \dfrac{n_1 n_3}{2n_2} - \dfrac{n_2 n_3}{2n_1}\right), & n_2 < 2n_1 \\ \dfrac{3n_1 n_3}{2n_2}, & n_2 > 2n_1 \end{cases}$
$n_3 < n_2 < n_1$	$n_1 + R\left(n_2 + n_3 - \dfrac{n_2 n_3}{n_1}\right) + R^2 \dfrac{n_2 n_3}{n_1}$	$n_2 + Rn_2 + R^2 \begin{cases} \left(2n_3 - \dfrac{n_2 n_3}{2n_1} - \dfrac{n_1 n_3}{2n_2}\right), & n_1 < 2n_2 \\ \dfrac{3n_2 n_3}{2n_1}, & n_1 > 2n_2 \end{cases}$	$n_3 + Rn_3 + R^2 \begin{cases} \left(2n_3 - \dfrac{n_1 n_3}{2n_2} - \dfrac{n_1 n_2}{2n_3}\right), & n_1 < 2n_2 \\ \dfrac{3n_2 n_3}{2n_1}, & n_1 > 2n_2 \end{cases}$

Source: Jin, X., *IEEE Sensors J.*, 2010. With permission.

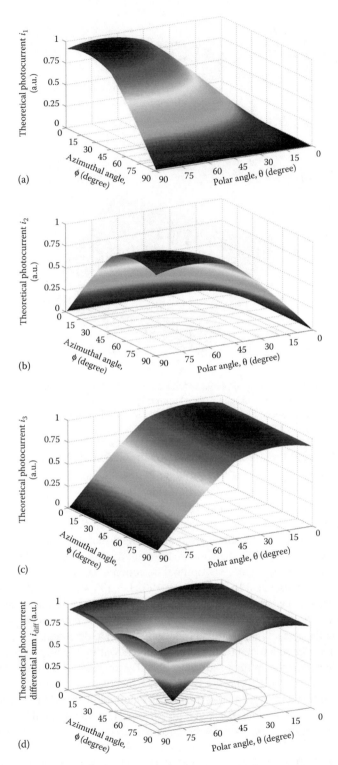

FIGURE 10.3 Theoretical photocurrents are shown as surfaces varying with ϕ and θ. Independent and normalized photocurrents are displayed for (a) P_1, (b) P_2, and (c) P_3. (d) Complete photocurrent differential sum.

(continued)

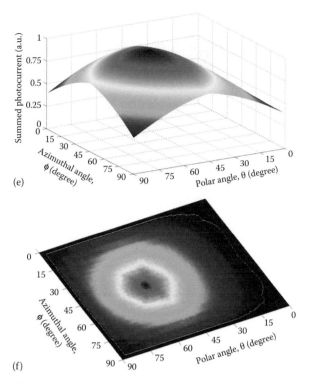

(e)

(f)

FIGURE 10.3 (continued)   Theoretical photocurrents are shown as surfaces varying with ϕ and θ. The normalized photocurrents are displayed for (e) absolute summed photocurrent, and (f) a 2D view of retroreflected power are also shown and produced by the analytical expressions derived in this chapter. (From Jin, X., *IEEE Sensors J.*, 2010. With permission.)

is reached. The systematic optimization introduced in Section 10.2 will bring the retrodetector alignment to the $\phi = 45°$ and $\theta = \cos^{-1}(1/\sqrt{3}) \approx 54.7°$ absolute minimum of Figure 10.3d surface with the incident light rays entering the structure along the $\hat{r} = -\hat{x} - \hat{y} - \hat{z}$ direction. This balanced orientation is aligned for optimal communications between the optical source and retrodetector.

From a signal detection standpoint, the summed photocurrent is recorded by combining the individual PD photocurrents. The absolute minimum of the differential photocurrent sum concavity in Figure 10.3d is then transferred to the absolute maximum of the summed photocurrent convexity of Figure 10.3e, and this orientation gives optimal detection conditions. In contrast to this, differential photocurrent sum maxima at $(\phi, \theta) = (0°, 0°)$, $(90°, 0°)$, $(0°, 90°)$, and $(90°, 90°)$ correspond to minima and poor detection conditions for the summed photocurrent minima at these extreme orientations. The retrodetection optimization procedure leading to an orientation of $\phi = 45°$ and $\theta \approx 54.7°$ will therefore bring the system into alignment for optimal signal detection.

From a signal retroreflection standpoint, the alignment optimization procedure will bring the system into an orientation for optimal retroreflection. The theoretical retroreflected power of this retrodetector is shown as a surface varying with ϕ and θ in a two-dimensional view in Figure 10.3f, and its signal strength is shown as a color map. It is readily apparent that the maximum retroreflected power (corresponding to the dark red regions) appears at the optimal orientation of $\phi = 45°$ and $\theta \approx 54.7°$. In contrast to this, the dark blue regions around the periphery at $(\phi, \theta) = (0°, 0°)$, $(90°, 0°)$, $(0°, 90°)$, and $(90°, 90°)$ correspond to near-zero retroreflection from the retrodetector. Ultimately, the described differential triangulation and optimization procedure can bring the system into optimal alignment for bidirectional FSO communication with enhanced detection and retroreflection.

10.4 SENSING APPLICATION

In the previous section, the theory and methodology for optical retroreflection, photodetection, and active control were described for differential triangulation in the proposed retrodetector. The retrodetector is now tested experimentally by way of a broadband LED optical broadcasting configuration and a monochromatic laser configuration with uniform illumination. In Sections 10.4.1 and 10.4.2, the FSO remote sensing capability and bidirectional communication capability of the proposed retrodetector architecture are systematically investigated.

10.4.1 FREE-SPACE OPTICAL REMOTE SENSING APPLICATION

Figure 10.1b setup is employed to test the retrodetector system. A broadband LED source uniformly illuminates all the retrodetector surfaces. A 5 m propagation distance is used. For angular testing purposes, the retrodetector is mounted in a gyroscope that allows for independent rotations in the ϕ and θ orientations. Phase-sensitive detection and/or electronic filtering can be employed as needed, to lock the PD signals into the modulation frequencies of interest. This filtering improves the sensitivity of the setup, removes extraneous/background optical signals, and compensates for the reduced reflectivity of the semiconductor surfaces (compared to, for example, metal mirror finishes). Interestingly, the cubic nature of the retrodetector architecture also lends itself to arrayed detector distributions with interleaved retrodetectors distributed across surfaces. Such periodic arrays can be scaled up to increase the overall signal amplitude without sacrificing the individual PD response (RC) times.

To validate the differential detection and optimization process inherent to the corner-cube PD assembly, the retrodetector is here tested for angular characteristics. The PD_1, PD_2, and PD_3 photocurrents are recorded by a digital acquisition system as the azimuthal and polar angles are scanned in $10°$ increments over the ranges $0° \leq \phi \leq 90°$ and $0° \leq \theta \leq 90°$, respectively. The results are shown in Figure 10.4a–c. The resemblance of these experimental results for PD_1, PD_2, and PD_3 to their theoretical counterparts in Figure 10.3a–c is immediately apparent. The results are in excellent agreement. The PD_1 photocurrent in Figure 10.4a rises from negligible levels at large azimuthal angles ($\phi \approx 90°$) and small polar angles ($\theta \approx 0°$) to a maximum of 70 μA when ϕ is small and θ is large. The PD_2 photocurrent in Figure 10.4b rises from negligible levels at small azimuthal angles ($\phi \approx 0°$) and small polar angles ($\theta \approx 0°$) to a maximum of 68.6 μA when ϕ and θ are both large. The PD_3 photocurrent in Figure 10.4c is largely independent of the azimuthal angle and rises from negligible levels at large polar angles ($\theta \approx 90°$) to a maximum of 72.9 μA when θ is small. These rotational features are all apparent qualitatively by visualizing the PD illumination for angular extremes in the Figure 10.1a coordinate system.

Having collected the independent photocurrents for PD_1, PD_2, and PD_3, values for the photocurrent differential sum can be recorded and the distribution can be used to optimize the alignment. The experimental photocurrent differential sum results are shown in Figure 10.4d as the azimuthal and polar angles are scanned in $10°$ increments over the ranges $0° \leq \phi \leq 90°$ and $0° \leq \theta \leq 90°$, respectively. The resemblance of this surface to the theoretical curve defined by (10.31) and displayed in Figure 10.3d is clear. The theoretical photocurrent analyses and related assumptions (e.g., angular-independent surface reflectivity) appear to be validated, as the experimental data points in Figure 10.4d deviate from those of Figure 10.3d with a standard deviation below 10 μA. The observed deviations are greatest for orientations with large, i.e., glancing, angles of incidence of one or more PDs, though such deviations at the angular extremes are well within the required accuracy for the retrodetector alignment process (as the overall surface concavity of the differential distribution is used minimize the photocurrent differential sum and optimize the structure's alignment).

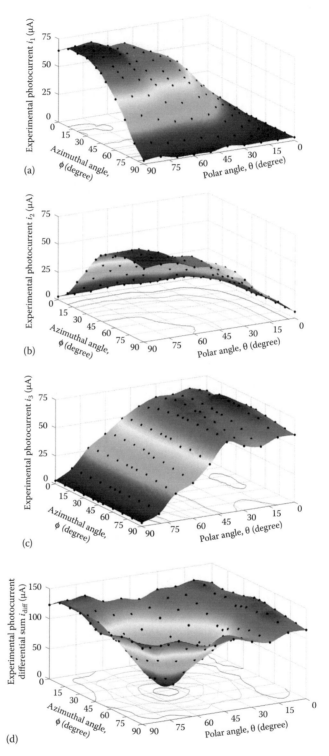

FIGURE 10.4 Experimental photocurrent surfaces varying with ϕ and θ are shown. Results for (a) P_1, (b) P_2, and (c) P_3, corresponding to PD_1, PD_2, and PD_3, respectively, are present and the complete photocurrent differential cum is shown in (d).

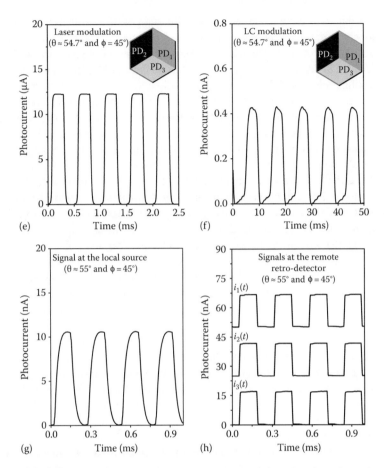

FIGURE 10.4 (continued) Experimental photocurrent surfaces varying with ϕ and θ are shown. The photocurrents for (e) active downlink mode and (f) passive uplink mode are shown as a function of time for an FSO laser setup, while the photocurrent results for (g) local and (h) remote detection (with $i_1(t)$ (shift up 50 nA), $i_2(t)$ (shift up 25 nA), and $i_3(t)$, respectively) are shown for an FSO LED setup. Both of those setups are toward the remote retrodetector at the optimal $\phi \approx 45°$ and $\theta \approx 55°$ orientation. (From Jin, X., *IEEE Sensors J.*, 2010. With permission.)

10.4.2   FREE-SPACE OPTICAL BIDIRECTIONAL COMMUNICATION APPLICATION

The proposed integrated retrodetector has experimentally demonstrated our desired ability for active real-time optimization of optical beam alignments in the previous subsection. This active control is also a useful extension for FSO bidirectional communication systems in general. The FSO bidirectional communication capability is therefore verified here for this control structure, and the balanced condition achieved by minimizing the photocurrent differential sum, at the azimuthal and polar angles of $\phi \approx 45°$ and $\theta \approx 54.7°$ with the incident light entering the structure along the $\hat{r} = -\hat{x} - \hat{y} - \hat{z}$ direction, leaves the structure in an orientation that is optimized for FSO bidirectional communication with a monochromatic laser configuration or a broadband LED optical broadcasting configuration.

The laser-based FSO bidirectional communication setup is built first here. The 1 mW, 650 nm monochromatic laser diode is employed and expanded to create sufficiently uniform collimated beams directing at the retrodetector so that the optical signals are recorded remotely. Both ϕ and θ angles are optimized with forementioned differential triangulation, and $(x, y, z) = (1, 1, 1)$ orientation becomes aligned toward the laser source. Both active downlink and passive uplink schemes are then demonstrated successively. The active downlink mode is investigated first with the laser modulation at 2 kHz.

The Pi-cell LC modulator is mounted at the entrance interface of the retrodetector device and remains in its on-state (open) at this investigation. The summed photocurrent, $i_1(t) + i_2(t) + i_3(t)$, is shown in Figure 10.4e as a function of time. The modulated characteristics of the laser transmitter are seen to be effectively detected by the retrodetector. The alternative form of FSO communication is that of the passive uplink configuration. As it is fundamentally bidirectional in nature, a biased silicon photodetector is located at the laser source side to monitor the retroreflected signal. The retrodetector is then illuminated by the continuous-wave laser with the entrance Pi-cell modulator operating from a biased Pi-state condition with a modulation frequency of 100 Hz. The resulting retroreflected signal level recorded by the photodetector is shown in Figure 10.4f as a function of time. Note that the Pi-cell LC modulator effectively maps the remotely encoded information onto the retroreflected beam. The signal level and modulation depth are seen to be appreciable with a signal-to-noise ratio (SNR) beyond 40 dB. This enhanced signal is due, in large part, to the previous beam alignment optimization procedure.

The retrodetector of interest to this investigation is also tested with a broadband LED optical broadcasting scheme. Carrying out the desired photocurrent differential sum minimization and optimization orients the silicon PD retrodetector at the azimuthal and polar angles of $\phi \approx 45°$ and $\theta \approx 54.7°$, respectively. Different from the previous subsection having individual transceiver (laser) and receiver (retrodetector with Pi-cell LC modulator) modulation, this LED-based FSO bidirectional communication investigation only employs transceiver (LED) modulation schemes with local and remote detection. A broadband LED local source is modulated at 4 kHz and broadcasted toward the retrodetector. After 4 m propagation, the light is reflected by the remote retrodetector and returned directly back to the local source, where it is sampled by a photodetector. The photocurrent signal recorded at the local source is shown in Figure 10.4g for the balanced orientation of $\phi \approx 45°$ and $\theta \approx 55°$. The 4 kHz retroreflected signal is immediately apparent in this figure. For a broadcasted optical power of 0.25 mW, a 10.6 nA local signal is detected with an SNR of ~40 dB. The low noise levels and large signal strength (well above the estimated 0.1 nA PD sensitivity) show the benefits of the electronic bandpass filter stage, as it filters external noise sources and removes the contribution from background light levels. The same signal modulation can be seen at the remote retrodetector in Figure 10.4h, where the PD_1, PD_2, and PD_3 photocurrent signals are displayed as a function of time. The remote signal levels here are well balanced with an SNR of ~50 dB, leading to a corresponding local signal that is maximized and relatively insensitive to extraneous reflections in the environment.

10.5 CONCLUSION

A new retrodetection technique was introduced through the merging of retroreflection and detection in an integrated optical package. The presented retrodetector met the requirements for FSO sensing in a bidirectional format. Moreover, it was shown that differential combinations of the local photocurrent signals could be used as a control mechanism to optimize the structure's alignment. The theory and methodology for the optical reflection, photodetection, and active control characteristics were presented and subsequently verified experimentally with a silicon PD retrodetection prototype. Excellent agreement was found between the theory and experimental angular characterizations, and an active remote sensing configuration was tested for the optimized alignment. The local and remote sensing capabilities of the retrodetector were found to be successful.

REFERENCES

1. Y. H. Lee and J. K. Lee, Passive remote sensing of three-layered anisotropic random media, in *Proceedings of IEEE IGARSS Conference*, Tokyo, Japan, 1993, vol. 1, pp. 249–251.
2. S. H. Yueh, R. Kwok, F. K. Li, S. V. Nghiem, W. J. Wilson, and J. A. Kong, Polarimetric passive remote sensing of wind-generated sea surfaces and ocean wind vectors, in *Proceedings of Oceans'93 'Engineering in Harmony with Ocean'*, Victoria, Canada, 1993, vol. 1, pp. 131–136.

3. P. Sharma, I. S. Hudiara, and M. L. Singh, Passive remote sensing of a buried object using a 29.9 GHz radiometer, in *Proceedings of Asia-Pacific Microwave Conference*, Suzhou, China, 2005, vol. 1, pp. 2–3.

4. R. D. Thom, T. L. Koch, J. D. Langan, and W. J. Parrish, A fully monolithic InSb infrared CCD array, *IEEE Trans. Electron Dev.*, 27(1), 160–170, 1980.

5. C. R. Sharma, C. Furse, and R. R. Harrison, Low-power STDR CMOS sensor for locating faults in aging aircraft wiring, *IEEE Sensors J.*, 7(1), 43–50, 2007.

6. X. Y. Kong, GPS modeling in frequency domain, in *Proceedings of IEEE Second International Conference on WBUWC*, Sydney, Australia, 2007, pp. 61–61.

7. Y. Zhou, J. Schembri, L. Lamont, and J. Bird, Analysis of stand-alone GPS for relative location discovery in wireless sensor networks, in *Proceedings of IEEE CCECE Conference*, St. John's, Canada, 2009, pp. 437–441.

8. J. K. Lee and J. A. Kong, Active microwave remote sensing of an anisotropic random medium layer, *IEEE Trans. Geosci. Remote Sens.*, GE-23, 910–923, 1985.

9. G. R. Allan, H. Riris, J. B. Abshire, X. L. Sun, E. Wilson, J. F. Burris, and M. A. Krainak, Laser sounder for active remote sensing measurements of CO_2 concentrations, in *Proceedings of IEEE Aerospace Conference*, Big Sky, USA, 2008, pp. 1–7.

10. M. A. Zuniga, T. M. Habashy, and J. A. Kong, Active remote sensing of layered random media, *IEEE Trans. Geosci. Remote Sens.*, 17(4), 296–302, 1979.

11. M. Steinhauer, H. O. Ruo, H. Irion, and W. Menzel, Millimeter-wave-radar sensor based on a transceiver array for automotive applications, *IEEE Trans. Microwave Theory Tech.*, 56(2), 261–269, 2008.

12. H. Dominik, Short range radar-status of UWB sensors and their applications, in *Proceedings of IEEE European Microwave Conference*, Munich, Germany, 2007, pp. 1530–1533.

13. T. Dorney, J. Johnson, D. Mittleman, and R. Baraniuk, Imaging with THz pulses, in *Proceedings of IEEE International Conference Image Processing*, Rochester, USA, 2002, vol. 1, pp. 764–767.

14. R. W. Mcgowan, R. A. Cheville, and D. Grischkowsky, Direct observation of the Gouy phase shift in THz impulse ranging, *Appl. Phys. Lett.*, 76(6), 670–672, 2000.

15. B. Warneke, M. Last, B. Liebowitz, and K. S. J. Pister, Smart dust: Communicating with a cubic-millimeter computer, *Computer*, 34(1), 44–51, 2001.

16. L. Zhou, J. M. Kahn, and K. S. J. Pister, Corner-cube retroreflectors based on structure-assisted assembly for free-space optical communication, *J. Microelectromech. Syst.*, 12(3), 233–242, 2003.

17. D. C. O' Brien, W. W. Yuan, J. J. Liu, G. E. Faulkner, S. J. Elston, S. Collins, and L. A. Parry-Jones, Optical wireless communications for micro-machines, *Proc. SPIE*, 6304, 63041A1-8, 2006.

18. W. S. Rabinovich, R. Mahon, H. R. Burris, G. C. Gilbreath, P. G. Goetz, C. I. Moore, M. F. Stell, M. J. Vilcheck, J. L. Witkowsky, L. Swingen, M. R. Suite, E. Oh, and J. Koplow, Free-space optical communications link at 1550 nm using multiple-quantum-well modulating retroreflectors in a marine environment, *Opt. Eng.*, 44(5), 056001-003, 2005.

19. D. A. Humphreys and A. J. Moseley, GaInAs photodiodes as transfer standards for picoseconds measurements, *IET Proc. J. Optoelectron.*, 135(2), 146–152, 1988.

20. D. H. Auston, Picosecond optoelectronic switching and gating in silicon, *Appl. Phys. Lett.*, 26, 101–103, 1975.

21. R. T. Howard, A. F. Heaton, R. M. Pinson, and C. K. Carrington, Orbital express advanced video guidance sensor, in *Proceedings of IEEE Aerospace Conference*, Big Sky, USA, 2008, pp. 1–10.

22. A. Makynen and J. Kostamovaara, Optimization of the displacement sensing precision of a reflected beam sensor in outdoor environment, in *Proceedings of 21th IEEE IMT Conference*, Como, Italy, 2004, vol. 2, pp. 1001–1004.

23. C. M. Collier, X. Jin, J. F. Holzman, and J. Cheng, Omni-directional characteristics of composite retroreflectors, *J. Opt. A. Pure Appl. Opt.*, 11, 085404(1–10), 2009.

24. X. Jin and J. F. Holzman, "Differential retro-detection for remote sensing applications," *IEEE Sensors J.*, vol. 10, no. 12, pp. 1875–1883, 2010.

Part II

Infrared and Thermal Sensors

11 Measurement of Temperature Distribution in Multilayer Insulations between 77 and 300 K Using Fiber Bragg Grating Sensor

Rajini Kumar Ramalingam and Holger Neumann

CONTENTS

11.1 INTRODUCTION

The branches of physics and engineering that involve the study of very low temperatures, how to produce them, and how materials behave at those temperatures are generally termed as cryogenics [1]. The word cryogenics comes from the Greek word "kryos," meaning cold, combined with a shortened form of the English verb "to generate," it has come to mean the generation of temperatures well below those of normal human experience. In a more operational way, it can also be defined as the science and technology of temperatures below 120 K [2]. The later definition depends on the characteristic temperatures of the cryogenic fluids. The limit temperature of 120 K comprehensively includes the normal boiling points of the main atmospheric gases, as well as of methane which constitutes the principal component of natural gas which begins to liquefy. Cryogenic technology has wide spectrum of application areas like

1. Gas technology
   a. Gas breakdown
   b. Liquefaction
   c. Production, storage, and transport of industrial gas
2. Electrotechnology
   a. Application of superconductive technologies in
      i. Energy management
      ii. Energy transformation, e.g., in turbo generators, motors, fusion reactors, magneto-hydrodynamic generators
      iii. Energy transport in superconductive cables
      iv. Superconductive magnet energy storage (SMES)
   b. Magnetic separation/conditioning (ore, minerals, contamination, e.g., in water)
   c. Traffic engineering (magnetic levitation train, electrodynamic systems)
   d. Nuclear magnetic resonance (NMR) technique (tomography, spectroscopy) for medical, biological, and chemical investigation
   e. Telecommunication engineering (e.g., low noise amplification, IR-radiation detectors)
   f. Microwave engineering
3. High-energy physics
   a. Particle accelerators
   b. Particle detectors
4. Vacuum technique
   a. Cryo vacuum pumps
5. Space technology
   a. Rocket systems
   b. Satellite, space telescopes, and ground stations
   c. Space simulation chambers
6. Fusion technology
   a. Superconductive magnets for plasma embedding
   b. Superconductive magnets for plasma heater technology with gyrotrons
   c. Cryo pumps for plasma heater technology with neutral injection
   d. Cryo vacuum pumps for evacuating the plasma
7. Biology and medicine
   a. Cryo probe in urology
   b. Cryo scalpel in surgery
   c. Tissue displacement, cancer therapy
   d. Conservation/storage of cell structures, blood plasma, sperm
   e. NMR spectroscopy for medical, biological, chemical, and biochemical investigation by the use of superconductive magnets
8. Recycling industry
   a. Separation process, recycling process by the use of LN2 (using different thermal contractions of the materials and the different brittleness)

9. Environmental protection
    a. SO_2 separation from waste gas by low temperature techniques
    b. Retention of radioactive substrates from nuclear power plants and recycling plants by low temperature rectification
    c. Waste air purification and solvent recovery by low temperature condensation
10. Liquid hydrogen technology
    a. Liquid hydrogen as energy carrier, e.g., for automotive applications
11. Electronics, microelectronics
    a. Sensitive electronic elements/measurement devices, low noise amplifier (e.g., space detectors for astronomy)
    b. Squids as extremely sensitive digital magnetometer, voltmeter (e.g., for magneto cardiogram, extremely sensitive material testing)
    c. Fast computer (alternative semiconductor by superconductive Josephson contacts)
    d. Miniaturized high voltage elements for satellites and mobile phones (antenna, resonators, filters, etc.)

All the earlier mentioned applications have to use cryogenic fluids which involve in storage, handling, and transferring from one point to another. In order to minimize the evaporation rate of the cryogenic fluids, an optimal thermal insulation has to be ensured. Hence, thorough knowledge of thermal insulation is a key part of enabling the development of efficient, low-maintenance cryogenic systems.

11.2   MULTILAYER INSULATION

Multilayer insulation (MLI) (also referred as superinsulation), which was developed in the 1950s by Peterson (Sweden) and first established in the 1960s by the space industry, is a key component in the reduction of heat load to cryogenic systems due to thermal radiation.

MLI consists of series of reflecting layers. Spacer elements with low heat conductivity are positioned between two reflecting layers. The whole MLI works under high vacuum conditions to prevent convection and minimize heat transfer by residual gas conduction. Generally speaking, for ideal systems the reduction in thermal radiation heat leak scales as $1/(N + 1)$ where N is the number of layers in the MLI. However, most systems are far from ideal and great attention to detail is required to use the MLI properly.

The reflecting layers consist mostly of aluminum-metalized Mylar films and sometimes of pure aluminum foils if the MLI should be not flammable. The spacer elements are mostly a net of glass fibers or foils. But also paper, polyester, tulle, or silk is used as spacer material. Some manufactures produces a unit of reflector and spacer with metalized Mylar films which are crinkled or embossed to reduce the contact surface between the reflecting layers without spacer element.

The high temperature difference between cryogenic and environmental temperatures, the very small latent heat value for most of the cryogenic fluids, and the need of very high-energy input for generating low temperatures necessitate the extremely high quality of thermal insulation. So an economical low temperature operation requires optimal thermal insulation which corresponds to minimal heat loads from room temperature to cryogenic temperature.

11.2.1   HEAT TRANSFER THROUGH MLI

The heat transfer modes through MLI have complex interactions and also dependent on variety of parameters such as number of layers, layer density, contact pressure and area, boundary temperatures, gas pressure within the insulation, emissivity of the shields, absorption and scattering coefficients of thermal spacers, etc. Further, due to the highly anisotropic characteristics of the radiation shields, there exists a coupling between longitudinal and lateral heat conduction in MLI [3].

As a performance index parameter, a Fourier law–type conductivity coefficient called "apparent" or "effective" thermal conductivity coefficient (λ) is often used to evaluate the MLI effectiveness and to optimize the insulation parameters for specific applications. Since the thickness of the insulation is small compared to physical dimensions of cryogenic system, the heat transfer in MLI can be approximated to one-dimensional for heat flow normal to the surface. Then

$$\lambda = \frac{Q \times \delta}{A \times (T_h - T_c)} \tag{11.1}$$

where
  Q is the heat flow through the insulation in the normal direction (W)
  δ is the insulation thickness (m)
  A is the area (m^2)
  T_h and T_c are the warm and cold boundaries (K)

11.2.2 Measurement of Heat Flux in MLI

Thermal effectiveness of MLI can be commonly evaluated using unsteady-state methods and steady-state methods. In unsteady-state methods sample does not require being in thermal equilibrium. The thermal diffusivity of the insulation sample can be estimated quickly from the cooling rate of the sample and this provides an indirect estimation of thermal conductivity with poor accuracy [4]. On other hand, the steady-state method ensures the thermal equilibrium. There are many major steady-state methods used in the estimation of heat flux/effective thermal conductivity. Boil-off calorimetry is the most commonly used method which is also used in the present work to estimate the heat flux. When the insulation is applied over the test section of a calorimeter containing a cryogenic fluid, the heat transport across the insulation can be estimated from the evaporation rate of the cryogen at steady-state conditions, provided secondary heat currents are eliminated. However to estimate heat flux with a good degree of accuracy, correction has to be made for the factors like unaccounted vapor fraction of evaporation [5], effect due to sensible heat increase of the vapor fraction [6,7], effects due to stratification and superheating [8], measurement error, edge effect, and lateral heat conduction [9].

11.2.3 Theoretical Estimation of Heat Transfer

The theoretical formulation of heat transfer through MLI is subject of many publications [10–13]. Due to unpredictable changes in parameters such as winding pressure, uniform contact pressure, and interstitial pressure, accurate theoretical prediction of MLI performance is very difficult [14].

In steady-state condition, the total heat transfer ($\dot{Q}_{total}$) is the sum of the radiation heat ($\dot{Q}_{radiation}$), the solid heat conduction ($\dot{Q}_{s,cond}$), and the residual gas conduction ($\dot{Q}_{g,cond}$). This total heat transfer is constant between the single layers but the amount of the single parts depends on the neighboring layers on either side:

$$\dot{Q}_{total} = \dot{Q}_{total_{i,j+1}}$$
$$\dot{Q}_{total} = \dot{Q}_{radiation_{i,j+1}} + \dot{Q}_{s,cond_{i,j+1}} + \dot{Q}_{g,cond_{i,j+1}} \tag{11.2}$$

For the numerical calculation of the temperatures of the single layers, the single heat transfer parts can be described as follows:

$$\dot{Q}_{radiation_{i,j+1}} = \frac{\sigma}{\frac{1}{\varepsilon_i} + \frac{1}{\varepsilon_{i+1}} - 1} \times \left(T_i^4 - T_{i+1}^4\right) \times (1-f) \times A_i \tag{11.3}$$

$$\dot{Q}_{s,\text{cond}_{i,j+1}} = \frac{\lambda_{i,i+1}}{s} \times (T_i - T_{i+1}) \times f \times C \times A_i \tag{11.4}$$

$$\dot{Q}_{g,\text{cond}_{i,j+1}} = \frac{K+1}{K-1} \times \frac{\alpha}{2-\alpha} \times p_i \times \sqrt{\frac{2 \times R}{8 \times \pi \times (T_i + T_{i+1})}} \times (T_i - T_{i+1}) \times (1-f) \times A_i \tag{11.5}$$

where

A is the area (m²)

C is an empirical constant considering contact heat transfer resistance

f is the ratio of contact area of adjacent superinsulation layers and whole heat transfer area,

$$f = \frac{A_{s,\text{cond}}}{A}$$

p is the residual gas pressure (N/m²)

s is the distance between adjacent superinsulation layer (m)

T is the temperature (K)

α is the thermal accommodation coefficient

ϵ is the emission coefficient

K is the ratio of specific heat

λ is the thermal conductivity (W/(m K))

σ is the Stefan–Boltzmann constant (5.67051 × 10⁻⁸ W/(m² K⁴))

i is the counter variable for the number of superinsulation layers

Figure 11.1 shows the calculated temperatures of 24 layers between a warm wall (300 K) and a cold wall (77 K) considering the different heat transfer mechanisms of Equations 11.3 through 11.5. The temperatures of the upper curve with circular symbols are calculated by restricting the heat transfer only on radiation heat, which means in Equations 11.3 through 11.5, that $f = 0$ and $p_i = 0$. The temperatures of the lower linear curve with square symbols are calculated by restricting the heat transfer only on solid heat conduction, which means in Equations 11.3 through 11.5, that $f = 1$. The middle curve with triangle symbols describes a realistic case considering all heat transfer mechanisms of Equations 11.3 through 11.5. For the cases with radiation heat transfer (triangle and circular symbols), the amount of the temperature gradient increases with a decrease of the temperature. Consequently, the amount of solid and gaseous heat conduction increases and the radiation heat transfer decreases approaching the cold wall.

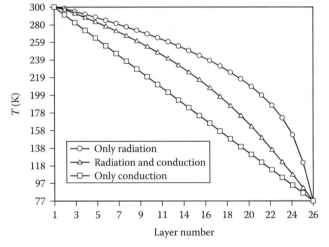

FIGURE 11.1   Calculated temperatures of 24 layers between the warm wall (300 K) and the cold wall K (77 K) considering different heat transfer mechanism.

11.2.4 PRESENT MEASUREMENT AND MODELING DRAWBACKS

As pointed out in the introduction, accurate information about the performance of the MLI is important to estimate the cooling budget requirements for a very large-scale system. Moreover, insulation is a key element in long-duration missions requiring cryogenic storage since relatively additional small heat fluxes can result in significant boil-off losses, increased tank pressure, and increased liquid saturation conditions. As already discussed, the theoretical estimation of heat flux in MLI using numerical modeling only helps us to understand the complex physical phenomenon associated with MLI [15]. They also try to estimate the temperature distributions in the intermediate layers analytically to give only upper bounds for parameters in MLI. Even though theoretical analyses seem to be very promising, they lack experimental validation.

The measurement of heat flux in MLI is not an easy task and it also includes some errors. In such a scenario, axial and transversal temperature distribution matrix can help the measurement technique to understand the results of the heat load in a better way. The major problem to measure the axial and transversal temperature distribution in MLI is the unavailability of a suitable sensor that can be integrated effectively during assembly of the MLI for accurate and reliable measurements. The electromechanical sensor technology cannot be used to measure the temperature distribution in MLI. Because these sensors need penetration of the electrical cabling inside the MLI which may influence the temperature characteristics of an MLI pack. Moreover, to measure the temperature distribution, a lot of sensors are required to be installed and this makes the sensing system more complex and costly. Furthermore, such standard temperature sensors are bigger in size and therefore they disturb the temperature profile through the layers. Use of fiber Bragg gratings (FBG) measurement system could be a solution for the earlier discussed problem as the thermal conductivity of the glass fiber is 0.04 W/m K at 298 K [16] and hence the heat conduction is very low compared to the other standard sensors. Also, the temperature profiles throughout the layers are not disturbed due to the very small dimensions (125 µm Ø) of the sensors which are nearly equal to the thickness of the spacer material. Many other advantages of using FBG measurement system have been discussed in the Section 11.3.2.

11.3 FIBER BRAGG GRATING

FBG consists of a periodic modulation of the index of refraction along the fiber core, as shown in Figure 11.2. Ultraviolet (UV) laser light can be used to write the periodic modulation directly into photosensitive fiber. When a broadband light signal is sent along the FBG sensor, it reflects back one particular wavelength (reflected light) generally called as Bragg wavelength and the remaining light signal will be transmitted via other side of the fiber as a transmitted light signal. Reference [17] explains the theory of the FBG sensors in detail.

11.3.1 FBG SENSORS PRINCIPLE

When an FBG is expanded or compressed, its grating spectral response is changed. The Bragg reflection wavelength λ_B of an FBG is given as [18]

$$\lambda_B = 2n_{eff}d \tag{11.6}$$

where
λ_B, the Bragg grating wavelength, is the free space center wavelength of the input light which will be back reflected from the Bragg grating
n_{eff} is the effective refractive index of the fiber core at the free space center wavelength
d is the grating spacing

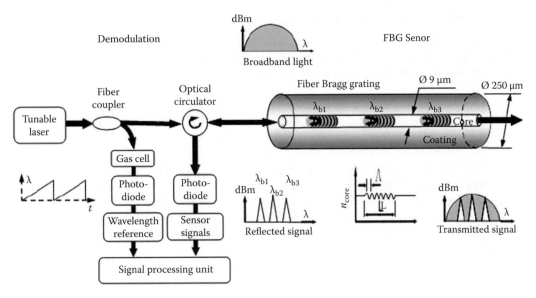

FIGURE 11.2 FBG sensor and demodulation technique.

The Bragg grating resonance, which is the center wavelength of back-reflected light from a Bragg grating, depends on the effective index refraction of the core and the periodicity of the grating. The effective index of refraction, as well as the periodic spacing between the grating planes, will be affected by changes in strain and temperature. Using Equation 11.6, the shift in the Bragg grating center wavelength due to strain and temperature changes is given by

$$\Delta\lambda_B = 2\left(d\frac{\partial n_{\text{eff}}}{\partial l} + n_{\text{eff}}\frac{\partial d}{\partial l} \right)\Delta l + 2\left(d\frac{\partial n_{\text{eff}}}{\partial T} + n_{\text{eff}}\frac{\partial d}{\partial T} \right)\Delta T$$

$$\Delta\lambda_B = \Delta\lambda_{B_s} + \Delta\lambda_{B_T} \tag{11.7}$$

The first term in Equation 11.3 represents the strain effect on an optical fiber. This corresponds to a change in the grating spacing and the strain–optic-induced change in the refractive index. For temperature measurement, this term is kept constant. The second term in Equation 11.3 represents the effect of the temperature on an optical fiber. A shift in the Bragg wavelength due to thermal expansion changes the grating spacing and the index of refraction. This fractional wavelength shift for a temperature change ΔT may be written as [18]

$$\Delta\lambda_{B_T} = 2\left(d\frac{\partial n_{\text{eff}}}{\partial T} + n_{\text{eff}}\frac{\partial d}{\partial T} \right)\Delta T$$

$$= \lambda_B\left(\alpha_d + \alpha_n\right)\Delta T \tag{11.8}$$

where

$\alpha_d = \left(\dfrac{1}{d}\right)\left(\dfrac{\partial d}{\partial T}\right)$ is the thermal expansion coefficient for the fiber ($\sim 0.55 \times 10^{-6}$ for silica)

$\alpha_n = \left(\dfrac{1}{n_{\text{eff}}}\right)\left(\dfrac{\partial n_{\text{eff}}}{\partial T}\right)$ is thermo–optic coefficient ($\sim 8.6 \times 10^{-6}$ for the germania-doped, silica-core fiber)

11.3.2 Features of Fiber Bragg Gratings

The Bragg wavelength shift response of the FBG sensors depends on both temperature and strain effect and hence it can in principle sense both parameters. In this chapter, the focus has been given to the features related to temperature measurement only.

- The FBG sensors can be used to measure the temperature in a quite large range from 2 to 973 K, provided choosing appropriate fiber type and the coatings. Therefore, FBGs are very well suited for temperature measurement applications.
- FBGs are small-sized and lightweight and hence it can be more suitable in measurement areas where the space is limited and in space applications where weight is one of the vital issues.
- FBGs are immune to electromagnetic interference and hence FBG can be one of the best sensors you could have in such environments like RF cavities [19].
- FBG temperature sensors written in photosensitive fiber, without any pre- or post-writing treatment, exhibit the higher radiation tolerance in both pure gamma and mixed gamma neutron environments [20].
- FBGs are intrinsically passive sensors and hence no electrical power required for its operation. Therefore, these sensors can be used in high voltage and potentially explosive atmosphere areas.
- Signal from FBG sensor can be transmitted to very long distance of nearly 100 km and hence it could be installed in very remote locations.
- Multiplexing scheme could be used to have up to 25 sensors (or more, depends upon the detection scheme and the bandwidth of the light source) in one single mode fiber.
- FBGs have high thermal stability and hence could be used in high temperature range.
- FBGs are resistance to corrosion.
- FBGs have very low thermal conductivity which is appropriate for cryogenic applications and in MLI measurements.
- FBGs are negligibly affected by high magnetic fields and hence could be a better sensor for superconductor application.
- FBGs have long lifetime of nearly 20 years.

11.3.3 Demodulation Technique

The block diagram of the interrogation unit used in this study is shown in Figure 11.2. Light from the tunable laser was split into two by a 50/50 fiber coupler and optical circulator. Half of the light was guided to a 12 FBG sensor array while the other half to a National Institute of Standards and Technology (NIST) traceable wavelength gas cell. This gas cell covers the wavelength range from 1520 to 1570 nm. When the wavelength of light emitted by the laser is continuously swept from 1520 to 1570 nm, reflections from FBGs are obtained at the photo detector at different instants. This happens because light reflected from each grating left the laser source at different times. In this way, the exact instant at which the reflection from a given FBG is obtained depends on its wavelength. Time can therefore be used for measuring wavelength. For absolute wavelength accuracy, on each laser sweep, the spectrum of a gas cell is also gathered. By comparing time-synchronized spectrum from both gas cell and FBGs, it is possible to monitor absolute Bragg wavelength changes as small as 1 pm. On each laser sweep, the spectrum of the FBGs as well as the one of the gas cells is acquired and digitalized using tenths of thousands of data points. All the points above a defined threshold are analyzed in order to evaluate if they correspond to a FBG signal or not (noise). A centroid method is then used to determine the central wavelength of the FBGs. This method is based on the geometrical determination of the spectrum centroid, in a way that it is dependent on the set of points used to calculate λ_B [17].

Technical specification of the detection unit used in this work is given in Table 11.1.

TABLE 11.1
Detection Unit Specification

Technical Parameters	Specification
Resolution	0.5 pm
Absolute accuracy	±2.0 pm
Repeatability	±1.0 pm
Optical output power	10 dBm
Line width	125 kHz
Optical isolation	45 dB

11.4 FBG SENSOR ARRAY FOR MLI MEASUREMENT

Corning single mode fiber (SMF 28) has been used to fabricate the FBG sensing arrays. Five FBG sensor arrays are fabricated with four individual sensing elements in each array. In A_1 and A_4, first sensor was damaged and hence it is not shown in Figure 11.3. These sensing elements are fabricated with different Bragg wavelength, which will enable the instrumentation setup to measure all the sensors simultaneously using wavelength division multiplexing (WDM) scheme. Figure 11.3 shows the scheme of the sensor design for MLI temperature measurements. The sensor elements are fabricated at equal distance in 1.5 m length fiber. This helps to locate the position of the sensors inside the MLI after installation.

The fabricated FBG sensors are then calibrated placing arrays in a copper plate, ensuring the sensors are not attached firmly with copper plate to nullify any stress cross effect. A calibrated PT 100 temperature sensor is used as reference. Then the sensors-attached copper plate is placed in the calibration chamber which is then filled with liquid nitrogen. The change in the Bragg wavelength was recorded and correlated with PT 100 sensor for corresponding temperature. The specification of the WDM FBG sensing array used for MLI measurement is shown in Table 11.2.

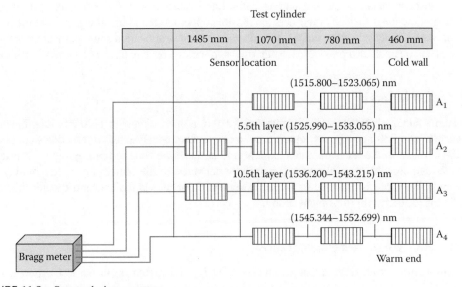

FIGURE 11.3　Sensor design concept.

TABLE 11.2
Specification of WDM FBG Sensing Array for MLI Measurement

Parameters	Specification	
Fiber type	Corning SMF 28	
Grating length	10 mm	
Number of sensors	14 sensors	
Bragg wavelength	1515–1553 nm	
Diameter	~200 μm	
PMMA recoating thickness	~190 μm	
Connector type	FC/APC	
Reflectivity	~97%	
Effective refractive index	~1.4682 at 1550 nm at 23°C	
Temperature sensitivity (calibrated with PT 100 as a reference)	Temperature range (K)	Sensitivity (pm)
	300–200	17
	200–100	14
	100–77	13

11.5   EXPERIMENTAL SETUP

11.5.1   Facility THISTA

The Thermal Insulation Test Facility (THISTA) at the Institute of Technical Physics was used to conduct the measurement. Figure 11.4 shows a cross section of THISTA. In principle, THISTA could be used for many test configurations. All test configurations can use the common outer vacuum tank, the same measurement chamber (LH$_e$/LN$_2$), intermediate guard chamber (LH$_e$/LN$_2$), and external guard (LN$_2$/RT), together with their respective fill and vent lines in thermal contact with the radiation baffles. The volume of the measurement chamber is nearly 50 L which enables to have a continuous test run of 7 days with no fill in between up to a test module heat load of 0.225 W at 4.2 K and 14 W at 77 K. Though refilling of the guard chamber does not affect the thermal equilibrium of the measurement chamber, adequate volumes are provided for comfortable management of the experimental program. The lining of the inner surface of the steel tanks with copper mesh and metal vapor outlet tubes fixed inside the measurement chamber reduces the adverse effect of stratification and superheating in the test and guard chambers. It also provides an isothermal surface irrespective of cryogen levels in the tanks.

11.5.2   Test Module

The test module is a cylinder with dimensions of 219 mm outer diameter, 1820 mm length, and the cold surface of 1.368 m^2. The test module has a concentric pipe system, which allows connection with the cryogen in the measurement chamber and at the same time inhibiting any geyser action. The inner demountable pipe allows cryogen to be delivered to the bottom of the test module and the evaporating vapor is returned to the measurement chamber via the concentric outer tube, thus providing an effective thermosiphon action.

11.5.3   FBG Sensor Array Installation

FBG sensor arrays with three sensing elements were first integrated on to the test cylinder. From Equation 11.7, it is clearly understood that the total Bragg wavelength shift is the result of both temperature

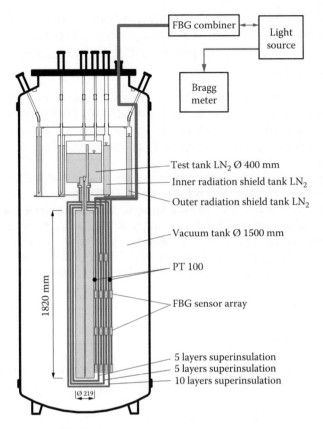

FIGURE 11.4 Cross section of THISTA.

and strain effects. To measure only the MLI temperature, the strain effect has to be cancelled. Hence during installation, the FBG sensors are left hanging free near the wall of the test cylinder to ensure that the sensors did not experience any physical stress. Similar method has been followed for integrating the remaining sensing arrays in between 10th and 11th MLI layer, 15th and 16th MLI layer, and in the outer layer as shown in Figure 11.5. After installation, the Bragg wavelength of the sensors remain unchanged

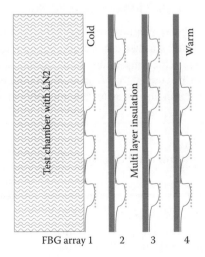

FIGURE 11.5 FBG sensor array installation.

and this ensures the stress-free installation of sensing array. The MLI blankets are integrated onto the test module after careful visual inspection for any damage and contamination.

11.5.4 INSTRUMENTATION

A laminar flow meter from the company Hyperschall-und Strömungstechnik is used. A calibration was done by measuring the volume flux with an exact given heating power into the LN_2 bath. According to this calibration, heat fluxes with an accuracy of $\Delta Q = \pm 20$ mW can be measured. Platinum temperature sensors are positioned on the cold and warm surfaces for calculating the heat conductance. Further temperature sensors are used in the shields and the LN_2 reservoir to control the liquid level. An Alcatel rotary vane pump with a pumping speed of 2060 m³/h is used as a fore pump and a Pfeiffer turbo pump with a pumping speed of 0.5 m³/s is used to achieve a vacuum of about 10^{-6} mbar. The final pressure in the vacuum vessel is measured by a Blazers IKR-020 cold cathode gauge head and a Pfeiffer HPT 100 transmitter which have a measurement range of 10^{-3} to 10^{-9} mbar. This vacuum measurement device is directly linked to the personal computer to record the vacuum levels together with the temperature values. In addition a fore vacuum gauge head Alcatel FA-111 with a measurement range of 1000 to 5×10^{-4} mbar and Alcatel CA-111 vacuum gauge head with a measurement range of 1000 to 10^{-3} mbar were used. Also, FBG sensing array to measure the temperature distribution in the intermediate layers was also integrated.

11.6 EXPERIMENTAL PROCEDURE

The determination of the heat transfer through MLI blankets between environment (300 K) and liquid nitrogen temperature (77 K) consists of the following steps:

- Integrating the test cylinder into THISTA, closing, and sealing the facility
- Evacuating
- Cooling down by filling test tank, test cylinder, and radiation shields with LN_2
- Starting measurements of volume flux, pressure, and temperature
- Waiting for stationary conditions and if necessary refilling of LN_2
- Data recording of measurement values for the evaluation during stationary conditions

11.7 RESULTS AND DISCUSSION

11.7.1 INSULATION EFFICIENCY AT 10^{-6} mbar AND AT 77 K

The relative Bragg wavelength shift was measured once the THISTA attains the steady-state condition. The measured Bragg wavelength shift was then correlated to the corresponding temperature change in the intermediate layers. The change in the temperature at 10^{-6} mbar and at 77 K is shown in Table 11.3.

As discussed in Section 11.2.3, it is understood that at lower temperature ranges the solid conduction and residual gas conduction dominate. As expected, this behavior is also exhibited in the results

TABLE 11.3
ΔT in the Intermediate Insulation Layers

Layers Number	Total Layers	Temperature Difference (ΔT)
ΔT warm (0.5)—10.5th layer	10 layers	34 K
ΔT 10.5th layer—15.5th layer	5 layers	67 K
ΔT 15.5th layer—Cold wall (20.5)	10 layers	106 K

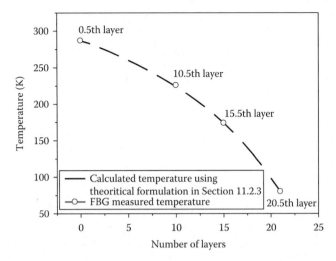

FIGURE 11.6   Comparison of measured temperature and calculated temperature distribution.

of FBG temperature measurement system. The temperature difference between the cold wall and the 15.5th layer is relatively large because of the domination of both solid conduction and residual gas conduction phenomenon. On the other hand, the temperature difference between the warm layer and the next neighboring layer becomes small as the major contribution of heat exchange happens only because of the residual gas conduction. The measurement, hence, coincides perfectly with the theoretical formulation. Figure 11.6 shows the comparison of the measured values with the calculated values from the theoretical formulation described in Section 11.2.3.

The temperature distributions in the intermediate layers have been calculated by giving the heat flux "Q." The heat flux given for the temperature distribution calculations was measured to be 1.022 W. The measured heat flux for the MLI without FBG sensors and the calculated heat flux from the installed FBG temperature distribution measurement data were found to be coinciding with each other. This ensures that the FBG sensors installed in the intermediate layers of the MLI did not disturb the insulation quality and the temperature distribution measurement made with FBG sensors is strain-free. As both the calculated and the measured values coincide well with each other, the FBG sensors can be a reliable and promising candidate for such applications.

11.7.2   Axial and Transverse Temperature Distribution in MLI

As discussed in Section 11.2.4, the FBG sensors can be a right choice to measure the axial and transverse temperatures as they have many advantages compared to other electromechanical sensors (see Section 11.3.2). Complete temperature profile in the insulation system of both axial and transverse directions has been measured using FBG sensor arrays and is shown in Figure 11.7.

From Figure 11.7, it can be observed that the axial temperature distribution varies within 3 K which increases linearly from bottom to top of the MLI. This could be due to the pressure gradient existence, in turn temperature gradient from neck of the test cylinder to bottom of the test cylinder.

11.7.3   Insulation Efficiency at Degraded Vacuum Levels

In general, the insulation quality gets worse when it is used at the poor vacuum levels. In this experiment, the temperature distribution in the intermediate layers was measured and the approximate estimation of decrease in the insulation quality in percentage was calculated. The vacuum level in the THISTA facility was decreased from 10^{-6} to 10^{-1} mbar and the corresponding relative Bragg wavelength shift from 10^{-6} to 10^{-1} mbar was recorded which is shown in Figure 11.8.

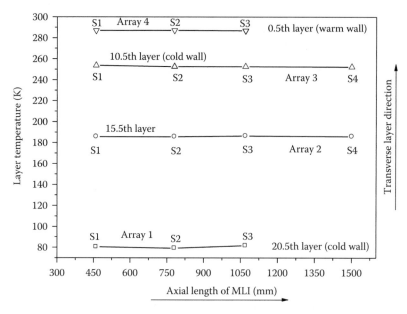

FIGURE 11.7   Axial and transverse temperature distribution in MLI.

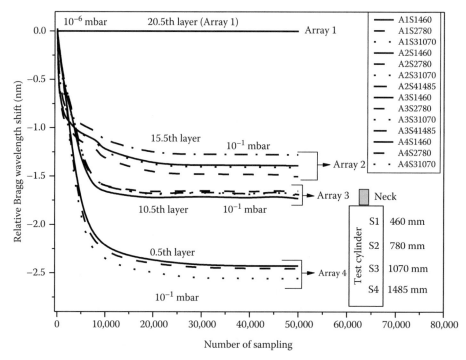

FIGURE 11.8   The FBG sensor wavelength shift when the vacuum levels are changed from 10^{-6} to 10^{-1} mbar at 77 K at cold wall (20.5th layer), 15.5th layer, 10.5th layer, and in warm end.

From Figure 11.8, it is seen that as the vacuum level get decreased to 10^{-1} mbar, the relative Bragg wavelength shift in the sensor array 4, 0.5th layer (warm end) becomes larger and the relative wavelength change at the 20.5th layer (cold wall), sensor array 1, is negligibly small. Large change in the Bragg wavelength means the large temperature gradient. This could be seen in Figure 11.9a.

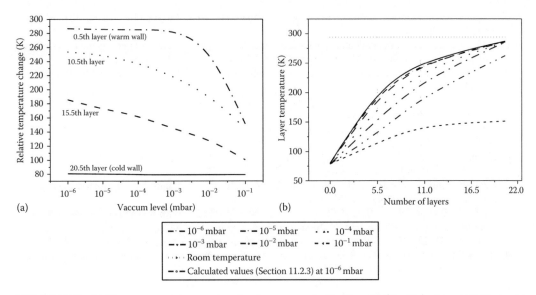

FIGURE 11.9   (a) The measured temperature for the vacuum levels from 10^{-6} to 10^{-1} mbar at 77 K at cold wall (20.5th layer), 15.5th layer, 10.5th layer, and in warm end and (b) degradation of the insulation as the vaccum level decreases.

Figure 11.9b shows the degradation of the insulation quality as the vacuum level decreased. It can be observed that below 10^{-3} mbar the insulation quality is decreased nearly to 90% at 10^{-2} mbar and to nearly 50% at 10^{-1} mbar. Hence, to get a good insulation, the vacuum level should be maintained in the order of 10^{-3} to 10^{-6} mbar.

11.8   CONCLUSION

In this work, the axial and transverse temperature distribution of the superinsulation intermediate layer has been measured successfully using the multiplexed FBG sensor arrays. The experimental data were then compared with calculated data. The measured and the calculated data are well coincided with each other. From the data, it has been understood that the insulation quality gets degraded as the vacuum level decreased. Below 10^{-3} mbar, the insulation quality is decreased nearly to 90% at 10^{-2} mbar and to nearly 50% at 10^{-1} mbar. Hence, to get a good insulation, the vacuum level should be in the order of 10^{-3} to 10^{-6} mbar. This infers that the solid conduction and residual gas conduction phenomenon dominate along with the radiation as the vacuum level decreases. The calculated heat flux for the MLI with the FBG sensors installation was found to be 1.022 W which coincides with the calculations. This ensures that the FBG sensors installed in the interlayer of the MLI did not disturb the insulation quality and the temperature distribution measurement made with FBG sensors is strain-free. From the earlier results, it is evident that the FBG sensors could be the right choice to study the thermal performance of the intermediate layers of the MLI.

REFERENCES

1. *Oxford English Dictionary*, 2nd edn., Oxford University Press, Oxford, U.K. (1989).
2. *New International Dictionary of Refrigeration*, 3rd edn., IIR-IIF, Paris, France (1975).
3. Kropschot, R.H., Schrodt, J.E., Fulk, M.M., and Hunter, B.J., Multi-layer insulation, *Adv. Cryog. Eng.* (1959), 5, 189–198.
4. Gibbon, N.C., Matsch, L.C., and Wang, D.I.-J., Thermal conductivity measurement of insulating materials at cryogenic temperatures, American Society of Testing Materials, Baltimore, MD (1967), STP 411, pp. 61–73.

5. Kaganer, M.G., Thermal insulation in cryogenic engineering, Israel Program for Scientific Translations, Jerusalem (1969), p. 164.
6. De Hann, J.R., Thermal conductivity measurements for insulating materials at cryogenic temperatures, American Society of Testing Materials, Baltimore, MD (1967), STP 411, pp. 95–109.
7. Kaganer, M.G., Thermal insulation in cryogenic engineering, Israel Program for Scientific Translations, Jerusalem (1969), p. 161.
8. Jacob, S., Multilayer. Insulation in Cryoequipment—A Study of Refrence Literarture, kernforschungszentrum, karlsruhe, kfk 5165, March 1993.
9. Kline, S.J. and McClintock, F.A., Uncertainty estimation in single sample experiments. *Mech. Eng.* (1953), 3, 75–78.
10. Zhitormirskij, I.S., Kislov, A.M., and Romanenko, V.G., A theoretical model of the heat transfer processes in multilayer insulation, *Cryogenics* (1979), 19, 265–268.
11. Chen, G., Sun, T., Zheng, J., Huang, Z., and Yu, J. Performance of multilayer insulation with slotted shields. *Cryogenics* (1994), 34, 381–384.
12. Jacob, S., Kasthurirengan, S., and Karunanithi, R., Investigations into the thermal performance of multilayer insulation (300–77 K). Part 2: Thermal analysis, *Cryogenics* (1992), 32(12), 1147–1153.
13. Chau, H. and Moy, H.C., Thermal characteristics of multilayer insulation, AIAA Paper, No. 70-850.
14. Bapat, S.L., Narayankhedkar, K.G., and Lukose, T.P., Experimental investigations of multilayer insulation, *Cryogenics* (1990), 30, 711–719.
15. Neumann, H., Concept for thermal insulation arrangement within a flexible cryostat for HTS power cables, *Cryogenics* (2004), 44, 93–99.
16. White, G.K. and Birch, J.A., Thermal properties of silica at low temperatures, *Physics Chem. Glasses* (1965), 6(3), 85–89.
17. Othonos, A. and Kalli, K., *Fiber Bragg Gratings—Fundamentals and Application in Telecommunications and Sensing*, Artech House Optoelectronics Library, Norwood, MA (1999).
19. Kashyap, R., *Fiber Bragg Gratings*, Academic Press, San Diego, CA (1999).
20. Albin, S., Fu, W., Zheng, J., Lavarias, A., and Albin, J., A cryogenic temperature sensor using fiber Bragg grating, *International Cryogenic Engineering Conference (ICEC 17)*, Bournemouth, U.K. (1998), pp. 695–698. US Patent 6,072,922, June 6, 2000.

12 Thin Film Resistance Temperature Detectors

Fred Lacy

CONTENTS

12.1 INTRODUCTION TO RESISTANCE TEMPERATURE DETECTORS

Resistance temperature detectors (RTDs) are temperature sensors that are composed of a metallic material. These devices operate via an inherent property in which their electrical resistivity is a linear function of temperature. To determine the temperature of an object, the electrical resistance of the metal is the parameter that is measured. RTDs are typically fabricated from platinum, but nickel and copper are sometimes used as substitute materials. The range of the linearity varies according to the metal used, but platinum exhibits the widest range of linearity and has been shown to be extremely linear from −200°C to 650°C [1–3].

In addition to using RTDs to sense or measure temperature, other popular sensors include thermistors, thermocouples, and infrared [1–3]. RTDs are generally considered to be more accurate compared to other temperature sensors [1–3]. Because the RTD is extremely linear with a highly correlated equation, the RTD will provide less measurement error than the other type of temperature sensors. In addition to its accuracy, RTDs are also very stable and repeatable or reproducible [1–3]. Conversely, the other temperature sensors are typically as good or better in terms of sensitivity and response time [1–3].

The electrical resistance of an RTD is directly proportional to its electrical resistivity as well as its length and it is inversely proportional to its cross-sectional area. Because metals are good conductors, their electrical resistivity is small and therefore to achieve a desired nominal resistance, a metal with a specified cross-sectional area must be fabricated with a certain length. To achieve an adequate length, RTDs are typically fabricated by winding a strand of wire into a coil or by depositing a thin layer of metal in serpentine form on a substrate as shown in Figure 12.1. Thin film RTDs can be created with much smaller cross-sectional areas than wire-wound devices, so it is easier to achieve devices with higher nominal resistances. However, if the films become too thin, other performance characteristics can be adversely affected [4–16]. Numerous studies have shown that various factors (e.g., surface effects, fabrication techniques) can affect the resistivity of thin films, so accurately characterizing thin films can be somewhat complex [17–33].

FIGURE 12.1   Top view of a thin film RTD constructed with a serpentine shape and pads for power input and measurement connections. (Reprint from Lacy, F., Characterizing nanometer sized platinum films for temperature measurements, *Current Themes in Engineering Technologies, Selected Papers of the World Congress Engineering and Computer Science*, Vol. 1007, Eds. S.I. Ao, M.A. Amouzegar, and S.S. Chen, Springer, New York, 2008, pp. 128–139. With permission.)

To determine the temperature-sensing limitations of thin film RTDs, films with various thicknesses were analyzed. Experimental measurements, computational analysis, and theoretical examination of thin film RTDs are presented to determine the limitations for these temperature-sensing devices.

12.2   TEMPERATURE MEASUREMENTS WITH RTDs

It is well known that the electrical resistance of a material is given by

$$R = \rho \frac{l}{A} \tag{12.1}$$

where
R is the electrical resistance
ρ is the electrical resistivity
l is the length
A is the cross-sectional area [34]

Additionally, RTDs exhibit a linear relationship between temperature and electrical resistivity given by

$$\rho = \rho_0 \left[1 + \alpha(T - T_0)\right] \tag{12.2}$$

where
T is the temperature
T_0 is a reference temperature (usually 0°C)
α is the temperature coefficient of resistance (TCR)
ρ is the resistivity at temperature T
ρ_0 is the resistivity at temperature T_0 [34]

Thus, determining the electrical resistance of the RTD will indicate the electrical resistivity which in turn will indicate the temperature of the material.

To determine the electrical resistance, a very common technique uses a power source to force current through the RTD while the voltage drop across the sensor is measured (see Figure 12.2). Once the voltage is measured, then Ohm's law (i.e., $V = IR$) can be used to determine the resistance.

Joule heating is produced in these devices when current flows through them; therefore, the current or current density in the sensor must be limited to prevent measurement error [1,6]. So for a given film thickness, the current flowing through the RTD should be lower than a threshold value such that any heat generated by the current will not significantly contribute to the temperature reading.

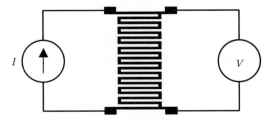

FIGURE 12.2   Resistance measurement technique in which current *I* is supplied to the RTD and the voltage drop *V* is measured.

12.3   COMPUTATIONAL ANALYSIS

Finite element analysis was performed on the RTD structure to determine how much the temperature of the sensor will rise for a given current density [6]. As a result, this computational analysis will determine if the aforementioned current density will cause a significant temperature elevation and thus lead to measurement error. A detailed description of the computational analysis is provided elsewhere [6], but a brief description is provided here.

The structure as shown in Figure 12.1 (without the bonding pads) was constructed using the 3D Builder module of the IntelliSuite software program. The dimensions of the computer-modeled structure were the same as the physical structure. After constructing the RTD computer model, the automatic mesh function was used to perfectly mesh the structure.

After meshing the structure, it was imported from the 3D Builder into the thermoelectrome-chanical module to perform simulations. Then material properties for platinum were loaded into the system for the structure. Initially, electrical voltage loads were applied to the appropriate faces of the structure and current densities were determined. Subsequent calculations revealed that the structures had resistances of ~1 MΩ (for the 46.3 nm film), 51 kΩ (for the 74.0 nm device), and 5.1 kΩ (for the 92.6 nm device). Then the finite element program was run to determine the surface temperature of the structure for various electrical currents.

Figure 12.3 illustrates the result for a 46.3 nm thin film with a current of 1.46 mA flowing through it. The reference temperature is set at 25°C, and as a result of the aforementioned current, most of the RTD has a temperature of ~25.025°C or a 0.1% increases. The inside corners of the RTD have a slightly higher temperature while the outside corners have a slightly lower temperature. The same analysis was performed to determine the level of current that would create a 0.1% temperature increase in 74.0 and 92.6 nm thin films. Finally, computational analysis was performed to determine the current that would create a 0.01% increase in temperature for all three film thicknesses.

A general relationship between the film thickness and the current limit (see Table 12.1) was obtained through curve fitting the data. Each curve fit has a correlation coefficient of $R^2 = 1$.

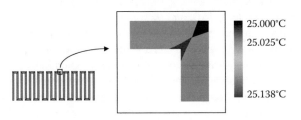

FIGURE 12.3   Illustration of the finite element output of the surface temperature (with an expanded view of a corner) for a 46.3 nm platinum film at 25°C with 1.46 mA of current.

TABLE 12.1

Values for Electrical Current Limits That Produce an Insignificant Increase in Temperature for Thin Film RTDs

Film Thickness	Current Causing a 0.01% Temperature Increase (mA)	Current Causing a 0.1% Temperature Increase (mA)
t	$C = 0.00465e^{0.1t}$	$C = 0.0144e^{0.1t}$

For thin films of thickness t, electrical currents less than $C = 0.00465e^{0.1t}$ or $C = 0.0144e^{0.1t}$ will produce temperature increases less than 0.01% or 0.1%, respectively.

12.4 FABRICATION ISSUES

The complete procedure for fabricating these thin film temperature sensors has been given elsewhere [4–6], but a brief overview of the procedure is given here. The thin film platinum temperature sensors were fabricated on glass substrates. Photoresist was deposited onto the substrates and then this photoresist was hardened by baking the substrate in an oven for several minutes. Then the substrate was patterned by placing a photomask (with the pattern shown in Figure 12.1) over the sample and exposing it to ultraviolet radiation. Next, the unwanted photoresist was removed by immersing the substrate in developer solution. Then, platinum was deposited over the entire glass substrates using a DC sputtering machine. The unwanted platinum was removed through liftoff by rinsing the substrate with acetone until only the patterned platinum remained.

Film thicknesses of 46.3, 74.0, and 92.6 nm were created in order to experimentally characterize thin film RTDs and thus determine and confirm their behavior under certain conditions. However, in fabricating metallic films, variations in the performance characteristics will result [35–38]. The fabrication procedure requires numerous processing steps with various times, temperatures, chemical concentrations, etc., so variations in any of these items will result in differences in the expected electrical resistance of the film. Furthermore, because all of the dimensions of the sensor are very small, slight variations in processing will have drastic effects on the device. Additionally, a few impurities in the processing environment that find their way into/onto the device will also have a significant impact on a nanometer size device. Therefore, because of dimension uniformity, processing impurities, fabrication inconsistencies, etc., caution must be exercised when a device is fabricated and experimental results are compared to computational or theoretical results.

12.5 EXPERIMENTAL RESULTS

Again, to determine the temperature-sensing limitations on thin film RTDs, platinum films were fabricated with thicknesses of 46.3, 74.0, and 92.6 nm. The resistance of the fabricated films was measured at 25°C and it was found that the 46.3 nm film had a value of ~1 MΩ, the 74.0 nm film had a value of ~51 kΩ, and the 92.6 nm film had a value of ~5.1 kΩ [4–6]. The electrical resistance values from the experimental measurements agree with the computational analysis, so using electrical current values from the computational analysis will ensure accurate temperature measurements.

The fabricated films were used to perform electrical resistance measurements at selected temperature values. The relationship between temperature and electrical resistance was nonlinear as shown in Figure 12.4 for the 46.3 nm film. As pointed out previously, bulk platinum RTDs are

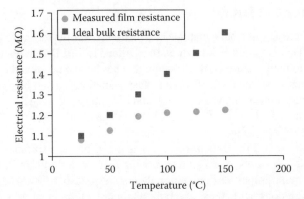

FIGURE 12.4 Resistance vs. temperature for the 46.3 nm platinum film compared to bulk platinum. (Reprint from Lacy, F., Characterizing nanometer sized platinum films for temperature measurements, *Current Themes in Engineering Technologies, Selected Papers of the World Congress Engineering and Computer Science*, Vol. 1007, Eds. S.I. Ao, M.A. Amouzegar, and S.S. Chen, Springer, New York, 2008, pp. 128–139.)

linear in the temperature range from −200°C to 650°C, so the response of bulk platinum is also displayed in Figure 12.4 to display the contrast between bulk and thin film RTDs.

The other films displayed a similar type response in which the resistance increased in a linear type manner and then the resistance did not stay linear, but became relatively flat as the temperature continued to increase [4–6]. These regions will be referred to as the linear and saturation regions. The demarcation between these two regions will be referred to as the transition point. The transition point varied according to the thickness of the film and the transition point occurs at higher temperatures for thicker films. In equation form, the saturation temperature is given by $T_{sat} = T_0 + 0.009e^{0.1t}$ where T_{sat} is the saturation temperature, T_0 is a constant which equals 84°C for these platinum films, and t is the film thickness. For the 46.3 nm thin film, the saturation point is measured to be ~85°C and the saturation temperature equation calculates the value to be 84.9°C. Similarly, the 74.0 and 92.6 nm films have calculated values of 98.7°C and 178.6°C, respectively.

The temperature coefficient of resistance (TCR) can be determined from the equation $\Delta R/R_0 = \alpha \Delta T$ where ΔR is the change in resistance, ΔT is the change in temperature, R_0 is the resistance at 0°C, and α is the TCR [1–6]. The TCR (also known as the sensitivity) for bulk platinum has a constant value of 0.00385/°C in the temperature range from −200°C to 650°C [1–3]. The bulk data shown in Figure 12.4 displays this TCR. However, since the slope of the thin film data does not match the slope of the bulk data, thin film RTDs will have a different TCR (and thus a different sensitivity) than the standard 0.00385/°C.

When the temperature is below the saturation point (i.e., $T < T_{sat}$), the ratio of the TCR for thin films to the TCR for bulk platinum is on average 0.54 or $TCR_{film}/TCR_{bulk} \approx 0.54$ [6]. Thus the thin film RTDs are about half as sensitive as their bulk counterparts. When thin film RTDs reach their saturation point, the slope changes and for temperatures beyond the saturation temperature, the resistance vs. temperature curve becomes flat and thus the TCR is approximately zero. Thus, for $T > T_{sat}$, $TCR_{film}/TCR_{bulk} \approx 0$ [6].

This analysis shows that the useful measurement range for thin film RTDs is limited based upon the thickness of the film. Thin film RTDs will be incapable of accurately measuring temperatures above their saturation level. However, in the linear region, thin film RTDs will be able to measure temperatures accurately even though the thin film will not be as sensitive as a corresponding bulk sensor.

12.6 THEORETICAL RESULTS

RTDs constructed from several bulk metals (i.e., platinum and nickel) have a well-defined characteristic equation. This equation is known as the Callendar–van Dusen equation and relates the electrical resistance to temperature [1,3]. This equation has been determined through experimental measurements. For platinum, the Callendar–van Dusen equation is given by $R = R_0 [1 + AT + BT^2 + C(T - 100)\ T^3]$ for temperatures between −200°C and 0°C, and by $R = R_0[1 + AT + BT^2]$ for temperatures between 0°C and 650°C where the coefficients will have values $A = 3.908 \times 10^{-3}$, $B = -5.77 \times 10^{-7}$, and $C = -4.183 \times 10^{-12}$ [1,3].

Research has shown that if the dimension(s) of a material become sufficiently small, its properties will change [4–16]. As a result, it is not surprising that these thin film RTDs exhibit different resistance vs. temperature properties than their bulk counterpart. In order to completely understand why thin film platinum does not behave like bulk platinum, a theoretical model that examines the microscopic properties of the RTD material is needed. In developing this model, various principles are already well established (e.g., band theory, phonons) [39–45] and therefore they are assumed to be valid while developing the model.

The specific details and derivation of this theoretical model has been given elsewhere [7], but a general description will be provided here. A two-dimensional model is used with the following assumptions: electrons are confined to energy bands, phonons or atom vibrations in the lattice will exist, and electrons will interact with phonons when they move through the material. The interaction between electrons and lattice atoms will impede the advancement of the electrons. An increase in temperature will increase the energy of the lattice atoms which will enhance or lattice vibrations and thus provide larger impedance for the electrons. Thus, when an electric field is applied to the material, the velocity and momentum of conduction electrons will be affected by temperature.

To create a realistic and yet simplified model, a two-dimensional lattice structure is used to develop an equation for the electrical resistivity as a function of temperature. This model, as shown in Figure 12.5a, uses a face-centered cubic structure (FCC), in which atom vibration is restricted to the y-direction and electron movement is restricted to the x-direction. As electrons travel through the material, they will either be in a region where there is a gap (where they will travel without being scattered by an atom), or they will be in a region where there is no gap (and they will eventually be scattered by an atom). This is illustrated in Figure 12.5b in which electrons will eventually collide with an atom if it is not lined up in a gap. As temperature increases, the gaps are reduced in size and thus the percentage of electrons that can travel unhindered is reduced and thus the conductivity (which is the inverse of the resistivity) is reduced.

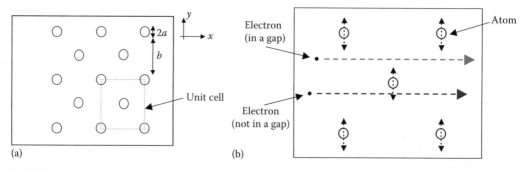

FIGURE 12.5 Two-dimensional structure for the theoretical model. (a) Lattice structure showing atoms, the diameter of an atom, the spacing between atoms, and a unit cell. (b) Electrons traveling through vibrating atoms in a unit cell (electrons that are in a gap will not be scattered while electrons that are not in a gap will be scattered).

As a result of this theoretical analysis, the resistivity of a conductor in general or an RTD in particular is given by

$$\rho = \rho_0 \left[\cfrac{1}{\cfrac{\left[2 \sqrt[\delta]{\cfrac{\gamma}{kT}} - b\right]\left(\cfrac{\tau_1}{\tau_2} - 1\right)}{2a + b} + 1} \right] \tag{12.3}$$

where

$$\rho_0 = \frac{m}{ne^2 \tau_2}$$

a is the atomic radius
b is the size of the opening between atoms (when the atoms are stationary)
τ_1/τ_2 is the ratio of travel times before scattering when an electron is in the gap to when an electron is not in a gap
δ is 1
k is Boltzmann's constant
T is temperature (in Kelvin)
γ is the proportionality term in the energy equation $U = \cfrac{\gamma}{r^\delta}$

Equation 12.3 represents the main equation from the theoretical model.

This model must first demonstrate that it is accurate in producing experimental data relating to bulk materials (i.e., the Callendar–van Dusen equation) and then the model can be used to explain the results for thin film conductors (i.e., the saturation effect). Depending upon the values of the parameters selected in Equation 12.3, the result can have a linear or nonlinear response. When Equation 12.3 is plotted with the following values: $\delta = 1$, $a = 1 \times 10^{-12}$ m, $b = 3.92 \times 10^{-10}$ m, $\gamma = 5.39 \times 10^{-33}$ J m, and $\tau_1/\tau_2 = 2.005$, the result is linear and is illustrated in Figure 12.6. This is symbolic of bulk platinum RTDs which are linear from 73 to 923 K (or equivalently from −200°C to 650°C).

Of all metals or conductors, platinum (and thus a platinum RTD) is generally recognized as having the widest range of linearity. However, all conductors will not be linear over this temperature range (i.e., from 73 to 923 K), but the result shown in Figure 12.6 is symbolic of the linear resistivity response to temperature that is typical of experimental measurements of all bulk conductors as

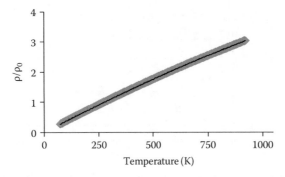

FIGURE 12.6 Linear response of data generated from the theoretical model for electrical resistivity as a function of temperature for bulk conductors.

given in Equation 12.2. It will be shown that Equation 12.3 can be used to simulate the response for platinum and nickel (which are the two prominent materials used for RTDs).

Again, when data is generated from Equation 12.3 to simulate platinum, $\delta = 1$, $a = 1 \times 10^{-12}$ m, $b = 3.92 \times 10^{-10}$ m, $\gamma = 5.39 \times 10^{-33}$ J m, and $\tau_1/\tau_2 = 2.005$, and a curve fit is applied to this data for positive Celsius values, the equation $R/R_0 = [1 + AT + BT^2]$ is obtained where the coefficients have values $A = 3.5 \times 10^{-3}$ and $B = -5.83 \times 10^{-7}$. This is very similar to the experimental results obtained for platinum in which the coefficients have values $A = 3.908 \times 10^{-3}$ and $B = -5.77 \times 10^{-7}$. The curve fit provides a very good match with a correlation coefficient of $R^2 = 0.9999999975$.

Similarly, data can be generated from Equation 12.3 to simulate nickel so when $\delta = 1$, $a = 1 \times 10^{-12}$ m, $b = 3.52 \times 10^{-10}$ m, $\gamma = 4.85 \times 10^{-33}$ J m, and $\tau_1/\tau_2 = 2.007$, and a curve fit is applied to this data for positive Celsius values, the equation $R/R_0 = [1 + AT + BT^2]$ is obtained where the coefficients have values $A = 4.49 \times 10^{-3}$ and $B = 6.63 \times 10^{-6}$. This is very similar to the experimental results for nickel in which the coefficients have values $A = 5.49 \times 10^{-3}$ and $B = 6.65 \times 10^{-6}$. The curve fit provides a very good match with a correlation coefficient of $R^2 = 0.9999952808$.

Therefore, the results obtained with the theoretical model are in very good agreement with experimental findings for bulk platinum and nickel RTDs. Now the model must be adjusted and/or modified in order to generate theoretical data that is in agreement with experimental data for thin films. Since it is well known that physical properties of materials change when their size is reduced to the nanometer scale [46–50], it is reasonable to expect that one of the parameters in Equation 12.3 is responsible for the difference between the bulk and thin film resistivity responses. For example, it has been shown that melting point depression (i.e., the phenomena in which the melting point of a material becomes lower when the size of that material is reduced) occurs in materials because the molecules gain enough energy at lower temperatures to change from the solid to the liquid state [51–53].

Equation 12.3 contains a term, γ, that is the proportionality variable relating the energy of atoms to the separation distance between them. This parameter was initially selected to be constant, and this led to a good match with the Callendar–van Dusen coefficients. However, after performing additional analysis, it was determined that by varying γ, a better match for the Callendar–van Dusen coefficients was obtained. As a result of this aforementioned analysis and since γ is a term that is related to the energy of the atoms, it is reasonable to think that further varying this term will have the desired effect of altering the resistivity from bulk values to nanoscale values. When γ is suitably varied, a match between experimental and theoretical data is obtained as shown in Figure 12.7 [8].

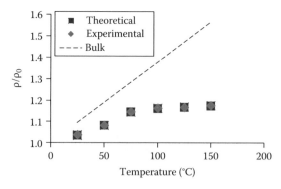

FIGURE 12.7 Resistivity vs. temperature graph showing that the theoretical model can be used to match experimental data for thin film conductors. (Reprint from Lacy, F., *Nanoscale Res. Lett.*, 6, 636, 2011. With permission.)

12.7 THIN FILM RTD DESIGN

Because of the saturation effect, if thin films with thicknesses of ~100 nm or less are used to measure temperatures, then extreme care must be taken if accurate results are to be obtained over an extended temperature range. Because of this limitation, the thickness of the films must be above a certain level if they are to be use for temperature sensing at high temperatures. To use a thin film platinum RTD for measuring temperatures below some maximum temperature T_{max}, the RTD would have to be fabricated such that $T_{max} < T_{sat}$, where T_{sat} represents the saturation temperature for the thin film. For platinum, calculations reveal that this would occur if the thin film had a thickness greater than

$$t = 10 \times \ln\left(\frac{T_{sat} - T_0}{0.009}\right) \tag{12.4}$$

where

t is the film thickness
T_0 represents the minimum saturation point for platinum films and equals 84°C

Although temperature measurements with RTDs may be limited, one beneficial use of the saturation feature of thin film RTDs could be their ability to limit current to a load. Because the resistance of this device increases as the temperature increases, this device can be used in a similar manner as a thermistor and prevent damage to circuit components. A thermistor is a ceramic material (or metal oxide) that can be fabricated with a PTC response (positive temperature coefficient) in which the resistance increases as temperature increases or with an NTC response (negative temperature coefficient) in which the resistance decreases as temperature increases. Many uses of the thermistor involve current limiting, but not many of these applications involve the resistance temperature characteristic of the PTC thermistor. The current limiting applications for the PTC thermistor involve either their voltage–current characteristic or their current–time characteristic. Because the voltage–current and current–time responses for the thermistor are more favorable than the RTD, the thermistor is typically preferred for these types of current limiting applications. However, the resistance temperature characteristic of the RTD (i.e., the saturation feature) could make it more favorable than the thermistor in limiting the electrical current delivered to a circuit component.

Consider the circuit shown in Figure 12.8a in which a load (e.g., a fan) is connected in parallel to the thin film RTD. At a particular temperature, the RTD will have a certain electrical resistance and thus a specific amount of current will flow through the load. As the temperature increases and remains below the saturation temperature, the resistance of the RTD will increase and allow more current to flow through the load. Once the temperature reaches the saturation point, the RTD resistance will reach its saturation resistance and thus the current flowing through the load will also reach its limit. Any further increases in temperature will not result in an additional increase in the load current. Depending upon the desired saturation temperature, the RTD can be fabricated with a certain thickness, and this will determine the temperature at which the load receives the maximum current (see Figure 12.8b).

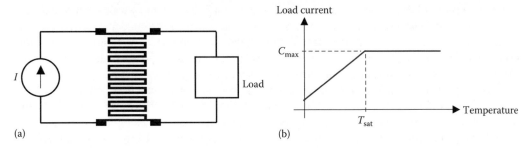

(a) (b)

FIGURE 12.8 (a) An electrical circuit using a thin film RTD to increase the current to a load when the temperature increases and (b) a graph of the load current profile (as a function of temperature) for this circuit.

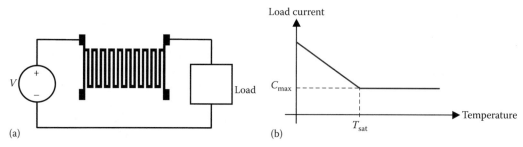

FIGURE 12.9 (a) An electrical circuit using a thin film RTD to limit the current to a load when the temperature increases and (b) a graph of the load current profile (as a function of temperature) for this circuit.

Conversely, if the current to a load (e.g., an electric motor) needs to be reduced when temperatures are elevated, an electrical circuit as shown in Figure 12.9a could be used. This circuit contains the RTD in series with the load. At a particular temperature, the RTD will have a certain electrical resistance and thus a specific amount of current will flow through the load. As the temperature increases and remains below the saturation temperature, the resistance of the RTD will increase and reduce the current flow through the load. Once the temperature reaches the saturation point, the RTD resistance will reach its saturation resistance and thus the current flowing through the load will also reach its lower limit. Again, the thin film RTD can be fabricated with a desired thickness to achieve the desired saturation temperature at which the load will receive the limiting current (see Figure 12.9b).

12.8 CONCLUSIONS

In summary, metallic materials (such as platinum, nickel, and copper) in bulk form have been successfully used as temperature sensors known as resistance temperature detectors. The electrical resistance of these devices changes linearly as temperature changes. Bulk materials have a well-established response which is characterized by the Callendar–van Dusen equation. However, because nanoscale materials in general behave differently than their bulk counterpart, thin film RTDs are expected to have different characteristics than bulk devices.

Thin film RTDs were fabricated and after measuring their resistance temperature response, it was confirmed that thin film RTDs are not linear throughout the same temperature region as RTDs fabricated in bulk form. The experimental findings reveal that the electrical resistance of thin film RTDs changes linearly with temperature at low temperatures; however, at high temperatures the electrical resistance saturates and does not change as the temperature increases.

To confirm these experimental results, a theoretical model was created. After appropriate parameter values were applied to this theoretical model, data was generated and it was found that the theoretical data matched the experimental values. Therefore, it is concluded that the theoretical model is trustworthy and that this model elucidates the underlying physics of RTDs.

Because thin film RTDs are not accurate when temperatures reach the sensor's saturation point (i.e., T_{sat}), caution must be exercised when using these devices to measure temperatures if those temperatures may exceed that saturation point. Although the saturation effect will limit the use of thin film conductors for temperature measurements, these thin films can be valuable and find extensive use in automatically limiting electrical current to a device in an electrical circuit.

This work has been performed at the experimental, computational, and theoretical levels. It has provided great information and tremendous insight into the characteristics of thin film conductors and their limitation for temperature sensing. Nevertheless, additional research is always needed to fully develop and expand the capacity of this work (e.g., fully characterizing the thin film RTD in the current limiting circuits, performing similar research with copper and/or nickel films, and comparing the results with platinum films).

ACKNOWLEDGMENTS

This work was supported in part by the U.S. Department of Energy, in part by the National Nuclear Security Agency under Grant DE-FG02-94EW11427. The author would like to thank Ms. April Page for collecting the experimental data, Dr. Ernest Walker for his continued support on this project, and Dr. Steven L. Richardson for his influence on this work.

REFERENCES

1. J. Fraden, *Handbook of Modern Sensors: Physics, Designs, and Applications*, New York: Springer-Verlag, 2004, pp. 461–477.
2. G. T. A. Kovacs, *Micromachined Transducers Sourcebook*, Boston, MA: McGraw Hill, 1998, pp. 559–561.
3. H. L. Trietley, *Transducers in Mechanical and Electronic Design*, New York: Marcel Dekker, 1986, pp. 27–41.
4. F. Lacy, Investigating thin films for use as temperature sensors, *Proceedings of the World Congress on Engineering & Computer Science*, San Francisco, CA, October 24–26, 2007, pp. 441–444.
5. F. Lacy, Characterizing nanometer sized platinum films for temperature measurements, *Current Themes in Engineering Technologies. Selected Papers of the World Congress Engineering and Computer Science*, Vol. 1007, Eds., S.I. Ao, M.A. Amouzegar, and S.S. Chen. New York: Springer, 2008, pp. 128–139.
6. F. Lacy, Using nanometer platinum films as temperature sensors (constraints from experimental, mathematical, and finite-element analysis), *IEEE Sensors J*, 9(9), 1111–1117, 2009.
7. F. Lacy, Evaluating the resistivity-temperature relationship for RTDs and other conductors, *IEEE Sensors J*, 11(5), 1208–1213, 2011.
8. F. Lacy, Developing a theoretical relationship between electrical resistivity, temperature, and film thickness for conductors, *Nanoscale Res Lett*, 6, 636, 2011.
9. F. Warkusz, The size effect and the temperature coefficient of resistance in thin films, *J Phys D Appl Phys*, 11, 689–694, 1978.
10. M. J. Lourenco, J. M. Serra, M. R. Nunes, A. M. Vallera, and C. A. Nieto deCastro, Thin-film characterization for high-temperature applications, *Int J Thermophys*, 19(4), 1253–1265, 1998.
11. E. V. Barnat, D. Nagakura, P. I. Wang, and T. M. Lu, Real time resistivity measurements during sputter deposition of ultrathin copper films, *J Appl Phys*, 91(3), 1667–1672, 2002.
12. S. U. Jen, C. C. Yu, C. H. Liu, and G. Y. Lee, Piezoresistance and electrical resistivity of Pd, Au, and Cu films, *Thin Solid Films*, 434(1–2), 316–322, 2003.
13. M. Kawamura, T. Mashima, Y. Abe, and K. Sasaki, Formation of ultra-thin continuous Pt and Al films by RF sputtering, *Thin Solid Films*, 377, 537–542, 2000.
14. X. Zhang, H. Q. Xie, M. Fujii, H. Ago, K. Takahashi, T. Ikuta, H. Abe, and T. Shimizu, Thermal and electrical conductivity of a suspended platinum nanofilm, *Appl Phys Lett*, 86(17), 171912-1–171912-3, 2005.
15. X. Zhang, H. Q. Xie, M. Fujii, H. Ago, K. Takahashi, T. Ikuta, H. Abe, and T. Shimizu, Thermal and electrical properties of a suspended nanoscale thin film, *Int J Thermophys*, 28(1), 33–43, 2007.
16. N. Stojanovic, J. Yun, E. B. K. Washington, J. M. Berg, M. W. Holtz, and H. Temkin, Thin-film thermal conductivity measurement using microelectrothermal test structures and finite-element-model-based data analysis, *J Microelectromech Syst*, 16(5), 1269–1275, 2007.
17. K. Fuchs, The conductivity of thin metallic films according to the electron theory of metals, *Proc Camb Philos Soc*, 34(1), 100–108, 1938.
18. E. H. Sondheimer, The mean free path of electrons in metals, *Adv Phys*, 1(1), 1–42, 1952.
19. Y. Namba, Resistivity and temperature coefficient of thin metal films with rough surfaces, *Jpn J Appl Phys*, 9, 1326–1329, 1970.
20. A. A. Cottey, The electrical conductivity of thin metal films with very smooth surfaces, *Thin Solid Films*, 1, 297–307, 1968.
21. C. R. Tellier and A. J. Tosser, Adequate use of the Cottey model for the description of conduction in polycrystalline films, *Electrocomp Sci Technol*, 6, 37–38, 1979.
22. H. D. Liu, Y. P. Zhao, G. Ramanath, S. P. Murarka, and G. C. Wang, Thickness dependent electrical resistivity of ultrathin (<40 nm) Cu films, *Thin Solid Films*, 384, 151–156, 2001.
23. S. M. Rossnagel and T. S. Kuan, Alteration of Cu conductivity in the size effect regime, *J Vac Sci Technol B*, 22, 240–247, 2004.
24. D. Dayal, P. Rudolf, and P. Wissmann, Thickness dependence of the electrical resistivity of epitaxially grown silver films, *Thin Solid Films*, 79, 193–199, 1981.

25. J. M. Camacho and A. I. Oliva, Morphology and electrical resistivity of metallic nanostructures, *Microelectron J*, 36, 555–558, 2005.
26. A. F. Mayadas and M. Shatzkes, Electrical-resistivity model for polycrystalline films: The case of arbitrary reflection at external surfaces, *Phys Rev B*, 1, 1382–1389, 1970.
27. J. R. Sambles, The resistivity of thin metal films—Some critical remarks, *Thin Solid Films*, 106, 321–331, 1983.
28. C. Durkan and M. E. Welland, Size effects in the electrical resistivity of polycrystalline nanowires, *Phys Rev B*, 61, 14215–14218, 2000.
29. J. Feder, P. Rudolf, and P. Wissmann, The resistivity of single-crystal copper films, *Thin Solid Films*, 36, 183–186, 1976.
30. S. X. Shi and D. Z. Pan, Wire sizing with scattering effect for nanoscale interconnection, *Proceedings of the 11th Asia and South Pacific Design Automation Conference*, Yokohama, Japan, January 24–27, 2006, pp. 503–508.
31. A. E. Yarimbiyik, H. A. Schafft, R. A. Allen, M. E. Zaghloul, and D. L. Blackburn, Modeling and simulation of resistivity of nanometer scale copper, *Microelectron Rel*, 46, 1050–1057, 2006.
32. P. Fan, K. Yi, J. D. Shao, and Z. X. Fan, Electrical transport in metallic films, *J Appl Phys*, 95, 2527–2531, 2004.
33. W. Zhang, S. H. Brongersma, O. Richard, B. Brijs, R. Palmans, L. Froyen, and K. Maex, Influence of the electron mean free path on the resistivity of thin metal films, *Microelectron Eng*, 76, 146–152, 2004.
34. A. Hudson and R. Nelson, *University Physics*, New York: Harcourt Brace Jovanovich, 1982, p. 586.
35. U. Schmid and H. Seidel, Influence of thermal annealing on the resistivity of titanium/platinum thin films, *J Vac Sci Technol A*, 24(6), 2139–2146, 2006.
36. J. A. Chiou and S. Chen, Thermal hysteresis and voltage shift analysis for differential pressure sensors, *Sensors Actuat A Phys*, 135(1), 107–112, 2007.
37. K. Sadek and W. Noussa, Studying the effect of deposition conditions on the performance and reliability of MEMS gas sensors, *Sensors*, 7(3), 319–340, 2007.
38. M.-J. Huang, P.-K. Chou, and M.-C. Lin, Thermal and thermal stress analysis of a thin-film thermoelectric cooler under the influence of the Thomson effect, *Sensors Actuators A Phys*, 126(1), 122–128, 2006.
39. C. Kittel, *Introduction to Solid State Physics*, New York: Wiley, 2005.
40. H. T. Stokes, *Solid State Physics*, Boston, MA: Allyn & Bacon, 1987.
41. M. Razeghi, *Fundamentals of Solid State Engineering*, New York: Springer, 2007.
42. L. Solymar and D. Walsh, *Electrical Properties of Materials*, Oxford: Oxford University Press, 2004.
43. J. R. Christman, *Fundamentals of Solid State Physics*, New York: Wiley, 1988.
44. G. Burns, *Solid State Physics*, Orlando, FL: Academic Press, 1985.
45. N. W. Ashcroft and N. D. Mermin, *Solid State Physics*, Belmont, CA: Brooks/Cole, 1976.
46. S. Link, M. B. Mohamed, and M. A. El-Sayed, Simulation of the optical absorption spectra of gold nanorods as a function of their aspect ratio and the effect of the medium dielectric constant, *J Phys Chem B*, 103, 3073–3077, 1999.
47. M. Fujii, X. Zhang, H. Xie, H. Ago, K. Takahashi, T. Ikuta, H. Abe, and T. Shimizu, Measuring the thermal conductivity of a single carbon nanotube, *Phys Rev Lett*, 95, 065502–065505, 2005.
48. X. W. Wang, G. T. Fei, K. Zheng, Z. Jin, and L. D. Zhang, Size dependent melting behavior of Zn nanowire arrays, *Appl Phys Lett*, 88, 173114, 2006.
49. G. Bilalbegovic, Structures and melting in infinite gold nanowires, *Solid State Commun*, 115, 73–76, 2000.
50. W. H. Qi, M. P. Wang, M. Zhou, and W. Y. Hu, Surface area difference model for thermodynamic properties of metallic nanocrystals, *J Phys D Appl Phys*, 38, 1429–1436, 2005.
51. M. Zhang, M. Y. Efremov, F. Schiettekatte, E. A. Olson, A. T. Kwan, S. L. Lai, T. Wisleder, J. E. Greene, and L. H. Allen, Size-dependent melting point depression of nanostructures: Nanocalorimetric measurements, *Phys Rev B*, 62, 10548–10557, 2000.
52. G. L. Allen, R. A. Bayles, W. W. Gile, and W. A. Jesser, Small particle melting of pure metals, *Thin Solid Films*, 144, 297–308, 1986.
53. Q. Jiang, S. Zhang, and M. Zhao, Size-dependent melting point of noble metals, *Mater Chem Phys*, 82, 225–227, 2003.

13 The Influence of Selected Parameters on Temperature Measurements Using a Thermovision Camera

Mariusz Litwa

CONTENTS

13.1 THEORY

13.1.1 HISTORY OF THERMOGRAPHY

Thermography, also called thermovision, dates its origins to the late eighteenth and early nineteenth century, namely the year 1800. During this period, Sir William Herschel (1738–1822), a famous astronomer, designer of telescopes, and composer, quite accidentally discovered the spectrum of infrared radiation.

In modern times, however, the majority of people who questioned: "Who was William Herschel?" will answer that he was an outstanding astronomer who had discovered the planet Uranus. Only few people will connect this name with infrared radiation. Herschel discovered infrared radiation while searching for an optical filter to reduce the brightness of the observed image in a telescope. He perceived that blue-colored glass transmits less heat radiation than red-colored glass. Herschel placed the mercury thermometers along the different colors of glass of the visible spectrum and observed temperature rise from blue to red. Herschel also noted a rise in temperature on the thermometer outside the red color. He concluded that outside the visible spectrum there is something more. Herschel newly discovered an area of the spectrum called "the invisible thermometrical spectrum." Perhaps in 1800, the great astronomer did not realize how great the discovery was, which we are now fully enjoying.

Thermography is currently a very rapidly growing field of science. But this has not always been so. During the war and the postwar period, work on the infrared radiation has been withheld because of its military use. Only in the second half of the twentieth century, thermovision became a widely available science. Infrared radiation is better known today than 10 years ago. This became possible because newspapers regularly published the latest developments in the field of thermography, apart from frequent international conferences [5–7].

13.1.2 Electromagnetic Radiation

Electromagnetic radiation is the radiation that is constant throughout its full range. There is no upper or lower limit of the scale of the frequency or wavelength of the electromagnetic wave. The overall distribution of the radiation spectrum depends on the wavelength λ. Electromagnetic radiation can be divided into smaller ranges:

- Cosmic rays and gamma rays ($\lambda < 10^{-5}$ μm)
- X-rays ($10^{-5} < \lambda < 10^{-2}$ μm)
- Ultraviolet ($10^{-2} < \lambda < 0.35$ μm)
- Visible radiation ($0.35 < \lambda < 0.75$ μm)
- Infrared radiation ($0.75 < \lambda < 10^{3}$ μm)
- Microwaves and radio waves ($\lambda > 10^{3}$ μm)

Graphical distribution of electromagnetic radiation is shown in Figure 13.1.

The boundaries of the division are contractual values, which may differ slightly in different literature works.

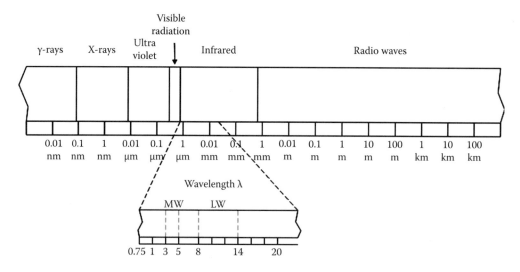

FIGURE 13.1 Distribution of electromagnetic radiation depending on the wavelength (MW—medium waves, LW—long waves).

To detect objects, thermography is performed using the infrared radiation, which is located behind the visible radiation. There are several ratings of infrared radiation. In thermography, the infrared band is usually divided into six smaller bands:

1. Near infrared (NIR): 0.7–1.4 μm
2. Short-wave infrared (SWIR): 1.4–3 mm
3. Mid-wave infrared (MWIR): 3 mm–8 μm
4. Long-wave infrared (LWIR): 8–12 μm
5. Very long-wave infrared (VLWIR): 12–25 μm
6. Far infrared wave (FWIR): 25–1000 μm

Infrared cameras operate in two bands: middle waves and long waves. This is because of the atmospheric attenuation, which will be referred to in a later section [2].

13.1.3   RADIOMETRIC QUANTITIES AND INFRARED RADIATION LAWS

Table 13.1 describes the basic radiometric quantities.

From Table 13.1, the luminous emittance, also known as emittance, is defined as the aerial density of radiation energy flux Φ from the surface A to half-space. This is the most important volume used in thermography, which schematically describes the amount of power radiated by an object (unit area) at a specified temperature. The mathematical form of the luminous emittance is as follows:

$$M = \frac{\partial \Phi}{\partial A} \quad [\text{W/m}^2] \tag{13.1}$$

13.1.3.1   Planck's Law

In 1900, the German physicist Max Planck made a claim which has become one of the fundamental theorems of modern physics. He described the spectral distribution of blackbody radiation, i.e., the density of spectral emittance:

$$M_\lambda(T) = \frac{2\Pi hc_0^2}{\lambda^5(e^{hc_0/\lambda kT} - 1)} \quad [\text{W/(m}^2\mu\text{m})] \tag{13.2}$$

TABLE 13.1

Radiometric Quantities

Symbol	Quantity	Unit	Definition
Q_e	Radiation energy	J	Energy emitted, transferred or incident on a surface
Φ_e	Radiant flux	W	Power emitted, transmitted or incident on a surface
I_e	Radiation intensity	W/sr	Radiant flux emitted in unit solid angle
E_e	Irradiance	W/m²	Radiant flux incident per unit area
M_e	Luminous emittance	W/m²	Radiant flux emitted by unit surface
L_e	Radiance	W/(m² sr)	Radiant flux emitted by unit surface in unit solid angle

Source: Bielecki, Z. and Rogalski, A., Detekcja sygnałów optycznych, WNT, Warszawa, 2001.

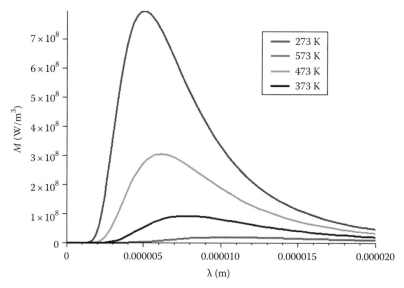

FIGURE 13.2   Distribution of blackbody radiation depending on wavelength for different temperatures.

where
   $M_\lambda(T)$ is blackbody spectral emittance for wavelength λ
   c is speed of light = 299792458 (m/s)
   h is Planck's constant = 6.626×10^{-34} (J $\cdot$ s)
   k is Boltzmann's constant = 1.38×10^{-23} (J/K)
   T is blackbody absolute temperature (K)
   λ is wavelength (μm)

   A graphical representation of the blackbody emittance depending on the wavelength for different temperatures is shown in Figure 13.2.
   Planck described his theory using the model of atomic processes occurring in the walls of the cavity. He assumed that the atoms in the wall of the cavity act like electromagnetic oscillators, which have a certain frequency of vibration. These oscillators simultaneously absorb and emit the electromagnetic radiation from the cavity. Planck's deliberations have forced one to accept certain assumptions regarding the atomic oscillators. The most important Planck's assumption suggests that the oscillators do not radiate energy continuously:

$$\Delta E = h\nu \qquad (13.3)$$

where
   ΔE is energy change
   h is Planck's constant
   ν is oscillator frequency

13.1.3.2   Stefan–Boltzmann's Law
The total power radiated by a blackbody can be obtained by integrating Planck's formula (13.2) within the entire electromagnetic spectrum from $\lambda = 0$ to $\lambda = \infty$:

$$M_\lambda(T) = \frac{2\pi^5 k^4}{15c^2 h^3} T^4 = \sigma \times T^4 \quad [\text{W/m}^2] \qquad (13.4)$$

where σ is Stefan–Boltzmann's constant, equal to 5.67×10^{-12} [W/(cm^{2}K^4)].

This relationship is called the Stefan–Boltzmann's law in honor of the explorers. The most important finding of this model is that the total power of radiation emitted by a blackbody is proportional to the fourth power of its temperature [12].

These rights and relationships are valid only for objects that are ideal emitters, i.e., perfect blackbody.

13.1.4 BLACKBODY

The body that completely absorbs the incident radiation, regardless of wavelength, is called the perfect blackbody. According to the Gustav Robert Kirchhoff law (1824–1887), a body capable of absorbing radiation at any wavelength is also capable of emitting this radiation.

This is an idealized model, because there is no perfect blackbody in nature. It can be used only for theoretical considerations. Models similar to the blackbody are used instead. A radiant cavity may be an example of such a model, which is used for calibration of infrared cameras. The radiation reflected from the inner walls is partially dispersed. The internal walls of the cavity should have the same temperature. Another example can be a sphere with a small hole, through which radiation can enter. Then, the radiation is dispersed as mentioned earlier.

At this point, it is necessary to mention the basic principle by which it is possible to use thermal imaging as a measurement tool.

Any body with a temperature above absolute 0, that is −273.15°C emits energy. The value of the energy emitted increases with increasing temperature of the object. Thus, the body temperature can be determined by measuring the energy, particularly in the infrared band. Objects whose temperature is lower than 500°C emit radiation only in the infrared band. While objects with a temperature T more than 500°C sends also part of the radiation in the visible band, it is possible to estimate the temperature of an object by its color.

13.1.4.1 Blackbody and Real Body

Until now, all considerations were related to idealized bodies that cannot be found in nature. Real bodies are capable of absorption, transmission, and reflection of the electromagnetic radiation. There are three factors:

1. Absorption coefficient is the ratio of the absorbed radiation to total radiation incident on a given surface:

$$a = \frac{\Phi_a}{\Phi}$$

2. Transmission coefficient is the ratio of transmit radiation to the total radiation incident on a given surface:

$$\tau = \frac{\Phi_\tau}{\Phi}$$

3. Reflection coefficient is the ratio of reflected radiation to total radiation incident on a given surface:

$$r = \frac{\Phi_r}{\Phi}$$

The sum of these three factors is always equal to 1:

$$a + \tau + r = 1 \tag{13.5}$$

For opaque objects, transmission coefficient is eliminated.

Equation 13.5 changes as follows:

$$a + r = 1 \tag{13.6}$$

The emitted thermal radiation flux Φ_e is usually different for many bodies, despite their temperature remaining same. This is because all the bodies have a certain ability to emit radiation, which is characterized by the factor called emissivity [1].

13.1.5 EMISSIVITY COEFFICIENT

Emissivity is a parameter characterizing the physical properties of radiation of the real bodies. Emissivity of the object depends on its physicochemical parameters [5]. In particular, they are the type of material, type of surface, surface geometry, the wavelength, the angle of incidence, and the surface temperature. There are also some less important parameters [13].

The emissivity of the object has a characteristic value. Knowledge of this parameter is necessary to make quantitative measurement of temperature using infrared camera.

The emissivity ε of the body for the whole radiation spectrum, called the total emissivity, is a ratio of emittance $M(T)$ from the given surface of the body to the emittance $M_B(T)$ of the blackbody at the same temperature:

$$\varepsilon = \frac{M(T)}{M_B(T)} \tag{13.7}$$

where
  T is temperature
  $M(T)$ is emittance of measured object
  $M_B(T)$ is emittance of blackbody

Blackbody emissivity is equal to 1 ($\varepsilon = 1$), while all other objects found in nature have emissivity in the range between 0 and 1. There is also a third kind of radiation source. They are selective radiators for which ε depends on the wavelength λ.

Objects with smooth surfaces and polished surfaces have very low emissivity, about 0.1. For comparison, the emissivity of human body is ~0.98 and is close to the value of blackbody emissivity [10].

Precise knowledge and correct choice of emissivity in the infrared measurement are very important. There are several ways of determining the emissivity of the object. The most popular of these methods is described in the following. Of course, it has also some limitations. Nevertheless, it is mostly used in practice. This method relies on determining the emissivity by comparing the temperature of the test object with the temperature of the sample of known emissivity (point of reference). This sample can be a piece of tape stuck on an object or paint applied to the object. The most serious limitation is the need to intervene in the structure of the object. The process of determining the emissivity coefficient with this method is as follows:

- Apply conductive paint or stick a tape on the surface of the test object.
- Increase the temperature of the test object to reduce the impact of ambient temperature on the measurement results.
- Set the measuring point of infrared camera on the object in the place where the benchmark of known parameters was applied and check the temperature value.

- Direct the measurement point on the test object next to the reference point.
- Set the value of emissivity for which the temperature indicated by the camera was equal to the temperature on the reference point [5].

The second section describes the influence of emissivity coefficient changing in camera on the result of temperature measurement.

Emissivity of the object not only is a function of its temperature but also depends on the wavelength λ, time t, and angle of view θ [9]:

$$\varepsilon = f(\vartheta, \lambda, t, \theta) \tag{13.8}$$

During the thermovision measurements, the knowledge about the changes of emissivity ε as a function of angle of view θ is not always known. Mostly, only the total emissivity for zero angle of view is known. The question is: for which values of the angle of view the temperature measurement using infrared camera is reliable? An attempt to answer this question is presented later in this chapter.

13.1.6 TRANSMITTANCE OF THE ATMOSPHERE

Earth's atmosphere is a mixture of different gases, which absorb electromagnetic radiation, including infrared radiation. The absorption is more or less depending on the wavelength (Figure 13.3). Molecules of water vapor, carbon dioxide, and ozone take part in this process. The number of molecules increases with increase in the distance between the camera and the object, which of course change the attenuation by the atmosphere. External factors such as fog, smoke, and fumes may also cause loss of signal coming from the object.

There are two spectral bands in the infrared band, in which atmospheric attenuation is negligible. These bands are called "atmospheric windows" and contain the ranges of wavelengths:

- $3 \div 5$ μm (medium-wave range)
- $8 \div 14$ μm (long-wave range)

Because of these "atmospheric windows," thermal imagers are divided into two groups: medium term and long term.

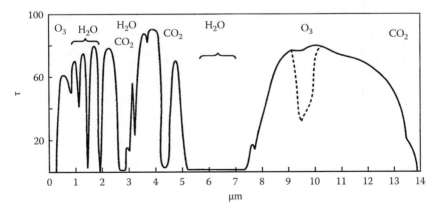

FIGURE 13.3 The transmittance of the atmosphere τ depending on the wavelength λ. (From Sabins, F.F., *Remote Sensing—Principles and Interpretation*, W.H. Freeman, San Francisco, 1978; Więcek, B., Wybrane zagadnienia współczesnej termowizji w podczerwieni, Politechnika Łódzka, Łódź, 2010.)

Several research centers around the world are studying the problem of atmospheric absorption. Several models of atmospheric transmittance were developed based on empirical data (different seasons and different height above sea level) and the theoretical analysis. These models include parameters that directly affect the number of molecules absorbed on the distance between the camera and the object (temperature, humidity, type, and size of molecules of water vapor, height, and many others). The most popular models are LOWTRAN and MODTRAN [5].

Some of the infrared cameras have the possibility of attenuation coefficient directly to the camera. A prerequisite is knowledge of this parameter, which is sometimes difficult to determine. Rough estimate of the value of attenuation affects the resultant error of temperature measurement using infrared camera. Other cameras calculate this parameter using a complicated algorithm, "hidden" inside the camera.

13.1.7 Mathematical Model of Temperature Measurement Using Infrared Camera

An infrared camera measures the radiation in a narrowband of wavelengths. Most cameras on the market register the radiation wavelength between 8 and 14 μm. These devices have arrays of uncooled bolometric detectors (focal plane array [FPA]), working at ambient temperature. This makes them much more popular than their counterparts with detectors cooled. Both types of cameras record the radiation coming from the object, as well as the radiation reflected from it and coming directly from the atmosphere. The last two types of radiation can be regarded as interference, which should be eliminated. Unfortunately, it is not always possible to measure the temperature of thermal imaging camera in ideal conditions, such as in the laboratory.

Infrared camera–trained operators must be aware of the disturbances affecting the measurement (sun, sky, earth, clouds, etc.). They should try to reduce some interference, so that their impact was minimal (taking measurements at a specific time of day, to cover the powerful sources of radiation, etc.).

As mentioned earlier, infrared camera receives radiation from several sources (Figure 13.4).

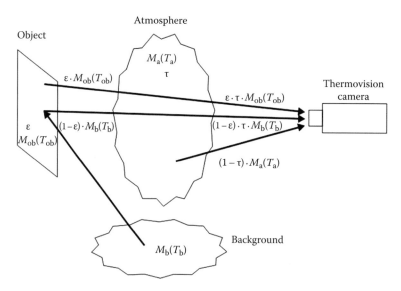

FIGURE 13.4 Components of the radiation measured by the infrared camera. (From Więcek, B., Wybrane zagadnienia współczesnej termowizji w podczerwieni, Politechnika Łódzka, Łódź, 2010. With permission.)

The following indications are used in Figure 13.4:

ε is emissivity of the object
τ is transmission coefficient of the atmosphere between the camera and object
$M_{ob}(T_{ob})$ is power emitted by an object
T_{ob} is temperature of the object
$M_a(T_a)$ is power emitted by the atmosphere
T_a is temperature of the atmosphere
$M_b(T_b)$ is power emitted by the background
T_b is background temperature

Finally, the total power that enters the thermal imager can be given as follows:

$$M = \varepsilon \times \tau \times M_{ob}(T_{ob}) + (1 - \varepsilon) \times \tau \times M_b(T_b) + (1 - \tau) \times M_a(T_a) \qquad (13.9)$$

The next step in thermal measurement is the conversion of the radiation to the electrical signal (voltage), which can be provided in a simple way:

$$U = A \times M \qquad (13.10)$$

where A is constant of the camera, describing the conversion of radiation in the detector and the amplification circuits inside the camera.

13.2 EXPERIMENTAL

13.2.1 MEASUREMENT CONDITIONS

Measurements were carried out under laboratory conditions using a setup consisting of a camera and an infrared radiator. The place of study was a laboratory of the Division of Metrology and Optoelectronics at University of Technology, Poland. The test conditions were strictly defined. The scheme of this set is shown in Figure 13.5. The main element of infrared radiator is

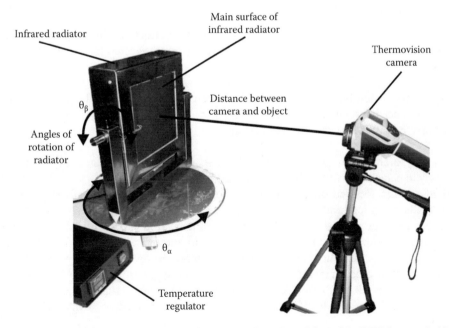

FIGURE 13.5 A laboratory setup used in experimental studies. (From Litwa, M., *IEEE Sensors J.*, 10, 1552, 2010. With permission.)

a flat surface. This surface has constant temperature ϑ_z and specific emissivity coefficient ε. The angle of view can be set in two perpendicular axes. The infrared radiator was made of aluminum. To obtain a homogeneous temperature field, the dimensions of radiator were equal to 210 mm × 210 mm × 40 mm. The block has been warming up with a flat heater. One-way flow of the radiation was forced to get sufficient homogeneity of the field of temperatures—the back and sides were isolated from the surroundings. The aluminum block was exposed to anodic oxidation process to get high value of emissivity coefficient. The emissivity coefficient is $\varepsilon = 0.98$. The value was determined by comparing the temperature value ϑ_x measured using the thermovision camera with the contact measurement ϑ_z using the sensor Pt1000. A PID controller was used to stabilize the surface temperature.

One of the main objectives of the study was to compare the largest possible amount of infrared cameras with different parameters. All cameras have worked in long-wave range. The limits of ranges slightly differed for several infrared cameras (Table 13.2). Each of them had a microbolometer array detectors, operating at ambient temperature. Unfortunately, it was impossible to use all cameras in all measurements.

The following infrared cameras were used in experimental studies:

- FLIR ThermaCAM E25 (model no longer available for sale)
- Fluke Ti20
- Fluke Ti32
- NEC TH7800
- VIGO v50
- Chauvin Arnoux RAYCAM CA 1888

Table 13.2 represents a comparison of the parameters of the earlier infrared cameras. In the experiment, infrared cameras were used with different spatial resolution, thermal resolution, matrix size, angle of view, and temperature ranges.

13.2.2 Description of Studies

The main aim of the experimental studies was to show the effect of selected parameters on temperature measurement using infrared camera. The parameters selected for experiments are listed as follows:

1. The angle of view between the axis of the camera and the normal of the object
2. The distance between the camera and the object
3. Emissivity inflicted in the infrared camera

The first two parameters relate to the attributes associated with an object, while the third factor relates only to parameter, which can be set into the camera.

The discussion focused on the determination of the limit of error of temperature measurement. The technical documentation of these cameras [14–19] describes the measurement uncertainty as the greater of value: ±2°C or ±2% of measured value. Only in the camera from Chauvin Arnoux company, uncertainty is defined in another way: ±(2°C + 2%). This slight difference in the mathematical notation generates large differences in the value of the error, e.g., inaccuracy of most cameras used in experiments for an object at 50°C is equal ±2°C, while for the camera RAYCAM CA 1888 is equal ±3°C. These values are deteriorating with increasing temperature (to 100°C, similarly to the earlier inaccuracies are ±2°C and ±4°C, respectively).

TABLE 13.2
Comparison of Cameras Used in Experimental Studies

	FLIR E25	VIGOCAM v50	NEC 7800	Fluke Ti20	Fluke Ti32	Chauvin Arnoux 1888
Temperature range	−20 to 250	−20 to 120 and −10 to 450	−20 to 100 and 0 to 250	−10 to 550	−10 to 600	−20 to 250 and 200 to 600
Accuracy	±2°C or 2% (the greater of value)	±2°C or 2% (the greater of value)	±2°C or 2% (the greater of value)	±2°C or 2% (the greater of value)	±2°C or 2% (at 25°C, the greater of value)	±(2°C + 2%)
Detector type, resolution	Uncooled focal plane array (microbolometer) (FPA), 160 × 120	Uncooled focal plane array (FPA), 384 × 288	Uncooled focal plane array (microbolometer) (FPA), 320 × 240	Uncooled focal plane array (microbolometer) (FPA), 128 × 96	Uncooled focal plane array (microbolometer) (FPA), 320 × 240	Uncooled focal plane array (microbolometer) (FPA), 384 × 288
Thermal sensitivity (NETD)	0.2°C at 30°C	<0.065°C at 30°C	0.1°C at 30°C	0.2°C	0.05°C at 30°C (50 mK)	0.08°C at 30°C (80 mK)
NUC	Automatic	Automatic or manual	Automatic	Automatic	Automatic	Automatic or manual
Spectral range / Visual camera	7.5 to 13 μm —	8 to 14 μm 640 × 480 pixels (0.3 mega)	8 to 14 μm 752 × 480 pixels (0.41 mega)	7.5 to 14 μm —	7.5 to 14 μm 2.0 megapixels	8 to 14 μm 640 × 480 pixels (0.3 megapixels)
Lens	Standard	Wide angle lens	Standard	Standard	Standard	Standard
Field of view	19° × 14°	30° × 23°	27° × 20°	20° × 15°	23° ×17°	24° × 18°
Spatial resolution (IFOV)	No data in documentation	1.4 mRad	1.5 mRad	2.8 mRad	1.25 mRad	1.3 mRad
Minimal focal distance	0.3 m	0.5 m	0.5 m	0.15 m	0.15 m	0.1 m
Focus regulation	Manual	Manual	Manual	Manual	Manual	Manual
File format	Radiometric JPEG	Images and sequences are recorded on the SD card or computer hard drive via Ethernet	1000 thermal images with digital photos	Radiometric 100 images	Non-radiometric (.bmp lub.jpeg) or full radiometric (.is2)	Radiometric
Battery operating time	2 h	2 h	2.5 h	3 h	4 h	3 h
Weight	0.7 kg	1.5 kg	1.3 kg	1.2 kg	1.05 kg	No data in documentation

(continued)

TABLE 13.2 (continued)
Comparison of Cameras Used in Experimental Studies

	FLIR E25	VIGOCAM v50	NEC 7800	Fluke Ti20	Fluke Ti32	Chauvin Arnoux 1888
Operating temperature range	−15°C to 45°C	−20°C to 40°C	−15°C to 45°C	0°C to 50°C	−10°C to 50°C	No data in documentation
Storage temperature range	−40°C to 70°C	−30°C to 70°C	−40°C to 70°C	−25°C to 70°C	−20°C to 50°C	No data in documentation
Case protection classification	IP54	IP54	IP54	IP54	IP54	IP54
Correcting functions						
Emissivity	Yes	Yes	Yes	Yes	Yes	Yes
Atmosphere temperature	Yes	Yes	Yes	No	No	Yes
Background temperature	No	Yes	Yes	No	Yes	No
Humidity	No	Yes	Yes	No	No	Yes
Distance	No	Yes	Yes	No	No	Yes
Atmosphere transmittance	No	No	No	No	Yes	No

13.2.2.1 Influence of Angle of View on the Temperature Measurements Using Infrared Camera

Experimental measurements were carried out in the system of Figure 13.5 as follows:

1. The camera and radiator were situated to obtain the established distance l and values of angles $\theta_\alpha = 0$ and $\theta_\beta = 0$.
2. The chosen temperature was set ϑ_z.
3. The angle θ_α has been changing ($\theta_\beta = 0$).
4. The angle θ_β has been changing ($\theta_\alpha = 0$).

Four infrared cameras were used in the measurements:

1. FLIR ThermaCAM E25
2. Fluke Ti20
3. NEC TH7800
4. VIGO v50

The influence of angle of view was estimated on the basis of the analysis of values of the measurement error. The relative error of the measurement δ_ϑ has been calculated using the following dependences:

- The absolute error of the measurement of the temperature Δ_ϑ

$$\Delta_\vartheta = \vartheta_x - \vartheta_{pop} \tag{13.11}$$

where

ϑ_x is the value measured with the thermovision camera for the given angle of view, suitably θ_α and θ_β

ϑ_{pop} is expected correct value of the temperature measured with thermovision camera for angles $\theta_\alpha = 0°$ and $\theta_\beta = 0°$

- The relative error of measurement of temperature δ_ϑ

$$\delta_{\vartheta\%} = \frac{\Delta_\vartheta}{\vartheta_{pop}} \times 100\% \tag{13.12}$$

The measurements were made for three values of temperature $\vartheta_z = 100°C$ and $200°C$ and for the distance $l_1 = 0.55$ m, $l_2 = 1.50$ m, and $l_3 = 3.60$ m. The graphical results of measurements for $l_1 = 0.55$ m made with camera FLIR ThermaCAM E25, Fluke Ti20, NEC TH7800, and with camera VIGO v50 are presented in literature [3,4].

In Table 13.3, the critical values of the angle of view θ_g for three tested distances l are shown. Appointed values θ_g include in the interval from 47° to 60°. Measurements of the temperature with thermovision cameras should be executed for the angles of view smaller than 45°. Measurements for larger angles of view can be burdened with additional considerable error.

13.2.2.2 Influence of Distance between the Camera and the Object on Temperature Measurement Using an Infrared Camera

According to the theory presented in Section 13.1.6, the atmosphere located between the camera and the object absorbs radiation. The attenuation is greater or smaller depending on the type of

TABLE 13.3

Critical Values Θ_g of the Angle of View

Length between the Camera and the Object	Camera Type			
	ThermaCAM E25	Ti20	TH7800	VIGO v50
	Critical Values of Angle of View Θ_g			
$l_1 = 0.55$ m	53°	58°	58°	53°
$l_2 = 1.50$ m	54°	57°	60°	47°
$l_3 = 3.60$ m	57°	57°	60°	47°

molecules in the atmosphere. It is evident even in the laboratory, where the measurement distances often do not exceed 10 or 15 m.

Experimental measurements were carried out in the system of Figure 13.5 as follows:

1. Infrared camera and radiator were placed to obtain the established distance l for the values of angles of view θ equal to 0 ($\alpha = 0$ and $\beta = 0$).
2. The chosen temperature was set ϑ_z.
3. The distance between the camera and the radiator was changed (without changing this parameter in camera—measurements at 1 m) maintaining constant external parameters (humidity, atmosphere temperature, background temperature, angles of view).
4. The distance between the camera and the radiator was changed (with change in this parameter in camera) maintaining constant external parameters (humidity, atmosphere temperature, background temperature, angles of view).

Four infrared cameras were used in the measurements according to point 3:

1. FLIR ThermaCAM E25
2. Fluke Ti32
3. Chauvin Arnoux RAYCAM CA 1888
4. VIGO v50

While in point 4, cameras that allow for correction of distance l are listed as follows:

1. Chauvin Arnoux RAYCAM CA 1888
2. VIGO v50

The measurements were carried out for two values of temperature: 50°C and 150°C. Graphical representation of the results of the influence of distance on the temperature measurement results is shown in Figures 13.6 through 13.9.

13.2.2.3 Influence of Emissivity ε Changing in the Infrared Camera on Temperature Measurement

Experimental measurements were carried out in the system of Figure 13.5 as follows:

1. Infrared camera and radiator were situated within a certain distance $l = 1$ m.
2. The chosen temperature was set ϑ_z.
3. The value of emissivity inside the camera was changing, maintaining constant external parameters (humidity, atmosphere temperature, background temperature, angles of view, distance between the camera and the object).

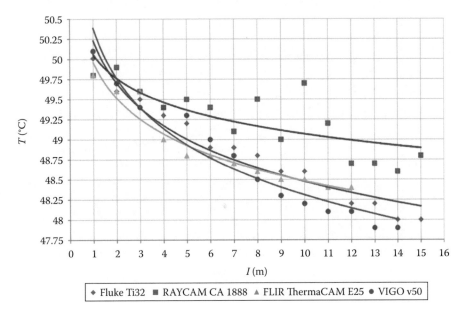

FIGURE 13.6 The influence of distance l between camera and object for object temperature $\vartheta_z = 50°C$. (From Litwa, M., *IEEE Sensors J.*, 10, 1552, 2010. With permission.)

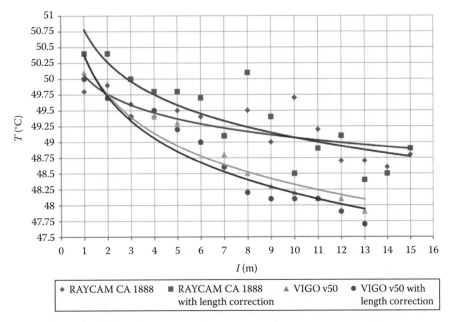

FIGURE 13.7 The influence of distance l between camera and object for object temperature $\vartheta_z = 50°C$ (for cameras with length correction). (From Litwa, M., *IEEE Sensors J.*, 10, 1552, 2010. With permission.)

Five infrared cameras used in the measurements are listed as follows:

1. FLIR ThermaCAM E25
2. Fluke Ti32
3. Chauvin Arnoux RAYCAM CA 1888
4. VIGO v50
5. NEC TH7800

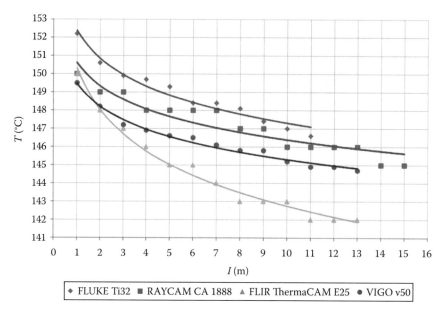

FIGURE 13.8 The influence of distance l between camera and object for object temperature $\vartheta_z = 150°C$. (From Litwa, M., *IEEE Sensors J.*, 10, 1552, 2010. With permission.)

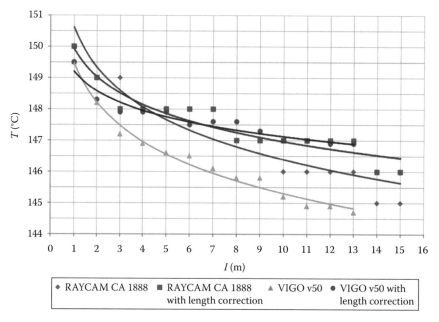

FIGURE 13.9 The influence of distance l between camera and object for object temperature $\vartheta_z = 150°C$ (for cameras with length correction).

Measurements were carried out for five selected values of temperature: 40°C, 80°C, 120°C, 160°C, and 200°C. Selected graphical representation of the results of the influence of distance on the temperature measurement results is shown in Figures 13.10 through 13.12.

A broad convergence of a series of measurements for different cameras can be observed during the analysis of the obtained graphs (Figures 13.10 through 13.12). Only for the lowest temperature (40°C), the curve representing the camera VIGO v50 differs from the others. However, the trend is

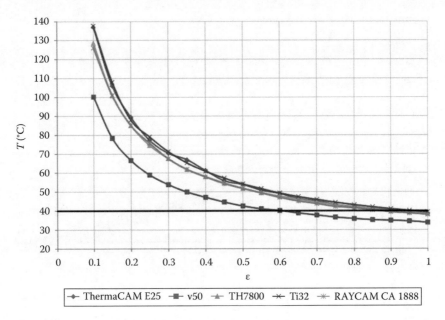

FIGURE 13.10 The influence of emissivity coefficient changing in camera on temperature measurements for object temperature $\vartheta_z = 40°C$.

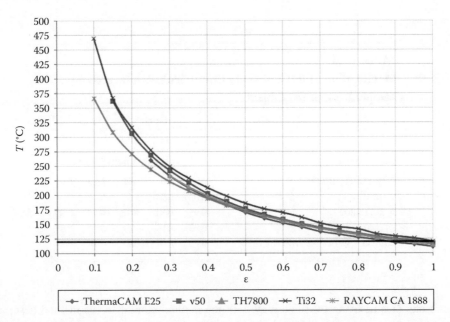

FIGURE 13.11 The influence of emissivity coefficient changing in camera on temperature measurements for object temperature $\vartheta_z = 120°C$.

similar to other curves. Inadequate assessment of the value of the test object emissivity can lead to large measurement errors.

An example showing how to determine the values of emissivity was presented in Section 13.1.5. This is the most popular method. Unfortunately, it has some limitations. It is impossible to simply apply the tapes on surface, where the temperature is more than 100°C. During thermovision measurements, a thermographer should be aware of the fact that the value of emissivity also changes with temperature of the object.

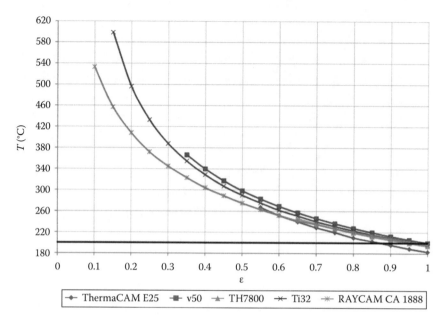

FIGURE 13.12   The influence of emissivity coefficient changing in camera on temperature measurements for object temperature $\vartheta_z = 200°C$.

TABLE 13.4
Modification of the Emissivity Coefficient with Increasing Temperature of the Object

	Temperature (°C)				
	40	80	120	160	200
Camera Model	Emissivity Coefficient ε				
FLIR ThermaCAM E25	0.9	0.9	0.88	0.88	0.87
VIGO v50	0.6	0.88	0.95	0.99	1
NEC TH7800	0.9	0.91	0.93	0.95	0.95
Fluke Ti32	0.95	0.97	1	0.99	0.97
Chauvin Arnoux RAYCAM CA 1888	0.88	0.97	0.98	0.98	0.97

The results of measuring the emissivity with the method similar to the method described in Section 13.1.5 are presented in Table 13.4. A Pt1000 sensor was used instead of sticking the tape to the object. It was placed inside a hole drilled in the emitting surface of the radiator. This eliminates the influence of external factors to a large extent. In the next step, the emissivity coefficient inside the camera was changing to display the temperature on the camera screen equal to the temperature indicated by the reference point (measured by the sensor).

By analyzing the values obtained in Table 13.4, it cannot be clearly determined how the emissivity is changing as a function of temperature. Some of them showed an increasing trend (VIGO v50 and NEC TH7800), one of them—decreasing (FLIR ThermaCAM E25), an initially increasing and then decreasing (Fluke Ti32), and the last one—a constant (Chauvin Arnoux RAYCAM CA 1888).

13.3 SUMMARY

The advantage of measuring the temperature using infrared camera lies in its simplicity. The thermovision camera is a perfect tool for noncontact measurement. However, the reliability of thermovision in measuring the temperature depends on essential factors including emissivity coefficient ε and angle of view. It occurs that the distance between the camera and the object (more precisely the composition of the atmosphere) is also very important. Not maintaining this condition may cause an additional, substantial measurement error. The problem with keeping suitably small value of the angle of view occurs during measurements under conditions that prevent proper location of the observed object, such as machine bearings, located near infrared assessment of buildings, the observation of objects with complex shapes.

This chapter presents the results of experimental studies on the influence of angle on temperature measurements using infrared camera. It was found that if the angle of view does not cross 45° the result of the measurement of the temperature is sufficiently reliable. It was also demonstrated how the changes indicate the camera with the increase in its distance from the object. The last stage of this study was to demonstrate the importance of proper valuation of the emissivity of the object.

All the earlier-described parameters have influence on the result of temperature measurement using infrared camera. Everyone who performs thermovision measurements should avoid generating unnecessary errors.

REFERENCES

1. Bielecki, Z. and Rogalski, A. Detekcja sygnałów optycznych, WNT, Warszawa, 2001.
2. Epperly, R.A., Heberlein, G.E., and Eads, L.G. Thermography, a tool for reliability and safety, *IEEE Industry Applications Magazine*, 5, 1999, 28–36.
3. Litwa, M. Influence of angle of view on temperature measurements using thermovision camera, *IEEE Sensors Journal*, 10, 2010, 1552–1554.
4. Litwa, M. and Wiczyński, G. Wpływ kąta obserwacji na wynik pomiaru temperatury kamerą termowizyjną, *ELEKTRONIKA—konstrukcje, technologie, zastosowania*, 6, 2008, 147–148.
5. Madura, H. Pomiary termowizyjne w praktyce, PAK, Warszawa, 2004.
6. Minkina, W., Rutkowski, P., and Wild, W. Podstawy pomiarów termowizyjnych, część I—Istota termowizji i historia jej rozwoju, część II—Współczesne rozwiązania *systemów* termowizyjnych, błędy metody, *Pomiary Automatyka Kontrola*, 46(1), 2000, 7–14.
7. Minkina, W. Pomiary termowizyjne—przyrządy i metody, Wydawnictwa Politechniki Częstochowskiej, Częstochowa, 2004.
8. Sabins, F.F., Jr. *Remote Sensing—Principles and Interpretation*, W.H. Freeman, San Francisco, 1978.
9. Sala, A. Radiacyjna wymiana ciepła, WNT, Warszawa, 1982.
10. Taimarov, M.A., Rusev, K.A., and Garifullin, F.A. Directional emissivity of structural materials, *Journal of Engineering Physics and Thermophysics*, 49(2), 1985, 939–942.
11. Więcek, B. Wybrane zagadnienia współczesnej termowizji w podczerwieni, Politechnika Łódzka, Łódź, 2010.
12. Więcek, B. and De Mey, G. Termowizja w podczerwieni. Podstawy i zastosowanie, Wydawnictwo PAK, Warszawa, 2011.
13. Xiru, X., Chen, L., and Zhuang, J. The passive measurements of object's directional emissivity in laboratory, *Science in China Series E: Technological Sciences*, 43(Suppl 1), 2000, 55–61.
14. Specification of thermovision camera TheramCam E25, Flir, http://www.instruments4hire.co.uk/ThermaCAM%20E25%20EN.pdf, accessed on June 10, 2011.
15. Specification of thermovision camera Ti20, Fluke, http://www.pqmeterstore.com/crm_uploads/fluke_ti20_thermal_imager_datasheet.pdf, accessed on June 10, 2011.
16. Specification of thermovision camera Ti32, Fluke, http://www.transcat.com/PDF/Ti32_Ti25_Ti10.pdf, accessed on June 10, 2011.
17. Specification of thermovision camera v50, Vigo, http://vigo.com.pl/index.php/en/content/download/2233/9235/file/VIGOcam%20v50.pdf, accessed on June 10, 2011.
18. Specification of thermovision camera TH7800, NEC—Avio, http://www.nec-avio.co.jp/en/products/ir-thermo/pdf/th7800-e.pdf, accessed on June 10, 2011.
19. Specification of thermovision camera RAYCAM C.A 1888, Chauvin—Arnoux, http://www.chauvin-arnoux.com/display.asp?10922, accessed on June 10, 2011.

14 Adaptive Sensors for Dynamic Temperature Measurements

Paweł Jamróz and Jerzy Nabielec

CONTENTS

14.1 INTRODUCTION

Rapid development of motorization at the beginning of the twentieth century was the reason for engineers to set more and more difficult tasks. They wanted to increase their knowledge of physico-chemical processes taking place in such hardly accessible places like interior of combustion engine and particularly in the combustion chamber. Nevertheless, the rate of such processes to be analyzed was the biggest difficulty standing in the way of development of system designed for such applications. An attempt to design a sensor with sufficiently fast measurement properties led to a sensor which had to be very thin and thus very prone to damage. A sensor which might have been sufficiently durable would have been very thick and thus completely useless for planned measurements.

A very inventive measurement method was proposed by Pfriem [1], which theoretically allowed for this contradiction to be overcome. By means of two different massive sensors, each of which does not meet the imposed requirements considering its transmission bandwidth, it is possible to obtain correct measurement results while measuring the rapidly changing signal. Nevertheless, the practical implementation of this idea requires the use of some of the very advanced numerical methods.

Unfortunately, such tools were not available at the beginning of the twentieth century. Despite these inconveniences, the scientists and engineers of that time were able to build adequate mathematical models describing the dynamics of investigated processes. Present instruments allow for practical implementation of the method, which was proposed dozens years earlier. A new area in which the measurement of high rate gas processes found its applicability was found in studies of turbo compressors driven by means of engine's exhaust gases just behind the exhaust valves.

Moreover, modern simulation tools revealed some new features of this method and indicated new areas of its applicability, including the measurements of rapidly changing time courses in power networks or in modern inverter electric drives. It may also be used in order to investigate the respiration ventilation when the composition of flowing gas is undeterminable due to changes in humidity between inspired and expired air.

14.2 TWO-SENSOR CORRECTION METHOD

In the literature, this method is also often referred to as the adaptive or the "blind" method. It allows for very difficult measurement of rapidly changing signal $u_M(t)$ to be performed utilizing the pair of sensors with different yet unknown dynamic properties, which are specified as their dynamic coefficients $f_M(t)$ and $g_M(t)$ for the first and the second sensors, respectively. The subscripts "M" denote the variables as measurands, the values of which are unknown. Variables without these subscripts stand for signals, which are used to estimate the measurands on the basis of results of measurements or data processing. Intentionally, the concept of "time constant" is not used here, since it refers only to quantities which are invariable in time. In the case under consideration, the dynamic properties of sensors are time-dependent and change as quickly as does the measured signal. These time-varying dynamic coefficients are determined directly at the measurement site and under the sensors' operational conditions by means of measured signal serving as the only excitation during the process of their identification. Recorded and processed are only the $x(t)$ and $p(t)$ signals which reproduce the output signals of the pair of sensors $x_M(t)$ and $p_M(t)$, denoted, respectively, for the first and the second sensors. These variables denote the instantaneous temperature of sensors, which are determined indirectly by means of measuring the instantaneous sensor resistances or its thermo-powers. The design of measurement system under consideration is presented in Figure 14.1. The $(f_M * u_M)(t)$ symbol stands for the convolution in time domain of measurand $u_M(t)$ and the dynamic coefficient $f_M(t)$ of the first measurement channel. Analogically, the $(g_M * u_M)(t)$ symbol stands for the convolution of $g_M(t)$ and $u_M(t)$ variables of the second measurement channel. The t_i symbol denotes the moments of sampling the measured signals by A/D converters.

The mathematical model of dynamic process of heat exchange between the sensor and the gas flowing around at variable velocity is described by means of the first-order ordinary differential equation comprising product of the time-dependent $f(t)$ and the first derivative of the sensor temperature $x(t)$. The second measurement channel is characterized by a differential equation of identical structure comprising the $g(t)$ term and $p(t)$. The common quantity of both equations is the sought estimate $u(t)$. Therefore, it is convenient to rewrite the system of these two equations in the form (14.1). Every differential equation should be supplemented by initial conditions in order to make it solvable. Nevertheless, in the case under consideration, we do not seek for the solution of the pair of equations with respect to $x(t)$ and $p(t)$, since these quantities are in fact available through the measurement. Thus, the solution of inverse problem in the form of unknown variables $f(t)$ and $g(t)$ and the $u(t)$ term in turn is what we are searching for:

$$f(t)\frac{dx(t)}{dt} + x(t) = u(t) = g(t)\frac{dp(t)}{dt} + p(t), \quad x(t), p(t) \tag{14.1}$$

In order to solve such system of equations, it becomes necessary to add an additional equation describing the dependence between two of these three unknowns. Considering the fact that no "a priori" information on measured signal $u(t)$ is available, the only obtainable information is a dependence relating the dynamic properties of two sensors used. In this relationship $g(f)$, time must not be expressed in explicit form.

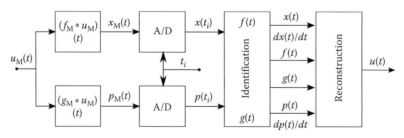

FIGURE 14.1 Structure of the system for the "blind" correction method.

$g(t) = Cf(t)$, where $C \in \Re^+$, is the most convenient form of such a relationship. Such a relation was described for the first time by Pfriem [1] and it expressed the ratio of heat transfer coefficients of two sensors.

Employment of the Wiener–Hammerstein models for measurement channels allows for extension of applicability of this method to sensors with nonlinear static characteristics.

14.3 STUDIES AND IMPLEMENTATIONS OF THE TWO-SENSOR METHOD PERFORMED UP TO DATE

At the turn of the twentieth and the twenty-first centuries, published were some new studies the leading point of which was to measure the variable temperature of gas flowing through at variable velocities. The main need for performing such measurements was raised by the car industry.

All these studies employ the method proposed by Pfriem [1] although introducing certain modifications into it. The essence of the proposed solution consists in employment of two first-order sensors (thin thermoresistors or thermocouples) of different dynamic properties with the assumption that coefficients occurring in differential equations at first-order derivatives are proportional to each other. In certain amount of earlier publications, this coefficient was considered a constant value and referred to as the time constant [2,3]. Such an assumption leads to the formal unambiguous solution of identification problem [4].

Nevertheless, such approach does not provide a faithful description of the process of heat exchange between the flowing gas and a sensor. This may present a satisfactory approximation of dynamic phenomena of heat exchange but only in the case of small range of gas velocity changes.

In more recent publications, these coefficients were considered as time varying and determine an instantaneous value of the heat transfer coefficient, which depends significantly on instantaneous velocity of gas flowing over the sensor [5].

Kar et al. [6] employed the Kalman filter in order to identify the pair of time-varying dynamic coefficients. However, omission of the relationship between these coefficients leads to negative identification results, despite the very good numerical convergence of the task.

The two-sensor method assumes the utilization of additional relationship describing the ratio of time-varying dynamic coefficients of sensors. The value of the coefficient of proportionality described in the literature so far is exponentially dependent only on the ratio of sensors' diameters. Exponent occurring in this dependence is strictly dependent on adopted mathematical description of dynamic properties of sensors and adopted estimates of dimensionless numbers in fluid dynamics and heat transfer [1,7–9].

On the other hand, the coefficient of proportionality does not depend on physical properties of the flowing medium. Nevertheless, Cambray [10] stressed that this condition is met as long as the fluctuations of physical parameters of gas are not extensively large, provided that the thermocouple sensors do not vary excessively between each other.

Such assumption of low diversification of dynamic properties of sensors and low fluctuations of measured signal leads to unfavorable numerical conditioning of identification task [11].

Vachon et al. [12] presented is also the method of experimental determination of the coefficient of proportionality at individual instants, for which the values of the temperature responses of two sensors are equal.

14.4 IDENTIFICATION OF DYNAMIC COEFFICIENTS

Solutions of the task of dynamic coefficients identification employing the "blind" method utilizing the proportional relationship between them have already been published in the literature. Then, Equation 14.1, allowing for the identification of the dynamic coefficient $f(t)$, is reduced to

$$f(t) = \frac{p(t) - x(t)}{(dx(t)/dt) - C(dp(t)/dt)} \tag{14.2}$$

It is crucial that the coefficient of proportionality C be determined also at the site and at the time of measurement by means of the "blind" method too.

This is a valuable feature, since the value C determined under laboratory conditions may not correspond to the actual value of this coefficient at the site where sensors are used. The physical properties of gas influence the value of the coefficient of proportionality C of the pair of differently shaped sensors (sphere–cylinder) [13].

The response of the faster sensor intersects the response of the slower one at least twice per period. Such an observation is correct provided that both recordings are performed with the same scale appropriate to the measured signal utilizing the static parameters of sensors. It happens that the constant component of recorded responses recalculated to the scale of measured signal does not coincide the constant component of measured signal. This phenomenon is caused by the fact that in the output signal of the sensor, signals the frequencies of which do not occur in the measured signal are generated. It is a result of multiplication of signals of the same frequencies. The velocity at which the combustion gases flow over the sensor is the periodic function with the basic harmonic depending on the rotational speed of the engine. So, the dynamic coefficient describing the process of heat exchange between the gas and the sensor is also a periodic function with the same basic frequency. Also, the instantaneous temperature of combustion gases flowing over the sensor is a periodic function with the very same basic harmonic. In an equation describing the dynamics of this process, a term occurs in which the dynamic coefficients and the derivative of sensor temperature are multiplied by each other. The result of their multiplication is a signal with zero pulsation, that is the constant component and a periodic signal with pulsation doubled compared to the basic one. The generated constant-component signal apparently perturbs the static parameters of the sensor.

For selected moments $t = t_=$ at which the intersection of responses of sensors $x(t_=) = p(t_=)$ occurs, Equation 14.1 is reduced to

$$C = \frac{dx(t)/dt}{dp(t)/dt}, \quad \text{for } t = t_= \tag{14.3}$$

The constant C determined according to relationship (14.3) reaches the same value irrespective of whether it was determined for rising or falling slopes of recorded responses of both sensors. Otherwise, there is no proportion between coefficients $f(x)$ and $g(x)$.

However, for instants $t = t_=$ (14.2) becomes the indeterminate symbol, since its numerator and denominator equal zero. Value of dynamic coefficients for such instants may be determined by means of the l'Hospital's rule.

Reliable determination of the value of dynamic coefficient $f(t)$ in the neighborhood of the point $t = t_=$ encounters also other limitations. In the neighborhood of such points, the denominator (14.2) adopts the values only slightly different from zero. Therefore, any small inaccuracies in determination of numerator in this relationship result in very huge deviations of value of this fraction from its real value. Additional results of our own simulation studies confirmed that the extension of l'Hospital's rule to the neighborhood of the discontinuity point allows for more precise estimation of $f(t)$ compared to the accuracy which can be obtained using (14.2). The second method for correction of dynamic coefficient estimation accuracy in the neighborhood of discontinuity points was also verified. It is based on approximation of $f(t)$ obtained according to (14.2) using a rational function.

14.5 MATHEMATICAL MODEL OF RELATIONSHIP BETWEEN THE DYNAMIC COEFFICIENTS OF TWO SENSORS

With respect to the fact that the instantaneous value of dynamic coefficient depends on instantaneous velocity of gas flowing over the sensor, the criterial numbers describing such process are of crucial importance when describing the value of this coefficient. Of particular importance is the Reynold's number Re, dependent on the velocity of gas, raised to the power of a constant exponent.

Over a span of years, many studies dealing with semiempirical expressions describing the Nusselt number, Nu, being a function of Reynold's and Prandtl numbers, were published. They concern various ranges of its applicability, under various measurement conditions. As a result of their employment in order to describe the dynamic properties of thermosensors, various values of exponent of the power occurring in the relationship between the dynamic coefficients of the pair of sensors were obtained.

In order to confirm the assumptions concerning the usage of the two-sensor method, it became necessary to reanalyze the relationships between dynamic coefficients of sensors on the basis of commonly used formulae describing the Nusselt number. Here just another crucial question emerges—which way to determine Nu number should be used in order to determine $f(t)$ and $g(t)$?

14.5.1 MATHEMATICAL MODEL OF THE SENSOR

Traditionally, discussed is the case of a temperature sensor realized as a thermocouple or the thermoresistive converter and treated as a homogeneous body without the protective shield (Figure 14.2).

The sensor is placed in the stream of gas flowing at the constant velocity. Transforming the equation of the heat balance leads to the differential equation (14.1), where coefficient $f(t)$ next to the first derivative of $x(t)$ is described by

$$f(t) = \frac{\rho \times c \times d^2}{s \times \mathrm{Nu}(t) \times \lambda_a} \tag{14.4}$$

The fraction, traditionally referred to as the "time constant," determines the dynamic properties of the sensor. The value of the this coefficient is directly proportional to sensor heat capacity (product of the sensor: material density ρ, the specific heat of the material c, and the volume determined by diameter d) and inversely proportional to the ability of heat receiving determined by means of the product of Nu, λ_a—the thermal conductivity and shape factor $s = 4$ for the cylinder-shaped thermoresistor or sphere-shaped thermocouple $s = 6$.

The traditional description holds true for analysis of processes where model coefficients are independent of time. When analyzing time-dependent cases, it is necessary to account for the influence of changes of sensor model parameters together with variable measurement conditions. The results of simulation studies performed up to date for certain parameters of unsteady flows [14] show that variable gas velocity has the most prominent effect on the value of dynamic coefficient. The effect of fluctuations in other flow parameters related to temperature or pressure changes on dynamic coefficient is insignificant and may be neglected in practice (in certain range of their magnitude). Such an approach requires that traditional time constant should be replaced by time-dependent coefficient $f(t)$, being in function of gas velocity.

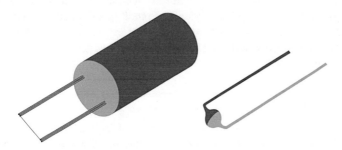

FIGURE 14.2 Sensors for temperature measurement.

TABLE 14.1

Relationships Describing the Nusselt Number Characterizing the Heat Transfer through the Surface of a Cylinder (Model of a Resistance Sensor)

Author of Equation	Range of Usability	Form of the Nusselt Number
King [15]	$0.055 < \mathrm{Re} < 55$	$0.318 + 0.69 \cdot \mathrm{Re}^{0.5}$
Kramers [16]	$0.01 < \mathrm{Re} < 10^4$	$0.42 \cdot \mathrm{Pr}^{0.2} + 0.57 \cdot \mathrm{Pr}^{0.33} \cdot \mathrm{Re}^{0.5}$
McAdams [17]	$0.1 < \mathrm{Re} < 10^3$	$0.32 + 0.43 \cdot \mathrm{Re}^{0.52}$
Van der Hegge Zijnen [18]	$0.01 < \mathrm{Re} < 10^4$	$0.35 + 0.5 \cdot \mathrm{Re}^{0.5} + 0.001 \cdot \mathrm{Re}$
Zukauskas [19]	$1 < \mathrm{Re} < 10^3$	$(0.43 + 0.5 \cdot \mathrm{Re}^{0.5}) \cdot \mathrm{Pr}^{0.38}$
Davis [19]	$0.1 < \mathrm{Re} < 200$	$0.96 \cdot \mathrm{Re}^{0.43} \cdot \mathrm{Pr}^{0.3}$

Source: Jamróz, P., *IEEE Sensors J.*, 11(2), 335, 2011. With permission.

TABLE 14.2

Relationships Describing the Nusselt Number Characterizing the Heat Transfer through the Surface of a Sphere (Model of a Round Thermocouple Junction)

Author of Equation	Range of Usability	Form of the Nusselt Number
Ranz and Marshall [20]	$\mathrm{Re} \leq 2 \times 10^5$	$2 + 0.6 \cdot \mathrm{Re}^{0.5} \cdot \mathrm{Pr}^{0.33}$
Hsu [21]	$\mathrm{Re} \leq 2 \times 10^5$	$0.921 \cdot (\mathrm{Re} \cdot \mathrm{Pr})^{0.5}$
Sideman [22]	$\mathrm{Re} \leq 2 \times 10^5$	$1.13 \cdot (\mathrm{Re} \cdot \mathrm{Pr})^{0.5}$
Whitaker [23]	$\mathrm{Re} \leq 8 \times 10^4$	$2 + (0.4 \cdot \mathrm{Re}^{0.5} \cdot 0.06 \cdot \mathrm{Re}^{0.67}) \cdot \mathrm{Pr}^{0.4}$
Kacnelson and Timofiejewa [24]	$1 \leq \mathrm{Re} \leq 3 \times 10^5$	$2 + 0.03 \cdot \mathrm{Re}^{0.54} \cdot \mathrm{Pr}^{0.33} + 0.35 \cdot \mathrm{Re}^{0.58} \cdot \mathrm{Pr}^{0.356}$

Source: Jamróz, P., *IEEE Sensors J.*, 11(2), 335, 2011. With permission.

Tables 14.1 and 14.2 present experimentally determined dependences describing the Nusselt number correlation depending on the assumed geometry of the sensor in the case of forced convection with external flow normal to the sensor axis. These dependences hold true within the basic ranges of the Reynold's number related among others to the sensor geometry and gas flow velocity. Owing to an experimental character of their determination, they do not introduce any time-dependent relationships which in the next part of this elaboration are determined according to the formula assessing the Re number under assumption of gas flow with time-varying velocity and temperature. The relationships proposed by authors such as King, McAdams, Van der Hegge Zijnen were determined for the flow of air, owing to which a 0.7 value of the Prandtl number was adopted. The other studies consider generalized form with the function Prandtl number dependence.

Dynamic properties of temperature sensors described in such a way may be used in order to determine the relationships between dynamic coefficients of two sensors utilized in the two-sensor method for dynamic error correction.

14.5.2 Relationships

Considering a special case in which a pair of sensors is used for measurement of time-varying temperature of gas flowing with time-varying velocity, it may be assumed that besides the measured temperature this velocity $v(t)$ presents a common quantity which connects such processes taking place at both sensors. It comprises a base for determination of theoretical relationships between dynamic properties of both thermometric sensors [13]. In the case of the use of sensors characterized by the same heat-exchange model, solving the system of equations describing the dynamic properties of sensors $f(t)$ and $g(t)$ written according to (14.4) leads to the following relationship:

$$g(t) = \frac{C}{(1/f(t)) - D} \tag{14.5}$$

where C and D are

$$C = \frac{\rho_g \times c_g \times d_g^2 \times d_f^{0.5}}{\rho_f \times c_f \times d_f^2 \times d_g^{0.5}} \quad D = \frac{4 \times 0.42 \times \lambda_a \times \mathrm{Pr}^{0.2} \times \left(d_g^{0.5} - d_f^{0.5}\right)}{\varepsilon_f \times d_g^{0.5}} \tag{14.6}$$

Subscripts f and g refer to properties characteristic of a given sensor, which characterize $f(t)$ and $g(t)$.

A dimensionless positive coefficient C specifies thus the ratio of their heat inertias during the process of heat exchange. The C constant depends exclusively on parameters of the sensors and does not depend on the type of gas provided that the sensors are of the same shape. The D coefficient of the [s^{-1}] dimension determines the nonlinearity causing deviation from the proportional relationship between heat transfer coefficients. Its value depends predominantly on the difference in square root of geometric dimensions between sensors, and thus on the difference in area for heat exchange between sensors. Moreover, it also depends on physical properties of the gas. The sign of the constant D depends on the difference in geometric properties of the sensors. An attention needs to be paid to the fact, that in the case of the same diameters of the two sensors, $D = 0$ [s^{-1}], although the sensors may be manufactured of different materials.

For a given pair of sensors, parameters C and D are constant within assumed ranges of fluctuations of flow parameters. The only reason for change of D may consist in flow of gas with different chemical composition. Nevertheless, these constants do not depend on the velocity of gas.

Interrelationships between the models of $f(t)$ and $g(t)$ functions are valid for the rest of relationships determining the Nusselt number, and were performed analogically (Table 14.3).

Of particular notice are the relationships between dynamic coefficients of sensors considering the forms of Nu number according to Davis, Hsu, and Sideman. They lead to a proportional relationship with the coefficient C.

Provided that the two sensors of the same density and specific heat of the material are used and considering the form of the Nu number proposed by Davis, Hsu, or Sideman, the constant C may be written as a sensors' diameter power ratio.

The dynamic properties of sensors depend exclusively on the ratio of their geometric dimensions. It conforms to the assumption introduced by Pfriem. They will not depend on physical properties of the gas, and thus will not be sensitive to the change of chemical composition of the medium flowing by. They will depend only on the power of the quotient of their diameters [1]. Should the sensors be manufactured of different materials, then the postulate determining the coefficient of proportionality would depend also on physical properties of these materials.

In analytical calculation of relationship between coefficients $f(t)$ and $g(t)$, the cases of models utilizing the Nu number forms proposed by Van der Hegge Zijnen, Whitaker, Kacnelson, and Timofiejewa were not considered, due to the fact that in these formulae, occurring on one side of the equation there is the Reynold's number raised to various powers. In such situations, it is not possible to eliminate the gas velocity from the equation.

In the literature which the authors are aware of, there are no results of experimental studies confirming the existence of any algebraic relationship between the dynamic coefficients, so far. This comprised the main rationale for undertaking a laboratory experiment aimed to confirm the existence of such relationship.

TABLE 14.3

Constants C and D for Twin Sensors and for the Same Nu Formula

Author of Equation	C	D
King	$\dfrac{\rho_g \times c_g \times d_g^2 \times d_f^{0.5}}{\rho_f \times c_f \times d_f^2 \times d_g^{0.5}}$	$\dfrac{4 \times 0.318 \times \lambda_a \times \left(d_g^{0.5} - d_f^{0.5}\right)}{\rho_f \times c_f \times d_f^2 \times d_g^{0.5}}$
Kramers	$\dfrac{\rho_g \times c_g \times d_g^2 \times d_f^{0.5}}{\rho_f \times c_f \times d_f^2 \times d_g^{0.5}}$	$\dfrac{4 \times 0.39 \times \lambda_a \times \left(d_g^{0.5} - d_f^{0.5}\right)}{\rho_f \times c_f \times d_f^2 \times d_g^{0.5}}$
Kramers	$\dfrac{\rho_g \times c_g \times d_g^2 \times d_f^{0.5}}{\rho_f \times c_f \times d_f^2 \times d_g^{0.5}}$	$\dfrac{4 \times 0.42 \times \lambda_a \times \mathrm{Pr}^{0.2} \times \left(d_g^{0.5} - d_f^{0.5}\right)}{\rho_f \times c_f \times d_f^2 \times d_g^{0.5}}$
McAdams	$\dfrac{\rho_g \times c_g \times d_g^2 \times d_f^{0.5}}{\rho_f \times c_f \times d_f^2 \times d_g^{0.5}}$	$\dfrac{4 \times 0.32 \times \lambda_a \times \left(d_g^{0.52} - d_f^{0.52}\right)}{\rho_f \times c_f \times d_f^2 \times d_g^{0.52}}$
Zukauskas	$\dfrac{\rho_g \times c_g \times d_g^2 \times d_f^{0.5}}{\rho_f \times c_f \times d_f^2 \times d_g^{0.5}}$	$\dfrac{4 \times 0.318 \times \lambda_a \times \left(d_g^{0.5} - d_f^{0.5}\right)}{\rho_f \times c_f \times d_f^2 \times d_g^{0.5}}$
Davis	$\dfrac{\rho_g \times c_g \times d_g^2 \times d_f^{0.5}}{\rho_f \times c_f \times d_f^2 \times d_g^{0.5}}$	0
Ranz and Marshall	$\dfrac{\rho_g \times c_g \times d_g^2 \times d_f^{0.5}}{\rho_f \times c_f \times d_f^2 \times d_g^{0.5}}$	$\dfrac{12 \times \lambda_a \times \left(d_g^{0.5} - d_f^{0.5}\right)}{\rho_f \times c_f \times d_f^2 \times d_g^{0.5}}$
Hsu	$\dfrac{\rho_g \times c_g \times d_g^2 \times d_f^{0.5}}{\rho_f \times c_f \times d_f^2 \times d_g^{0.5}}$	0
Sideman	$\dfrac{\rho_g \times c_g \times d_g^2 \times d_f^{0.5}}{\rho_f \times c_f \times d_f^2 \times d_g^{0.5}}$	0

Source: Jamróz, P., *IEEE Sensors J.*, 11(2), 335, 2011. With permission.

14.6 EXPERIMENT

In order to experimentally verify the existence of the earlier derived relationships, a measurement system and a measurement workstation generating variable flow of gas with variable temperature was developed (Figure 14.3).

The measurement system consists of two sensors with different dynamic properties and a pseudo-reference sensor with the dynamic response to changing temperature signal $u_M(t)$ considerably faster compared to remaining two sensors. In the experiment, it was originally assumed that the pseudo-reference sensor would allow for determination of actual temperature changes of the flow, nevertheless with respect to its non-zero inertia it was supposed that its output signal designated $y(t)$ may in fact not reproduce the measured signal $u_M(t)$ reliably. Therefore, it is called the "pseudo-reference sensor." Performed simulation studies enabled to determine: the parameters of the pseudo-reference sensor, whose diagram of time response to variable dynamic excitation matches the assumed temperature excitation, and parameters of the sensor wires which cause sensors various dynamic properties.

The generation of variable flow was realized by means of channel blower, heated chamber, ball valve, stepper motor. The control is realized by means of the stepper motor controller, cooperating with the programmable counters of the National Instruments USB 6221 control-measurement card and application created in LabVIEW. For the measurement of instantaneous voltage across thermoresistors as well as across the manganate calibration resistor, a four-channel 24-bit NI WLS 9237 device was used. The equipment ensures synchronically recording the dynamic response of the sensors $y(t)$, $x(t)$,

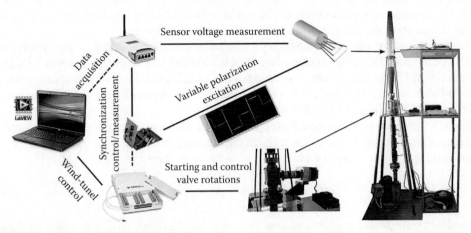

FIGURE 14.3 The experiment setup.

and $p(t)$ under the conditions of periodical changes of temperature and flow velocity with frequency about 5 Hz, whereas the sampling frequency was set to 50 kS/s. It makes it easier to do numerical analysis and allows to filtrate at high frequency perturbations by averaging per period.

Reed relays realizing the synchronous to valve rotation switching polarity power supply of the sensor with the analog channel auto-zero phase were developed. Such a solution enables in particular to get rid of effects related to offset shift during measurement and to level the effect of eventual parasitic thermoelectric forces existing in analog part of the measurement channel.

As a result of experiments' realization, obtained was a number of measurement data involving a several thousand of measurement cycles, out of which selected were the individual cycles characterized by a convergence according to assumption criteria. First of all, it was checked whether the recorded sequence of signals complied to the basic physical laws and whether the signal $y(t)$ intersected the extreme values of $x(t)$ and $p(t)$. One exemplary time period of signal showing the time course of temperature is presented in Figure 14.4.

The procedure of determination of dynamic coefficients $f_y(t)$ and $g_y(t)$ in (14.7) is based on the use of pseudo-reference signal $y(t)$ instead of the no measurable signal $u(t)$ and the signals of time response of these sensors $x(t)$, $p(t)$ together with time courses of their evaluated derivatives.

Two methods of derivative determination and noise reduction were employed. The first one was based on determination of frequency spectrum of the recorded signal. Subsequently, the very high

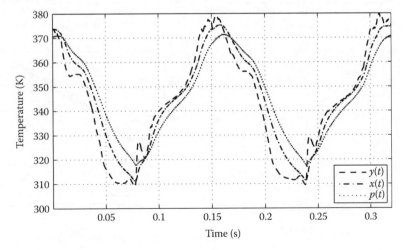

FIGURE 14.4 Exemplary signals of sensors' responses.

frequency components were discharged. The rest of the spectrum was utilized for the synthesis of analytical courses of estimated courses as well as of their derivatives.

The second method consisted in moving a window involing 200 samples along the recorded signal. The recorded time course contained within the window was approximated by a straight line by means of the LSQ method. The instantaneous value of estimated time course was determined for a selected time instant on the basis of this line's equation. The slope of this straight line determined the value of the derivative of estimated signal.

These two procedures allow for determination of instantaneous values of estimated signals for any time instant, not only for moments at which the sampling was performed. This allows for these signals to be treated as continuous and analytical as has been shown in Figure 14.1:

$$f_y(t) = \frac{y(t) - x(t)}{dx(t)/dt} \quad g_y(t) = \frac{y(t) - p(t)}{dp(t)/dt} \tag{14.7}$$

In the case of periodical changes of temperatures, there may be time intervals, in which the solution of identification task will not be possible to be obtained. These intervals refer to moments, in which the low variability of reference signal occurs, due to which the responses of sensors will match them and their derivatives will be close to zero. This results in unfavorable numerical conditioning of the task of identification of dynamic coefficients of individual sensors.

Considering the occurrence of intervals resulting in the earlier mentioned doubts regarding the reliability of obtained sensors' dynamic coefficient identification which appear in presented measurement results, the responses of sensors in time intervals in which the derivatives of $x(t)$ and $p(t)$ signals adopt the maximal values were selected for further analysis.

Figure 14.5a contains a fragment of time response of individual sensors to variable temperature excitation. Figure 14.5b presents the instantaneous values of derivatives of analyzed

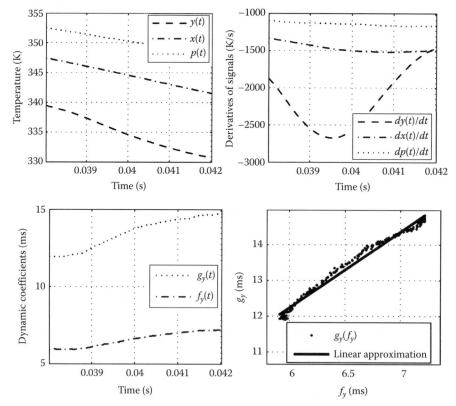

FIGURE 14.5   The experimental results.

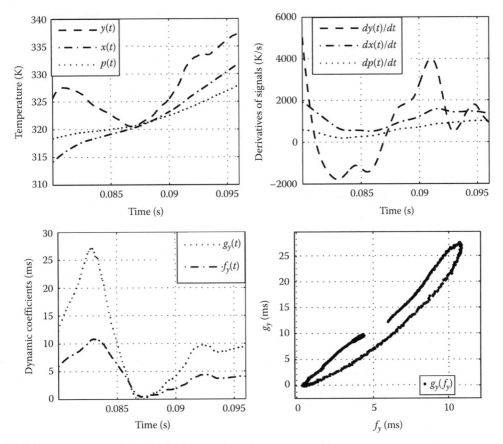

FIGURE 14.6 Deformed experimental result.

signals the moduli of which adopt high values, what enables to conclude on the correctness of selected measurement data with respect to proper numerical conditioning of the identification task. Figure 14.5c presents the instantaneous values of determined dynamic coefficients of analyzed measurement sensors, the interrelationship between which is presented in Figure 14.5d. Obtained characteristic of the $g_y(f_y)$ dependence is an example of typically obtained experimental results, close to linear characteristic. For many processed time courses of recorded signals, linear approximation was performed. A set of straight lines with slopes ranging from 2.12 to 2.35 and shifts ranging from −1.8 to −0.5 ms were obtained. This slope determined in such a way quantifies the sought C constant.

Analyzing the results of individual fragments of obtained dependences, intervals were found in which despite high values of derivatives of response signals of sensors, the obtained results were characterized by deformation of solution. Obtained characteristics were hallmarked by characteristic bends resulting ultimately in high deformation of the dependence Figure 14.6.

The occurrence of areas with high deformation in characteristic of dependences between dynamic coefficients is closely related to occurrence of extrema in the reference temperature signal $y(t)$. In order to determine the reason of such deformation characteristics, additional simulation studies were performed.

14.7 DEFORMATION OF SOLUTIONS

The simulation analysis was performed for a simple example, in which the temperature and velocity changes were described by means of the sine wave function. The simulated values of wire properties complied with used in the experiment.

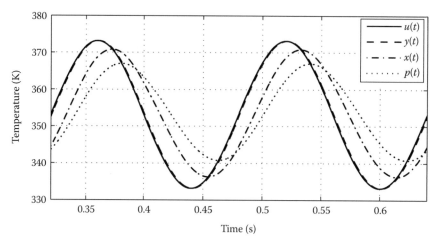

FIGURE 14.7 Simulated temperature changes and responses of sensors.

The value of Nusselt number was determined according to the relationship provided by Davis. The proportional dependence between the dynamic coefficients was employed with $C = 2$.

Figure 14.7 presents the time course of simulated temperature changes $u(t)$ and the responses of wire sensors $y(t)$, $x(t)$, and $p(t)$. It needs to be emphasized that signals $u(t)$ and $y(t)$ match with each other so much that they are indiscernible from each other.

In the next phase of the simulation, the dependences between dynamic coefficients $g_u(f_u)$ (the $y(t)$ was substituted by $u(t)$ in (14.7)) in relation to simulated reference signal $u(t)$ and $g_y(f_y)$ in relation to the pseudo-reference signal $y(t)$ were determined. During the simulation, a case in which the moduli of derivatives of $y(t)$, $x(t)$, and $p(t)$ signals reached their maximal values was analyzed (Figure 14.8).

Obtained results revealed significant differences in the case of relationship between dynamic coefficients of sensors when determined with the use of reference signal in the form of temperature excitation $u(t)$ which was known in the simulation, and when the reference signal was comprised by the response of the sensor with even the lowest inertia (the pseudo-reference one), whose response $y(t)$ did not keep the pace with quickly occurring changes of simulated measurand $u(t)$. In intervals within which the extremes of derivatives of signals $x(t)$ and $p(t)$ occur, characteristic deformation in diagrams of dependence $g_y(f_y)$ was obtained, which shapes comply into the ones obtained in the

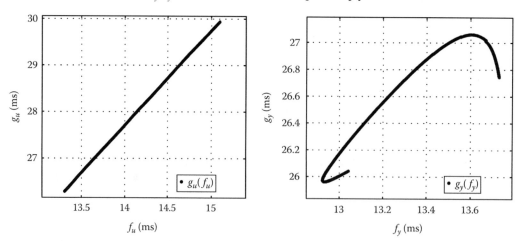

FIGURE 14.8 Deformed result of simulation.

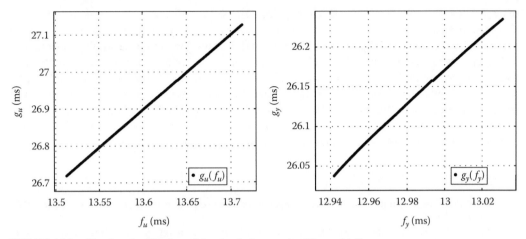

FIGURE 14.9   Simulated validation of the simulation results (Figure 14.5).

case of analysis of actual measurement data. In the next phase, similar simulations were performed, following the selection of the fragment of signal of sensors' dynamic response, in which the derivative of the response signal of the pseudo-reference sensor changes within the assumed range from 60% to 80% of its maximal value (Figure 14.9).

For such a case, the relationship between dynamic coefficients determined by both approaches adopts the form of function which is close to the linear one.

In the case, when this relationship is determined on the basis of signal of known temperature changes, this characteristic may be approximated by a linear function with the slope $C = 2.0359$ and a shift value of -0.8 ms. A similar approximation, in the case of the relationships obtained on the basis of the response of the pseudo-reference sensor, may possibly be described by a relationship with the slope equal to 2.0197 and a shift equal to -0.1 ms. Performed simulation analysis revealed that low inertia of reference sensor, which has been neglected so far, has a huge influence on the shape of developed dependence between the dynamic properties of two slower sensors. It is quite surprising effect, considering the fact that the dynamic coefficient of the pseudo-reference sensor is many times lower compared to analyzed sensors, 30 and 60 times, respectively. The disturbances in solutions research details were presented in Nabielec [25].

14.8   CONCLUSIONS

The presented results of experimental and concurrent simulation analysis confirm that the dependence between the dynamic coefficients determining the process of heat exchange between the thin wire thermoresistor and the gas with time-varying temperature flowing with variable velocity pass this thermoresistor is well approximated by the proportion. The results of experiments confirm that employed simplified mathematical model of such process describes the process of heat exchange within the analyzed range of measurement conditions with satisfactory accuracy. Assumptions concerning the omission of heat exchange with the surrounding taking place through supports of thermoresistor wires and through the radiation were proven to be correct for the experimental condition. Influence of these phenomena is negligibly small when compared to errors caused by the pseudo-reference sensor's inertia.

The sought proportional dependence occurs within certain range of temperature and gas velocity fluctuations. Experimental results do not exclude that given other experimental conditions (e.g., wider ranges of temperature and gas velocity changes) the dependence between the dynamic coefficients takes nonlinear forms. The equipment should be to rebuild for the new circumstances of experiment like the blast wave which may occur in the case of sudden gas breakouts in mines. The procedure of solution of the task of identification when the dynamic coefficients $f(t)$ and $g(t)$ are mutually connected a hyperbola ($D \neq 0$ [s^{-1}]) will be presented in the separate publication.

The presented method allows for determination of rapidly changing time courses of temperature, which were not able to be measured so far. Further development of this method should result in creation of new generation of sensors apart from the ability to record the values of physical quantities and should be able to perform complex mathematical operations upon the recorded signal.

The measurement systems using that method should be considered as a standard in dynamic calibration of sensors, but specific rules of the method utilization are required.

REFERENCES

1. H. Pfriem, Zur Messung veränderlicher Temperaturen von Gasen Und Flüssigkeiten, *Gen. Ingen.*, 7(2), 85–92, 1936.
2. M. Tagawa and Y. Ohta, Two-thermocouple probe for fluctuating temperature measurement in combustion—Rational estimation of mean and fluctuating time constants, *Combustion and Flame 109*, Elsevier Science Inc., pp. 549–560, 1997.
3. A. Olczyk, Problems of unsteady temperature measurements in pulsating flow of gas, *Meas. Sci. Technol.*, 19, 1–11, 2008.
4. J. Nabielec, An outlook on the DSP dynamic error blind correction of the analog part of the measurement channel, *Proceedings of the 16th IEEE Instrumentation and Measurement Technology Conference*, Venice, Vol. 2, pp. 709–712, 1999.
5. P. Jamróz, Investigation of polynomial models of the dynamic properties of temperature sensors in case of unsteady flow, *Przegląd Elektrotechniczny*, pp. 196–203, 2008.
6. K. Kar, S. Roberts, R. Stone, M. Oldfield, and B. French, Instantaneous exhaust temperature measurements using thermocouple compensation techniques, *SAE J. Fuels Lubr.*, 4, 652–673, 2004.
7. S. McLoone, P. Hung, G. Irwin, and R. J. Kee, Exploiting a priori time constant ratio information in difference equation two-thermocouple sensor characterization, *IEEE Sensors J.*, 6(6), 1627–1637, 2006.
8. I. Warshawsky, On-line dynamic gas pyrometry using two-thermocouple probe, *Rev. Sci. Instrum.*, 66, 2619–2624, 1995.
9. L. J. Forney and G. C. Fralick, Multiwire thermocouple in reversing flow, *Rev. Sci. Instrum.*, 66(10), 5050–5054, 1995.
10. P. Cambray, Measuring thermocouple time constants: A new method, *Combust. Sci. Technol.*, 45, 221–224, 1986.
11. J. Nabielec and J. Nalepa, The 'blind' method of dynamic error correction for the second order system, *Proceedings of XVII IMEKO World Congress*, Dubrovnik, Croatia, pp. 841–846, June 22–27, 2003.
12. M. Vachon, P. Cambray, T. Maciaszek, and J. C. Belet, Temperature and velocity fluctuation measurements in a diffusion flame with large buoyancy effect, *Combust. Sci. Technol.*, 48, 223–240, 1986.
13. P. Jamróz, Relationship between dynamic coefficients of two temperature sensors under nonstationary flow conditions, *IEEE Sensors J.*, 11(2), 335–340, 2011.
14. A. Olczyk, Modelling of dynamic properties of unsteady gas temperature sensors, *XV Symposium on Measuring Systems—Modelling and Simulation*, Krynica, Poland, 2005.
15. L. V. King, On the convection of heat from small cylinders in a stream of fluid: Determination of the convection constants of small platinum wires with applications to hot-wire anemometry, *Philos. Trans. Roy. Soc. A*, 214, 373–432, 1914.
16. H. Kramers, Heat transfer from spheres to flowing media, *Physica*, 12, 61–80, 1946.
17. W. H. McAdams, *Heat Transmission*, McGraw-Hill Book Company Inc., New York, 1954.
18. B. G. Van der Hegge Zijnen, Modified correlation formulae for the heat transfer by forced convection from horizontal cylinders, *Appl. Sci. Res. Ser. A*, 6, 129–140, 1956.
19. G. E. Andrews, D. Bradley, and G. F. Hundy, Hot wire anemometers calibration for measurements of small gas velocities, *Int. J. Heat Mass Transfer*, 15, 1765–1786, 1972.
20. W. E. Ranz and W. R. Marshall, Evaporation from drops: Part 1, *Chem. Eng. Prog.*, 48, 141–150, 1952.
21. C. Hsu, Heat transfer to liquid metals flowing past spheres and elliptical rod bundles, *Int. J. Heat Mass Transfer*, 8, 303–315, 1965.
22. S. Sideman, The equivalence of the penetration theory and potential flow theories, *Ind. Eng. Chem. Res.*, 58(2), 54–58, 1966.

23. S. Whitaker, Forced convection heat transfer correlations for flow in pipes, past flat plates, single cylinders, single spheres, and for flow in packed beds and tube bundles, *AIChE J.*, 18(2), 361–371, 1972.
24. B. Melissari and S. A. Argyropoulos, Development of a heat transfer dimensionless correlation for spheres immersed in a wide range of Prandtl number fluids, *Int. J. Heat Mass Transfer*, 48(21–22), 4333–4341, 2005.
25. J. Nabielec, Error in the method of sensor dynamics parameter identification, *Przegląd Elektrotechniczny*, 87(12A), 99–103, 2011 [in Polish].

15 Dual-Band Uncooled Infrared Microbolometer

Qi Cheng, Mahmoud Almasri, and Suzanne Paradis

CONTENTS

15.1 INTRODUCTION

This chapter describes the design and modeling of a smart uncooled infrared detector with wavelength selectivity in the long-wavelength infrared (LWIR). The design takes advantage of the smart properties of vanadium dioxide (VO_2): it can switch reversibly from an IR-transparent to an IR-opaque thin film when properly triggered. This optical behavior is exploited here as a smart mirror that can modify the depth of the resonant cavity between the suspended thermistor material and a patterned mirror on the substrate, thereby altering wavelength sensitivity. VO_2 belongs to a family of materials referred to as chromogenic which are capable of reversibly modifying their optical and electrical properties upon specific stimulation such as temperature, pressure, or electrical field. The phase transition of VO_2 which is triggered at 68°C is responsible for its optical and electrical modulation. At room temperature, the film behaves as a semiconductor material with an energy band gap of ~0.7 eV. When the temperature is increased to 68°C, the energy band gap vanishes, the lattice becomes tetragonal, and the film behaves as a metal. The electrical resistivity of the film can be varied by 3 to 4 orders of magnitude, and the film can be changed from IR-transparent in the semiconducting phase to IR-reflective in the metallic phase [1–7]. These features have prompted many groups to use VO_2 at the phase transition for electro-optical switches [6,7]. We have used the VO_2 stoichiometry to design the smart mirror because no other oxide of vanadium has the stoichiometry needed to engender the desired chromogenic behavior. VO_X (non-stoichiometric film) in the semiconducting phase is selected as an IR-sensitive material because of its high temperature coefficient of resistance (TCR), low $1/f$ noise, and low resistance, which is compatible with readout electronics. Several other materials can be used for IR detection, including amorphous silicon (a-Si) [8], yttrium barium copper oxide (YBaCuO) [9], silicon germanium (SiGe) [10], silicon germanium oxide (SiGeO) [11,12], and metals [13].

The capability of an IR detector to switch wavelength sensitivity between two bands enhances the probability of detecting and identifying objects in a scene. Commercially, multiband IR cameras typically use two separate focal plane arrays (FPAs) and electronics for each spectral band, or a single FPA and an optical element that selects the appropriate spectral band such as filter wheel or spectral dispersion techniques [14,15]. The use of separate FPAs has several drawbacks, including significant added complexity and cost, and the challenge of alignment between the FPAs. These approaches are effective for certain applications, but they have proven to be bulky and costly [16]. Cooled multiband FPAs are based on mercury cadmium telluride (MCT), quantum-well IR photodetectors using III–V materials, and quantum-dot IR photodetectors [17–20]. These detectors offer many advantages such as high sensitivity and very high speed, but require expensive and heavy cryogenic equipment to operate. These are used for military applications where ultrahigh sensitivity and speed are required. For a wide range of applications that do not require ultrahigh sensitivity, however, the high cost of cooling becomes a problem. The development of an uncooled dual-band detector would retain the advantages of uncooled detectors over cooled detectors, which are much lower fabrication cost since no exotic and difficult to produce semiconductor materials are used, reduced weight, size, and energy consumption since no bulky and noisy cooling system are required. The commercial and defense sectors would both benefit from uncooled dual-band cameras: highway safety, medical diagnostics and surgery, surveillance, law enforcement, and security. Dual-band cameras also offer a possible solution for real-time detection and identification of carefully concealed targets even in cluttered environments—environments with unwanted parasite signals. Having two bands instead of one will ease clutter rejection, and therefore plays a significant role in discriminating the actual temperature and unique signature of objects in the scene [21-29]. An IR camera offering wavelength selectivity in the LWIR band has shown promise in the detection of such difficult targets as recently buried mines [30]. Recently, monolithic detectors offering dual-band tunability have been investigated: wavelength-selective IR microbolometers using planar multimode detectors [31] and two-color thermal detectors [32] have been designed, and adaptive microbolometers have been designed, fabricated, and characterized [33]. These detectors were designed to operate in the LWIR band. In the first two designs, the cavity was tuned by using movable micromirrors underneath the microbolometer pixel, while in the third design the cavity was tuned by moving the microbolometer pixel via electrostatic actuation. The first approach has not been realized, while the second approach has achieved low responsivity and detectivity [33]. These designs are relatively complicated since they require movable components. In this chapter, a dual-band microbolometer model is presented. It can be tuned to two spectral bands with some overlap without using a movable structure, and hence the same microfabrication procedures developed for single-band microbolometers can be used to fabricate the proposed device.

15.2 BACKGROUND AND THEORY

A microbolometer is a thermal sensor whose resistance changes with temperature, associated with the absorption of IR radiation. Its performance is characterized by several figures of merit such as responsivity (R_v), TCR, detectivity (D^*), and noise equivalent temperature difference (NETD) [34]. Responsivity is the output voltage divided by the input radiant power falling on the detector. It is given by

$$R_v = \frac{I_b R \eta \beta}{G\left(1 + \omega^2 \tau_{th}^2\right)^{1/2}} \tag{15.1}$$

where
   I_b is the bias current (A)
   R is the bolometer electrical resistance (Ω)
   η is the fraction of the incident radiation absorbed
   G is the total thermal conductance to the substrate (W/K)
   ω is the radiation modulation frequency
   τ_{th} is the thermal response time

β is used to quantify the temperature dependence of resistance to measure the heating effect of the absorbed IR radiation and is given by

$$\beta = \frac{1}{R}\frac{dR}{dT} \tag{15.2}$$

The detectivity D^* is a figure of merit that measures the signal-to-noise ratio and normalizes the performance of the detector with respect to the detector size:

$$D^* = \frac{R_v\sqrt{A\Delta f}}{\Delta V_n} \tag{15.3}$$

where

Δf is the amplifier frequency bandwidth (Hz)
A is the detector area (cm^2)
ΔV_n is the total noise voltage of the system (V)

Hence, for good performance, a microbolometer must have large values of β, R_v, and D^* [34,35]. The NETD is a camera figure of merit. It is given by

$$\text{NETD} = \frac{4F^2}{\tau_0 A^{1/2} D^* (\Delta M/\Delta T)_{\lambda_1 - \lambda_2}} \tag{15.4}$$

where

F is the focal ratio of the optics
τ_0 is the transmittance of the optics
$(\Delta M/\Delta T)_{\lambda_1 - \lambda_2}$ is the change in optical power emitted with respect to temperature per unit area radiated by a blackbody at temperature T measured within the spectral band from λ_1 to λ_2

Note that total noise must be minimized and the responsivity must be maximized to achieve the best sensitivity.

The microbolometer's thermal response time high sensitivity can be achieved by maximizing R_v, and D^* and TCR. The thermal time constant, τ_{th}, was calculated from the ratio of the device's thermal mass to its thermal conductance, C/G_{th}. The thermal mass of the microbolometer is given by

$$C = \sum_n W_n L_n t_n c_n + \frac{1}{3}\sum_m W_m L_m t_m c_m \tag{15.5}$$

where

W_n, L_n, and t_n correspond to width, length, thickness, and specific heat of each thin film layer
W_m, L_m, t_m, and c_m correspond to width, length, thickness, and specific heat of each layer of the electrode arms

The total thermal conductance of the microbolometer is given by

$$G_{\text{tot}} = \sum_m \frac{K_m W_m t_m}{L_m} \tag{15.6}$$

where

K_m is the thermal conductivity of each component
W_m, L_m, and t_m correspond to width, length, and thickness of each component

The thermal time constant, τ_{th}, was calculated from the ratio of the device's thermal mass to its thermal conductance, C/G_{th}.

The microbolometer presented here consists of an IR-sensitive element deposited on top of a microbridge and suspended above the substrate by support arms. The gap between the IR-sensitive element and the IR-reflective mirror patterned on the substrate plays an important role in the absorption of IR radiation for the suspended microbolometer since it is wavelength-dependent. With the appropriate design, this gap behaves as a Fabry–Perot resonant cavity which maximizes IR absorption and hence maximizes responsivity and detectivity. The resonant cavity can be created between the incoming and reflected waves if the cavity depth is tuned accordingly:

$$d_n = \left[(2n-1) - (\varphi_1 - \varphi_2) \right] \frac{\lambda}{4} \qquad (15.7)$$

where
    d_n is the air gap depth
    n is an integer
    λ is the wavelength
    φ_1 and φ_2 are the phase differences between the incoming and reflected light [36]

In operation, the microbolometer requires either current or voltage biasing. The infrared cameras that are based on microbolometer FPA do not require chopper for operation. However, for modeling and testing the individual devices, a chopper is used in order to enable the calculation of the cutoff frequency and the video frame rate.

15.3 MICROBOLOMETER DESIGN

15.3.1 Optical Design

The actual temperature and unique signature of objects in the scene can be discriminated if spectral response of the uncooled microbolometer is changed before each exposure to IR power. This is achieved by taking advantage of the smart properties of VO_2: it can switch reversibly from an IR-transparent to an opaque thin film when properly triggered. This optical behavior is exploited here as a smart mirror that can modify the Fabry–Perot resonant-cavity depth between the suspended thermistor material (VO_x) and a mirror patterned on the substrate, thereby altering wavelength sensitivity and creating a dual-band microbolometer. A schematic of the dual-band microbolometer along with the structural view and operational view are shown in Figure 15.1. The fixed reflective mirror is switching between the smart material (VO_2) and Au layer. In the first case, the VO_2 is used in the metallic phase by heating the detector to 68°C. Thus, the IR radiation is reflected back to the suspended VO_x enhancing the IR absorption in the first spectral band (9.4–10.8 μm). In the second case, VO_2 is switched to the semiconducting phase by cooling the detector to room temperature. Thus, most of the IR radiation is transmitted through the transparent VO_2 layer and through a SiO_2 layer used as a spacer material, and then reflected by the gold (Au) thin film mirror layer, which is patterned underneath the SiO_2 spacer layer on the substrate. Thus, the gap underneath the microbolometer pixel, which initially contained only vacuum, is increased by the thicknesses of the SiO_2 and VO_2 transparent materials. Therefore, the IR absorption is shifted to a second spectral band (8–9.4 μm). The resonant-cavity depths are optimized to maximize IR absorption, thereby increasing the responsivity and detectivity of the microbolometer in the 8–9.4 μm and 9.4–10.8 μm spectral bands. The two-band design enhances the probability of detecting and identifying objects in a scene. These two windows were chosen for demonstration of concept: they split the 8–12 μm LWIR band roughly into two. In this design, VO_2 is switched between the two semiconductor–metallic phases at a rate double the camera frame rate (i.e., 60 Hz), thus changing a resonant-cavity depth and providing a response to

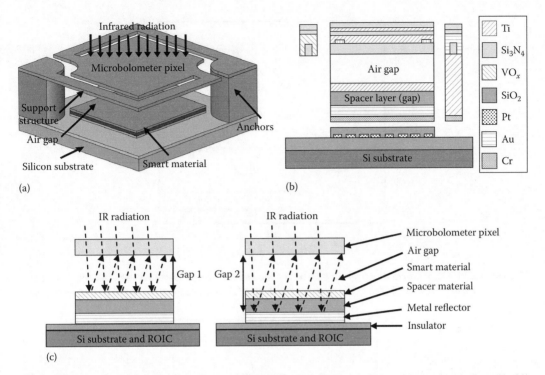

FIGURE 15.1 The schematics show the dual-band microbolometer in (a) three-dimensional view, (b) side view, and (c) operational view. IR radiation is reflected by the smart material when it is used in the metallic phase (c—left), and is reflected by the gold layer when the smart material is used in the semiconducting phase.

a two-wavelength spectral bands. To separate the response spectral bands, the readout has to be performed twice at the frame rate level to achieve a two-color picture at the standard 30 Hz frame rate.

Thin film Wizard software (Film Wizard™) developed by Scientific Computing International (SCI-Soft) is used to determine the absorption of the microbolometer for the case of normal incidence of the radiation using the refractive index $((n(\lambda))$ and absorption coefficient $(k(\lambda))$ of the various layers, which include the VO_x IR-sensitive layer, Si_3N_4 bridge, Ti absorber, Si_3N_4 passivation layers, VO_2 reflective/transparent layer, SiO_2 spacer layer, Au reflective mirror, chromium (Cr) adhesive layer, SiO_2 insulating layer, and silicon substrate. The thin film layer thicknesses and the gap depths were adjusted until the maximum absorption in the specified wavelength windows is determined. The result was absorption with an average value of 65% and 59% for cavity depths of 3.9 and 4.63 μm, respectively. In the first case, the absorption was maximized for the 9.4–10.8 μm band, while in the second case the absorption was maximized for the 8–9.4 μm band. The absorption was plotted as a function of wavelength for both cavity depths as shown in Figure 15.2. In addition, the difference in absorption value between 8–9.4 μm and 9.4–10.8 μm bands for metallic and semiconductor phases, respectively, were 40.6%, and 32%. This difference in absorption can be seen as the contrast between the two modes. The $n(\lambda)$ and $k(\lambda)$ of the VO_2 and Si_3N_4 thin film layers were found in the literature [37] for the former, and in FilmStar Optical Thin Film Software for the latter. When VO_2 with a thickness of 130 nm is used in the metallic phase, the IR radiation was reflected back to the suspended VO_x IR-sensitive layer. The absorption results were not affected by changing the SiO_2 spacer layer thickness. This suggests that the selected VO_2 thickness is satisfactory for reflecting most of the IR radiation (Figure 15.3). It is important to note that the electrical, mechanical, thermal, and optical properties are optimized simultaneously in order to design a high-performance microbolometer. The thickness of VO_2 was fixed at 130–150 nm in the model. This thickness was optimized by the coauthor's laboratory [1,33].

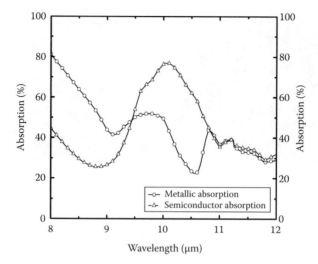

FIGURE 15.2 The calculated optical absorption for the metallic phase (9.4–10.8 μm) and semiconducting phase (8–9.4 μm) of VO_2 are plotted as a function of wavelength for cavity depths of 3.9 and 4.63 μm, respectively.

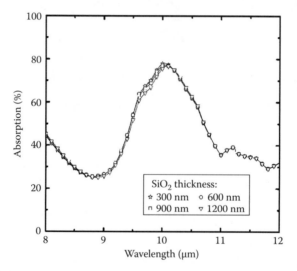

FIGURE 15.3 The calculated optical absorption are plotted as a function of wavelength. VO_2 is used as a reflector. The air gap is fixed at 3.9 μm while the SiO_2 spacer layer is variable, and all other films are of fixed thickness.

15.3.2 STRUCTURAL DESIGN

The microbolometer is designed with a pixel size of 25×25 μm², enabling the fabrication of megapixel format arrays. Each microbolometer consists of a Si_3N_4 bridge with a thickness of 350 nm suspended above a silicon substrate. Si_3N_4 is used because of its excellent thermal properties, processing characteristics, and high infrared absorption.

Encapsulated in the center of the Si_3N_4 bridge is a layer of VO_X infrared-sensitive material and thin Ti absorber with thicknesses of 150 and 10 nm, respectively. A layer of Si_3N_4 with a thickness 50 nm was used to prevent electrical contact between the VO_X and the Ti absorber. The VO_X thickness is optimized experimentally to achieve high TCR, low resistivity, and low noise properties. The Ti was chosen as an absorber with this thickness since it maximizes the absorption difference between metallic and semiconducting phases in each spectral band. Thus, it increases IR absorption in the 9.4–10.8 μm band and decreases it in the 8–9.4 μm band when the underlying

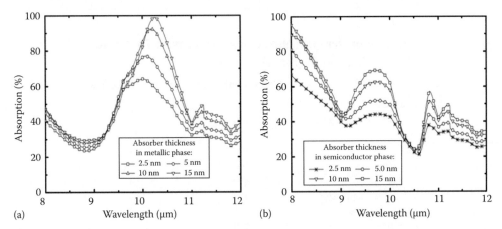

FIGURE 15.4 Microbolometer optical absorption is plotted versus wavelength. The Ti absorber thickness is variable while all other films are of fixed thickness (see Table 15.1). VO_2 is used in (a) metallic phase and (b) semiconducting phase.

VO_2 is used in the metallic phase. When the VO_2 is switched to the semiconducting phase, IR absorption is decreased in the 9.4–10.8 µm band and increased in the 8–9.4 µm band. The effects of the Ti absorber thickness on absorption are shown in Figure 15.4. Several other metals such as nickel–chromium (Ni–Cr) and platinum (Pt) could be used as an absorber. Atop the Ti absorber, a third layer of Si_3N_4 is deposited and patterned with a thickness of 50 nm in order to passivate and hence protect it from oxidation during the removal of the polyimide sacrificial layer using oxygen plasma to release the microbolometer. The bridge arms are made of Si_3N_4 and Ti layers with an area of 54×4 µm². The arms serve as support structures, conductive legs, and thermal isolation legs. Au was used as the contact (area 23×4 µm²) because it makes excellent contact with the VO_X IR-sensitive layer resulting in linear behavior. We selected Ti as the electrode material because it has a much lower thermal conductivity than Au, thus providing better thermal isolation. Vacuum cavities with depths of 3.9 and 4.63 µm are used to create the wavelength resonant cavities at two separate windows. This difference in absorption can be seen as the contrast between the two modes. The first cavity is between the suspended VO_X and the underlying VO_2 in the metallic phase. The vacuum cavity also provides thermal isolation from the substrate. The second cavity is between the suspended VO_X and the underlying Au mirror. This cavity consists of vacuum and SiO_2 spacer material and VO_X layer. Finally, a silicon substrate with a resistivity of 10 Ω cm is used. A complete list of the calculated film thicknesses and their electrical and thermal properties are shown in Table 15.1. The thermal mass and thermal conductance were calculated for the microbolometer with pixel, support arm, and Au contact areas of 25×25 µm², 54×4 µm², and 23×4 µm² as shown in Table 15.2.

The pixel was designed with low thermal mass (4.48×10^{-10} J/K) and the supporting arms with relatively low thermal conductance (1.71×10^{-7} W/K), to reduce the thermal time constant (2.6 ms) and maximize responsivity and detectivity. The thermal conductance through the electrode arms is made as small as possible to meet response time requirements. We aimed for a low thermal time constant for high-speed camera applications. The thermal mass of the microbolometer was calculated from the specific heat capacity and volume of each layer forming the microbridge and one-third of each Au/Si_3N_4 electrode arm. The thermal conductance was estimated from the calculated thickness, width, and length of the electrode arms that were obtained from the electromechanical model results. It was also obtained from the thermal simulation. Since the thermal conductance of the two components making up the electrode arms were in parallel, the total thermal conductance was determined by summing the thermal conductance of the different layers. Thermal conduction occurs mainly through the electrode arms and the surrounding air. Therefore, the electrode arms were carefully designed to

TABLE 15.1

Microbolometer Pixel and Support Arm Layers Thicknesses That Are Used in the Model

Layers	Pixel Thickness (nm)	Arm Thickness (nm)
Ti absorber	10	—
VO$_X$ (semiconducting phase)	150	—
Si$_3$N$_4$	400	350
Au contact	100	—
Ti electrodes	100	100
Air gap	3900	—
VO$_2$ mirror/transparent	130	—
SiO$_2$ spacer layer	600	—
Au mirror	200	—
Chromium adhesion layer	150	—
SiO$_2$ insulating layer	200	—
Si substrate	Thick	—

Note: The pixel, support arm, and Au contact sizes are 25×25 μm^2, 54×4 μm^2, and 23×4 μm^2, respectively.

TABLE 15.2

Specific Heat and Thermal Conductivity Used in the Model, and the Calculated Thermal Mass and Thermal Conductance of Each Layer

Layers	Heat Capacity (J/K cm^3)	Thermal Conductivity (W/cm K)	Thermal Conductance G_{th} ($10^7 \times$ W/K)	Thermal Mass C_{th} ($10^{10} \times$ J/K)
Ti	2.266	0.217	—	0.14
Si$_3$N$_4$	5.413×10^{-4}	0.003	0.0889	1.35
VO$_X$	2.316	0.307	—	1.88
Au	2.482	3.15	—	0.17
Ti	—	0.217	1.62	—

reduce the thermal conductance path and to meet the thermal time response requirements. The pixel array will be packaged in a vacuum in order to thermally isolate the IR-sensitive element from its surroundings to reduce the rate of heat loss, thereby increasing the sensitivity of the detector.

15.4 STRUCTURAL AND THERMAL SIMULATION

Finite element analysis using CoventorWare™ was employed to provide an accurate prediction of the microbolometer performance. The simulations were performed using published material properties of all thin film layers. The resulting stress distribution (Von Mises) of the electrode arms and the pixel due to the internal forces acting on the microbolometer, which includes the axial force, shear, moment, and torque, were calculated and plotted in Figure 15.5 for the case when the heater is in the off-state. In general, the result shows that the microbolometer with a 25×25 μm^2 pixel and 54×4 μm^2 support arms has flat surfaces and low stress distribution. The deflection of the electrode arms and the pixel was negligible, at 2.77×10^{-5} μm. A detailed

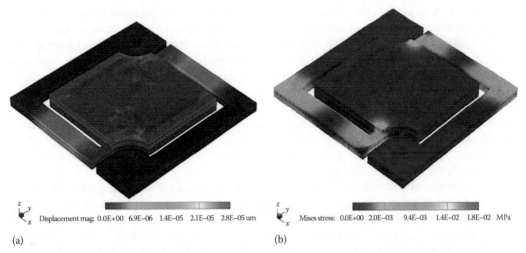

Displacement mag: 0.0E+00 6.9E–06 1.4E–05 2.1E–05 2.8E–05 um

(a)

Mises stress: 0.0E+00 2.0E–03 9.4E–03 1.4E–02 1.8E–02 MPa

(b)

FIGURE 15.5 (a) The plots show an optimized microbolometer structure, with pixel and support arm size of 25×25 μm^2 and 54×4 μm^2, with relatively little deflection. The largest deflection occurs at the corners of the square pixel with a value of 2.77×10^{-5} μm. (b) Von Mises stress distribution of the microbolometer with flat surface. The greatest stress -1.8×10^{-2} MPa occurs at the joints between support arms and fixed anchors. The simulation was performed using Coventor finite element tool.

thermal simulation of the dual-band microbolometer was performed using CoventorWare. The simulation was performed by assuming an input heat power of 3×10^7 W/m^3, which corresponds to a constant power of 0.264 μW absorbed by the VO$_X$ IR-sensitive layer. This power increases the microbolometer temperature from a reference temperature of 300 K (room temperature) to 301.53 K on the pixel, which is reached in steady-state simulation. The applied input power and the corresponding temperature rise of the microbolometer are used to calculate the thermal conductance as 1.72×10^{-7} W/K. The temperature gradient across the microbolometer was simulated in vacuum condition with respect to the reference temperature (300 K) and is shown in Figure 15.6a. The heat flux distribution is shown in Figure 15.6b, with the heat flux highest at the support arm. The simulated and calculated thermal conductance were essentially the same, which confirms the accuracy of the model. The stress distribution and the temperature gradient across

Temperature: 300.00 300.38 300.76 301.15 301.53 K

(a)

Heat flux mag: 4.56E–07 1.06E+06 2.12E+06 3.19E+06 4.25E+06 pWum^2

(b)

FIGURE 15.6 (a) Temperature gradient across the microbolometer structure with pixel and support arm size of 25×25 μm and 54×4 μm. The highest temperature (301.53 K) occurs in the pixel in steady-state simulation. (b) Heat flux distribution across the microbolometer structure.

the microbolometer for the case when the heater is in the on-state are expected to be similar to the off-state. Voltage application on the dedicated microheater which is located underneath the VO_2 induces heating via Joule effect. The response time for the transition by Joule effect is a function of the quality of the contact and the surface area to be heated. The response time and down time are expected to be in microseconds range. However, this needs to be confirmed experimentally. On the other hand, the VO_X switches between the two metallic and semiconductor phases at a rate double the camera frame rate (60 Hz). Therefore, the resonant-cavity depth (gap) changes between two positions and hence provides a response to two spectral windows. In other words, the total thermal response time for both spectral windows is 2.6 ms. This response time will be divided equally between the two spectral window, each 1.3 ms. This time is much higher than the microheater's response time and down time.

The dual-band microbolometer can be realized by employing a series of photolithography, surface micromachining, and polyimide sacrificial layer processes in the following sequence (see Figure 15.1b). (1) The wafer is thermally oxidized to grow a thick SiO_2 layer for insulation. (2) The Pt layer is deposited and patterned to form the heater. (3) A layer of oxide is deposited using PECVD for insulation. (4) A thin layer of chromium (Cr) and a thin layer of gold (Au) are sputter-deposited and patterned to form the reflective mirror, trace line, and bonding pads. (5) A thick layer SiO_2 spacer material is deposited and patterned. (6) A layer of VO_2 is deposited and patterned. This layer will be used as a reflecting mirror when operated in the metallic phase, and will transmit IR radiation through it when switched to the semiconducting phase. (7) A polyimide sacrificial layer is spin-coated, patterned the gap, and a mold for fabricating the anchors. (8) A photoresist layer is spin-coated and patterned at locations corresponding to the microbolometer anchors. (9) Ti is then sputter-deposited and patterned using a liftoff technique to create the microbolometer anchors. (10) The photoresist is removed with acetone. (11) A thick layer of Si_3N_4 is deposited and patterned to form the bridge structure. (12) The VO_X IR-sensitive material is sputter-deposited and patterned. (13) Au is deposited and patterned to create the support arms and electrode contacts. (14) A thin layer of Si_3N_4 is deposited and patterned for insulation, followed by sputtering of a thin Ti absorber. (15) The absorber is then patterned. (16) A third layer of Si_3N_4 is deposited and patterned to protect the absorber. (17) All Si_3N_4 layers are patterned and dry-etched to expose the polyimide to oxygen ashing in the final step. (18) The polyimide sacrificial layers are removed by oxygen plasma ashing.

15.5   MICROBOLOMETER PERFORMANCE

Prior to calculating performance, the microbolometer is assumed to have a TCR and a resistance of 100 kΩ and 2.7%/K, respectively, based on published data in the literature. It will be dc-biased with a bias current of 25 μA. Performance is calculated for the microbolometer with a support arm size of 54×4 μm^2. Initially, the total noise, which includes Johnson noise, temperature fluctuation noise, and background fluctuation noise, was calculated as a function of chopper frequency and plotted in Figure 15.7. It is important to note that uncooled thermal cameras since it will allow the detector to reach the background limited noise performance and further improve NETD. The $1/f$ noise was not accounted for in the calculation of the total noise because the detector will be biased in the Johnson noise level region and the $1/f$ noise depends on many process technology factors related to the microbolometer fabrication. In addition, VO_X has low $1/f$ noise. Therefore, the $1/f$ noise does not affect the capabilities of the proposed design. The voltage responsivity and detectivity were calculated for the two spectral bands and plotted versus chopper frequency in Figure 15.8. The highest achieved values at 60 Hz frame rate were 1.23×10^5 V/W and the reduction of noise is crucial to the next generation of 1.62×10^9 cm sqrt (Hz/W). The responsivity was maximized by enhancing the absorption to 61% and 63% for the 8–9.4 μm and 9.4–10.8 μm bands, and by achieving relatively low thermal mass (J/K) and low thermal conductance (W/K) in the detector. The corresponding thermal time constant was

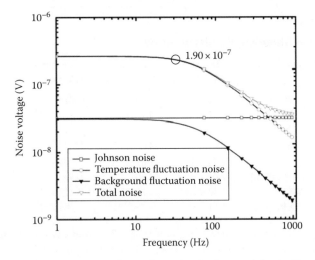

FIGURE 15.7 Johnson noise, temperature fluctuation noise, background fluctuation noise, and total noise were calculated as a function of chopper frequency.

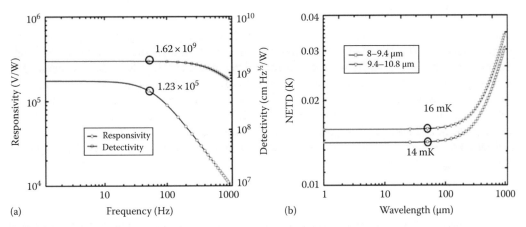

FIGURE 15.8 (a) Responsivity and detectivity and (b) NETD as a function of chopper frequency. The D^* and R_v values for both spectral bands are almost same.

2.6 ms. The high values of responsivity and detectivity will allow for improved NETD to a value as low as 14 and 16 mK for 8–9.4 μm and 9.4–10.8 μm windows, respectively. These NETD values are higher than the current state of the art; for example, the Raytheon group's published results on VO_X microbolometers with a size of 25 × 25 μm² have achieved NETD less than 50 mK [38,39]. We expect the calculated NETD value will be higher if the readout electronic and system noise is included. The NETD is plotted versus chopper frequency in Figure 15.8. The microbolometer performance was also calculated for narrower electrode arms and for shorter support arms. A complete list of the microbolometer performance results is shown in Table 15.3. We have chosen the support arm dimensions as 54 × 4 μm² in all the simulation results reported here because it represents an optimized value.

We also performed the simulation for three more designs. In all these dimensions, the deflection of the microbolometer pixel was minimal. It is important to note that the performance of the microbolometer with shorter and narrower arms (25 × 2 μm²) was similar to that achieved with a support arm size of 54 × 4 μm² and better fill factor. However, we selected the longer and wider arm because it can be fabricated with our current microfabrication capability.

TABLE 15.3

Summary of Modeling Results at 60 Hz

Parameter Support Arm Dimensions	Design 1 ($54 \times 4 \ \mu m^2$)	Design 2 ($54 \times 2 \ \mu m^2$)	Design 3 ($25 \times 4 \ \mu m^2$)	Design 4 ($25 \times 2 \ \mu m^2$)
Responsivity (V/W)	1.23×10^5	1.66×10^5	7.512×10^4	1.303×10^5
Detectivity (cm Hz$^{1/2}$)/W	1.62×10^9	2.27×10^9	1.103×10^9	1.563×10^9
NETD 8–9.4 μm (mK)	14	10	21	15
NETD 9.4–10.8 μm (mK)	16	11	23	16
Total noise (V)	1.9×10^{-7}	1.827×10^{-7}	1.703×10^{-7}	2.083×10^{-7}
Thermal time constant (ms)	2.6	4.86	0.99	1.91
Thermal mass (J/K)	4.48×10^{-10}	4.16×10^{-10}	3.672×10^{-10}	3.525×10^{-10}
Thermal conductivity (W/K)	1.71×10^{-7}	8.56×10^{-8}	3.696×10^{-7}	1.848×10^{-7}

Notes: Support arms are designed with lengths of 54 and 25 μm. Performance was also calculated for two different electrode widths: 2 and 4 μm. In all cases, deflection of the microbolometer pixel was minimal.

15.6   CONCLUSION

A dual-band microbolometer is designed based on semiconducting vanadium oxide (VO_x) infrared-sensitive material. The microbolometer's bottom mirror is switched between two metals to create two resonant vacuum cavities between the mirror and the suspended VO_x IR-sensitive material. The cavity depths maximize optical absorption in the 8–9.4 μm and 9.4–10.8 μm bands. The maximum optical absorption values observed were 65% and 59% for cavity depth of 3.9 and 4.63 μm, respectively. The achieved voltage responsivity was 1.23×10^5 V/W and the detectivity was 1.62×10^9 cm sqrt (Hz)/W. In addition, the microbolometer NETD was between 14 and 16 mK. The calculated thermal time constant was 2.6 ms for each spectral band. Hence, these detectors could be used for 60 Hz frame rate applications.

REFERENCES

1. S. Paradis, P. Merel, P. Laou, and D. Alain, Vanadium oxide films for optical modulation applications, *Proc. SPIE* 6343, 63433U, 2006.
2. S. Paradis, Tailoring electro-optical properties of RF sputtered vanadium dioxide with dopant for smart infrared modulation, *Proceedings of International Workshop Smart Materials and Structures*, October 23–24, 281–290, 2008.
3. G. Golan, A. Axelevitch, B. Sigalov, and B. Gorenstein, Investigation of phase transition mechanism in vanadium oxide thin films, *J. Optoelectron. Adv. Mater.*, 6(1), 189–195, 2004.
4. P. P. Boriskov, A. L. Pergament, A. A. Velinchko, G. B. Stefanovich, and N. A. Kuldin, Metal–insulator transition in electric field: A viewpoint from the switching effect. E-print arXiv:cond-mat/0603132, 2006.
5. L. A. Luz de Almeida, G. S. Deep, A. M. Nogueira Lima, and H. Neff, Modeling of the hysteretic metal–insulator transition in a vanadium dioxide infrared detector, *Opt. Eng.*, 41(10), 2582–2588, 2002.
6. X. Chena and J. Daib, Optical switch with low-phase transition temperature based on thin nanocrystalline VO_x film, *Optik*, Advance copy, 2009.
7. M. Soltani and M. Chaker, 1 × 2 optical switch devices based on semiconductor-to-metallic phase transition characteristics of VO_2 smart coatings, *Meas. Sci. Technol.*, 17, 1052–1056, 2006.
8. T. Schimert, C. Hanson, J. Brady, T. Fagan, M. Taylor, W. McCardel, R. Gooch, M. Gohlke, and A. J. Syllaios, Advances in small-pixel, large-format-Si bolometer arrays, *Proc. SPIE*, 7298, 72980T, 2009.
9. M. Almasri, D. P. Butler, and Z. Çelik-Butler, Self-supporting semiconducting Y-Ba-Cu-O uncooled IR microbolometers with low-thermal mass, *IEEE/JMEMS*, 10(3), 469–476, 2001.

10. A. H. Z. Ahmed, R. N. Tait, T. B. Oogarah, H. C. Liu, M. W. Denhoff, G. I. Sproule, and M. J. Graham, A surface micromachined amorphous $Ge_xSi_{1-x}O_y$ bolometer for thermal imaging applications, *Proc. SPIE*, 5578, 298–308, 2004.

11. Q. Cheng and M. Almasri, Silicon germanium oxide ($Si_xGe_{1-x}O_y$) infrared material for uncooled infrared detection, *Proc. SPIE*, 7298, 72980K, 2009.

12. M. M. Rana and D. P. Butler, Radio frequency sputtered $Si_{1-x}Ge_x$ and $Si_{1-x}Ge_xO_y$ thin films for uncooled infrared detectors, *Thin Solid Films*, 514, 355–360, 2006.

13. M. V. S. Ramakrishna, G. Karunasiri, U. Sridhar, and G. Chen, Performance of titanium and amorphous germanium microbolometer infrared detectors, *Proc. SPIE*, 3666, 415–420, 1999.

14. R. Breiter, W. A. Cabanski, K-H. Mauk, W. Rode, J. Ziegler, H. Schneider, and M. Walther, Multicolor and dual-band IR camera for missile warning and automatic target recognition, *Proc. SPIE*, 4718, 280–288, 2002.

15. P. Ljungberg, R. G. Kihlen, S. H. Lundqvist, P. Potet, and S. Berrebi, Multispectral imaging MWIR sensor for determination of spectral target signatures, *Proc. SPIE*, 3061, 823–832, 1997.

16. L. Becker, Multicolor LWIR focal plane array technology for space- and ground-based applications, *Proc. SPIE*, 5564, 1–14, 2004.

17. L. Becker, Novel quantum well, quantum dot, and superlattice heterostructure based infrared detectors, *Proc. SPIE*, 7298, 729805, 2009.

18. M. Nagashima, M. Kibe, Mi. Doshida, H. Yamashita, R. Suzuki, Y. Uchiyama, Y. Matsukura, H. Nishino, T. Fujii, and S. Miyazaki, High-performance 256 × 256 pixel LWIRQDIP, *Proc. SPIE*, 7298, 72980D, 2009.

19. P. Bensussan, P. Tribolet, G. Destéfanis, and M. Sirieix, Fifty years of successful MCT research and production in France, *Proc. SPIE*, 7298, 72982N, 2009.

20. A. Dutta, R. Sengupta, A. Krishnan, S. Islam, P. S. Wijewarnasuriya, and N. Dhar, Broadband image sensors for biomedical, security, and automotive applications, *Proc. SPIE*, 6769, 67690C, 2007.

21. D. K. Breakfield and D. Plemons, The application of microbolometers in 360° ground vehicle situational awareness, *Proc. SPIE*, 7298, 72981K, 2009.

22. R. Breiter, J. Wendler, H. Lutz, S. Rutzinger, K. Hofmann, and J. Ziegler, IR-detection modules from SWIR to VLWIR: Performance and applications, *Proc. SPIE*, 7298, 72981W, 2009.

23. M. Münzberg, R. Breiter, W. Cabanski, K. Hofmann, H. Lutz, J. Wendler, J. Ziegler, R. Rehm, and M. Walther, Dual color IR detection modules, trends and applications, *Proc. SPIE*, 6542, 654207, 2007.

24. W.-B. Song and J. J. Talghader, Design and characterization of adaptive microbolometers, *J. Micromech. Microeng.*, 16, 1073–1079, 2006.

25. A. K. Sood, Y. R. Puri, R. Richwine, L. Becker, N. Dhar, S. Sivanathan, and J. Zimmerman, Advances in multi-color large area focal plane array sensors for standoff detection, *Proc. SPIE*, 5881, 58810N, 2005.

26. P. W. Norton, S. Cox, B. Murphy, K. Grealish, M. Joswick, B. Denley, F. Feda, L. Elmali, and M. Kohin, Uncooled thermal imaging sensor and application advances, *Proc. SPIE*, 6206, 620617, 2006.

27. D. Murphy, M. Ray, A. Kennedy, J. Wyles, C. Hewitt, R. Wyles, E. Gordon, T. Sessler, S. Baur, D. Van Lue, S. Anderson, R. Chin, H. Gonzalez, C. Le Pere, S. Ton, and T. Kostrzewa, Expanded applications for high performance VO_x microbolometer FPAs, *Proc. SPIE*, 5783, 448, 2005.

28. D. Murphy, M. Ray, A. Kennedy, J. Wyles, C. Hewitt, R. Wyles, E. Gordon, T. Sessler, S. Baur, D. Van Lue, S. Anderson, R. Chin, H. Gonzalez, C. Le Pere, S. Ton, and T. Kostrzewa, Two-color HgCdTe infrared staring focal plane arrays, *Proc. SPIE*, 5209, 1–13, 2003.

29. L. S. R. Becker, Multicolor LWIR focal plane array technology for space- and ground-based applications, *Proc. SPIE*, 5564, 1, 2004.

30. A. C. Goldberg, T. Fischer, and Z. I. Derzko, Application of dual-band infrared focal plane arrays to tactical and strategic military problems, *Proceedings of SPIE: Infrared Technology and Applications XXVIII*, 4820, pp. 500–514, 2003.

31. S. Han, J. Jung, and D. P. Neikirk, Wavelength selective bolometer design, *Int. J. High Speed Electron. Syst.*, 18(3), 569–574, 2008.

32. V. N. Leonov and D. P. Butler, Two-color thermal detector with thermal chopping for infrared focal-plane arrays, *Appl. Opt.*, 40(16), 2601–2610, 2001.

33. Y. Wang, B. J. Potter, and J. J. Talghader, Coupled absorption filters for thermal detectors, *Opt. Lett.*, 31(13), 1945–1947, 2006.

34. P. W. Kruse and D. D. Skatrud, Uncooled infrared imaging arrays and systems, *Semiconductors Semimetals*, vol. 47, pp. 17–42, Academic Press, New York, 1997.

35. P. W. Kruse, The design of uncooled infrared imaging arrays, *SPIE*, 2746, 34–37, 1994.
36. S. J. Ropson, J. F. Brady, III, G. L. Francisco, J. Gilstrap, R. W. Gooch, P. McCardel, B. Ritchey, and T. R. Schimert, a-Si 160 × 120 micro IR camera: Operational performance, *Proc. SPIE*, 4393, 89–98, 2001.
37. F. Guinneton, L. Sauques, J.-C. Valmalette, F. Cros, and J.-R Gavarri, Optimized infrared switching properties in thermochromic vanadium dioxide thin films: Role of deposition process and microstructure, *Thin Solid Films*, 446, 287–295, 2004.
38. S. Black, M. Ray, C. Hewitt, R. Wyles, E. Gordon, K. Almada, S. Baur, M. Kuiken, D. Chi, and T. Sessler, RVS uncooled sensor development for tactical applications, *Proc. SPIE*, 6940, 694022, 2008.
39. D. Murphy, M. Ray, J. Wyles et al., 640 × 512 17 μm microbolometer FPA and sensor development, *Proc. SPIE*, 6542, 65421Z, 2007.

16 Sensing Temperature inside Explosions

Joseph J. Talghader and Merlin L. Mah

CONTENTS

16.1 INTRODUCTION

The interior of an explosion is one of the harshest environments on earth. The violent chemical reactions and rapid changes in pressure and temperature make sensing with conventional electronics virtually impossible. One can obtain significant information using remote optical techniques, and indeed the spectroscopy of explosions is a major field of research [1,2]; however, the presence of debris or opaque chemical reactions can make the fireball impossible to probe, even with the most advanced equipment. In this chapter, we describe temperature and thermal history sensors based on microparticle or nanoparticle luminescence. These particles are embedded in explosive material and disperse with it but are undamaged by the explosion since they have no mechanical or electronic "parts." The luminescence of the particles can be examined in the debris or measured in a lab and can ascertain the distribution of temperatures seen in the periphery of an explosive fireball. This method is currently under development at the University of Minnesota and Oklahoma State University and shows good promise as a thermal diagnostic where common sensing methods fail.

Any sensor that is placed inside a fire or explosion will be subject to conditions so violent that they are likely to destroy any mechanical components or conventional electronics. A sensor that could work within these limitations is a block of material with some nonvolatile, temperature-dependent property. Many oxides, fluorides, and other materials have embedded traps with a distribution of trap energies. If the material has traps filled with electrons or holes that are significantly deeper than an electron-volt, the traps will usually be stable over many years or even centuries. High temperatures can excite and empty these traps, a process that is described more fully in the next section. Generally, higher temperatures empty deeper traps. The thermoluminescence (TL) of these particles is an indication of trap population and thus gives one insight into the thermal history of the particle. In an explosion, these particles could be embedded into or near the explosive material and could thus probe temperature distributions around the explosive area, or post-detonation environment, without fear of sensor damage. Since the particles can be developed with a variety of sizes and shapes, they can mimic the aerodynamic behavior of components of an explosive (e.g., metal nanoparticles) or its target (e.g., a bioweapon containing viruses or bacteria). Figure 16.1 shows

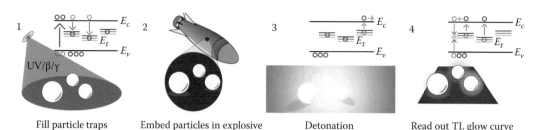

Fill particle traps   Embed particles in explosive   Detonation   Read out TL glow curve

FIGURE 16.1 Conceptual diagram of measuring thermal history using microparticles. Initially, the traps inside thermoluminescent particles are filled using ionizing radiation. The particles are embedded in an explosive, which detonates and empties the shallower traps. By examining the luminescence, one can determine the trap population distribution and therefore the temperature-time relationship of the explosive event.

a conceptual diagram of how such a process might work. A group of particles, numbering in the thousands to millions, is exposed to deep ultraviolet (UV) light or some other ionizing radiation source to fill the particle traps. After this, the particles are embedded in a flammable or explosive material (or placed around it) so that they will be dispersed during the reaction. Finally, the debris or ashes are examined in the field using a handheld luminescence reader or taken to a lab to assess the temperature and/or thermal history distribution.

This approach has several advantages:

1. The particles themselves have no parts that can be damaged by a fire or explosion; they are merely doped glasses and other materials that can easily withstand the harshest environments and procedures.
2. The materials are usually very cheap, and millions of the particles can be fabricated using existing processes.
3. The particles can be micromachined to take on just about any shape or size and thus will have essentially identical aerodynamics and thermal experience to any nanoparticles used in the explosion itself.
4. Fluctuations in the thermal history can be seen from the differences in luminescence from region to region or even particle to particle.
5. The luminescence data are exponentially weighted toward the highest temperatures experienced, which are the most critical in understanding post-detonation effects.
6. Extreme thermal profiles can be simulated using micromachined heaters that increase temperature to hundreds of degrees and cool on timescales on the order of a millisecond, or longer if desired. This also allows one to isolate temperature effects from pressure effects.

16.2 THERMOLUMINESCENCE AS A THERMOMETER

A luminescent material is one that emits light in response to some external stimulus such as heat or optical pumping. Many applications of luminescent materials depend on the presence of very long-lived deep traps that can store charge for years or even millennia. Perhaps, the most widely known application of this type is radiation dosimetry, where luminescent particles are embedded in an identification badge or other object. If this object is exposed to ionizing radiation, then electron–hole pairs are generated that can fall into traps within the gap. These trapped charge carriers can be released upon stimulation by heat in TL or by light in optical stimulated luminescence (OSL), after which they may recombine and emit light. The number of filled traps is, for many materials, directly proportional to the dose of ionizing radiation received; therefore, the amount of light emitted upon heating or optical excitation is a direct indication of the amount of radiation to which the object has been exposed.

The difference in using luminescent materials for dosimetry and temperature sensing is that in radiation dosimetry one is interested in how the trapping states are filled by exposure to radiation, whereas in temperature sensing one is interested in how the trapping states are emptied by a certain temperature profile. Therefore, for temperature sensing, the trapping states need to be filled prior to use of the materials as a temperature sensor. This can be easily accomplished using ionizing radiation or, more conveniently for some materials, deep UV light.

16.2.1 BASICS OF TRAP LUMINESCENCE

The energy of the trap can be understood using a bandgap model for the charge carrier states of the solid, as shown in Figure 16.2. A valence band filled with electrons is separated from an empty conduction band by an energy gap of several electron volts. The size of the gap for the thermoluminescent materials under discussion is large enough that the number of thermally excited valence electrons that reach the conduction band is effectively zero. However, there can be a number of traps distributed throughout the bandgap. These traps are not necessarily isolated levels but generally have a density of states that clusters at one or more energies. Some materials, such as carbon-doped aluminum oxide (Al_2O_3:C), have a single dominant trap cluster, while others, such as LiF:Mg,Ti, have a number of trap clusters that contribute to the luminescence.

The simplest model of the population statistics of a trap is governed by an Arrhenius type equation. Assuming first-order kinetics [3], the probability per unit time of an electron being released from an electron trap via thermal energy is as follows:

$$p = s \exp\left(\frac{-E_t}{kT}\right) \tag{16.1}$$

where
p is the probability of emission in s^{-1}
E_t is the depth of the trap below the conduction band (or above the valence band)
k is Boltzmann's constant
T is the temperature
s is a frequency factor that can be roughly understood to be related to the thermal phonon interactions that an electron undergoes with the surrounding lattice

The parameter s usually has a value between 10^{12} and 10^{14} s^{-1}, but in the literature, the parameter is frequently used for curve-fitting, and highly nonphysical values of s can be reported. This does not mean the values are "wrong"; it just indicates that the aforementioned simple equation is insufficient to describe the complexity of most traps. Traps may be interacting with one another, requiring a kinetic model that includes multiparticle interactions. Traps may also have tunneling-assisted transitions with surface or boundary states that require explicitly quantum mechanical treatments. Finally, defects in many materials have behaviors yet to be fully explained. So while the simple Arrhenius expression is very useful for understanding concepts and developing simple models, we should recognize that it rigorously applies to only a very limited number of traps and materials. A discussion of more complex models is beyond the scope of this review but is covered in excellent texts [3–5].

It is not enough for a material to have traps—it must also luminesce. Visible radiation from non-volatile traps usually does not arise from direct transitions from conduction band to valence band (which would be deep in the UV) but rather through a luminescence defect in the material. This is

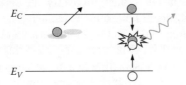

FIGURE 16.2 Bandgap model of charge traps. An electron (or hole) is excited to the conduction (valence) band where it travels until recombining at a luminescence center, emitting a photon.

typically an impurity such as Cr^{3+}, Eu^{2+}, or Tb^{3+} or a vacancy in the lattice that attracts and recombines oppositely charged carriers, emitting a photon of characteristic energy. The terms "F-center" or "color center" are often used to describe luminescence centers that derive from vacancies in a crystal lattice [3]. The recombination process can create an excited luminescence center, which relaxes to the ground state by photon emission.

In TL, the trap characteristics are obtained from a plot of emitted light intensity versus temperature. The temperature is usually ramped linearly with time, say at a heating rate of 1°C/s, but there are certain applications where other profiles, such as short thermal [6] or laser [7] pulses, are used. The resultant TL glow curve usually has one or more peaks, each corresponding to a different trap distribution. These trap distributions can be assigned an activation energy E_t and a frequency factor s as described earlier. To measure temperature, one fills the traps using ionizing radiation prior to measurement. In most applications, from radiation dosimetry to archaeological dating, one fills the traps using radioactive decay from gamma or beta sources, since radiation dose is the parameter to be measured or calibrated against. However, deep UV light will also work with many materials such as Al_2O_3:C and Mg_2SiO_4:Co,Tb. This may seem to be somewhat of a contradiction since it is well known that daylight, which contains UV light, removes carriers from certain types of traps, and the last exposure to daylight is considered the "zeroing" point from which a date is reckoned in archaeological TL. The difference involves the wavelength of the UV light. If a photon is extremely energetic, it will be able to excite an electron from the valence band all the way to the conduction band, giving that band a very large number of carriers. Excited carriers then fall into the traps, filling them. Shallow UV photons will be unable to bridge the bandgap in TL materials. Occasionally, there are other absorptions, such as the F-center absorption of Al_2O_3:C at 205 nm [8], which also fill the traps effectively. In most cases, however, near-UV and visible photons will empty the traps.

16.2.2 THERMOLUMINESCENCE FOR TEMPERATURE AND THERMAL HISTORY MEASUREMENT

The use of mineral or ceramic luminescence to determine temperature has been utilized previously in the space sciences and archaeology but not extensively because of complications in assessing initial (prior to the thermal event) trap populations, among other issues. In the early 1980s, there was interest in determining the former orbital characteristics of meteorites that had fallen to earth [3,9–11]. One way to determine this was to look at the TL of mineral particles within the meteorite. Since the TL would be characteristic of the highest temperature commonly reached by the meteorite while it was in orbit, it was assumed that the luminescence would indicate, via the temperature, the approximate distance of the perihelion of its orbit around the sun. This particular application was much more complex in many ways than what is proposed here because of complications such as heating during entry into Earth's atmosphere, determining cosmic ray exposure that would give a steady-state trap population, predicting the albedo (reflectivity) of the overall meteor, and other issues. Despite this, the spread in the perihelia of known meteorite sources could be roughly correlated to the spread in TL measurements [11]. Trap luminescence has also been used to obtain information about the firing temperatures of pottery [12]. However, in this case, it is not the change in trap population that is measured, but rather how the firing temperature changes the sensitivity of the materials to radiation dose.

While in theory, it is possible to obtain temperature measurements in rapid events using only the absolute intensity of the luminescence (as used in radiation dosimetry, archaeological dating, and the meteorite measurements just described), in reality, the complexity of the combustion environment makes this extremely difficult. For example, temperature-sensing particles will be mixed in with debris and soot, attenuating the signal to an unknown and uncontrollable extent. A better method of temperature measurement relies on examining the ratios of the intensities of two or more peaks. The population of the traps corresponding to a lower temperature peak will empty at lower temperatures than those corresponding to a higher temperature peak. Since the higher temperature peak can be used as a partial reference, the ratios of the intensities of the two peaks will be a more reliable indicator of temperature than the intensity of just one.

In order to understand the details of obtaining a thermal history from TL, we must develop an expression for trap population as a function of temperature, where temperature can vary with time. Consider a TL particle with a single peak that undergoes a rapid heating followed by a (relatively) slow cooling:

Rapid heating

$$T(t) = \beta_{\text{expl}} \cdot t + T_{\text{init}} \tag{16.2}$$

Slow cooling

$$T(t) = T_{\text{max}} e^{-\frac{t}{\tau}} + T_{\text{init}} \tag{16.3}$$

where
   β_{exp} is the (very large) slope of the temperature vs. time curve during heating
   T_{init} is the ambient temperature before the explosion
   T_{max} is the peak temperature to which the particle is exposed
   τ is a cooling time constant
   t is the time in seconds
   T is the temperature of the particle at a time t

A graph of this heating profile is shown in Figure 16.3.

The population of a specific set of traps can be derived by multiplying the probability that a trap is filled (Equation 16.1) by the number of traps in that group and taking the derivative. At this point, the temperature is a single number. If we wish to analyze trap population as a function of time, the temperature must become a function of time and then the entire expression must be integrated:

$$\frac{dn}{dt} = -p \cdot n = -nse^{-E/kT(t)} \tag{16.4}$$

$$n(t) = n_0 \exp\left(-\int_0^t se^{-E/kT(t)} dt\right) \tag{16.5}$$

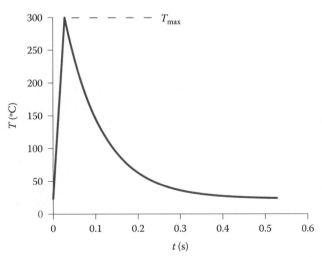

FIGURE 16.3 Temperature profile used to simulate an explosion. There is an extremely fast initial rise (usually considered to be instantaneous) followed by slow asymptotic cooling.

In these expressions, n is the trap population at time t, n_0 is the initial trap population before the event, and p is the probability of a single trap being emptied as described in Equation 16.1.

Now, consider two sets of traps whose energies, E_{t1} and E_{t2}, cluster about 1.277 and 1.30 eV, respectively. For simplicity, we will assume that the frequency factor, s, of both traps is the same at $s = 10^{12}$ s^{-1}. If we make a plot of the population ratio, n_1/n_2, as a function of cooling time for a variety of maximum temperatures, we will obtain Figure 16.4. (Note that the cooling time is related to the thermal time constant, τ, defined earlier.) Here, we can see that even small changes in maximum temperature can have a dramatically different effect on the populations of two sets of traps, even when those trap sets are separated by only relatively small energies. On the other hand, Figure 16.5 shows a plot of n_1/n_2 as a function of T_{max} for a variety of cooling times. The cooling time is much

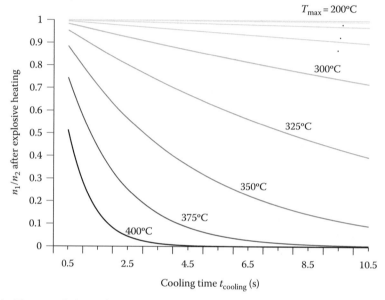

FIGURE 16.4 Trap population ratio as a function of cooling time for a variety of maximum temperatures. Note the cooling time only has a large impact once the temperature has reached a high value.

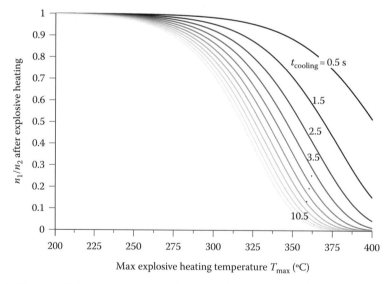

FIGURE 16.5 Trap population ratio versus maximum heating temperature for a variety of cooling times. The transitions from high ratio to low ratio all cluster near similar temperatures.

less important in making quantitative changes to the population ratio. This can be understood to be a result of the exponential dependence on temperature of the detrapping rate of Equation 16.1. A maximum temperature increase, even if held for a short period of time, is more effective at causing carriers to escape traps than long periods of exposure to lower temperatures. It is a valid question to ask whether the heating rate plays a large role in the trap population ratios. The answer depends on the speed of the heating. In the simulations leading to Figures 16.4 and 16.5, we have assumed a more or less infinitely fast heating rate. So long as the heating taking place over time scales much shorter than the cooling, then we may neglect the effect of β_{exp} on the population profile. However, if the cooling and heating times become comparable, then both must be included. Attempts to extract both temperature and cooling times will be discussed in more detail in Section 16.4.

Now that we have established that the population ratio of two traps is intimately related to the thermal history to which the traps have been exposed, it is useful to examine how the luminescence might indicate these ratios. Ideally, a TL curve will show us multiple peaks that are spaced far enough apart that their populations can be individually determined without reference to any others. However, if two traps are extremely close together in energy, such as the traps described in Figures 16.4 and 16.5, the TL from each trap will overlap and be difficult to separate, even if there are significant population differences between the two. This phenomenon is shown in Figure 16.6, where a trap with $E_{t1} = 1.4$ eV and $s_1 = 2 \times 10^{12}$ s^{-1} is plotted with another trap with $s_2 = 1 \times 10^{13}$ s^{-1} where E_{t2} changes from 1.3 to 1.7 eV in steps of 0.1 eV. (We used different initial trap populations to scale the two peaks.) The darkest (top) curve is the sum of the two luminescences, which is what would be measured by a detector. It can be seen that an easy distinction between the two really only occurs at $E_{t2} = 1.3$ and 1.7 eV. This does not mean that we cannot use the information in the merged TL curve, but it does mean that our modeling must do more than just measure peak heights (see Section 16.4).

So far, we have considered that we are using a single type of particle, even if there may be millions of them in a fire or explosion. It is theoretically possible to enhance the amount of information that we can extract from the luminescence using multiple types of particles. Perhaps the most attractive feature is that each type of particle could have traps that luminesce at different wavelengths than other types of particles. Many, but not all, TL materials have fixed emission characteristics, even if they have multiple trap populations. The reason for this is that the luminescence center is usually a completely separate entity from the trap itself. Carriers are excited from traps to the conduction or valence band and from there travel to recombine at the luminescence center. This means that within a single particle, one cannot count on being able to distinguish traps by wavelength,

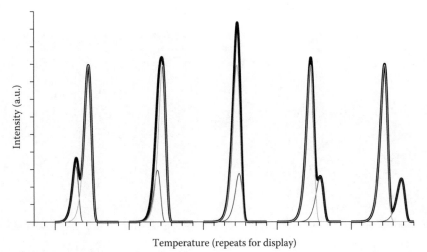

FIGURE 16.6 Overlapping thermoluminescent glow curves as the energy between two traps is changed. The parameters of the highest peak trap are $E_{t1} = 1.4$ eV and $s_1 = 2 \times 10^{12}$ s^{-1} while the parameters of the variable trap are $s_2 = 1 \times 10^{13}$ s^{-1} and E_{t2} changes from 1.3 to 1.7 eV in steps of 0.1 eV.

but with multiple particles, where there are different luminescence centers with different emission wavelengths, this should be possible. However, one must be aware that the spectral width of the luminescence can be very large. For example, Al_2O_3:C has a peak emission near 420 nm with a spectral width of over 100 nm [13]. The emission of LiF:Mg,Ti [14] has peaks that overlap those of Al_2O_3:C in wavelength, so it would not be straightforward to use emission wavelength selectivity to distinguish the luminescence of particles of these two materials. On the other hand, there are materials, such as MgO:Tm,Li and MgO:Eu,Li, that have relatively narrow emission spectra [15], such that wavelength selectivity among multiple particles would be viable.

16.3 MICROHEATERS

Obviously, developing TL temperature measurements will be easier if one does not have to perform large scale explosions just to obtain a data point, particularly in a university environment. In order to eliminate the need to explode particles, we perform much of our thermoluminescent testing on micromachined heaters. The heating element of a traditional TL measurement is usually a hotplate or planchet heater, in which a sample-bearing platform is heated by an electrical current passing through a resistive element [16]. While these conventional devices serve their purpose well enough for TL dosimeters and the like, their size and thermal mass makes them ponderous to heat and cool, vulnerable to contamination, power-hungry, and difficult to control.

Many of these problems can be solved by shrinking the heating element into the micro-regime. Using VLSI microfabrication techniques, semiconductor materials can be patterned into heating elements with dimensions of tens of microns. Examples are shown in Figure 16.7. At these scales, the thermal mass of a heater is so small that temperature excursions of many hundreds of degrees can be made on time scales of milliseconds. It should be explicitly noted that arbitrary temperature waveforms can be simulated with a microheater. It is extremely easy, using a pulse generator or other high speed current source, to drive the heater to heat within hundreds of microseconds to hundreds of degrees and then cool over several seconds (or any rate within the limit of the thermal time constant). Further, heat transfer through convection is limited, making control and analysis relatively simple. The microheaters can often eschew the complex control schemes needed to deal with response delay and ringing in conventional hotplates. The highly concentrated application of heat greatly reduces unwanted TL signals from sample residue or any other contamination that may be nearby, as well as background light and distortions from radiative heating and convection, respectively. This selectivity of heating means that thermoluminescent films can be deposited

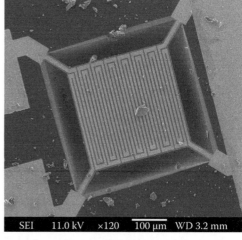

FIGURE 16.7 Scanning electron microscope images of microheaters used in luminescent particle studies.

directly onto the heater surfaces during fabrication if desired, providing for a sample with integrated heating facilities on a single chip. Finally, their miniature size, power, and processing requirements make microheaters ideal for integration into portable TL readers.

On a microheater, measurement with a commercial thermistor or thermocouple is out of the question, as the heater would be dwarfed by the measurement device. However, platinum can be easily integrated into the heater and exhibits an extremely linear thermal coefficient of resistance (TCR). Since platinum is also well suited for a heating resistor, the sensing and driving elements can even be one and the same, which greatly simplifies the device and eliminates concerns about sensor placement.

The advantages of microheaters do not come without complications. The miniscule surface area and highly focused heating of the heaters inherently translate to a smaller volume of TL material and thus lower signal intensity. If the microheaters are poorly designed, the temperature of one section of the microheater may differ enough from another that particles on the two sections will have TL glow curves that are displaced along the temperature axis from one another. Also, differing film stress characteristics can cause heater platforms to take on curved shapes, and these shapes can buckle rapidly under extreme temperatures. One batch of our early microheaters would abruptly transform from concave to convex (or vice versa) and thus launch any loosely adhered particles into the air!

Ultimately, the previously listed complications are a matter of proper engineering and have seldom caused trouble after initial development. However, one issue is more fundamental: the speed of the heat transfer from the microheater to the particles. The small time constants that allow temperature profiles that mimic explosions very accurately can be too short for large particles. This problem is magnified when dealing with TL powders with highly irregular shapes, which may result in poor thermal contact between heater and material, resulting in inaccurate temperature readings and smeared-out TL curves. Whereas a conventionally sized heater never heats fast enough to make these issues relevant, a microheater makes the calculation of the time constants of the particles important to the analysis. Thus one must estimate the thermal mass of any large particle on the microheater to insure that the temperature at the platinum thermistor is actually the temperature of the particle. It should be noted, however, that this will also be an issue in explosions (post-detonation environments) and particles on the order of 10 μm or larger should be avoided.

Our latest-generation microheaters use a serpentine resistor of sputtered platinum a few hundred nanometers thick, patterned by liftoff photolithography and underlain by a 10 nm titanium adhesion layer. These electrical parts are sandwiched by silicon nitride films, with a plasma enhanced chemical vapor deposition (PECVD) layer deposited on top for wet etch protection and a low-stress low pressure chemical vapor deposition (LPCVD) layer underneath. A plasma etch through both nitride layers to the silicon wafer below allows the use of anisotropic wet etching to carve a pit below the heater, resulting in a suspended platform of silicon nitride. The sizable air gaps below and to all sides of the heater platform limit heat conduction paths to four support beams of high aspect ratio; this degree of isolation, when combined with the minuscule thickness—in the hundreds of nanometers—and mass of the support structure, ensures a rapid response to joule heating and a time constant of around 30 ms. For applications where a much faster time constant is desired, the release etch can be skipped to leave the underlying silicon substrate intact; this provides a massive heat conduction path from the heater platform's entire underside to a relatively enormous block of material, and thus drives the time constant down to the 200 μs range. To date, we have not seen any background TL signals from any of the materials used to fabricate the heaters. Experience has shown that the released heaters are mechanically robust enough to handle powder samples being removed by spraying with pressurized nitrogen. Even with single-wafer processing, hundreds of devices can be derived from a single batch, allowing the option of a single-use per heater and the practical elimination of sample-to-sample cross-contamination.

Before operation, the heaters are placed on a conventional hotplate and their resistance measured as the die is heated; this yields the linear TCR, which is used to measure and control their temperature in use. Stress testing has shown that these microheaters can easily and repeatedly ramp to temperatures of over 500°C without change in characteristics.

16.4   EXAMPLE PROCESS

Using the rapid heating technology of microheaters, we can study the effects of an explosion-type temperature profile on TL materials in conditions much more easily produced, measured, and controlled than actual explosions [17–19]. For an initial test, microparticles of Mg_2SiO_4:Tb,Co on the order of 10–20 μm in diameter were prepared [20] by Dr. M. Prokic, Institute of Nuclear Sciences. This terbium- and cobalt-doped magnesium silicate formulation produces at least two easily distinguishable TL peaks below 350°C upon irradiation with broad-spectrum 180–400 nm UV light, which allows us to refill sample traps inside our TL setup without the need for more hazardous radiation sources or heavier shielding. The entirety of the experiment can thus be run without removing the sample from the TL chamber, eliminating the risk of photobleaching (to which the closely related Mg_2SiO_4:Tb is known to be vulnerable [21]) during handling and ensuring that all data are taken from the same sample with a consistent thermal contact.

Microheaters similar to the variety previously described, with platforms 300 μm on a side, were used for both explosive and linear (glow curve) heating. Four calibrations of the microheater resistors on hot plates produced closely matching linear TCR results that, at the highest temperature used in the experiment, produced a maximum temperature deviation of 4.4%. The microheaters were also run through a temperature ramp to record the nonlinear relationship of voltage and temperature, which would be used to target the output power during pulse heating. (The actual pulse temperature reached was still determined via resistance measurement, however.) A computer-controlled sourcemeter was used to drive the microheater in both modes of heating; while this was observed to work very well for slow heating, later revisions of the experiment use faster control programs and dedicated pulsed generation devices to achieve better accuracy during rapid operation. TL readout is handled by a photomultiplier tube, which is protected during irradiation by a motorized metal shutter. This setup is shown in Figure 16.8.

Mg_2SiO_4:Tb,Co microparticles were sprinkled onto the heater die and repositioned with microprobe manipulators; the microheaters are thermally isolated from their substrates, so the excess particles and any other contaminants that land near the device will not be heated enough to luminesce. Once laden with samples, the microheater was placed inside the darkened TL chamber, and its microparticles irradiated with UV light from a 30 W deuterium lamp. A temperature pulse of total duration 200 ms (±13 ms) and variable maximum temperature (i.e., variable current amplitude) was applied in noncontrol runs. A TL glow curve was then collected using a linear temperature ramp of 2°C/s up to 350°C, where the magnesium silicate glow is observed to die away while a strong background intensity—most

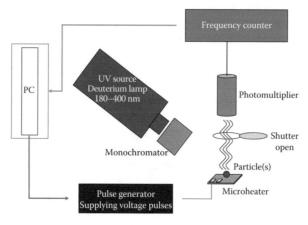

FIGURE 16.8 Experimental set-up used to test the response of thermoluminescent Al_2O_3:C microparticles to rapid thermal profiles. Included are devices to fill the traps, simulate explosive heating, and measure TL glow curves.

likely thermal blackbody emission—rises to prominence. Although no ill effect from repeated rapid cooling was observed in stress testing, the microheater temperature was gradually ramped down to ambient to avoid any thermal gradient bimorph stresses that might jostle the particle. The entire experiment run was automated and repeated as necessary without manual handling of the sample.

To reduce photomultiplier tube (PMT) noise and fluctuations in overall intensity, each raw data point was averaged with its 20 closest neighbors (corresponding to approximately 10°C) and each curve multiplicatively scaled so that their heights are equal at 266°C, the apex of the second peak. The blackbody emission background, which becomes dominant at ramp temperatures above 310°C after the Mg_2SiO_4:Tb,Co TL peaks (where present) have died away, was modeled as an exponential function curve-fitted to the control curve above 310°C and subtracted from the intensity data of each subsequent pulsed run. The resulting glow curves are shown in Figure 16.9.

In actual use, our technique must compare the TL glow curves of different samples of the same material, some scarred and sullied by explosion, possibly using detectors and read-out systems with slightly different characteristics. To compensate for the changes in overall intensity that will follow, we must preserve one of our two distinguishable TL peaks undiminished as an intensity reference; this can be done by choosing a TL material with a high temperature peak that will not be impacted by the expected heating level, which we may set by observing the effects of slow preheating (using a conventional planchet heater) on the glow curve of Mg_2SiO_4:Tb,Co. In such intensity curves, gathered with a calibrated thermoluminescent dosimetry (TLD) reader, the two peaks occur at slightly different temperatures than in the microheater readings, and a small third peak is visible between them in some samples; this discrepancy, and the different temperatures at which the peaks occur, is likely due to differences in materials preparation and the sensing instruments used. As the maximum preheating temperature increases, the height of the lowest-temperature peak quickly falls, but the highest temperature peak stays relatively untouched for much longer. This implies that we can almost entirely avoid depopulating the traps responsible for the highest-energy peak by choosing a material whose highest peak has a temperature significantly above that created by the heat pulse. Different glow curves can then be easily compared, regardless of overall intensity, using the ratio of the height of the first peak to that of the second.

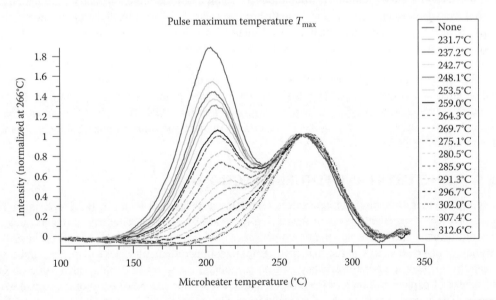

FIGURE 16.9 The thermoluminescence glow curves of Mg_2SiO_4:Tb,Co microparticles after a 190 ms explosive heating pulse. The legend at right indicates the peak temperature of the pulse corresponding to each curve. The intensities shown are normalized at 266°C, the top of the second observable peak. The ratios of the intensities of the peaks strongly correlate with the maximum temperature to which the particles had been exposed. (Reprinted from Mah, M. L. et al., *IEEE Sens J*, 10(2), 311, February 2010.)

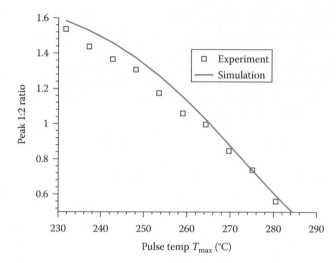

FIGURE 16.10 The ratio of the height of the first TL peak of Figure 15 to the height of the second as a function of pulse temperature for simulated and experimental data. (Reprinted from Mah, M. L. et al., *IEEE Sens J*, 10(2), 311, February 2010.)

Theory predicts that a pulse with a higher peak temperature will have a progressively stronger depopulation effect on the relatively shallow electron traps. Our data suggest that the deeper traps responsible for the second peak are relatively unaffected by the pulse temperatures and durations used here, but will begin to empty at higher excitation temperatures as can be seen by slow preheating. Increasing the maximum pulse-heating temperature [17] applied by 5°C at a time yields easily discernible decreases in the height of the low-temperature peak until it dwindles to an indistinguishable intensity when the maximum preheating pulse temperature reaches around 300°C, as shown in Figure 16.9.

To compare these results to theoretical prediction, we chose trap parameters to closely fit our nonpulsed control curve; numerical simulations based on first-order kinetics theory were then used to produce postpulse glow curves for these theoretical particles. Figure 16.10 compares the relation between the intensity peak height and temperature pulse maxima of the resulting simulations to that of the experimental data obtained using microheaters. As expected, a clear one-to-one decrease occurs in the intensity ratio of the first peak to second peak heights as the pulse temperature climbs [17]. The slope of the curve is reasonably stable, indicating that the sensitivity of this Mg_2SiO_4:Tb,Co will be fairly steady over the usable pulse temperature range. In the range we observed, model and data differed by an average of 4.4%, with no errors greater than 9.1%. (Even better accuracy could be obtained by simply curve-fitting the TL peak ratio in much the same manner that the resistance of commercial thermistors is calibrated.)

16.5 CONCLUSION AND FUTURE DIRECTIONS

We conclude that thermoluminescent micro- and nanoparticles with multiple, light-insensitive TL peaks provide an excellent means of measuring thermal history in extreme environments such as fires and explosions, where no traditional sensors could function. These TL materials have the advantages of no separable parts, low cost, customizable aerodynamics, easy dispersal, and high sensitivity to peak temperatures. They can be stimulated many times per day using microheaters instead of explosions and survive postdetonation even when placed directly on high-explosive charges. The concept has been experimentally proven using MgSiO4:Tb,Co particles on microheaters with rapid heating and cooling over approximately 232°C–313°C.

We feel future work in the area will proceed in four directions. The first, obviously, is to test TL thermal history sensing in real explosions, and in fact, this work is already proceeding in a collaboration between our group and NSWC in Indian Head, Maryland. Second, there is a definite need for materials research. Almost all TL materials have been developed and characterized for radiation dosimetry, a technology where materials with a single well-behaved peak and high radiation sensitivity are paramount. For temperature measurement, multiple light-insensitive peaks are preferred and radiation sensitivity is largely irrelevant. Many materials that were rejected for dosimetry may be viable for sensing. In addition, new materials are being developed at Oklahoma State University and Clemson University using solution combustion synthesis [15]. Third, one can develop specially engineered particles and materials to enhance the collection of thermal history beyond what we have described here. For example, one might construct a thin film stack of alternating TL materials to engineer a "particle" with a desired set of traps. One may also create thick core-shell particles where the shell will follow temperatures at the surface very rapidly while the core will only follow slower average events. By examining the luminescence as a function of depth in the particle (say by using core/shell materials with different emission wavelengths) one can extract temperature information in much the same manner that millennia of climate history can be obtained from deep caves and boreholes drilled into the ice in Antarctica or Greenland [22]. Finally, it may be possible to measure parameters other than temperature. It will be interesting to see if the extreme pressures in the center fireball of an explosion cause any changes in trap populations. From one point of view, one would like to ignore such a complication. On the other hand, if there is a dependence, then certain particles are likely to be more sensitive to this than others, and the development of pressure sensors could begin. In any case, it will be interesting to see if other applications of microparticle thermoluminescent sensors arise!

ACKNOWLEDGMENTS

We would particularly like to thank the Defense Threat Reduction Agency (DTRA) for funding our work under grants HDTRA1-07-1-006 and HDTRA1-10-0007. We would also like to thank Su Peiris of the DTRA for ongoing guidance and support. This article is an abridged version of our article that appeared in Optical, Acoustic, Magnetic, and Mechanical Sensor Technologies, Ed. Kris Iniewski, CRC Press, 2012. Portions of this work are based on research in our group from prior publications, particularly Section IV using references [17–19]. We would like to thank Eduardo Yukihara of Oklahoma State University and Jim Lightstone of the Naval Surface Warfare Center for their fine research in the area and for their continuing collaboration and assistance. We would also like to acknowledge our colleagues at the University of Minnesota: Michael Manfred, Sangho Kim, and Nick Gabriel, who have greatly contributed to the work in this chapter.

REFERENCES

1. W. K. Lewis and C. G. Rumchik, Measurement of apparent temperature in post-detonation fireballs using atomic emission spectroscopy, *Journal of Applied Physics*, 105(5), 056104, 1 March 2009.
2. J. Wilkinson, J. M. Lightstone, C. J. Boswell, and J. R. Carney, Emission spectroscopy of aluminum in post-detonation combustion, *AIP Conference Proceedings*, 955(1), 1271–1274, 2007.
3. S. W. S. McKeever, *Thermoluminescence of Solids*, Cambridge University Press: Cambridge, U.K., 1985.
4. Y. S. Horowitz, *Thermoluminescence and Thermoluminescent Dosimetry*, CRC Press: Boca Raton, FL, 1984.
5. L. Botter-Jensen, S. W. S. McKeever, and A. G. Wintle, *Optically Stimulated Luminescence Dosimetry*, Elsevier: Amsterdam, the Netherlands, 2003.
6. M. E. Manfred, N. Gabriel, E. Yukihara, and J. J. Talghader, Thermoluminescence measurement technique using millisecond temperature pulses, *Radiation Protection Dosimetry*, 139(4), 560–564, 2010.
7. J. Gasiot, P. Braunlich, and J. P. Fillard, Laser heating in thermoluminescence dosimetry, *Journal of Applied Physics*, 53, 5200, 1982.
8. K. H. Lee and J. H. Crawford, Jr., Luminescence of the F center in sapphire, *Physical Review B*, 19(6), 3217–3221, 1979.

9. G. Valladas and C. Lalou, Thermoluminescence of the Saint-Severin Meteorite, *Earth and Planetary Science Letters*, 18(1), 168–171, 1973.

10. S. W. S. McKeever and D. W. Sears, Thermoluminescence and terrestrial age of the Estacado meteorite, *Nature*, 275(5681), 629–630, 1978.

11. S. W. S. McKeever and D. W. Sears, The natural thermoluminescence of meteorites: A pointer to meteorite orbits? *Modern Geology*, 7(3), 137–145, 1980.

12. G. S. Polymeris, A. Sakalis, D. Papadopoulou, G. Dallas, G. Kitis, N. C. Tsirliganis, Firing temperature of pottery using TL and OSL techniques, *Nuclear Instruments and Methods in Physics Research A*, 580, 747–750, 2007.

13. E. G. Yukihara, V. H. Whitley, J. C. Polf, D. M. Klein, S. W. S. McKeever, A. E. Akselrod, and M. S. Akselrod, The effects of deep trap population on the thermoluminescence of Al_2O_3:C, *Radiation Measurements*, 37, 627–638, 2003.

14. P. D. Townsend, Analysis of TL emission spectra, *Radiation Measurements*, 23(2/3), 341–348, 1994.

15. V. R. Orante-Barrn, L. C. Oliveira, J. B. Kelly, E. D. Milliken, G. Denis, L. G. Jacobsohn, J. Puckette, and E. G. Yukihara, Luminescence properties of MgO produced by solution combustion synthesis and doped with lanthanides and Li, *Journal of Luminescence*, 131(5), 1058–1065, 2011.

16. I. Bibicu and S. Calogero, High-efficiency heater for a thermoluminescence apparatus, *Journal de Physique III*, 6, 475–480, 1996.

17. M. L. Mah, M. E. Manfred, S. S. Kim, M. Prokić, E. G. Yukihara, and J. J. Talghader, Measurement of rapid temperature profiles using thermoluminescent microparticles, *IEEE Sensors Journal*, 10(2), 311–315, February 2010.

18. M. L. Mah, M. E. Manfred, S. S. Kim, M. Prokic, E. G. Yukihara, and J. J. Talghader, Sensing of thermal history using thermoluminescent microparticles, *IEEE/LEOS International Conference on Optical MEMS and Nanophotonics 2009*, Clearwater, FL, pp. 23–24, August 2009.

19. J. J. Talghader, Micro- and nano-particles for the distributed sensing of thermal history, *Seventh International Conference on Networked Sensing Systems*, Kassel, Germany, pp. 169–170, June 15–18, 2010.

20. J. C. Mittani, M. Prokić, and E. G. Yukihara, Optically stimulated luminescence and thermoluminescence of terbium-activated silicates and aluminates, *Radiation Measurements*, 43(2), 323–326, February 2008.

21. M. Prokić and E. G. Yukihara, Dosimetric characteristics of high sensitive Mg2SiO4:Tb solid TL detector, *Radiation Measurements*, 43(2), 463–466, February 2008.

22. W. S. B. Patterson, *The Physics of Glaciers*, 3rd edn., Butterworth-Heinemann: New York, 1994.

Part III

Magnetic and Inductive Sensors

17 Accurate Scanning of Magnetic Fields

Hendrik Husstedt, Udo Ausserlechner,
and Manfred Kaltenbacher

CONTENTS

17.1 INTRODUCTION

17.1.1 MOTIVATION

Magnetic sensors are used in versatile automotive applications such as the position detection of throttle valves, cam- and crankshafts, pedals, wipers, winders, etc. Before the assembling of these sensors in a car, each device is extensively tested not only electrically but also magnetically. Consequently, dedicated test equipment is needed to generate magnetic reference fields emulating the fields during application. These reference fields are used to decide which parts are sorted out and which are free from defects. Thus, it is essential that the magnetic field vector at the position of the device under test (DUT) exactly meets the desired properties, such as direction and absolute value of the magnetic field. Errors of the reference field affect the testing of millions of parts and, therefore, the equipment is not assembled in the test environment before it is well analyzed. For this purpose, a highly accurate three-dimensional (3D) magnetic scanning system is needed to measure the magnetic field in the volume where the DUT is positioned.

17.1.2 APPROACH

The magnetic field is described with three degrees of freedom in every point in space, which is represented by a vector field. Today, there are integrated magnetic sensors available that measure precisely the strength of the magnetic field in one direction. Combining three one-dimensional sensors on one silicon die or in separate packages allows for measurements of all three components of the magnetic field. Attaching such a 3D magnetic field sensor to moving axes realizes the positioning of the sensor to arbitrary points near the field source. Furthermore, an optical probe is added to the moving axes so that the geometry of the field source can be measured. We call such a system a magnetic and coordinate measuring machine (MCMM), which is able to measure the magnetic field of an arbitrary field source in all three directions and relate the results to the geometry.

17.1.3 OUTLOOK

The rest of the chapter is organized as follows. First, the scanning of magnetic fields is explained, and the approach of standard measurement systems is discussed. Then, the measurement principle and the realization of an MCMM are presented, which consist of moving axes, a magnetic sensor, and an optical probe. This setup allows for precise measurements of the magnetic field relative to the geometry of the field source. This feature is especially important for the measurement of strongly inhomogeneous magnetic fields where small mechanical tolerances lead to crucial measurement errors. To relate the measurement data of the magnetic sensor with the data of the optical probe, the setup needs to be calibrated. Here, the calibration of geometrical parameters of the optical probe and the magnetic sensor are depicted. Finally, the working principle of an MCCM is illustrated with a measurement example.

17.2 MAGNETIC SCANNING

17.2.1 STRAIGHT FORWARD APPROACH

In common measurement setups, a 3D magnetic sensor is attached to moving axes (see Figure 17.1), e.g., [1,2]. With this setup, the magnetic sensor can be positioned at arbitrary points near the field source, which roughly provides the field vectors relative to the reference frame of the field source x^d, y^d, and z^d. However, due to assembly tolerances, the exact position and alignment of the magnetic sensor and the field source are unknown.

In detail, the magnetic sensor is usually assembled in a ceramic or plastic package. Inside this package, the position of the silicon die varies in the range of ±100 μm [3,4]. Furthermore, the accuracy of the orientation of the silicon die inside the package is usually specified with ±3° [5], whereas internal investigations with x-ray photography at Infineon Villach show typical values of ±0.5°. Then, the package is mounted to an attachment that connects the magnetic sensor with the moving axes, which also causes mechanical errors. Overall, the position of the magnetic sensor w.r.t. the coordinate system of the moving axis can be determined to an accuracy of ±300 μm, and typical alignment errors are in the range of ±3°. Furthermore, not only the magnetic sensor but also the positioning of the field source in the measurement setup is accompanied by mechanical tolerances. Thus, the position and orientation of the field source vary if it is placed several times in the setup. This reduces not only the absolute accuracy but also the repeatability.

In short, due to mechanical tolerance, the position and alignment of the magnetic sensor w.r.t. the coordinate system of the field source are unknown. The impact of translational tolerances may

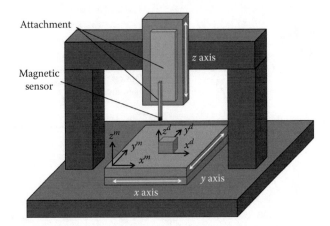

FIGURE 17.1 Schematic drawing of a common setup to measure the spatial characteristic of magnetic fields. The reference frame of the moving axes has the coordinates x^m, y^m, and z^m, and the coordinate system of the field source or DUT has the coordinates x^d, y^d, and z^d.

be neglected for homogeneous magnetic fields. However, if strongly inhomogeneous magnetic fields have to be measured, e.g., with a gradient of 1%/10 μm as used in automotive applications [6], mechanical tolerances are the crucial source of error.

17.2.2 MAGNETIC AND COORDINATE MEASURING MACHINE

To overcome problems caused by mechanical tolerances, the approach is using not only a magnetic sensor and moving axes but also a probe to measure the geometry as shown in Figure 17.2 [7,8]. Since this system represents a coordinate measuring machine (CMM) [9,10] that includes a magnetic sensor, it is denoted as MCMM. A schematic drawing of an MCMM and a photograph of a real setup are shown in Figure 17.2.

The idea of this concept is the following. First, the field source (DUT) is placed in the measurement setup, whereas its alignment and position are not exactly known due to mechanical

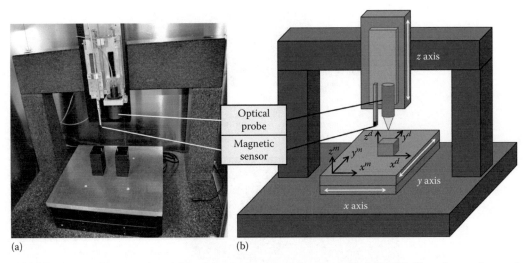

(a) (b)

FIGURE 17.2 (a) Photograph and (b) schematic drawing of the setup of an MCMM. The reference frame of the moving system has the coordinates x^m, y^m, and z^m, and the coordinate system of the field source (DUT) are denoted as x^d, y^d, and z^d.

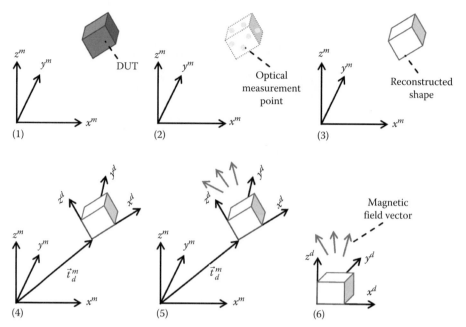

FIGURE 17.3 Measurement principle of an MCMM.

tolerances (Figure 17.3-1). Then, the optical probe detects several points on the surface of the DUT (Figure 17.3-2) so that its shape can be reconstructed (Figure 17.3-3). If the shape of the DUT is known, the position and orientation of the reference frame of the DUT can be determined (Figure 17.3-4). All these steps are equal to the working principle of a CMM, whereas an MCCM also allows one to detect the magnetic field. Since the alignment and position of the DUT is known, the magnetic sensor can be moved to points w.r.t. the coordinate system of the DUT (Figure 17.3-5). Finally, the data of the geometry and of the magnetic field can be transformed into the coordinate system of the DUT so that assembly tolerances have no influence on the measurement result (Figure 17.3-6).

However, moving the magnetic sensor to points relative to the DUT requires that the distance between the optical probe and the magnetic sensors $(\vec{P}_B)$ is known (see Figure 17.4). Moreover, the

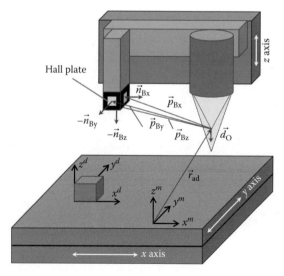

FIGURE 17.4 Schematic drawing of the measurement setup including the parameters of calibration. $\vec{r}_{ad}$ represents the position adjusted with the moving axes.

field vectors measured can only be transformed to the coordinate system of the field source if the alignment of the magnetic sensors w.r.t. the coordinate system of the moving axes ($\vec{n}_B$) is known. In addition, also the alignment of the optical probe ($\vec{d}_o$) has to be calibrated because misalignments cause distortions of the geometry measured.

To conclude, an MCMM allows one to measure the magnetic field w.r.t. the geometry of the field source so that the impact of mechanical tolerances is suppressed. Nevertheless, the measurement principle requires an accurate calibration of the optical probe and the magnetic sensor, which is discussed in Section 17.4.

17.2.3 COORDINATE SYSTEMS

Two orthonormal coordinate systems are used to define the position and orientation of a sensor, a field source, etc. The axes of these coordinate systems are denoted as x^k, y^k, z^k, which are distinguished with the superscript index $k \in \{m,d\}$, where m represents the coordinate system of the MCMM and d the reference axes of the field source (DUT). In addition, the superscript index denotes in which coordinate system an arbitrary point $\vec{r}^k$ is represented. Furthermore, the standard basis vectors of the coordinate system k are denoted as $\vec{x}_k^l = \left(x_{kx}^l \ x_{ky}^l \ x_{kz}^l\right)^T$, $\vec{y}_k^l = \left(y_{kx}^l \ y_{ky}^l \ y_{kz}^l\right)^T$, and $\vec{z}_k^l = \left(z_{kx}^l \ z_{ky}^l \ z_{kz}^l\right)^T$, where the subscript index k denotes to which coordinate system the basis vectors belong, and the superscript index l denotes in which coordinate system they are represented. If $k = l$, the standard basis vectors are written as $\vec{x}_k^k = \vec{e}_x = (1 \quad 0 \quad 0)^T$, $\vec{y}_k^k = \vec{e}_y = (0 \quad 1 \quad 0)^T$, and $\vec{z}_k^k = \vec{e}_z = (0 \quad 0 \quad 1)^T$. In addition, if a basis vector, point, or axis has no superscript index, it is defined in the coordinate system of the MCMM (m). Since all coordinate systems describe real geometries without scaling, a transformation of a point $\vec{r}^k$ to the coordinate system l can be realized only with a translation vector $\vec{t}_l^k$ and a rotation matrix $\mathbf{R}_{kl}$:

$$\vec{r}^l = \mathbf{R}_{kl}\left(\vec{r}^k - \vec{t}_l^k\right). \tag{17.1}$$

Moreover, if both parameters ($\vec{t}_l^k$, $\mathbf{R}_{kl}$) of the transformation from the coordinate system k to l are known, the inverse transformation is written as $\vec{r}^k = \mathbf{R}_{lk}\left(\vec{r}^l - \vec{t}_k^l\right)$ with $\mathbf{R}_{lk} = \mathbf{R}_{kl}^{-1}$; $\vec{t}_k^l = -\mathbf{R}_{kl}\vec{t}_l^k$.

17.3 REALIZATION OF AN MCMM

17.3.1 COMPONENTS

The realization of an MCMM is depicted in Figure 17.5. This setup consists of a moving system, made by *Feinmess Dresden*, with a resolution of 0.1 μm, an absolute accuracy of ±2 μm, and a measurement range of 150 mm in all directions. The alignment between the moving axes is accurate to 0.004° [11,12].

As probe to measure the geometry, a one-dimensional white light sensor, made by *Precitec Optronik GmbH*, has been chosen. This sensor consists of a nonmagnetic, passive probe that is connected to the sensor electronics with an optical fiber. The probe is attached to the setup so that it points in negative z direction. Three-dimensional measurements are realized by moving the probe at different positions near the DUT. Moreover, there are two probes available: (1) the probe *RB200031* with a measurement range of 300 μm, a resolution in z direction of 10 nm, and a resolution in lateral direction of 2.5 μm and (2) the probe *RB200050* with a measurement range of 3000 μm, a resolution in z direction of 100 nm, and a resolution in lateral direction of 6 μm. Both probes use the same sensor electronics, *CHRocodileE*, which has a digital interface, and allows one to sample measurement values with a rate up to 4 kHz.

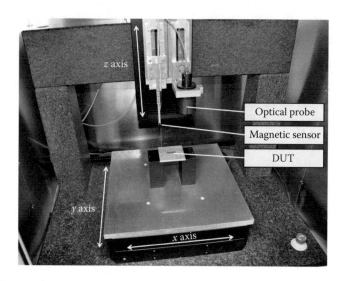

FIGURE 17.5 Photograph of the realization of an MCMM.

The magnetic sensor of the setup is a 3D integrated Hall sensor from *Senis GmbH* with a measurement range of 200 mT, an accuracy of 0.1%, and a bandwidth of 2.5 kHz [4,13]. Two different types of this 3D Hall sensor can be attached to the measurement. The type *T-H3A* has a package that does not surround the whole silicon die so that the sensor can be moved close to the DUT. Thus, this sensor is very fragile. If the distance to the DUT can be larger, the sensor of type *C-H3A* is used where the silicon die is fully surrounded by the package.

17.3.2 MECHANICAL ADJUSTMENT OF ALIGNMENT

The alignment of the magnetic sensor and the optical probe can be separately adjusted in the measurement setup around the x and y axes. Each adjustable angle is realized with an aluminum frame pivoted, a screw defining the alignment, and a spring fixing the position. The screws have a thread pitch of 0.5 mm per rotation, and experience shows that they can be adjusted with an accuracy of an eighth turn (45°). Furthermore, the distance between the pivot and the tip of the screw is 155 mm for the alignment around the y axis and 180 mm for the alignment around the x axis, which leads to an adjustability of ±0.020° or ±0.023°, respectively. The frame for the alignment around the y axis is mounted on the frame for the alignment around the x axis. Thus, the orientation of the rotation axis of this frame depends on the alignment of the inner frame. However, since only small angles are adjusted, this effect is neglected for the calculation of the adjustability.

17.3.3 TEMPERATURE STABILITY

A change of temperature causes thermal expansion so that the positions of all parts, such as the attachment, the magnetic sensor, the optical probe, and the field source, are shifted relative to each other. It is complex to accurately simulate the impact of a temperature change because this requires the temperature distribution of the setup. Nevertheless, the maximal shift of two points relative to each other can be estimated by the maximal distance of these two points, the linear coefficient of thermal expansion, and the maximal temperature change. The structural parts of the setup are made of aluminum and steel with a maximal linear coefficient of thermal expansion of $23 \cdot 10^{-6}$/K, and the maximal distance is about 0.5 m, which results in a maximal change of the distance between two points of approximately 11.5 µm/K.

The calibration of the MCMM as explained in Section 17.4 provides all important geometrical parameters of the setup so that the absolute temperature is not of importance. Only a temperature

FIGURE 17.6   Photograph of the entire measurement setup including the chamber for thermal insulation.

change between the start of the calibration and the end of the actual measurement impairs the measurement accuracy. To reduce this temperature change, a chamber for thermal isolation surrounds the entire measurement setup (see Figure 17.6).

This chamber mainly consists of 10 cm Styrofoam plates, wood, a window, and aluminum plates covering the inner faces of the chamber. Except for the motors of the moving axes, all active devices such as the controller and the electronics of the MCMM, the magnetic sensor, and the optical probe or other devices such as multimeters or current sources are outside of the chamber. Furthermore, the motors of the moving axes have passive breaks so that they can be turned off when the desired measurement position is reached.

An extensive analysis of the temperature at multiple points inside and outside of the chamber shows that the temperature between the start of the calibration and the end of the actual measurement is stable in the range of ±0.35 K for typical measurements with a duration of a few days [14]. Thus, according to the maximal change of position between two points of 11.5 μm/K, the maximal position error due to a change of temperature is approximately ±4 μm.

17.4   CALIBRATION OF AN MCMM

17.4.1   OPTICAL CALIBRATION

17.4.1.1   Distortions due to Misalignments of the Optical Probe

If all moving axes are set to zero ($\vec{r}_{ad} = (0 \quad 0 \quad 0)^T$), the starting point of the measurement range of the optical probe is defined to be at the origin of the coordinate system of the MCMM (see Figure 17.4). To detect points, the position of the DUT has to be known roughly, in advance. With this information, the axes of the MCMM can be adjusted so that the surface of the DUT is inside the measurement range of the optical probe. Then, the output signal of the probe o is saved, and the coordinates of the measurement point on the surface of the DUT are calculated with

$$\vec{r}_{om} = \vec{r}_{ad} + o\vec{d}_o. \tag{17.2}$$

In the considered setup, the optical probe is attached so that the direction of focus $\vec{d}_o$ is orientated in the negative z direction ($\vec{d}_o = -\vec{z}_m$). Nevertheless, attaching the optical probe to the moving axes is always accompanied by alignment errors. This misalignment of the probe is described with the two angles

$$\tan \alpha_{o1} = \frac{d_{ox}}{d_{oz}}; \quad \tan \alpha_{o2} = \frac{d_{oy}}{d_{oz}}, \tag{17.3}$$

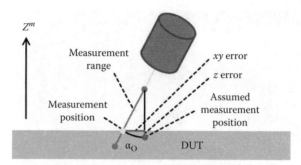

FIGURE 17.7   Two-dimensional cut plane through the axis of focus of the optical probe and the z^m axis.

which represent the angles between the z axis and the projection of $-\vec{d}_o$ on the xz plane and on the yz plane. A misalignment of the optical probe impairs all three components of the measurement point $\vec{r}_{om}$. First, the error of the scaling is regarded. In this case, only the angle α_o between the z axis and $-\vec{d}_o$ is of importance, which calculates from α_{o1}, α_{o2} as

$$\tan^2 \alpha_o = \tan^2 \alpha_{o1} + \tan^2 \alpha_{o2}. \qquad (17.4)$$

If the probe measures a point on a plane surface, it seems that the measurement point is further away (see z error in Figure 17.7). The absolute error is proportional to the measurement value o so that the measurement result is distorted. For instance, if the probe measures two points at different heights and the difference in height is calculated, the result is higher compared to the real difference.

Another important impact is the shift of the xy position on the surface of the DUT. If the distance of the optical probe to the DUT is changed, the xy position of the measurement point also changes (see xy error in Figure 17.7). If the geometry is a plane normal to the z axis (as in Figure 17.7) and the focus is not beside the borders, the shift has no impact on the measurement result. However, for an arbitrary geometry, this shift causes additional errors, which strongly depend on the geometry of the DUT [14].

17.4.1.2   Avoiding Impacts due to Misalignments of the Optical Probe

In the foregoing, the impact due to misalignments of the optical probe is explained. One simple way to avoid distortions is to adapt the height for each measurement point so that the measurement value o stays constant. With this technique, the optical probe is not used to measure the distance but to move the probe at a constant distance along the surface of the DUT. Therefore, the misalignment causes no distortion. However, a disadvantage of this method is the long measurement time due to the adaptation of the height at each measurement point.

Another approach is measuring the misalignment of the optical probe and compensating for it mathematically or mechanically. If $\vec{d}_o$ is known, the correct measurement point can be calculated with (17.2). Nevertheless, the misalignment changes the perspective of the optical probe so that the accessible points may also be changed. Therefore, the misalignment of the optical probe is measured and then mechanically corrected in the measurement setup with the attachment explained in Section 17.3.2.

17.4.1.3   Measurement Technique for Misalignment

To measure the misalignment, the idea is scanning a part where the geometry is exactly known and comparing the measurement results with the ideal geometry. A sphere is chosen as reference geometry because accurate reference spheres are commercially available, and a sphere does not have to be aligned in the measurement setup. In addition, the distortion of the measurement result due to misalignments of the optical probe can be described analytically [14].

First, the sphere is scanned with the optical probe, which results in measurement values including the distortion due to misalignments. Then, an optimization problem is formulated where the mean squared error between the measurement values and the analytical equation, describing the distortion caused by misalignment, is minimized. Finally, solving this minimization problem leads to the unknown angles of misalignment α_{o1} and α_{o2}. Moreover, the exact position of the center of the sphere $\vec{m} = (m_x \, m_y \, m_z)^T$ in the measurement setup is unknown so that these parameters also have to be fitted.

17.4.2 Magnetic Calibration

17.4.2.1 Magnetic Parameters of Calibration

For reasons of clarity, the magnetic sensor is often referred to as a single 3D magnetic sensor. However, at least three one-dimensional magnetic sensors are necessary to measure all three magnetic field components, which are Hall plates in the setup considered. The ideal Hall plate measures the component of the magnetic induction $\vec{B}$ normal to the Hall plate $\vec{n}_B$ and weights it with the sensitivity S, which results in the output signal:

$$M = S \, \vec{n}_B \cdot \vec{B}. \qquad (17.5)$$

After calibrating the optical probe, the geometry of an arbitrary field source can accurately be scanned, which results in a 3D image. Moreover, the optical measurement data are used to reconstruct the ideal shape of the field source (see Figure 17.3) and to calculate the rotation matrix $\mathbf{R}_{dm}$ and the translation vector $\vec{t}_m^d$ (see (17.1)). With this information, any measurement point defined in the coordinate system of the field source can be transformed to the coordinate system of the MCMM. If the magnetic field should be scanned at an arbitrary position $\vec{r}_p^d$, first, this position is transformed to the coordinate system of the MCMM $\vec{r}_p^m$ (see Figure 17.8a). Then, each Hall plate is moved to $\vec{r}_p^m$, and the output signals $\tilde{M}_i(\vec{r}_p^m)$ with $i \in \{x,y,z\}$ are saved. In order to move a Hall plate to the position $\vec{r}_p^m$, the MCMM has to be adjusted to $\vec{r}_{ad} = \vec{r}_p^m - \vec{p}_B$, which is exemplarily shown in Figure 17.8b for the sensor orientated normal to the y axis.

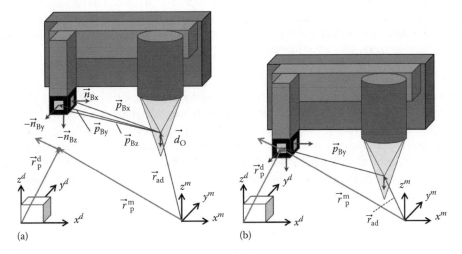

(a)　　　　　　　　　　　(b)

FIGURE 17.8 Taking magnetic measurement values with the MCMM.

Taking the model of the Hall probe of (17.5) into account, the magnetic field vector at the position $\vec{r}_p^m$ w.r.t. the coordinate system of the MCMM is given by

$$\vec{B}^m\left(\vec{r}_p^m\right) = \mathbf{A}_{nB}^{-1} \begin{pmatrix} \tilde{M}_x(\vec{r}_p)/S_x \\ \tilde{M}_y(\vec{r}_p)/S_y \\ \tilde{M}_z(\vec{r}_p)/S_z \end{pmatrix} \quad \text{with} \quad \mathbf{A}_{nB} = \begin{bmatrix} n_{Bxx} & n_{Bxy} & n_{Bxz} \\ n_{Byx} & n_{Byy} & n_{Byz} \\ n_{Bzx} & n_{Bzy} & n_{Bzz} \end{bmatrix}. \tag{17.6}$$

Finally, using the rotation matrix $\mathbf{R}_{md} = \mathbf{R}_{dm}^{-1}$ (see section 17.2.3), the magnetic field w.r.t. the coordinate system of the field source is written as follows:

$$\vec{B}^d\left(\vec{r}_p^d\right) = \mathbf{R}_{md}\vec{B}^m\left(\vec{r}_p^m\right). \tag{17.7}$$

Thus, as long as the positions, alignments, and sensitivities of all three Hall plates are known $\vec{P}_{Bi}, \vec{n}_{Bi}, S_i$ with $i \in \{x,y,z\}$), the magnetic field at an arbitrary position $\vec{r}_p^d$ can be measured w.r.t. the coordinate system of the field source. That means assembly tolerances of the magnetic sensors and of the field source do not impair the measurement result.

In general, the Hall plates could have any orientation, whereas (17.6) requires that the inverse matrix of $\mathbf{A}_{nB}$ exists. This is only the case if the Hall plates are not parallel to each other. Furthermore, it is preferable to align the Hall plates orthogonal to each other so that the matrix $\mathbf{A}_{nB}$ is well-conditioned.

To summarize, if the position and alignment of the Hall plates are known, the magnetic field can be scanned w.r.t. the field source. However, the accuracy of the measurement depends on how accurately the position and alignment of the magnetic sensors are known. The calibration of one Hall plate requires the measurement of three degrees of freedom for the position vector $\vec{P}_B$, and two degrees of freedom for the alignment $\vec{n}_B$ (see Figures 17.4 and 17.8). Thus, in general, 15 degrees of freedom have to be calibrated for all three directions. Nevertheless, if a multidimensional sensor is assembled, and the alignment and position of the sensing elements are exactly known relative to each other, not all distances and normal vectors have to be determined. Moreover, besides the calibration of assembly tolerances, important parameters of Hall plates such as offset and sensitivity have to be calibrated too. The offset can simply be measured without any field source or in a zero gauss chamber, whereas the calibration of the sensitivity S_i with $i \in \{x,y,z\}$ requires an external reference field.

In the following, a method is presented that determines all parameters of calibration such as the alignment, position, and sensitivity of the Hall plates. Moreover, a method has been developed that calibrates only the alignment of magnetic sensors with high precision by using multiple Hall plates on one silicon die. This method is not presented since the method discussed in the following also calibrates the alignment. Interested readers are referred to [8,15].

17.4.2.2 Calibration Principle

The idea of calibration is using an inhomogeneous reference field source that has a strong relation between its magnetic field and its geometry. Moreover, the magnetic reference field $\vec{B}_r(\vec{r})$ can be calculated analytically, if the geometry of the field source is known. First, the optical sensor measures this geometry so that the magnetic reference field $\vec{B}_r(\vec{r})$ can be computed. With this information, according to (17.5), the output signal M of any Hall plate can be calculated as a function of the position $\vec{P}_B$, alignment $\vec{n}_B$, and sensitivity S of the Hall plate, and of the position adjusted $\vec{r}_{ad}$:

$$M = S\,\vec{n}_B \cdot \vec{B}_r\left(\vec{r}_{ad} + \vec{p}_B\right). \tag{17.8}$$

Then, the Hall plate scans the field, which results in the output signal $\tilde{M}$. Finally, an optimization problem is formulated where the squared error between the values calculated and the values measured is minimized:

$$\min_{\vec{p}_B, \vec{n}_B, S} \| \tilde{M} - M \|^2 . \tag{17.9}$$

Solving this minimization problem results in the unknown parameters of calibration.

As reference field, the magnetic field of a current supplied, straight conductor with a circular cross section is used. This simple geometry can be manufactured accurately, and the magnetic field can be calculated with an analytical equation, if the position and alignment of the conductor are known [16]. Since this information is provided by the optical probe, the minimization problem of (9) can be formulated. However, the magnetic field of a straight conductor is homogenous along its rotation axis so that not all degrees of freedom of the desired calibration parameters $(\vec{P}_{Bi}, \vec{n}_{Bi}, S_i$ with $i \in \{x,y,z\})$ can be determined within one measurement cycle. To overcome this problem, the conductor is attached in different orientations in the measurement setup as shown in Figure 17.9 [7,14].

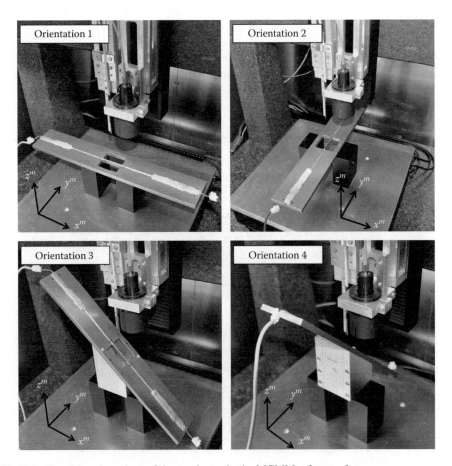

FIGURE 17.9 Possible orientations of the conductor in the MCMM reference frame.

17.5 MEASUREMENT EXAMPLE

17.5.1 FIELD SOURCE

As an example, the magnetic field of a neodymium permanent magnet is scanned with the Hall sensor of type *T-H3A*. According to the data sheet, the permanent magnet has a remanence of 1.240 T, it is plated with nickel, and it has the shape of a cube with the dimensions of $2 \times 2 \times 2$ mm³. To easily refer to the faces of the magnet, they are numbered equally to the sides of dices (see Figure 17.10). For the measurement, it is usually tried to orientate the field source toward the axes of the moving system so that only tolerances of the positioning have to be corrected (Figure 17.10a). However, in this example, the magnet is totally misaligned toward the axes of the moving system to demonstrate that an MCMM does not require any prealignments or exact positioning of the field source (see Figure 17.10b).

17.5.2 OPTICAL MEASUREMENT

The optical scan results in several measurement points of the surface of the permanent magnet. Outliers due to measurement errors of the white light sensor are filtered out, which cause holes in the surface plot of the measurement results (see Figure 17.11). To compare the real with the ideal geometry, it is necessary to define how the reconstructed geometry is uniquely created from the measurement points. To this end, several points of side 1 (blue), 2 (red), and 3 (yellow) are selected (see Figure 17.11b), and for each side, the least squares solution for the best fitting plane is calculated.

Moreover, the intersection lines of these plane equations are computed, which represent the edges of the reconstructed shape (see blue lines in Figure 17.11b). Moreover, a unique coordinate system has to be defined for the field source, which also takes mechanical tolerances into account. Usually, it is desired that the reference axes of the DUT consist of three orthonormal vectors. Therefore, it is not possible to define the axes of the reference coordinate system to be the edges of the magnet, because they are in general not orthogonal to each other. In this example, the orthogonality error

(a)

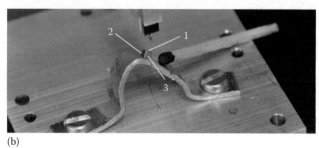

(b)

FIGURE 17.10 Photograph of the cubic permanent magnet and the magnetic sensor aligned (a) and totally misaligned (b).

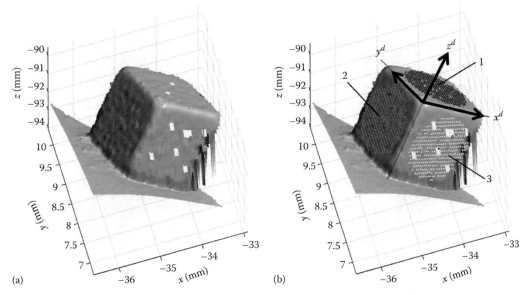

FIGURE 17.11 Optical measurement results of the scan of the permanent magnet. The holes in the surfaces occur due to outliers of the measurement points that are filtered out. The blue points are taken for the fit of plane 1, the red points for plane 2, and the yellow points for plane 3.

between surface 1 and 2 is $-0.18°$, between surface 1 and 3 it is $0.35°$, and between surface 2 and 3 it is $0.11°$. Furthermore, the coordinate system of the magnet is defined by the three orthonormal axes $\vec{x}_d^m$, $\vec{y}_d^m$, and $\vec{z}_d^m$ and the origin $\vec{t}_d^m$, which are defined as follows: (1) $\vec{z}_d^m$ is normal to the plane equation of surface no. 1 and points away from the cube; (2) the origin $\vec{t}_d^m$ is the intersection point of the three planes 1, 2, and 3; (3) $\vec{x}_d^m$ is parallel to the intersection line of the plane 1 and 3 and points from the origin to the intersection point of the planes 1, 3, and 5; and (4) $\vec{y}_d^m$ follows from $\vec{z}_d^m \times \vec{x}_d^m$. This coordinate system is also plotted on the right side of Figure 17.11. Although it is difficult to see, the y^d axis is not parallel to the edge between surface 1 and 2, and the z^d axis is not parallel to the edge between surface 2 and 3.

17.5.3 MAGNETIC MEASUREMENT

In this example, the magnetic field should be scanned in x^d and y^d direction in the range of 0 to 2 mm with steps of 0.5 mm and for a height z^d of 1 mm, 2 mm, and 3 mm. Before the magnetic measurement is started, the system is calibrated as explained in Section 17.4 so that the position and the alignment of the Hall plates are known (see $\vec{P}_B, \vec{n}_B$ in Figure 17.4). With this information, the magnetic sensor can be moved to points in the coordinate system of the field source (see Figure 17.8), which results in the magnetic field vectors at the desired positions but w.r.t. the coordinate system of the MCMM (see left part of Figure 17.12).

However, since the optical measurement provides the reference axis of the magnet, the geometry and the magnetic field vectors can be transformed to the coordinate system of the magnet as shown in the right part of Figure 17.12. This plot demonstrates that the original orientation and position of the magnet and the magnetic sensor have no influence on the measurement, and thus, assembly tolerances do not impair the measurement results. In addition, the absolute value of the magnetic flux density is evaluated on the line $x^d = y^d = 1$ mm against z^d. At $z^d = 1$ mm, the relative change is about $1.2\%/10$ μm, which demonstrates the strong inhomogeneity of the magnetic field and the crucial impact of small positioning errors.

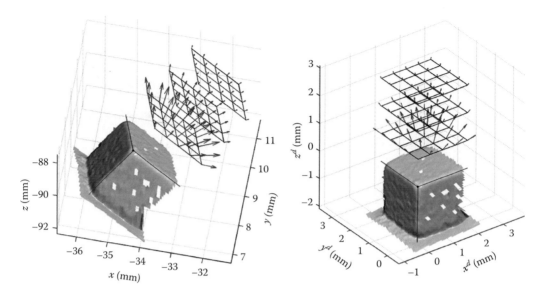

FIGURE 17.12 Magnetic field over the surface 1 of the permanent magnet in the coordinate system of the moving axes (left) and in the coordinate system of the magnet (right). Moreover, in both plots, a black grid visualizes the measurement points of the magnetic scan.

17.6 CONCLUSION

A measurement principle is explained to accurately scan magnetic fields even if the fields are strongly inhomogeneous as in automotive applications. The corresponding measurement setup is denoted as MCMM, which consists of three moving axes, a 3D magnetic sensor, and an optical probe. Standard measurement systems consist of moving axes and a magnetic sensor only, so that mechanical tolerances of the field source and of the measurement setup itself crucially impair the measurement accuracy, if inhomogeneous magnetic fields are measured. With an MCMM, the impact of assembly tolerances is significantly reduced, because the geometry is measured with the optical sensor first. Then, the magnetic field vectors can be measured at points relative to the field source, and the field vectors can be transformed to the coordinate system of the measurement object. This measurement principle requires a calibration of several geometrical parameters of the optical probe and the magnetic sensor. Here, solutions are presented for the calibration of the alignment of the optical probe as well as for the position and alignment of the magnetic sensor.

Finally, a real measurement setup of a MCMM is presented that consists of an integrated 3D Hall sensor, a white light sensor, and moving axes. With this setup, the magnetic field of a cubic permanent magnet is analyzed, demonstrating the working principle of an MCMM.

To conclude, a MCMM is a tool to measure the geometry and the magnetic field of an arbitrary field source. After a calibration of the setup, the magnetic field can be measured in the coordinate system of the field source so that assembly tolerances do not impair the measurement. Furthermore, if the magnetic results do not conform to the expectations, it is possible to distinguish between deviations due to changes of the geometry or changes of the magnetic material parameters such as permeability and magnetization.

REFERENCES

1. Dr. Brockhaus Messtechnik GmbH & Co. KG, Gustav-Adolf-Str. 4, 58507 Lüdenscheid, Germany. *XYZ Field Scanner*. Available at http://www.brockhaus.com
2. SENIS GmbH, Technoparkstrasse 1, CH-8005, Zürich. *Magnetic Field Mapping System MMS-1-R*, 3rd edn., January 2011. Available at http://www.senis.ch

3. Infineon Technologies AG, Am Campeon 1–12, 85579 Neubiberg, Germany. *TLE4998S3C: Linear Hall sensor, Data Sheet Rev 1.0*, 2009.

4. SENIS GmbH, Technoparkstrasse 1, CH-8005, Zürich. *Integrated 3-Axis Hall Probe C-H3A-xx*, 8th edn., 2010. Available at http://www.senis.ch

5. Infineon Technologies AG, Am Campeon 1–12, 85579 Neubiberg, Germany. *TLE5011: Angle Sensor, Preliminary Data Sheet V0.91*, 2009.

6. U. Ausserlechner and M. Holliber, U.S. Patent Application US 2009/0066465 A1, 2009.

7. H. Husstedt, U. Ausserlechner, and M. Kaltenbacher, Accurate measurement setup for strongly inhomogeneous magnetic fields. In: R. Lerch and R. Werthschützky, eds., *Sensor+Test Conferences 2011 Proceedings*, pp. 343–348. AMA Service GmbH, June 2011.

8. H. Husstedt, U. Ausserlechner, and M. Kaltenbacher, *Optical, Acoustic, Magnetic, and Mechanical Sensor Technologies*. CRC Press, Boca Raton, FL, January 2012.

9. A.H. Slocum, *Precision Machine Design*. Society of Manufacturing Engineers, Dearborn, MI, 1992.

10. J.A. Bosch, ed., *Coordinate Measuring Machines and Systems*. Marcel Dekker Inc., New York, 1995.

11. Feinmess Dresden GmbH, Fritz Schreiter Str. 32, 01259 Dresden, Germany. *Linear Stage PMT 160-DC*, 2009.

12. Feinmess Dresden GmbH, Fritz Schreiter Str. 32, 01259 Dresden, Germany. *Compact-XY-Stage KDT 380*, 2009.

13. D.R. Popovic, S. Dimitrijevic, M. Blagojevic, P. Kejik, E. Schurig, and R.S. Popovic, Three-axis teslameter with integrated hall probe. *IEEE Transactions on Instrumentation and Measurement*, 56:1396–1402, August 2007.

14. H. Husstedt, *Measurement of Magnetic Fields for the Testing of Automotive Sensors*. PhD thesis, Alps-Adriatic University of Klagenfurt, Klagenfurt, Austria, 2012.

15. H. Husstedt, U. Ausserlechner, and M. Kaltenbacher, Precise alignment of a magnetic sensor in a coordinate measuring machine. *Sensors Journal, IEEE*, 10(5):984–990, May 2010.

16. H. Hofmann, *Das elektromagnetische Feld*. Springer-Verlag, Berlin, Germany, 1974.

18 Low-Frequency Search Coil Magnetometers

Asaf Grosz and Eugene Paperno

CONTENTS

18.1 SINGLE-AXIS SEARCH COIL MAGNETOMETERS

Low-frequency search coil magnetometers are widely used for geophysical prospecting, space research, magnetic anomaly detection [1–17], etc. Their advantages compared to other magnetometers are high resolution and low power consumption, defined only by the power consumption of the preamplifier. The inherent reduction of the search coils' resolution with frequency can be compensated by increasing their size. Thus, large enough low-frequency search coils can compete with and even outperform fluxgates [9–11,15–17].

18.1.1 INTRODUCTION: PHYSICAL BACKGROUND

Search coils are passive magnetic sensors converting magnetic field into voltage. A typical search coil comprises a pickup winding and a ferromagnetic core. An additional winding for magnetic feedback can also be used [18].

The operation principle of a search coil is based on Faraday's law of induction: the voltage generated by an induction coil is proportional to the time derivative of the magnetic flux flowing through the coil area. For sinusoidal magnetic fields,

$$V(f) = 2\pi f N S \mu_a B_0. \tag{18.1}$$

The ferromagnetic core of a typical search coil has a cylindrical shape (see Figure 18.1a). Flux concentrators, for example, in the shape of thin disks (see Figure 18.1b), can be attached to the core [9–11]. The distributions of the apparent permeability along the cores with and without flux concentrators are given in Figure 18.1c. This figure illustrates that, despite significant shortening of the core length, the flux concentrators allow maintaining the same apparent permeability at the core center. Moreover, the distribution of the permeability along the core becomes nearly uniform.

Apparent permeability for cylindrical cores without flux concentrators [8,10] can be approximated by that of a prolate ellipsoid [8,10,19].

Apparent permeability for cylindrical cores with thin-disk flux concentrators can be approximated by [8] where H_D is the demagnetizing factor of an ellipsoid with the diameter equal to that of the flux concentrators and with the length equal to the total length of the search coil, including the flux concentrators.

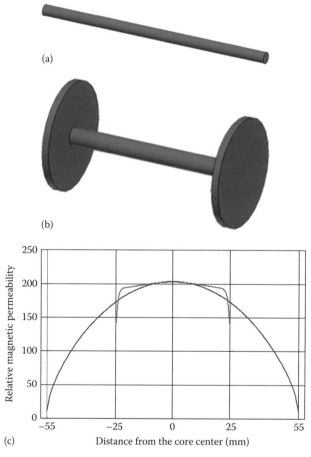

FIGURE 18.1 Types of search coil cores. Search coil cores (a) without and (b) with flux concentrators. (c) Distribution of the apparent permeability along the axis of a long (110 mm) core without flux concentrators (the dark gray curve) and a short (50 mm) core with flux concentrators (the light gray curve). The core diameters are 5 mm. The diameter of the flux concentrators is 30 mm and their thickness is 2 mm. The relative permeability of the cores and the flux concentrator material is 2000.

18.1.2 OPTIMIZATION OF SEARCH COIL MAGNETOMETERS

To reach the maximum resolution for a given volume and power consumption, a search coil magnetometer should be optimized. Conventional approaches to finding the best possible configuration are based on designing the magnetometer part by part [1–5]. For example, the search coil core is designed first to provide the maximum apparent permeability, then the coil distribution over the core is selected and, finally, a preamplifier is developed with low enough noise. Naturally, such a part by part design does not yield the best possible magnetometer configuration.

The optimization of the *entire* magnetometer is suggested in [7–12]. This optimization is based on an analytical model that includes the apparent permeability of the core and the flux concentrators, the coil winding, and the noise of the preamplifier. To find the optimum magnetometer configuration for given constraints, for example, for the maximum volume, weight, given power consumption, and the noise of the preamplifier, the analytical model is solved numerically for a large set of the parameters, and the configuration providing the best resolution is chosen.

This new approach has advanced the state-of-the-art low-frequency search coil magnetometers by substantially reducing their size, power consumption, and weight for the same resolution [7–12].

For a given search coil volume, aspect ratio, relative permeability of the core and the flux concentrators, and the noise of the preamplifier, the optimization of search coil magnetometers employing disk-shape flux concentrators (see Figure 18.2) includes the following six variables: the diameter and thickness of the flux concentrators, the outer diameter of the search coil winding, the diameters of the core and the wire, and the number of turns.

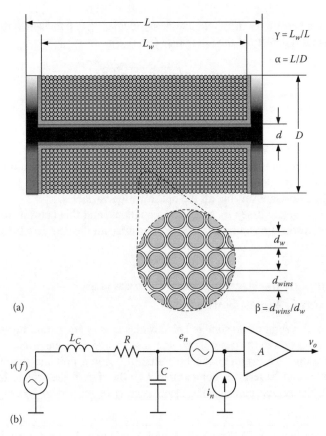

FIGURE 18.2 Search coil magnetometer. (a) Search coil structure and dimensions and (b) equivalent electrical circuit.

Our numerical simulations, performed with Maxwell finite element software, have shown that increasing the thickness of the flux concentrators beyond 5% of the total length of the search coil does not increase the apparent permeability enough to compensate for the decrease in the magnetometer sensitivity due to the corresponding reduction of the volume of the winding.

18.1.2.1 Analytical Model of the Magnetometer

The search coil structure and dimensions are shown in Figure 18.2a, and the equivalent electric circuit of the magnetometer is shown in Figure 18.2b. The amplitude of the voltage induced in the search coil by an applied sinusoidal magnetic field can be given by (18.1), where the number of turns

$$N = N_l N_h \quad \text{and} \quad S = \frac{\pi \cdot d^2}{4}. \tag{18.2}$$

At low frequencies, the search coil reactance can be neglected, and the spectral density of the total noise of the search coil and the preamplifier (see Figure 18.2b) referred to the coil input can be given as follows:

$$e_{tot} = \sqrt{e_R^2 + e_n^2 + (Ri_n)^2} \quad \text{with} \quad e_R = \sqrt{4kTR} \quad \text{and} \quad R = 4N \frac{d + d_{wins} N_h}{d_w^2} \rho. \tag{18.3}$$

The magnetometer sensitivity threshold can be defined as $B_{st}(f)$, for which the induced voltage (18.1) equals the spectral density of the total noise (18.3) [9]:

$$B_{st}(f) = \frac{\sqrt{e_R^2 + e_n^2 + (Ri_n)^2}}{2\pi f N S \mu_a'}. \tag{18.4}$$

18.1.2.2 Optimization of the Magnetometer

The aim of the optimization is to find the values of N, d, d_w, and D that minimize B_{st} in (18.4) for a given search coil size, type of preamplifier, and minimum acceptable wire diameter. This can be done numerically by discretely varying all the optimization variables.

Our optimization is solely based on analytical equations, and this helps us to find the best coil configuration. Other works, for example [8], do not optimize all the coil parameters regarding some of them as fixed factors.

18.1.3 INTEGRATION OF THE ELECTRONICS AND BATTERIES INSIDE THE HOLLOW CORE OF A SEARCH COIL*

Reducing the size and weight of sensors and electronics is very important for many applications. This is especially important in the cases where the sensors and their electronics are not integrated within a single housing. An example can be miniature search coil magnetometers, where the size of the electronics and batteries is comparable with the size of the sensor. In typical modular designs [1–9], both the search coil and the electronics have individual housings and individual

---

* Reprinted with permission from Grosz, A., Paperno, E., Amrusi, S., and Liverts, E., Integration of the electronics and batteries inside the hollow core of a search coil, *J. Appl. Phys.*, 107, 09E703-1–09E703-3, 2010. Copyright 2010, American Institute of Physics.

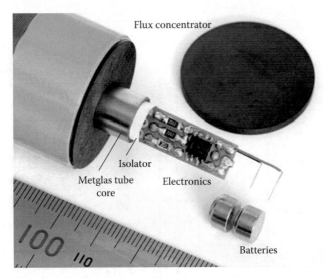

FIGURE 18.3 Experimental model of the search coil magnetometer. The magnetometer is disassembled to show the integration of its electronics and batteries inside the hollow core. (The alignment of the assembled flux concentrators and core is shown in the inset of Figure 18.4a.)

electrostatic shields. The electric wiring between the search coil and the electronics should also be shielded. All this increases the size, weight, and the complexity of the entire magnetometer. In designs where the search coil and the electronics share the same housing, the latter is larger than needed for the search coil alone.

A novel approach, suggested in [10], allows integrating all the magnetometer components, including the search coil, its electronics, and batteries within a single housing that is not larger than what is needed just for the search coil. Moreover, the search coil itself serves as both the housing and the electromagnetic shield for the electronics and batteries. This is obtained by replacing the solid rod core of the search coil with a thin-wall tube core and integrating the electronics and batteries inside this tube core (see Figure 18.3). It is found that a thin-wall tube core with a high enough permeability can keep the same average magnetic flux through the search coil winding. It is also found that increasing the outer diameter of the tube core beyond its optimum value provides more space for the electronics and batteries and does not strongly affect the search coil resolution.

18.1.3.1 Method

18.1.3.1.1 Magnetic Fluxes Averaged over the Length of the Rod and Tube Cores

Let us start from a comparison of the magnetic fluxes averaged over the length of the rod and tube cores of a search coil that has a 50 mm length and 30 mm diameter and employs disk-shape flux concentrators of a 30 mm diameter and 2 mm thickness.

To find the relative difference between the fluxes as a function of the wall thickness, t, of the tube core (see Figure 18.4a), we have used a Maxwell SV program based on the finite element method. In our simulations, the relative permeability of the rod core and the flux concentrators is fixed at a typical for ferrites value of $\mu_r = 10^3$, and the relative permeability of the tube core, μ_r'', is a parameter. Figure 18.4a shows that the permeability of a thin-wall tube core should be high enough to keep the same average magnetic flux as that inside the solid rod core of the same diameter and length. For example, a 10 mm outer diameter tube core has to have at least about 10×10^3 relative permeability if its thickness is 0.2 mm.

Considering the results of Figure 18.4a, we can conclude that it is possible to replace a rod core with a thin-wall tube one that has a high enough permeability and keeps the same average magnetic flux through the search coil winding and to use the interior of the tube core for the electronics and batteries.

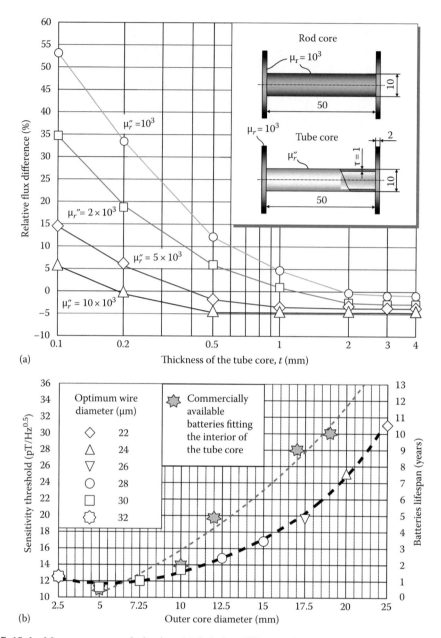

FIGURE 18.4 Magnetometer optimization. (a) Relative difference between the magnetic fluxes averaged over the length of the rod and tube cores as a function of the tube core thickness. (b) Theoretical sensitivity threshold, calculated at 1 Hz, and the magnetometer batteries' lifespan as a function of the outer diameter of the tube core.

18.1.3.1.2 Optimum Sensitivity Threshold as a Function of the Tube Core Diameter

It is important now to examine the magnetometer sensitivity threshold as a function of the outer diameter of the tube core.

The search coil sensitivity threshold (or resolution) can be found according to (18.4), where the apparent permeability of the tube core can be found by assuming that the relative permeability of the tube core material is high enough and its average magnetic flux equals that of the equivalent rod core. This leads us to

$$\mu_a' \frac{\pi}{4} D^2 = \mu_a'' \frac{\pi}{4} \left[D^2 - (D - 2\tau)^2 \right] \quad \text{and} \quad \mu_a'' = \mu_a' \cdot \frac{D^2}{D^2 - (D - 2\tau)^2}. \tag{18.5}$$

The aim of the optimization is to find the maximum possible resolution as a function of the outer diameter of the tube core and the diameter of the wire for a given search coil size and the type of preamplifier.

The results of the optimization are shown in Figure 18.4b for an OPA333 preamplifier. One can see from this figure that increasing the outer diameter of the tube core does not too strongly affect the magnetometer resolution. For example, a twofold increase, from 5 to 10 mm, decreases the resolution only by 20%: from 10.73 to 12.81 pT/Hz$^{0.5}$ at 1 Hz.

On the other hand, a twofold increase in the outer diameter of the tube core is very substantial in terms of increasing its interior space, which can be occupied by the electronics and batteries. Considering that the electronics can be miniaturized and most of the interior can be devoted to the batteries, we have estimated in Figure 18.4b the batteries' lifespan as a function of the outer diameter of the tube core. One can see from this figure that for a 10 mm outer core diameter, the batteries' lifespan can be as long as 2 years, instead of half a year for a 5 mm diameter. For a 17 mm diameter, the batteries' lifespan can be as long as 9 years, at the expense, however, of an about 60% decrease in the magnetometer resolution.

18.1.3.2 Magnetometer Model

To verify our theoretical results, we have constructed an experimental magnetometer model shown in Figure 18.3. This model is optimized for a 40 μm wire diameter, instead of the 30 μm wire, providing the best possible resolution, to alleviate on the coil winding. The resulting loss in the theoretical value of the resolution is about 10%: 14.1 against 12.8 pT/Hz$^{0.5}$ at 1 Hz. The wire has been wound on a 49 mm long plastic bobbin with a 10 mm diameter opening for the core. The outer winding diameter is 30 mm, the total number of turns is 229,515, and the wire resistance is 219 kΩ.

To manufacture the tube core, we have used a 50 mm wide and 22 μm thick Metglas 2075 M tape. The tape has been rolled to form a 50 mm long tube with a 10 mm outer diameter and 1 mm wall thickness (see Figure 18.3). Disk-shape ferrite flux concentrators of a 30 mm diameter and 2 mm thickness were attached to the core. The search coil inductance and self-resonant frequency are 4654 H and 177 Hz, respectively, and the corresponding stray capacitance is 174 pF. The search coil sensitivity at 1 Hz is 6.4 nV/pT. An additional single-layer, 31-turn winding is used to apply magnetic feedback.

Although the anisotropy of the core is circumferential, our experiments have shown that its permeability is high enough to keep the average flux through the tube core not below that of the rod core. A thin-wall ferrite core with a 10^3 relative permeability (the worst case for the material used for the flux concentrators) would not keep the magnetometer resolution at the same level. Although, the resolution decrease would be low, about 5% according to Figure 18.4a.

The magnetometer electronics comprises a preamplifier similar to [9]. It is already quite a miniature (see Figure 18.3), but can easily be miniaturized still further, leaving enough space inside the core for four pairs of the GP393 button cell batteries. These batteries would have an about 2-year lifespan, supplying the OPA333 preamplifier at its quiescent current of 17 μA. (For simplicity, we have used in our experimental model only a single pair of the batteries.)

Our experimental results are in very close agreement with the theoretical calculations. At 1 Hz, the magnetometer equivalent noise is 14.3 pT/Hz$^{0.5}$, which is very close to its theoretical estimation of 14.1 pT/Hz$^{0.5}$. The noise frequency behavior also very well fits its theoretical prediction.

18.2 THREE-AXIAL SEARCH COIL MAGNETOMETERS*

In this section, we integrate three orthogonal single-axis search coils, similar to the one described earlier, to design a three-axial magnetometer. Our aim is to minimize the total magnetometer volume. The volume is a limiting factor in various applications, for example, where many sensors are arranged in a network, or where an inhomogeneous magnetic field should be measured with a high accuracy.

To reach this aim, we optimize the coil parameters, considering this new constraint. We also investigate the effect of the integration on the magnetometer accuracy, taking into account the distribution of the applied magnetic flux between the magnetometer coils.

The designed three-axial magnetometer has a similar to fluxgates size and resolution at 1 Hz and significantly outperforms them in power consumption. Compared to conventional three-axial search coil magnetometers with a similar resolution (see Figure 18.5), the bulk of the suggested magnetometer is by an order of magnitude smaller.

18.2.1 Magnetometer Design

Much more compact design of search coils [9,10], where they are thick but short, as compared to traditional designs [1–4,6–8], where the coils are thin but very long, allows integrating three orthogonal coils in a single assembly [11] with a very high volume utilization factor (see Figure 18.5). One can see from Figure 18.5 that three orthogonal coils with an aspect ratio of 2 very efficiently fill in the total cubic volume of the magnetometer.

The two remaining small cubic volumes (see Figures 18.6a) are also utilized: one for accommodating the electronic board and the other for the batteries.

Considering the aspect ratio of 2, we have optimized the magnetometer coils. The optimization goal was to find the minimal magnetometer size, enabling the sensitivity threshold comparable to that of fluxgates, namely, about 10 pT/Hz$^{0.5}$ at 1 Hz.

The optimization has been done in accordance with Section 18.1.1.2. The optimization results are listed in Table 18.1 for an OPA333 preamplifier and an MnZn ferrite with a 2000 relative magnetic permeability, which has been selected as the material for the flux concentrators and the coil cores.

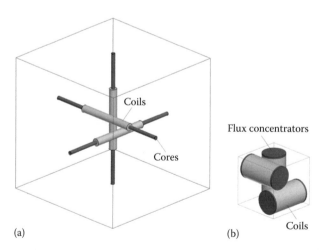

FIGURE 18.5 Integration of orthogonal search coils: (a) a conventional [1] and (b) the new design. While having a similar sensitivity and weight, the new design occupies a 22 times smaller volume. (Reprinted from Paperno, E., Grosz, A., Amrusi, S., and Zadov, B., Compensation of crosstalk in three-axial induction magnetometers, *IEEE Trans. Instrum. Meas.*, 60, 3416–3422, 2011. With permission. Copyright 2011 IEEE.)

* Reprinted from Grosz, A., Paperno, E., Amrusi, S., and Zadov, B., A three-axial search coil magnetometer optimized for small size, low power, and low frequencies, *IEEE Sens. J.*, 11, 1088–1094, 2011. With permission. Copyright 2011 IEEE.

TABLE 18.1

Parameters of the Magnetometer Coils

Coil Parts	Parameters
Flux concentrators	30 mm diameter, 4 mm thickness
Cores	5 mm diameter, 50 mm length[a]
	The shortest distance between the cores' axes is 32 mm
μ_r	2000
Copper wire	35 μm diameter (39 μm with isolation)
Winding	330,000 turns, 6 mm inner diameter, 29 mm outer diameter, and 47 mm length
Coil	
dc resistance	314.6 kΩ
Inductance	12.04 kH
Capacitance	95 pF
Self-resonant frequency	149 Hz

[a] The total length of each search coil is 60 mm, including the flux concentrators and conical springs to keep them attached to the coil cores.

The optimization results are related to a single magnetometer channel and do not take into account the applied flux distribution between the coils in the three-axial assembly. To address this, we have performed the following numerical simulations.

18.2.2 COIL INTEGRATION EFFECT ON THE MAGNETOMETER SENSITIVITY AND ACCURACY

18.2.2.1 Coil Integration Effect on the Magnetometer Sensitivity

To examine the sensitivity of the three-coil assembly (see Figure 18.6a), we have computed the apparent permeability of the coil cores with the help of commercially available three-dimensional finite element method (FEM) software, Maxwell 13.0. The results are shown in Figures 18.7 and 18.8.

The cores' apparent permeability averaged along their lengths on that the coils are wound equals 212. Compared to a single coil, it is only by 4% lower.

18.2.2.2 Crosstalk

We have also found in the simulations that the applied magnetic flux is flowing not only in the longitudinal coil but also in the transverse coils (see Figures 18.7 and 18.8). This is because the transverse cores provide a bypass for the applied flux and a part of it is flowing through them around the longitudinal core. For the same reason, a part of the secondary flux, generated in the longitudinal core by the electric current flowing in its coil, is also conducted by the transverse cores (see Figure 18.7).

We refer to the aforementioned effect as a crosstalk between the cores. The crosstalk can be described by the total magnetic flux flowing through the transverse coils relative to the primary flux Φ_p, caused by the applied field in the longitudinal coil,

$$\phi_{c\Sigma} = \frac{\Phi_{c\Sigma}}{\Phi_p} = \phi_{ca1} + \phi_{cs1} \frac{\Phi_s}{\Phi_p},$$

(18.6)

where

$\Phi_{c\Sigma}$ is the total flux in the transverse coils

ϕ_{ca1} is the flux due to the applied field that causes a unit Φ_p

ϕ_{cs1} is the flux due to a unit secondary flux Φ_s

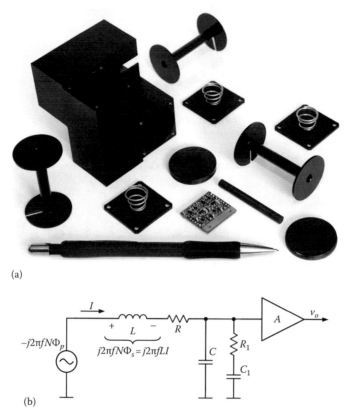

(a)

(b)

FIGURE 18.6 Magnetometer structure. (a) Magnetometer components (the windings, electrostatic shield, and the magnetometer housing are not shown). (b) Equivalent electrical circuit of a magnetometer channel. The damper $R_1 C_1$ is used to stabilize the circuit and to tune the coil resonant frequency to about 11 Hz. The channel electronics represented by block A is described in [11]. (From Grosz, A. et al., *IEEE Sensors J.*, 11, 1088, 2011. With permission.)

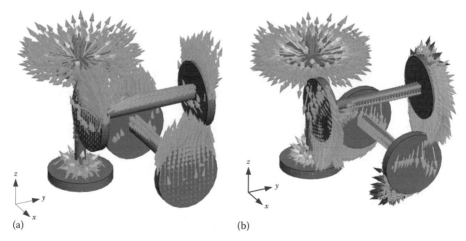

(a)   (b)

FIGURE 18.7 Magnetometer crosstalk. (a) Crosstalk due to the applied flux: magnetic induction within the cores for a uniform magnetic field applied along the Z-core. Note that there is a net magnetic flux within the transverse cores. Relative to the primary flux in the longitudinal coil, the crosstalk $\phi_{cp1} = 2.2\%$. (b) Crosstalk due to the secondary flux: magnetic induction within the cores for an electric current flowing in Z-coil. (The coils are not shown.) Note that there is a net magnetic flux within the transverse cores. Relative to the secondary flux in the vertical coil, the crosstalk $\phi_{cs1} = 6.3\%$.

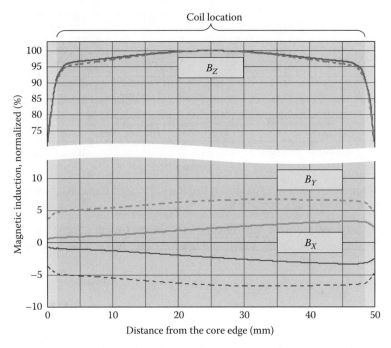

FIGURE 18.8 Crosstalk due to the applied and secondary fluxes: normalized magnetic induction along the coil cores' axes. A uniform magnetic field is applied along the Z-core (the solid curves, the maximum value of B_Z is 219 T), and an electric current is applied to Z-coil (the dashed curves). Note the enlarged scale for the B_X and B_Y magnitudes.

The values of the ϕ_{ca1} and ϕ_{cs1} factors found with the help of the FEM simulations are 2.2% and 6.3%, respectively. To find these factors, we have averaged the magnetic induction along the transverse cores' lengths on that the coils are wound (see Figure 18.8) for the applied and secondary fields, respectively.

Considering Figure 18.6b, the secondary flux can be found as follows:

$$\Phi_s = \frac{1}{N} LI = \frac{1}{N} L \frac{-j2\pi f N \Phi_p}{j2\pi fL + R + \dfrac{1}{j2\pi fC} \parallel \left(R_1 + \dfrac{1}{j2\pi fC_1} \right)}, \qquad (18.7)$$

where R_1 and C_1 are the resistance and capacitance of the damper used to tune the coil self-resonant frequency and to stabilize the circuit. From (18.6) and (18.7), the crosstalk can be found as a function of frequency (see Figure 18.9). One can see from this figure that the crosstalk nearly equals to ϕ_{ca1} below resonance and rapidly increases at resonance.

The crosstalk between the magnetometer channels can be reduced by applying magnetic feedback [13]. It can also be compensated by processing the magnetometer outputs as described in [14].

18.2.2.3 Effect of the Crosstalk on the Magnetometer Accuracy

To evaluate the effect of the crosstalk on the magnetometer accuracy, we have calculated the magnetometer outputs for a field vector rotating in such a way that its tip draws in space a spherical spiral.

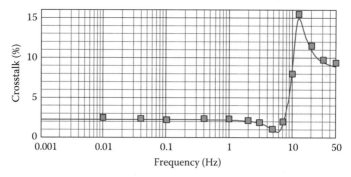

FIGURE 18.9 Magnetic crosstalk as a function of frequency. The solid curve and the squares represent the theoretical and experimental results, respectively.

The following approximations have been obtained for the maximum uncertainty in the measured field direction (in degrees) and for the maximum relative uncertainty in the field magnitude:

$$\varphi_{max} \approx 85\phi_{c\Sigma} - 40\phi_{c\Sigma}^2 + 35\phi_{c\Sigma}^3 \quad \text{and} \quad \delta R_{max} \approx \begin{cases} +\phi_{c\Sigma}100\% \\ -2\phi_{c\Sigma}100\% \end{cases} \tag{18.8}$$

Considering (18.8), φ_{max} and δR_{max} below resonance are 1.9° and {+2,2%, −4.4%}, respectively. This accuracy meets the requirements of our applications, where maximum uncertainties in the measured field direction and magnitude should be below ±2.5° and ±5% for 20 mHz–7 Hz frequencies.

18.2.3 FLATTENING THE FREQUENCY RESPONSE

The magnetometer frequency response is shown in Figure 18.10. One can see from this figure that frequency response is nearly flat in the 20 mHz–12 Hz frequency range. This is obtained due to the compensation of the rise of the coils' outputs with frequency by the roll-off of the

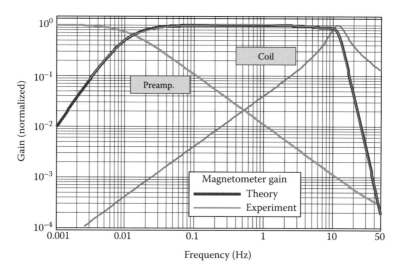

FIGURE 18.10 Shaping the magnetometer frequency response: the rise of the coils' outputs with frequency is compensated by the roll-off of the preamplifier gain. The steep drop-off of the magnetometer gain beyond 12 Hz is obtained with two second-order high-pass filtering stages. The magnetometer gain (sensitivity) is normalized to 7.7 μV/pT.

preamplifier gain. The upper limit of the frequency range is set at 12 Hz to shift the coils' resonance frequency away from the maximum frequency (7 Hz) at which the crosstalk should still be low (see Figure 18.9).

The steep drop off of the magnetometer gain beyond 12 Hz is obtained with two second-order high-pass filtering stages. It is needed to suppress the magnetometer sensitivity to interferences at frequencies of 50 and 100 Hz. This allows avoiding saturation and enables more efficient usage of the dynamic range of a 16-bit A/D converter.

Even after the filtering, magnetic interferences exceed the signal in our applications by about 80 dB. A 16-bit A/D converter meets this dynamic range. The magnetometer sensitivity (7.7 µV/pT) is high enough to bring the minimum noise density ($\sim$2 pT/Hz$^{0.5}$) at the A/D converter (AD7682) input above its intrinsic noise.

18.2.4 EQUIVALENT MAGNETIC NOISE

The magnetometer equivalent magnetic noise 12 pT/Hz$^{0.5}$ measured at 1 Hz is very close to the desired value. The minimum noise ($\sim$2 pT/Hz$^{0.5}$) is obtained at about 12 Hz [11].

18.2.5 POWER CONSUMPTION

To minimize the magnetometer power consumption, the OPA333 zero-drift, ultra-low power operational amplifiers with low white and no flicker noise have been chosen as preamplifiers. Their supply current is 17 µA.

To keep the power consumption as low as possible, we, instead of employing magnetic feedback, have flattened the magnetometer frequency response by reducing the preamplifier gain with frequency.

In [11], the OPA379 and OPA2369 operational amplifiers have been chosen due to their very low supply currents (2.6 and 0.7 µA, respectively) for the filtering stages and also for the circuit producing artificial ground. The supply current of the AD7682 A/D converter is also very low, about 1 µA. Thus, the total supply current is only 70 µA.

The magnetometer is powered by 3.6 V lithium batteries: four 1/2AA batteries, 1.1 A·h each, connected in parallel. Thus, its total power consumption is only 252 µW. Such ultra-low power consumption enables a 7-year continuous operation.

18.2.6 TESTING THE MAGNETOMETER

Experiments with a prototype have confirmed the theoretical evaluation of the magnetometer accuracy, frequency response, and noise. To measure these characteristics, a three-shell cylindrical magnetic shield was used. The magnetometer digital output was acquired by an external microcontroller.

To measure the crosstalk, we have used a three-shell magnetic shield and a solenoid to apply a magnetic field. A very good agreement between the measurements and the theoretical prediction has been obtained.

To measure the magnetometer accuracy, we used three-axial Helmholtz coils (the BH600-3-B type, manufactured by Serviciencia). A rotating magnetic field was applied to the magnetometer, sequentially in xy-, xz-, and yz-planes, and the channel outputs were recorded. These measurements have shown a less than 2.5° maximum uncertainty in the field direction and a less than 5% maximum relative uncertainty in the field magnitude.

The magnetometer supply current and the power consumption do not exceed the theoretical values.

NOMENCLATURE

α	Search coil aspect (length to diameter) ratio
β	Ratio of the wire diameters with and without the insulation: d_{wins}/d_w
γ	Ratio of the winding length and the total length of the search coil: L_w/L
μ_r	Relative permeability of the solid core and the flux concentrators
μ_r''	Relative permeability of the tube core
μ_a	Apparent permeability of the solid core without flux concentrators
μ_a'	Apparent permeability of the solid core with flux concentrators
μ_a''	Apparent permeability of the tube core with flux concentrators
ρ	Resistivity of the wire
τ	Wall thickness of the tube core
Φ	Average magnetic flux through the coil core
B	Magnetic field induction
B_0	Amplitude of the magnetic field induction
B_{st}	Sensitivity threshold (equivalent magnetic noise, resolution)
$B_{st\,min}$	Optimum sensitivity threshold (optimum equivalent magnetic noise, optimum resolution)
C	Search coil strain capacitance
d	Core diameter
d_{opt}	Optimum core diameter
d_w	Diameter of the wire, not including the insulation
$d_{w\,opt}$	Optimum diameter of the wire, not including the insulation
d_{wins}	Diameter of the wire, including the insulation
D	Diameter of the flux concentrators
e_n	Spectral density of the amplifier voltage noise
e_R	Spectral density of the coil thermal noise
e_{tot}	Spectral density of the total coil noise referred to the input
f	Frequency
H_D	Demagnetizing factor of a prolate ellipsoid
i_n	Spectral density of the amplifier current noise
k	Boltzmann constant
L	Total length of the search coil, including the flux concentrators
L_C	Search coil inductance
L_w	Length of the search coil winding
N	Number of turns
N_l	Number of turns in a layer
N_h	Number of layers
R	Search coil winding resistance
R_{opt}	Optimal resistance of the search coil winding
S	Cross-sectional area of the solid core
t	Time
T	Absolute temperature in kelvin.
$V(f)$	Amplitude of the voltage induced in the search coil by an applied sinusoidal magnetic field
Vol	Volume of the search coil

REFERENCES

1. H. C. Séran and P. Fergeau, An optimized low-frequency three-axis search coil magnetometer for space research, *Rev. Sci. Instrum.*, 76, 044502-1–044502-10, 2005.
2. R. J. Prance, T. D. Clark, and H. Prance, Compact broadband gradiometric induction magnetometer system, *Sens. Actuat. A: Phys.*, 76, 117–121, 1999.

3. R. J. Prance, T. D. Clark, and H. Prance, Compact room-temperature induction magnetometer with super-conducting quantum interference device level field sensitivity, *Rev. Sci. Instrum.*, 74, 3735–3739, 2003.
4. R. J. Prance, T. D. Clark, and H. Prance, Ultra low noise induction magnetometer for variable temperature operation, *Sens. Actuat. A: Phys.*, 85, 361–364, 2000.
5. K. Abe and J. I. Takada, Simulation and design of a very small magnetic core loop antenna for an LF receiver, *IEICE Trans. Commun.*, E90–B, 122–130, 2007.
6. Z. Chen, S. Zhou, and A. Jiang, Miniaturization design on magnetic induction sensors, *International Conference on Electronic and Mechanical Engineering and Information Technology (EMEIT)*, Harbin, Heilongjiang, China, 12–14 August, 2011.
7. C. Coillot, J. Moutoussamy, R. Lebourgeois, S. Ruocco, and G. Chanteur, Principle and performance of a dual-band search coil magnetometer: A new instrument to investigate fluctuating magnetic fields in space, *IEEE Sens. J.*, 10, 255–260, 2010.
8. C. Coillot, J. Moutoussamy, G. Chanteur, and A. Roux, Improvements on the design of search coil magnetometer for space experiments, *Sens. Lett.*, 5, 167–170, 2007.
9. E. Paperno and A. Grosz, A miniature and ultralow power search coil optimized for a 20 mHz to 2 kHz frequency range, *J. Appl. Phys.*, 105, 07E708-1–07E710-3, 2009.
10. A. Grosz, E. Paperno, S. Amrusi, and E. Liverts, Integration of the electronics and batteries inside the hollow core of a search coil, *J. Appl. Phys.*, 107, 09E703-1–09E703-3, 2010.
11. A. Grosz, E. Paperno, S. Amrusi, and B. Zadov, A three-axial search coil magnetometer optimized for small size, low power, and low frequencies, *IEEE Sens. J.*, 11, 1088–1094, 2011.
12. D. G. Lukoschus, Optimization theory for induction-coil magnetometers at higher frequencies, *IEEE Trans. Geosci. Electron.*, GE-17, 56–63, 1979.
13. A. Grosz, E. Paperno, S. Amrusi, and T. Szpruch, Minimizing crosstalk in three-axial induction magnetometers, *Rev. Sci. Instrum.*, 81, 125106-1–125106-9, 2010.
14. E. Paperno, A. Grosz, S. Amrusi, and B. Zadov, Compensation of crosstalk in three-axial induction magnetometers, *IEEE Trans. Instrum. Meas.*, 60, 3416–3422, 2011.
15. Magnetic field induction sensor BF-7, data sheet, Schlumberger, Ltd, Houston, USA. [Online]. Available: http:/www.slb.com/~/media/Files/rd/technology/product_sheets/emi_bf_7_sensor.ashx.
16. Broad band induction coil magnetometer MFS-06e, technical notice, Apex Tool Group, LLC, Maryland, USA. [Online]. Available: http://178.63.62.205/mtxgeo/index.php/sensors/mfs-06e.
17. Induction magnetometer for geophysical applications LEMI-121, technical notice, Lviv Centre of Institute of Space Research, Lviv, Ukraine. [Online]. Available: http:/www.isr.lviv.ua/lemi121.htm.
18. P. Ripka, *Magnetic Sensors and Magnetometers*, Artech House, Norwood, MA, 2001.
19. J. A. Osborn, Demagnetizing factors of the general ellipsoid, *Phys. Rev.*, 67, 351–357, 1945.

19 Inductive Coupling–Based Wireless Sensors for High-Frequency Measurements*

H.S. Kim, S. Sivaramakrishnan, A.S. Sezen, and R. Rajamani

CONTENTS

19.1 INTRODUCTION

Wireless sensor systems have been widely adopted in a variety of fields including manufacturing, structural health monitoring, transportation, security control, and in vivo medical devices [1–5]. Based on the presence of an on-board power source for the sensor, wireless sensor systems can be classified into active and passive wireless systems. An active wireless sensor system incorporates a battery to supply the power for the operation of an active circuit (microchip). While an active wireless sensor system provides relatively long telemetry distance up to dozens of meters with high accuracy and sensitivity, it inevitably becomes large and its lifetime is considerably affected by the capacity of the on-board battery. On the other hand, a passive wireless sensor system can be constructed in a relatively small size as well as low cost and offers potentially an unlimited life span. Although the passive wireless sensor system suffers from poor signal to noise ratio (SNR) and its telemetry distance is intrinsically restricted, the capability to eliminate the batteries from wireless sensor systems has significantly

---

* © 2010 IEEE. A significant portion of this chapter is reprinted, with permission, from IEEE Sensors, Vol. 10, No. 10, pp. 1647–1657, Oct 2010 [16].

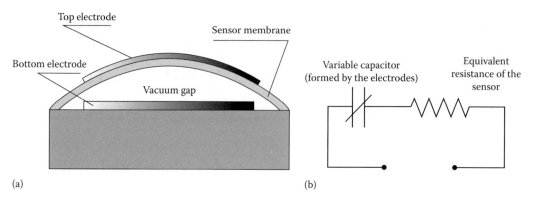

FIGURE 19.1 (a) Basic schematic of a capacitive pressure sensor and (b) simplified electrical model.

expanded its prospective applications from biomedical implants to industrial applications in hazardous environments where physical access to the sensor is highly restricted so that the battery cannot be exchanged or recharged [1,4]. In order to remotely transmit power to a passive sensor, a wireless link between an external interrogator and the sensor is typically adopted by virtue of inductive coupling.

A schematic of a basic microfabricated capacitive pressure sensor and its equivalent simplified electrical model are shown in Figure 19.1a and b, respectively. A sensing membrane is suspended over the sensor substrate. The sensing membrane and the bottom substrate are both equipped with metallic electrodes. The electrical capacitance between these electrodes is a function of the gap between the membrane and the substrate. Pressure on the diaphragm causes a change in the gap, changing the capacitance, and this change in capacitance needs to be measured (or estimated in a wireless fashion in the case of this chapter). The concept of the aforementioned sensing mechanism is well established, and many results have been published documenting the performance of microsensors based on this concept [6–8]. Capacitive sensors can be used to measure fluid pressure, interface or contact pressure, temperature, humidity, force, acceleration, pH value, and many other variables [2,9].

Two well-developed techniques that have been intensively explored to read a passive sensor in telemetry are (a) inductive coupling–based sensor system using resonant frequency determination and (b) a surface acoustic wave (SAW)-based sensor system. These techniques incorporate either an *LC* circuit or a SAW device into a sensor in order to detect a variation of sensor property such as the resonant frequency of the sensor or the amplitude and phase of the reflected SAW on the sensor. The interrogators in both techniques are excited by a radio frequency (RF) signal. Compared to SAW sensors, which typically use an RF signal frequency higher than 30 MHz, the inductive coupling–based sensor can adopt a lower frequency RF signal where the skin effect and parasitic resistance can be considerably attenuated [6,10]. However, as described in Section 19.2, the traditional inductive coupling–based technique requires a sweep of excitation frequencies in order to determine the resonant frequency, and hence this technique is mostly useful for "static" measurements. A "grid-dip" approach is a typical example of the resonant frequency determination–based technique [11–13]. This approach measures the "phase-dip" of the sensor impedance, which occurs whenever the frequency of the RF excitation signal precisely matches the resonant frequency of the sensor because the phase of the sensor impedance is shown to be more sensitive to the impedance change in comparison to its magnitude.

This chapter presents a novel real-time method called "cascaded filtering" that enables wireless estimation of high-frequency changes in capacitance by algebraically manipulating two measurements (the magnitude and the phase of the reflected sensor impedance). Compared to previous estimation techniques, the proposed capacitance estimation algorithm has been developed to obtain the following advantages. First, the proposed algorithm makes it possible to continuously estimate the rapidly varying capacitance of a capacitive sensor as well as static or slowly varying ones. Second, it is quite simple and inexpensive to practically implement the proposed algorithm since it is derived without the need for frequency sweeping operation on the primary circuit.

19.2 OPERATING PRINCIPLE OF CLASSICAL PASSIVE WIRELESS SENSOR SYSTEMS

19.2.1 INDUCTIVE COUPLING–BASED SENSOR SYSTEMS

An inductive coupling–based sensor system consists of a transceiver circuit (or interrogator) and a sensor circuit, which can be modeled as the primary and the secondary sides of a loosely coupled, air-cored transformer, respectively. A schematic circuit diagram of an inductive coupling–based sensor system is shown in Figure 19.2, where the transceiver consists of an RF signal generator and a primary coil L_1, and the sensor circuit consists of a secondary coil L_2, a capacitor C_2 representing a varying capacitance of the sensor, and a parasitic resistor R_2, respectively.

When a voltage input V_{in} of an RF frequency ω is applied to the transceiver circuit, an oscillating current i_1 is generated in the transceiver circuit, and a corresponding magnetic field is built up around the primary coil of the transceiver circuit, which activates the sensor circuit by virtue of inductive coupling. As a result, a current i_2 is induced in the sensor circuit. The current i_2 in turn, affects the transceiver circuit through the mutual inductance in the same manner as mentioned previously, which leads to a voltage change in the transceiver circuit. Since the induced voltage change appears in the opposite direction to the voltage input V_{in}, a load impedance $Z_L(\omega)$ is introduced on the transceiver circuit given by

$$Z_L(\omega) = \frac{(\omega M)^2}{R_2 + j\left(\omega L_2 - \dfrac{1}{\omega C_2}\right)} \tag{19.1}$$

where $j = \sqrt{-1}$ and $M\left(= k\sqrt{L_1 L_2}\right)$ is the mutual inductance between the primary and the secondary coils defined with the coupling coefficient $k(0 \leq k \leq 1)$ that changes according to the distance and the angle between the primary and the secondary coils, and their geometric dimensions, etc. As a result, the total impedance seen from the transceiver circuit due to the inductive coupling can be obtained by simply adding individual impedances of passive elements to the load impedance $Z_L(\omega)$ in Equation 19.1.

It is worthwhile to note that the mutual inductance between the primary and the secondary coils enables the transceiver circuit not only to remotely transmit power to the passive sensor but also to receive the information of load impedance $Z_L(\omega)$, including the sensor impedance. Therefore, by monitoring the change of total impedance on the transceiver circuit, it is possible to remotely detect a change of sensor property such as capacitance change, which is reflected on the transceiver circuit in the form of the load impedance $Z_L(\omega)$.

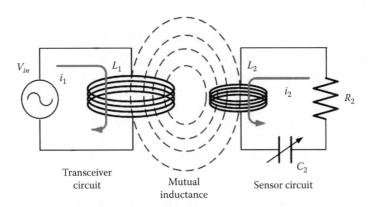

FIGURE 19.2 Schematic circuit diagram of an inductive coupling–based sensor system.

19.2.2 CLASSICAL RESONANT FREQUENCY DETERMINATION APPROACH

Among existing approaches to remotely monitor the capacitance change of the sensor, the resonant frequency determination approach is widely adopted, where the resonant frequency of the sensor circuit is determined by applying a predetermined set of excitation frequencies or a sweep of excitation frequencies near the natural frequency of the sensor circuit and checking the dip in the phase of the total impedance that occurs at the frequency-matching condition.

At the resonant frequency of the sensor $\left(\omega_0 = 2\pi f_0 = 1/\sqrt{L_2 C_2}\right)$, the reflected load impedance $Z_L(\omega_0)$ in Equation 19.1 becomes $(\omega_0 M)^2/R_2$. This sudden increase in the reflected load impedance causes a dip in the phase φ of total impedance in the transceiver circuit and the corresponding variation may be approximated by $\Delta\varphi_{dip} \approx \tan^{-1}\left(k^2\omega_0 L_2/R_2\right)$, which is maximized when the parasitic resistor R_2 in the sensor circuit is minimized and the inductance of the secondary coil L_2 is maximized [6]. However, when the inductance of the secondary coil is increased by increasing the number of turns in the coil, its parasitic capacitance increases so that the self-resonant frequency of the coil decreases, where the inductor ceases to function purely as an inductor and electrically looks like a capacitor. Therefore, since another dip is undesirably observed close to the self-resonant frequency of the coil, it becomes difficult to discriminate the resonant frequency of the sensor circuit from the self-resonant frequency of the coil for some cases [6,14].

Besides, it is reported that the frequency f_{min} corresponding to the phase dip does not exactly match the resonant frequency f_0 of the sensor circuit so that even if the telemetry distance between the primary and the secondary coils is relatively small, this discrepancy turns out to still exist because the minimum frequency is a function of both the coupling coefficient k and the quality factor Q [15]. Since the coupling coefficient k changes with distance, the calibration between f_{min} and f_0 keeps changing, and this frequency determination approach works correctly only if the telemetry distance is fixed and the self-resonant frequency of the coils is much higher than f_0 [14].

In order for the transceiver circuit to sweep the frequency range of interest, additional equipment such as a voltage controlled oscillator (VCO) is typically required [3]. Alternatively, network/impedance analyzers are typically used to interrogate the sensor circuit and to measure the total impedance on the transceiver circuit. These inevitably increase the cost and complexity of the transceiver circuit and further cannot avoid the increased processing time necessary to apply a predetermined sweep of excitation frequencies.

19.3 NOVEL CAPACITANCE ESTIMATION METHOD FOR INDUCTIVE COUPLING

19.3.1 MATHEMATICAL MODEL OF PASSIVE WIRELESS SENSOR SYSTEM

A novel method to read a passive capacitive sensor in telemetry by using inductive coupling is proposed in this section. Unlike the traditional resonant frequency determination approach to measure the sensor capacitance by identifying the resonant frequency of the sensor with a sweep of RF signals, the proposed new method estimates the capacitance change in real time by algebraically manipulating the measured magnitude and the phase of the reflected sensor impedance. This is possible because only one RF signal is used in the proposed method instead of a predetermined sweep of RF frequencies. A detailed circuit diagram of the telemetry system for the proposed method is very similar to Figure 19.2. However, an additional resistor R_1 and a tuning capacitor C_1 are connected to the transceiver circuit in order to measure a voltage output V_{out} across the resistor R_1 and to transmit the maximum power to the sensor circuit by tuning the capacitor C_1 appropriately for the given primary coil L_1, respectively. Based on the discussion in Section 19.2.1, the inverse of the frequency-domain transfer function from the RF voltage input $V_{in}(j\omega)$ to the voltage output $V_{out}(j\omega)$ measured across the resistor R_1 is given by

$$\frac{V_{in}(j\omega)}{V_{out}(j\omega)} = \frac{R_0 + R_1}{R_1} + \frac{j}{R_1}\left(\omega L_1 - \frac{1}{\omega C_1}\right) + \frac{(\omega M)^2/R_1}{R_2 + j\left(\omega L_2 - \dfrac{1}{\omega C_2}\right)} \tag{19.2}$$

where R_0 (not shown in Figure 19.2) is the series internal resistance of the power supply [16].

19.3.2 DEVELOPMENT OF NOVEL CAPACITANCE ESTIMATION METHOD

For the convenience of notation, the real and imaginary portions of Equation 19.2 are denoted by $z_1 = \mathrm{Re}\{V_{in}(j\omega)/V_{out}(j\omega)\}$ and $z_2 = \mathrm{Im}\{V_{in}(j\omega)/V_{out}(j\omega)\}$, respectively. These can be readily measured with a commercial gain/phase envelop detector chip. By expanding Equation 19.2 into real and imaginary parts, the two measurements z_1 and z_2 turn out to be related with the unknown variables x_1 and x_2 in the following manner:

$$\begin{bmatrix} z_1 \\ z_2 \end{bmatrix} = \begin{bmatrix} \dfrac{R_1 + R_0}{R_1} + \dfrac{\omega^2 x_2 R_2/R_1}{R_2^2 + (\omega L_2 - x_1/\omega)^2} \\[3ex] \dfrac{\omega L_1 - 1/(\omega C_1)}{R_1} - \dfrac{\omega^2 x_2 (\omega L_2 - x_1/\omega)/R_1}{R_2^2 + (\omega L_2 - x_1/\omega)^2} \end{bmatrix} \tag{19.3}$$

where $x_1 = 1/C_2$ and $x_2 = M^2$, respectively.

In order to find out the unknown variables x_1 and x_2, the closed form solution of Equation 19.3 is required but it does not seem possible to directly obtain a closed form solution of Equation 19.3 without any prior knowledge of both unknowns because this is a nonlinear algebraic problem involving two measurements z_1 and z_2 and two unknowns x_1 and x_2. The key idea to cope with this difficulty is to observe that if the known terms from the right-hand side of Equation 19.3 are separated, then the unknown terms remaining in the right-hand side of Equation 19.3 have not only $\omega^2 x_2/R_1$ in the numerator but also $R_2^2 + (\omega L_2 - x_1/\omega)^2$ in the denominator in common [16].

With new measured variables defined as $y_{c1} = z_1 - \dfrac{R_0 + R_1}{R_1}$ and $y_{c2} = z_2 - \dfrac{\omega L_1 - 1/\omega C_1}{R_1}$, Equation 19.3 can be rewritten as

$$\begin{bmatrix} y_{c1} \\ y_{c2} \end{bmatrix} = \begin{bmatrix} z_1 - \dfrac{R_1 + R_0}{R_1} \\[3ex] z_2 - \dfrac{\omega L_1 - 1/\omega C_1}{R_1} \end{bmatrix} = \begin{bmatrix} \dfrac{\omega^2 x_2 R_2/R_1}{R_2^2 + (\omega L_2 - x_1/\omega)^2} \\[3ex] -\dfrac{\omega^2 x_2 (\omega L_2 - x_1/\omega)/R_1}{R_2^2 + (\omega L_2 - x_1/\omega)^2} \end{bmatrix} \tag{19.4}$$

After dividing y_{c2} by y_{c1} and simplifying, the following closed form expression for the inverse of the unknown capacitance $x_1 = 1/C_2$ is compactly derived:

$$x_1 = \frac{1}{C_2} = \omega \cdot \left(R_2 \cdot \frac{y_{c2}}{y_{c1}} + \omega \cdot L_2\right) \tag{19.5}$$

Thus, the unknown $x_1 = 1/C_2$ can be linearly calculated by algebraically manipulating two measurements z_1 and z_2 and known parameters without any prior knowledge of the mutual inductance M. Since there is no restriction on the frequency bandwidth of the aforementioned capacitance sensing, the proposed estimation method turns out to be suitable for high-frequency sensing with inductive coupling–based sensors. The mutual inductance M is subsequently obtained by substituting $x_1 = 1/C_2$ into Equation 19.4. Hence, the proposed algorithm is called "cascade filtering" [16]. It should be noted that since the mutual inductance does not seem to provide any information of the sensor capacitance in Equation 19.5, it may be ignored for the proposed passive wireless sensing strategy. However, it will play an important role in compensating the error in the estimated sensor capacitance stemming from parameter uncertainties, which will be explained in the following section in detail.

19.4 ERROR ANALYSIS AND CALIBRATION OF THE PROPOSED ESTIMATION METHOD

19.4.1 EFFECTS OF PARAMETER ERRORS WITH DIFFERENT TELEMETRY DISTANCES

The performance of the proposed estimation method may deteriorate with errors in physical parameter values such as resistance, capacitance, and inductance or with noise in the measurements z_1 and z_2. These lead to inevitable deviations from the correct values of y_{c1} and y_{c2}. Then, a discrepancy between the estimated and the real capacitances occurs because it is impossible to completely cancel the common terms in Equation 19.4. For simplicity of analysis, the following simulations are carried out assuming that parameter errors in R_1, R_2, and L_2 are bounded within ±1%, which seems acceptable from the viewpoint of allowable errors in components of the electronics. In order to evaluate the performance of the proposed method in estimating high-frequency and multi-frequency capacitance changes, the parameter values used for the simulation are the same as in [17]. Multi-frequency components 150 and 420 Hz with the same peak-to-peak capacitance change of 15 pF are used over a constant nominal capacitance of 1010 pF. The effect of the parameter errors on the estimated capacitance and mutual inductance are shown in Figure 19.3a and b, respectively. The bold dotted, dash-dotted, and continuous lines denote the estimated capacitances with 1%, –1%, and no parameter errors, respectively, and other continuous lines between them denote the estimated ones obtained by increasing the parameter error from –1% by 0.2% each.

Although the proposed estimation method easily enables measurement of high-frequency and multi-frequency components of sensor capacitance, a discrepancy between the estimated and reference capacitances occurs for given parameter errors, as shown in Figure 19.3a. As a result, the corresponding estimation of the mutual inductance also suffers from a similar offset deviation as shown in Figure 19.3b. This simulation implies that parameter errors in electronic components cause an undesirable mismatch between the estimated and reference capacitances.

Recall that the coupling coefficient k is a function of the telemetry distance so that even if there exist no errors in the physical parameters, y_{c1} and y_{c2} may change according to the telemetry distance. Therefore, if x_2 is not completely cancelled due to parameter errors in R_1, R_2, and L_2, the discrepancy between the estimated and the reference capacitances will also change according to the telemetry distance. This is because the dependency of the mutual inductance x_2 on the telemetry distance undesirably appears in this case while estimating the capacitance.

19.4.2 DETERMINATION OF CALIBRATION ORDER FOR THE PROPOSED ESTIMATION METHOD

In order to overcome the distance-dependent deviation in the capacitance estimation from parameter errors, the coupling coefficient obtained from the estimated mutual inductance may be tactfully exploited since this deviation is mainly due to the fact that two mutual inductance terms containing x_2 in Equation 19.4 cannot be completely cancelled by division due to the parameter errors. Suppose that y_{c1} and y_{c2} in Equation 19.4 are replaced with $y'_{c1} = y_{c1} - \alpha$ and $y'_{c2} = y_{c2} - \beta$, where α and β

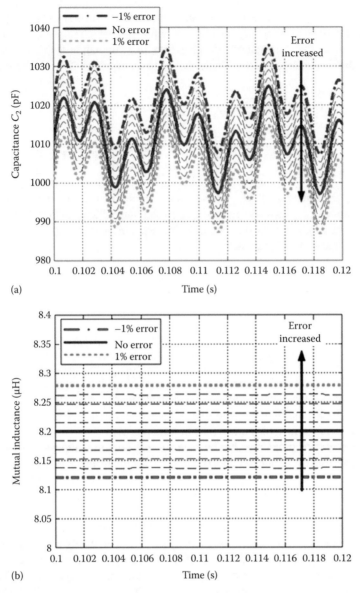

(a)

(b)

FIGURE 19.3 Error effect on (a) the proposed capacitance estimation and (b) the resulting mutual inductance estimation.

correspond to the functions of unknown errors of the parameters R_1, R_2, and L_2. Assuming $|\alpha/y_{c1}| < 1$ without loss of generality, y_{c2}'/y_{c1}' can be described in terms of y_{c2}/y_{c1} in the following form:

$$
\frac{y_{c2}'}{y_{c1}'} = \frac{y_{c2} - \beta}{y_{c1} - \alpha} = \frac{y_{c2} - \beta}{y_{c1}} \left(\frac{1}{1 - \alpha/y_{c1}} \right) = \frac{y_{c2} - \beta}{y_{c1}} \left(1 + \frac{\alpha}{y_{c1}} + \left(\frac{\alpha}{y_{c1}} \right)^2 + \cdots \right)
$$

$$
= \frac{y_{c2}}{y_{c1}} - \frac{\beta}{y_{c1}} + \alpha \frac{(y_{c2} - \beta)}{y_{c1}^2} + \alpha^2 \frac{y_{c2} - \beta}{y_{c1}^3} + \alpha^3 \frac{y_{c2} - \beta}{y_{c1}^4} + \cdots \qquad (19.6)
$$

Note that even though all terms on the right-hand side except for the first term y_{c2}/y_{c1} are functions of $1/x_2$, it seems unreasonable to simply describe y_{c2}'/y_{c1}' in such a form of the sum of y_{c2}/y_{c1} and a certain function of $1/x_2$ as $y'_{c2}/y'_{c1} = y_{c2}/y_{c1} + h(1/x_2)$. This is because the second term may produce another term proportional to the first term y_{c2}/y_{c1} in the case of $\beta = \gamma y_{c2}$, which implies that the first term also seems to depend on another function of $1/x_2$. Therefore, in combination with some parameter errors $(R'_2$ and $L'_2)$ and the corresponding error functions (α and β), Equation 19.5 must be recast as follows:

$$\frac{1}{C_2^c} = \omega\left[(R_2 + R'_2) \cdot \frac{y'_{c2}}{y'_{c1}} + \omega(L_2 + L'_2) \right]$$

$$= \omega\left[(R_2 + R'_2) \cdot \left\{ g\left(\frac{1}{x_2}\right)\frac{y_{c2}}{y_{c1}} + f\left(\frac{1}{x_2}\right) \right\} + \omega(L_2 + L'_2) \right] \tag{19.7}$$

where $g(\cdot)$ and $f(\cdot)$ are functions of $1/x_2$ whose values are defined as "1" and "0", respectively, for no parameter errors and C_2^c corresponds to the estimated capacitance affected by some parameter errors and the corresponding error functions. By replacing $R_2 + R'_2$ and $L_2 + L'_2$ with $e_R R_2$ $(e_R > 0)$ and $e_L L_2$ $(e_L > 0)$, Equation 19.7 becomes

$$\frac{1}{C_2^c} = \omega\left[e_R R_2 \left\{ g\left(\frac{1}{x_2}\right)\frac{y_{c2}}{y_{c1}} + f\left(\frac{1}{x_2}\right) \right\} + e_L \omega L_2 \right]$$

$$= e_R g\left(\frac{1}{x_2}\right)\frac{1}{C_2^o} + \omega^2 L_2 \left\{ e_L - e_R g\left(\frac{1}{x_2}\right) \right\} + e_R \omega R_2 f\left(\frac{1}{x_2}\right) \tag{19.8}$$

where C_2^o denotes the true capacitance without parameter error. It is noted that even for no error in R_2 and L_2, the capacitance estimation still suffers from the detrimental effects stemming from other parameter errors like R'_1, multiplied by the selected parameter values R_2 and L_2, and the operating frequency ω, and included in $g(1/x_2)$ and $f(1/x_2)$ in Equation 19.8.

Equation 19.8 seems to provide a clue for determining how to compensate for the undesirable mismatch resulting from the parameter uncertainties so that the inverse of the estimated capacitance C_2^c and the corresponding error functions is linearly approximated by the inverse of the true capacitance C_2^o as $1/C_2^c = a/C_2^o + b$, where $a = e_R g(1/x_2)$ and $b = \omega^2 L_2 \{e_L - e_R g(1/x_2)\} + e_R \omega R_2 f(1/x_2)$.

Equation 19.8 also explains why a lower operating frequency ω must be chosen to mitigate the effects of parameter uncertainties. Compared to the scaling deviation, the offset difference between the estimated and reference capacitances is relatively susceptible not only to the parameter errors but also to the parameters themselves as already confirmed through the simulation study in the previous section. Recall that both a and b depend on the telemetry distance because they are functions of the mutual inductance as shown in Equation 19.8. As a result, the scaling and offset deviations from the reference capacitances are subject to the influence of the telemetry distance. Therefore, since the telemetry distance between two coils may change continuously for certain applications, it is required to discriminate for the variations of the telemetry distance. Even if not calibrating by determining the present telemetry distance precisely, it may be possible to directly use the mutual inductance information subsequently obtained by combining the estimated capacitance with Equation 19.4.

19.5 EXPERIMENTS AND DISCUSSION

19.5.1 EXPERIMENTAL SETUP

Figure 19.4 shows a schematic diagram and a photograph of the experimental setup to evaluate the performance of the proposed algorithm for the inductive coupling–based wireless sensor system, which consists of the transceiver and sensor circuits. An RF reference signal of a constant frequency from an RF signal generator is divided into two signals of equal amplitude by using a coupler.

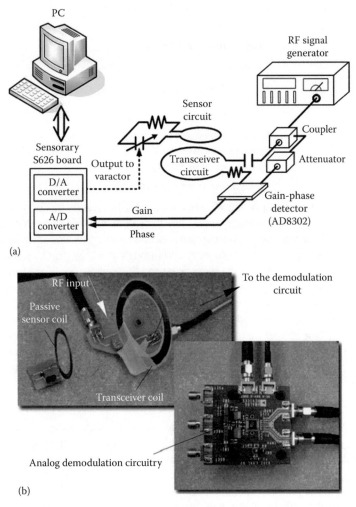

FIGURE 19.4 (a) Schematic diagram and (b) photograph of the experimental setup.

One signal from the coupler is guided to the transceiver circuit to interrogate the sensor circuit. The other signal is fed through an attenuator into a commercial gain/phase detector (AD8302 evaluation board, Analog Devices) together with the voltage output measured across the resistor on the transceiver circuit. With a sampling rate of 4 kHz, the experimental data are collected by using a PCI-type Sensorary S626 data acquisition board installed on a PC. In this study, Microsoft Visual Studio is used to create and compile C-code, which is then executed in Windows XP to handle the data acquisition and implement the proposed estimation method in real time.

Instead of a real capacitive sensor, a voltage-controlled capacitor (varactor) is adopted to produce a variety of sensed-signal conditions. Recall that for a given control voltage output of a fixed single frequency, the resulting capacitance of a varactor inevitably shows multiple high-order harmonics of the fundamental frequency since the relation between the control voltage output to the varactor and the corresponding capacitance variation is completely nonlinear [18].

As for the primary and secondary coils, both are made in the spiral planar type due to the simplicity of fabrication. Both are fabricated with a 30 American Wire Gauge (AWG) copper wire and fixed upon a paper tape so that their spiral shapes remain steady. Since the telemetry distance between the sensor and the transceiver is assumed to be at least 5 cm for practical applications, the diameter of the secondary coil is then determined to be 2 cm, while the size of the primary coil is not restricted. Of course,

TABLE 19.1
Properties of Parameters

Parameters	Values
Reader coil inductance L_1	14.4 μH
Reader resistance R_1	3.3 Ω
Reader capacitance C_1	15.9 pF
Sensor coil inductance L_2	5.24 μH
Sensor parasitic resistance R_2	0.1 Ω
Series internal resistance R_0	30 Ω

the size of the secondary coil can be significantly reduced if the telemetry distance becomes small, for example, the diameter of the secondary coil is less than 1 cm for a telemetry distance of 1 cm. It is worthwhile to note that the operating frequency (=resonant frequency) as well as the telemetry distance can have an influence on the size of the secondary coil through $f = 1/2\pi\sqrt{L_2C_2}$, where f is the operating (resonant) frequency, L_2 and C_2 are the inductance and the capacitance of the sensor, respectively. Therefore, even for an operating frequency of a few hundred kHz, the size of the secondary coil might increase significantly compared to the one presented in this study. The external diameter of the primary coil is 6 cm, and the corresponding inductances of the primary and the secondary coils are 14.4 and 5.24 μH, respectively. Considering that the skin effect and the parasitic series resistance are reduced as the operating frequency decreases, several frequencies of the reference RF signal lower than 30 MHz are chosen. Other parameters used for the experiments are summarized in Table 19.1.

19.5.2 NOMINAL PERFORMANCE OF THE PROPOSED ESTIMATION METHOD

The maximum telemetry distance based on the experimental setup explained in Section 19.5.1 is observed to be 5 cm. In order to evaluate the nominal and robust performances of the proposed estimation algorithm, experiments are conducted under a variety of conditions where the telemetry distance and the angle between the two coils are changed from 1 to 5 cm and from 0° to 50°, respectively. As discussed in the previous section, it is highly desired to adopt a lower RF signal in order to diminish its effect on the offset problem and the skin effect as well as parasitic resistance resulting in parameter errors, and so we tried to decrease the operating frequency down to around 10 MHz. During the initial stage of experiments, several operating frequencies around 13.56 MHz are chosen so that the capacitance estimations in Figure 19.5 are obtained with the operating frequencies of 13.53 and 13.43 MHz, respectively. On the other hand, other experimental results in Sections 19.5.2 and 19.5.3 are obtained with the operating frequency of 10.44 MHz.

The nominal performance of the proposed estimation procedure at a fixed telemetry distance is first evaluated for two cases of single- and multiple-frequency changes in capacitance. The experiments are conducted at a telemetry distance of 2 cm with the parameter values shown in Table 19.1. For the single- and multiple-frequency changes of the capacitances, the peak-to-peak magnitudes of the capacitance change are chosen as 0.15 and 0.7 pF, respectively. It is observed that the frequencies of the estimated capacitance changes are exactly the same as those of the real capacitance changes but the estimated capacitances show some discrepancies in the amplitude and offset compared to the real capacitance changes for both cases. Based on the discussion in Section 19.4.2, the scaling and offset coefficients a and b are then used to compensate for the capacitance estimation error in such a linear manner as $C_2^c = a\left|C_2^o - b\right|$, where C_2^c and C_2^o are the compensated and originally estimated capacitances, respectively. First, the offset coefficient b can be easily calculated since the real capacitance is absolutely determined by $C_2^c = 1/(\omega^2 L_2)$ without the capacitance change ($a = 1$). After subtracting b from the originally estimated capacitance C_2^o, the scaling coefficient a is then properly chosen to cope with the error in the amplitude of the estimated capacitance. Of course, this procedure

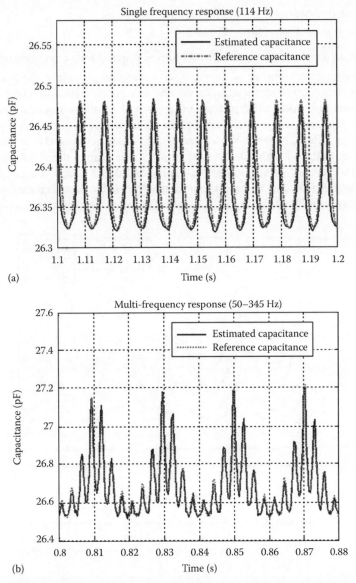

FIGURE 19.5 Comparison of the estimated and reference capacitances (a) with single-frequency component at 2 cm and (b) with multiple-frequency components at 2 cm.

to find out the scaling and the offset coefficients is off-line and dependent upon the telemetry distance. However, after obtaining the coefficients once to predict the real capacitance change precisely, no additional procedure is required so that the real capacitance change can be estimated in real time as well as for multiple-frequency components. These are the main contributions of the proposed technique to explicitly be different from the conventional approach based on the "grid-dip" technique.

Figure 19.5a and b compare the estimated capacitances that are calibrated according to the linear equation with the real capacitances for single-frequency change (114 Hz) and multiple-frequency change (35 and 345 Hz) at a telemetry distance of 2 cm, respectively. As shown in Figure 19.5, the estimated capacitance change in both cases matches the real capacitance changes with high fidelity. Recall that due to the nonlinearity of the varactor discussed in the previous section, both the estimated and real capacitance changes contain frequency components other than the frequencies of 35 and 345 Hz in Figure 19.5b.

19.5.3 ROBUST PERFORMANCE OF THE PROPOSED ESTIMATION METHOD WITH FIRST-ORDER CALIBRATION

19.5.3.1 Estimation of the Single/Multi-Frequency Capacitance Component

To evaluate the robust performance of the proposed estimation procedure, the telemetry distance is changed from 1 to 5 cm with capacitance varying at a single frequency of 115 Hz. Since the resonant frequency ω is 10.44 MHz and the inductance of L_2 of the secondary coil is 5.24 μH, the actual mean capacitance is around 44.305 pF in the ideal situation. As shown in Figure 19.6a, the estimated capacitances, however, show some discrepancies in the amplitude and the DC value appears according to the telemetry distance. Contrary to the case of the fixed telemetry distance considered in the previous section, where the scaling and offset coefficients remain unchanged, they must be adjusted according to the varying telemetry distance. Based on the estimated capacitance changes shown in Figure 19.6a, the corresponding coupling coefficient k obtained from the

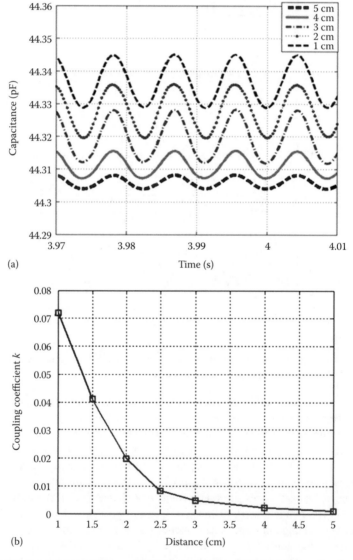

FIGURE 19.6 (a) Estimated capacitances with different telemetry distances from 1 to 5 cm and (b) corresponding coupling coefficient k measured according to the telemetry distance.

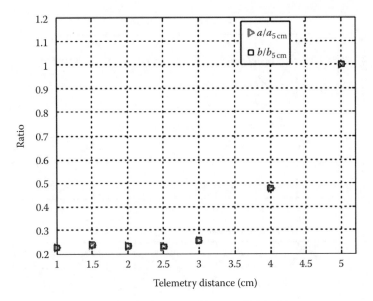

FIGURE 19.7 Relative magnitudes of coefficients a and b normalized with respect to $a_{5\,cm}$ and $b_{5\,cm}$.

estimated mutual inductance through the proposed "cascade filtering" algorithm is shown in Figure 19.6b. It is observed that the coupling coefficient k significantly changes according to the telemetry distance so that it is concluded that this parameter can sufficiently recognize the variations of the telemetry distance between two coils.

The relative magnitudes of the scaling and offset coefficients a and b normalized with respect to $a_{5\,cm}$ and $b_{5\,cm}$ calculated at 5 cm telemetry distance are shown in Figure 19.7, where it is observed that the behaviors of the normalized scaling and offset coefficients are very similar to each other and at the small telemetry distance less than 3 cm, the relative magnitudes remain almost the same but at the large telemetry distance, it considerably increases to compensate for the estimation error.

Figure 19.8 compares the reference capacitance to the new estimated capacitances with single/multiple-frequency components at 2 and 5 cm telemetry distances, which are simply compensated by the linear manner as discussed previously. For the estimation of single-frequency capacitance in Figure 19.8a, the magnitude and the frequency of the capacitance change are 4.5 pF and 115 Hz. The root mean square (RMS) errors of the resulting capacitances obtained at 2 and 5 cm with respect to the reference capacitance turn out to be 3% and 5% over the time interval (5 s), respectively. The estimated capacitance at 5 cm seems to still suffer from a slight drift in comparison to one obtained at 2 cm, which is attributed to the weak coupling effect between the primary and the secondary coils at this distance as shown in Figure 19.6b. Unlike the traditional approaches based on the "grid-dip" methodology, the proposed algorithm successfully detects the capacitance change even at this relatively large telemetry distance with the help of the linear calibration and, furthermore, the whole estimation procedure is carried out in real time after completing the calibration.

For the estimation of multiple-frequency capacitances in Figure 19.8b, the frequencies of the capacitance changes are 50 and 310 Hz and the corresponding magnitude variations are 15 and 30 pF, respectively. The RMS errors of the resulting capacitances obtained at 2 and 5 cm with respect to the reference capacitance turn out to be 11% and 24% over the time interval (5 s), respectively. Compared to the estimated capacitance at 5 cm, the estimated capacitance at 2 cm has the excellent capability of tracking the trend of the harmonic peaks with high fidelity. Although the RMS error is larger than that in the single-frequency capacitance estimation, the proposed algorithm seems to be sufficiently competitive to the traditional approaches, considering that the proposed algorithm is able to measure a huge number of frequency components in the capacitance change.

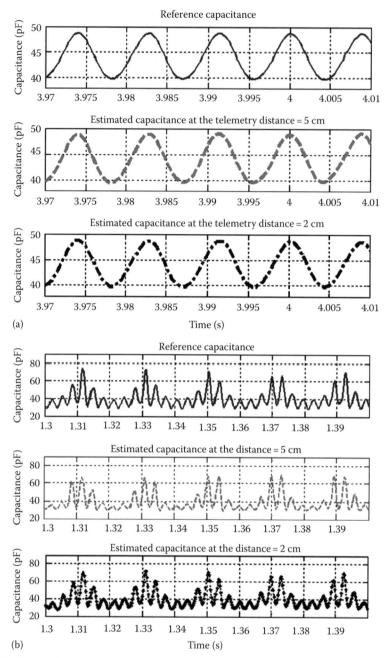

FIGURE 19.8 Comparison of reference and estimated capacitances (a) with single-frequency component and (b) with multiple-frequency components at 2 and 5 cm.

19.5.3.2 Estimation with Varying Telemetry Distance and Orientation

With the same offset and scaling coefficients, additional experiments are carried out to examine the robust performance of the proposed estimation algorithm from the viewpoint of the angle between the primary and the secondary coils.

Figure 19.9 shows the capacitance estimation results corresponding to the single- and multiple-frequency changes, while the angles between the primary and the secondary coils are 50°, 20°, and

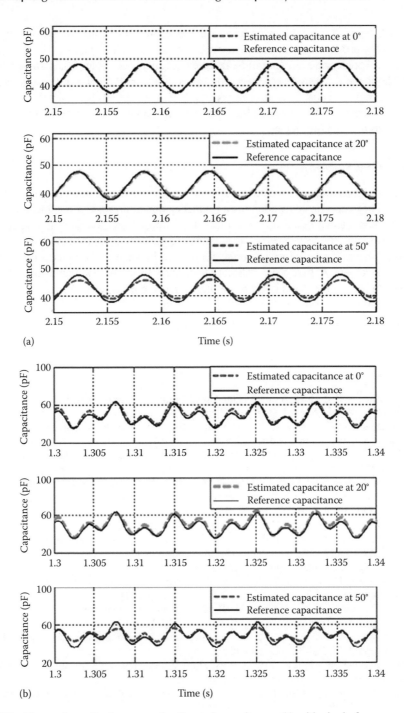

FIGURE 19.9 Comparison of reference and estimated capacitances (a) with single-frequency component and (b) with multiple-frequency components at the angles of 50°, 20°, and 0°.

0° (parallel), respectively, and the telemetry distance for both cases is fixed as 3 cm. The frequencies of the reference capacitance are chosen as 165 Hz for the single frequency change, and 115 and 285 Hz for the multiple-frequency changes, and the magnitudes of the capacitance change for the single- and the multiple-frequency changes are 4 and 15 pF, respectively. As shown in Figure 19.9, the proposed algorithm successfully detects both the single- and the multiple-frequency capacitance changes in real time although the magnitude of the estimated capacitance decreases slightly as the angle between the primary and the secondary coils becomes small. Actually, it is observed that the frequency of the capacitance change can be detected accurately at angles less than 30° even though the magnitude of the estimated capacitance becomes quite small. However, in some applications, detecting the capacitance change is more important than measuring it. For example, in a heart monitor counting the number of heartbeats [19–21], these experimental results establish the viability of the proposed algorithm at oblique angles.

19.6 CHAPTER SUMMARY

A novel real-time capacitance estimation methodology for inductive coupling–based wireless sensor systems has been developed by adopting a frequency domain transfer function approach. Unlike the traditional approaches based on "grid-dip," the proposed methodology makes it possible to continuously estimate a rapidly varying capacitance of a capacitive sensor as well as static or slowly varying capacitances. Furthermore, it is quite simple and inexpensive to practically implement the proposed methodology. A first-order calibration derived from the estimated mutual inductance significantly eliminates estimation errors stemming from parameter uncertainties. The superior performance of the proposed methodology has been successfully established through extensive experiments, thus making it potentially suitable for a variety of real-time wireless applications, including biomedical sensing.

REFERENCES

1. S. F. Pichorim and P. J. Abatti, A novel method to read remotely resonant passive sensors in biotelemetric systems, *IEEE Sensors Journal*, 8(1), 6–11, January 2008.
2. A. D. Dehennis and K. D. Wise, A wireless microsystem for the remote sensing of pressure, temperature, and relative humidity, *Journal of Microelectromechanical Systems*, 14(1), 12–22, February 2005.
3. Sajeeda and T. J. Kaiser, Passive telemetric readout system, *IEEE Sensors Journal*, 6(5), 1340–1345, October 2006.
4. A. S. Sezen, S. Sivaramakrishnan, S. Hur, R. Rajamani, W. Robbins, and B. J. Nelson, Passive wireless MEMS microphones for biomedical applications, *Journal of Biomechanical Engineering*, 127(6), 1030–1034, November 2005.
5. Y. Jia, K. Sun, F. J. Agosto, and M. T. Quinones, Design and characterization of a passive wireless strain sensor, *Measurement Science and Technology*, 17(11), 2869–2876, September 2006.
6. O. Akar, T. Akin, and K. Najafi, A wireless batch sealed absolute capacitive pressure sensor, *Sensors and Actuators A*, 95(1), 29–38, December 2001.
7. A. V. Chavan and K. D. Wise, Batch-processed vacuum-sealed capacitive pressure sensors, *Journal of Microelectromechanical Systems*, 10(4), 580–587, December 2001.
8. S. Guo, J. Guo, and W. H. Ko, A monolithically integrated surface micromachined touch mode capacitive pressure sensor, *Sensors and Actuators A*, 80(3), 224–232, March 2000.
9. T. J. Harpster, B. Stark, and K. Najafi, A passive wireless integrated humidity sensor, *Sensors and Actuators A*, 95(2–3), 100–107, January 2002.
10. A. Pohl, A review of wireless SAW sensors, *IEEE Transactions on Ultrasonics, Ferroelectrics and Frequency Control*, 47(2), 317–332, March 2000.
11. J. Goosemans, M. Catrysse, and R. Puers, A readout circuit for an intra-ocular pressure sensor, *Sensors and Actuators A*, 110(1), 432–438, February 2004.
12. R. Puers, G. Vandevoorde, and D. De Bruyker, Electrodeposited copper inductors for intraocular pressure telemetry, *Journal of Micromechanics and Microengineering*, 10(2), 124–129, June 2000.
13. S. Lizon-Martinez, R. Giannetti, J. L. Rodriguez-Marrero, and B. Tellini, Design of a system for continuous intraocular pressure monitoring, *IEEE Transactions on Instrumentation and Measurement*, 54(4), 1534–1540, August 2005.
14. D. Marioli, E. Sardini, M. Serpelloni, and A. Taroni, A new measurement method for capacitance transducers in a distance compensated telemetric sensor system, *Measurement Science and Technology*, 16(8), 1593–1599, August 2005.

15. M. A. Fonseca, J. M. English, M. von Arx, and M. G. Allen, Wireless micromachined ceramic pressure sensor for high temperature applications, *Journal of Microelectromechanical Systems*, 11(4), 337–343, August 2002.

16. H. S. Kim, S. Sivaramakrishnan, A. S. Sezen, and R. Rajamani, A novel real-time capacitance estimation methodology for battery-less wireless sensor systems, *IEEE Sensors Journal*, 10(10), 1647–1657, October 2010.

17. K. J. Cho and H. H. Asada, A recursive frequency tracking method for passive telemetry sensors, in: *IEEE Proceedings of the American Control Conference*, Denver, CO, June 4–6, 2003, pp. 4943–4948.

18. Zetex semiconductors. Data sheet of ZC930/ZMV930 series. [Online]. Available: http://www.datasheetcatalog.org/datasheet/zetexsemiconductors/930series.pdf

19. J. A. Potkay, Long term, implantable blood pressure monitoring systems, *Biomedical Microdevices*, 10(3), 379–392, June 2008.

20. M. A. Fonseca, M. G. Allen, J. Kroh, and J. White, Flexible wireless passive pressure sensors for biomedical applications, in: *Solid-State Sensors, Actuators, and Microsystems Workshop*, Hilton Head Island, SC, June 4–8, 2006, pp. 37–42.

21. S. Chatzandroulis, D. Tsoukalas, and P. A. Neukomm, A miniature pressure system with a capacitive sensor and a passive telemetry link for use in implantable applications, *Journal of Microelectromechanical Systems*, 9(1), 18–23, March 2000.

20 Inductive Sensor for Lightning Current Measurement Fitted in Aircraft Windows

A.P.J. van Deursen

CONTENTS

20.1 INTRODUCTION

Lightning is a real threat in flight [1], in particular, during take-off and landing. Statistically averaged, every aircraft has a chance of one lightning strike per year. The European project ILDAS [2] aimed to develop and validate a prototype of a system capable to measure and reconstruct lightning current waveform in-flight, to localize the attachment points of the lightning and the current trajectory after strike, and to create a database of events. The main goals are to assess the severity shortly after a lightning hit and to reduce maintenance time after landing. The acronym ILDAS is derived from "In-flight Lightning Strike Damage Assessment System." Many partners contributed to the project [2]. Special mention deserves NLR, the Dutch National Aerospace Laboratory that acted as the project leader and built the on-board data acquisition system. An extensive description is given in [3].

In order to retrieve the lightning attachment points, 12 sensors distributed over the aircraft continuously monitor the local electric and magnetic field. The selection of the number and positions of the sensors is described in [4]. The ILDAS is triggered by sudden changes in the field strength in the aircraft. The sensors then determine the local magnetic fields, retaining pre- and post-trigger data over the time span of about 1 s. After landing, the data are compared with possible field distributions derived for a number of current path scenarios [3], and the most likely entry and exit regions are determined.

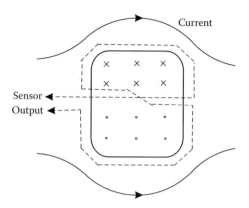

FIGURE 20.1 Principle of sensor windings.

Different sensor types were studied to see whether these would suit the aircraft environment: sensors based on Hall effect, giant magnetoresistance, or anisotropic magnetoresistance. The simple inductive coil emerged as most convenient for this application. Three inductive magnetic field sensors were developed [5,6]. Each sensor is followed by a combined passive/active integrator to redress the frequency characteristic and extend the usable bandwidth to lower frequencies. The first sensor is a small high-frequency (HF) coil. Its bandwidth with an integrator ranges from 100 Hz up to 20 MHz to measure the strokes. This sensor is similar to the one used in [7]. The second sensor is a low-frequency (LF) coil to measure the small continuous current part of the lightning current over the bandwidth of 0.16–100 Hz. It is mounted inside a properly designed shield to discriminate the intense but fast magnetic field of the strokes. Both HF and LF coils with their integrators and data storage units were intended for application outside the aircraft cabin, at the wings or tail airfoils under covers or in fairings.

A new type of sensor was proposed that can be fitted in a window [8] inside the cabin; see Figure 20.1. This sensor, the main topic of this contribution, avoids protrusions that would obstruct the airflow around the fuselage. No signal feed-through passing the hull is needed. The sensor has good sensitivity and large bandwidth. Because of the integrator chosen [5], the overall bandwidth is limited to 100 Hz at the LF side. In this contribution, the analysis of the window sensor proceeds in steps. First, the mathematical model is presented for the sensor mounted in a circular window opening. Second, a numerical method of moments (MoM) model provided the sensitivity of the sensor when placed around an opening in the bottom of a large rectangular transverse electromagnetic field (TEM) cell. Finally, the actual window replaced the circle in the MoM model, and the window mounting flanges were included.

Also, the experimental verification of the window sensor was done in steps. Introductory measurements have been performed in the laboratory: on a large tube with a hole in the wall. Preliminary tests have been carried out on a Nimrod aircraft [9], followed by a series of tests on a fuselage mock-up at Cobham, UK. Both tests will not be discussed here. The ground tests of the full ILDAS system have been performed in June 2009 on an A320 Airbus in Toulouse (France). Seventeen of 18 tests with current injection and return at different positions were correctly identified by ILDAS [3]. Comparison of the magnetic field data from the window sensor with the MoM model showed that an accuracy within 3% is attainable. ILDAS requirements were 10%. A reduced version of the ILDAS with two window sensors has been flown in March 2011 on board of an A340 test aircraft. The system faithfully recorded a large number of lightning strokes.

20.2 OPERATING PRINCIPLE

Figure 20.1 shows a window as a rounded rectangle and a lightning current pattern in the fuselage around the window. If the current pattern is symmetrical around the window, a part of the corresponding magnetic field enters the top half of the window and leaves through the lower half.

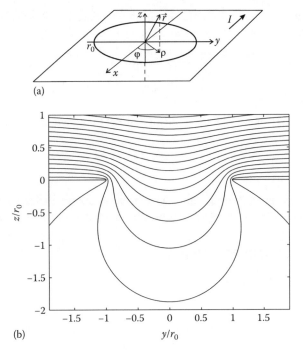

(a)

(b)

FIGURE 20.2   (a) Cartesian and cylindrical coordinate system with respect to a circular hole of radius r_0 in a plane. (b) Sketch of field penetration through the circular hole. Above the plane, the magnetic field is homogeneous and equal to H_0 at a large distance r.

A sensor coil is wound in the shape of a flattened figure "8." The sensor wire follows the central bar twice. The outer perimeter is close to the fuselage at some distance from the window. Such a coil captures the magnetic flux entering and leaving the window and sums the induced voltages in each half. The sensor output is proportional to the magnetic field H_0 that would exist parallel to the fuselage without a window. For a (large) circular fuselage of circular cross section with radius R_f and a lightning current I, one has the simple relation

$$H_0 = \frac{I}{2\pi R_f}. \tag{20.1}$$

Since R_f is much bigger than the window dimensions, the curvature of the fuselage can be neglected. An infinite conductive plane with a circular hole can be assumed to have a homogeneous magnetic field H_0 at one side for a large distance, as shown in Figure 20.2.

20.3   MATHEMATICAL APPROACH

Let us consider a circular hole with the radius r_0 instead of the rectangular window. The fuselage is regarded as a perfect conductor, i.e., no magnetic field penetrates the wall. The coordinate system employed is shown in Figure 20.2a. The current flows in the $-x$-direction at large distance from the hole. Bethe [10] analyzed the behavior of quasi-static field near a circular hole in an infinitely large wall of infinite conductivity but zero thickness. The configuration has been reanalyzed by many authors; see, e.g., [11]. Applications range from hole coupling between waveguides to lenses for charged particle beams. Earlier, F. Ollendorff gave an analytical solution for the scalar magnetic potential X [12, Sect. XIV] in terms of elliptical coordinates, as mentioned in [13, Ch. H]. Ollendorff's book is not easily available and—to the author's knowledge—has not been translated into English. In order to remedy the situation, a compact derivation has been reproduced in [8, Sect. III]. In the plane of the hole ($\rho < r_0$, $z = 0$), the resulting perpendicular component H_z of the magnetic field varies as

$$H_z = \frac{2H_0}{\pi} \frac{\rho}{\sqrt{r_0^2 - \rho^2}} \sin\varphi. \tag{20.2}$$

The in-plane component is homogenous, oriented parallel to the y-axis, and of magnitude

$$H_y = \frac{H_0}{2}. \tag{20.3}$$

With (20.2) and (20.3), one can study how the sensor sensitivity varies with the position of the central wire of the figure "8" with respect to the hole, in a first order approximation.

20.4 SELECTED SENSOR POSITIONS

Three positions of the figure "8" coil will be analyzed:

1. A flat sensor in the plane of the fuselage
2. A sensor with the central bar shifted inside the aircraft over the distance d; Figure 20.3a
3. A flat sensor completely shifted over the distance d; Figure 20.3b

The different shapes allow us to study the changes of the sensor sensitivity when obstructions prohibit the optimal shape 1.

20.4.1 FLAT SENSOR IN THE PLANE

With the sensor against the fuselage plane, there is no flux between the outer perimeter of the figure "8" and the fuselage. The magnetic flux through a sensor of radius $\rho_s < r_0$ is obtained by integrating H_z over the quarter circle and multiplying by 4:

$$\Phi_{\rho_s} = \frac{4\mu_0 H_0 r_0^2}{\pi} \left[\arcsin\frac{\rho_s}{r_0} - \frac{\rho_s}{r_0} \sqrt{1 - \frac{\rho_s^2}{r_0^2}} \right]. \tag{20.4}$$

For a sensor with $\rho_s \geq r_0$, one arrives at

$$\Phi_0 = 2\mu_0 H_0 r_0^2. \tag{20.5}$$

The square root term in (20.4) causes a fast decay in sensitivity for ρ_s slightly smaller than r_0. This result holds for an infinitely thin plane. Changes at the edge, for instance, a small rim, have large influence on Φ_0. On the other hand, Φ_0 does not depend on ρ_s as long as it is larger than r_0 and the coil wire is against the fuselage. The sensitivity does not strongly vary with the position of the central bar near the line $y = 0$ in the $z = 0$ plane because the H_z is zero there. For a displacement of the central bar by the amount e in the y direction, the sensitivity varies as

$$\frac{\Phi(e)}{\Phi_0} = 1 - \left(\frac{e}{r_0} \right)^2, \tag{20.6}$$

which is easily verified by rewriting (20.2) in Cartesian coordinates.

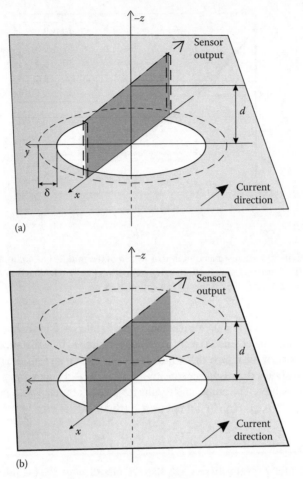

FIGURE 20.3 (a) The sensor (dashed line) against the fuselage, with the central bar shifted over the distance d and extended over the length δ. (b) The sensor fully lifted. The drawings show the inside of the fuselage, and the coordinate system is placed upside down. Please note that the coordinate system is reversed with respect to Figure 20.2.

20.4.2 CENTRAL BAR SHIFTED

As shown in Figure 20.3a, the central bar can be shifted to the position $z = -d$. Assume that vertical leads at $\rho = r_0$ connect the wires in the bar with the perimeter of the sensor. A quick estimate of the reduction in sensitivity is possible. The horizontal field H_y near the central bar is equal to $H_0/2$ as mentioned in (20.3). For small d, the flux through the rectangle formed by the central bar at $z = -d$ and at $z = 0$ is $\mu_0 H_0 r_0 d$, which is missed twice by the sensor. A first approximation of the sensitivity reduction is

$$\frac{\Phi_{1a}(d)}{\Phi_0} \approx 1 - \frac{d}{r_0}, \tag{20.7}$$

which overestimates the actual reduction at larger d. When the central bar is shifted over the distance d, the flux Φ_m not seen by the figure "8" sensor is

$$\frac{\Phi_m(d,\delta)}{\mu_0} = 4\int_0^d dz \int_0^{r_0+\delta} H_\varphi d\rho, \tag{20.8}$$

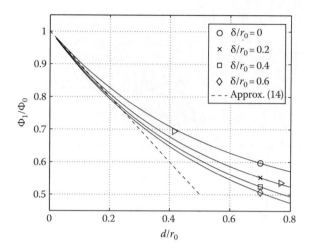

FIGURE 20.4 Sensitivity of the sensor as function of shift d of the middle bar, for four different extensions δ of the bar. When $d \to 0$, all curves merge with the approximation (20.7). The markers at $d/r_0 = 0.7$ are indicators for the legend. The markers $\triangleright$ are discussed in Section 20.5.

where $H_\varphi = \partial X / \partial \varphi$ evaluated at $\varphi = 0$. We used an upper limit $r_0 + \delta$ in the integration over r, since the sensor central bar may be extended over the window radius r_0 by a distance δ; see Figure 20.3a. The outer perimeter wire of the figure "8," which is always placed against the fuselage, is increased similarly. Figure 20.4 shows the resulting sensitivity $\Phi_1 = \Phi_0 - \Phi_m$ as a function of distance d for four values of the extension δ. The numerical results have been obtained with mathematica [14]. The divergence of the field at $(\rho, z) = (r_0, 0)$ did not pose particular problems in the integration.

20.4.3 SHIFTED SENSOR

If the full sensor is shifted over the distance d, also the flux through the cylinder between the sensor outer perimeter and fuselage is missed. The sensitivity can now be partially restored by increasing the outer perimeter to the radius $r_0 + \delta$. It is convenient to calculate the flux through the sensor directly via the quarter circle:

$$\frac{\Phi_2(d,\delta)}{\mu_0} = 4 \int_0^{\pi/2} d\varphi \int_0^{r_0+\delta} H_z \rho \, d\rho. \qquad (20.9)$$

The resulting Φ_2 is shown in Figure 20.5. The sensitivity of the sensor decreases approximately twice as fast with increasing d as the previous sensor with only the central bar shifted (Figure 20.4) because of the intense local field near the circle edge, midway between the central bar ends; see Figure 20.2b. The field divergence causes a substantial loss for a sensor that remains even close to the edge. The flux missed by the sensor is approximately proportional to $\sqrt{d}$, as shown by the steep increase of the curve for small d and $\delta = 0$, which is similar to (20.4).

20.5 TUBE TESTS

A brass tube of $r_t = 0.45$ m diameter and 1.5 m length acted as fuselage mock-up. The tube is used inside out, as it is excited internally by a thin wire near the axis with the tube as a return. A 82 mm diameter hole was made at mid-length in the tube. A single turn version of the window sensor was mounted; see Figure 20.6. The sensor wire was at the outside and was mounted midway over the window at

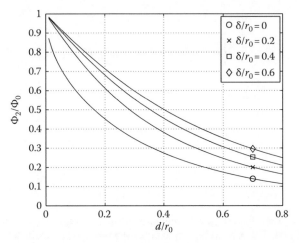

FIGURE 20.5 Sensitivity of the sensor as function of shift d of the whole sensor, for four different extensions δ of the sensor radius. Please note that the sensitivity drops faster than in Figure 20.4, in particular for $\delta = 0$.

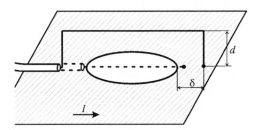

FIGURE 20.6 Single turn version of the window sensor, made by a coaxial signal cable extending the inner lead over the hole.

three distances d: 0, 17, and 32 mm or $d/r_0 = 0$, 0.42, and 0.76, respectively. The distance δ was 8.5 mm or $\delta/r_0 = 0.2$. The tube also acted as return for the measuring circuit. We measured the induced sensor voltages caused by an excitation current of $I = 2.1$ A at 500 kHz through the inner lead: U_0, U_{17}, and U_{32}. The ratios of the voltages U_{17}/U_0 and U_{32}/U_0 are plotted in Figure 20.4 by the markers $\triangleright$.

In a second set of measurements, we determined the coupling between the inner circuit and the sensor by a network analyzer HP 4396A with the S-parameters set HP 85046A over the frequency range up to 10 MHz. The sensor wire was at the distance $d = 1.5$ mm from the tubes cylindrical surface. Figure 20.7 shows the coupling as transfer impedance Z_{12}. The experimental coupling inductance M_e has been determined by a fit to $|Z_{12}| = \omega M_e$ between $f_1 = 0.9$ MHz and $f_2 = 5$ MHz. This range was chosen because of the noise below f_1 and the steeper increase above $2f_2$ caused by the onset of traveling wave effects in the tube. Approximate analytical expression of M_a follows from the inside field $H_0 = I/2\pi r_t$ without the hole:

$$M_a = \frac{\mu_0}{2\pi r_t}\left(r_0^2 - d \times r_0\right), \tag{20.10}$$

which assumes that the tube wall can be considered flat near the window. The resulting Z_{12} is displayed by the solid straight line in Figure 20.7. Alternatively, one can express the sensitivity in effective flux capturing area A: $V = \mu_0 A \partial H_0 / \partial t$. The agreement between M_e and M_a is within 7%; see Table 20.1. This is acceptable if one considers the limitations of the simple model, the mechanical tolerances on the actual tube, and the accuracy of the analyzer and S-parameters set (0.4 dB, equivalent to 5%).

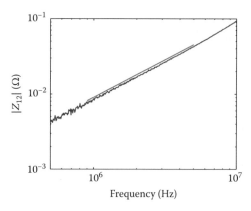

FIGURE 20.7   Coupling impedance $|Z_{12}|$ between the excitation circuit inside the tube and the sensor, measured by a network analyzer. The straight line represents the analytical approximation ωM_a.

TABLE 20.1

Values for M in nH and A in Units r_0^2

M_e	A_e	M_a	A_a	M_m	A_m
1.34	0.90	1.440	0.933	1.455	0.974

Subscript e stands for experimental, a for analytical, and m for MoM model.

The tube wall thickness was 1.5 mm; the finite thickness rounds the edge divergence and reduces the flux. We approximated the actual wall surface near the edge by an equi-flux surface with a smallest radius of curvature equal to half the wall thickness. The sensor sensitivity was reduced by about 15% for $d = 0$. This approach clearly overestimates the edge effect.

Also, the position of the current lead inside the tube turned out to be influential. An offset o from the axis in the direction of the sensor causes the field variation $\Delta H_0/H_0 = 2o/r_t$ at the inside wall of the tube. A 1 mm larger distance from the hole would cause approximately a 1% smaller M.

In order to determine the influence of the limited diameter of the tube, we modeled the tube with hole in the FEKO MoM approach [15]. On the tube, the triangular element sides were 3 cm, which made the set of triangles deviate from the cylindrical surface by 0.5 mm at most. The triangle sides decreased to 1 mm near the edge of the hole to deal with the diverging current and charge density there. As excitation, we used two voltage sources of opposite phase at the ends of the inner wire. This arrangement ensured a predominant inductive E-field near the sensor. A fixed frequency of 2 MHz was selected. Figure 20.8 shows the amplitude of E_z and of current density over the relevant surfaces. Of course, the E and H fields are $\pi/2$ out of phase. The sensor voltage was then determined by trapezoidal integration of the electric field over the line at 1.5 mm from the hole center, extending over 8.5 mm as in the measurements. Table 20.1 also shows the resulting coupling inductance $M_m = |V/I\omega|$. The 1% agreement with M_a shows that the curvature of the tube had little influence on the window sensor. As an additional check, we observed that the parallel component of the magnetic field H_y is near to constant and equal to $H_0/2$ to within 1% over the central line of the hole, in agreement with (20.3).

20.6   A320 TESTS ON THE GROUND

An A320 aircraft stood above a metal netting on the floor of a hangar. The netting had the shape of a cross under fuselage and wings and acted as ground plane (GP) and return for the current. Polyethylene sheets insulated the GP with respect to the floor. The discharge of a high-voltage

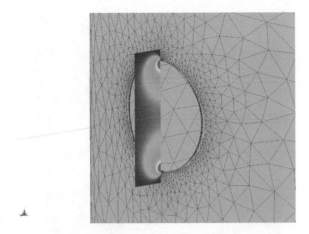

FIGURE 20.8   Meshing, current density in the tube and z-component of the electric field in the rectangular portion of the $y = 0$ plane near the hole. The tube axis extends along the z-direction; the x-axis is normal to the plane of the hole.

capacitor provided the current. The current waveshape approximated the stroke A—component mentioned in [16]. The amplitude was reduced to about 3 kA because the purpose of the test was to demonstrate the correct operation of the ILDAS system and not to establish the aircraft safety. A Pearson probe recorded the injection current. The aircraft was instrumented (Figure 20.9) with a mixture of ILDAS sensors and magnetic field sensors provided by the ILDAS partner Cobham Technical Services [17]. The magnetic field data were recorded by the ILDAS system and by digital scopes.

Eighteen scenarios were tested, with different current entrance and exit points; the successful outcome of the current pattern recognition has been presented in [3]. In this contribution, two measurements with the window sensors in the main current path are analyzed: current injected at the nose and retrieved at the vertical tail fin as in records 10 and 11 of the series taken on June 9, 2009. Figure 20.10 shows the window sensor as used.

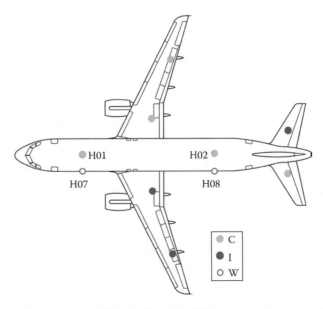

FIGURE 20.9   A320 with 12 sensor positions indicated. C: Cobham inductive sensors, H01 is on top, H02 at the bottom of the aircraft. I: Ildas LF and HF sensors. W: window sensor H07 and H08.

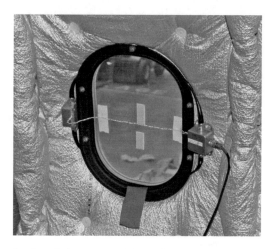

FIGURE 20.10 Photograph of a window sensor. Two twisted wires form the central bar. The boxes provide connections and termination of the coaxial cables that act as outer perimeter of the figure "8." The cables are mostly laid in a region of low magnetic flux, between fuselage and the top rim of the window pane mounting flange.

20.6.1 Waveform

The inductive sensors, coils, and window deliver a signal proportional to the time derivative of the magnetic field. The signals of the Cobham inductive sensors are integrated numerically in the scopes. A combination of passive and active integrators restored the waveform of the inductive ILDAS sensors. The passive integrators were the first stage. These also filtered frequencies higher than 20 MHz, the upper limit of the band of interest [9]. Active integrators extended the frequency range down to 100 Hz. The signal acquisition chain contained three high pass filters, two at 100 Hz in the integrator and one at 85 Hz before the analogue to digital converter (ADC). Because the frequency band of interest for the A-component is far above the three cross-over frequencies, the response of the window sensor was only slightly affected by the high-pass filters, and a simple procedure redressed the signal. Three time-domain filters, each with a frequency domain representation $1 + 1/j\omega\tau_i$, corrected the recorded window sensor data. Figure 20.11 shows the original response for window sensor H08, the corrected response, and the current scaled to the same amplitude and shifted in time for optimal coincidence. The current and corrected response coincide within the noise of the data, even after smoothing has been applied to reduce digital noise. At the end of the record, the difference amounts to 1% of the maximum; this can be readily attributed to the limited resolution of the input before correction. The data for sensor H07 give similar results but are not shown here. At frequencies of interest here, the metal parts around the window sensor can be considered impenetrable for magnetic fields. This allows the analysis described in the next section.

20.6.2 Sensitivity

The interpretation of the window sensor data requires two steps. First, the magnetic field pattern outside the aircraft placed above the GP needs to be known. The nearby return concentrates the current distribution at the bottom of the aircraft fuselage (ACF). The concentration depends on the width of the GP. EADS modeled the aircraft by an finite difference time domain (FDTD) method, which included the aircraft metallic and non-conducting parts. The FDTD mesh could be rather course compared to the window size, because the magnetic field H_0 at the window position was required, without the window opening being present. The FDTD input data are EADS proprietary information, and this calculation is not discussed here further. A simpler method

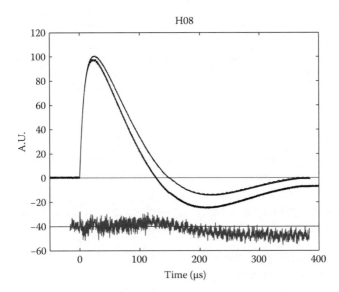

FIGURE 20.11   Output of the integrator for window sensor H08 in arbitrary units, before (lower curve, heavy due to digital noise) and after (upper thin curve) correction for the time constants, in comparison with the current. Corrected output and current are barely distinguishable on this scale. The difference between these is shown, multiplied by a factor of 10 and shifted over − 40.

relies on conformal mapping (CM); for details, the reader is referred to the Appendix of [18]. In this method, the ACF is represented by a well-conducting tube above a conducting GP of finite width. The diameter of the ACF is 3.95 m, the GP under the fuselage was 13.2 m wide, and the distance between ACF bottom and GP was 1.70 m.

Second, the penetration of the magnetic field through the window opening and its mounting need to be calculated accurately for reasons mentioned in Section 20.4. The sensor is thought to be mounted at the outside of the bottom of an existing TEM cell [19], which replaced the aircraft. The TEM cell limits the calculation effort. It is a closed structure except for the window opening and it is excited internally. This guarantees that the effects of the field penetration through the window model prevail. The rectangular measuring section of the TEM cell is 0.91 m long in the direction of current flow, 1.40 m deep, and 0.90 m high. The current carrying septum is asymmetrically placed at the height of 0.72 m, and it is designed to give the TEM cell a wave impedance of 50 Ω. The TEM cell has been modeled by the FEKO MoM [15]. Two voltage sources at 2 MHz excite the TEM cell at both ends of the septum. The choice of frequency depends on two factors: wave length effect should not interfere in the quasi-static regime, and the ratio of scalar to vector potential contributions must remain manageable with double precision arithmetics. The phases of the voltage sources are chosen opposite in order to enhance the induced electric field over the static field. The variation of H_0 in the x direction is negligible within the window area. The presence of the side walls causes a parabolic variation of the H_0 field in the y direction parallel to the long axis of the window. At the window edge, the field is 6% smaller than at the center. Increasing the size of the TEM cell by a factor of 1.5 decreases this variation to 2%, a factor of 2 even down to 0.7%. The 2% variation is acceptable in view of the measurement accuracy. The corresponding scale factor of 1.5 also keeps out-of-core computation time manageable, within 22 h on a 3 GHz, quad core PC with 8 GB memory. The magnetic field H_0 at the bottom is then 199 A/m per kA of current, in the absence of the window as shown in Figure 20.12.

A circular window of radius $r_0 = 0.165$ m is calculated first and acts as "calibration" because the analytical expression (20.5) for the sensitivity is available. Actual windows have the shape of a rounded rectangle of 0.33 m × 0.23 m. The long axis stands perpendicular to the current

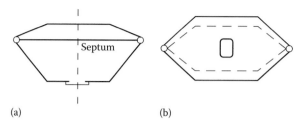

(a)   (b)

FIGURE 20.12 (a) Mid-body cross section of TEM cell, with path of the sensor lead indicated by the thin line at the bottom. (b) Bottom view of the TEM cell with window opening. The septum is indicated by the dashed lines. The circles at the septum ends represent the current sources.

flow, as in the fuselage. Its length is twice the circle radius r_0. Several approximations of the conducting parts of the window mounting have been studied:

a. A window-sized opening in a flat panel
b. With additional U-shaped mounting flange
c. With mounting flange approximating the actual shape

The cases a, b, and c refer to the shapes shown in Figure 20.14. The sides of the triangular MoM elements are 5 cm up to 7 cm on the body of the TEM cell and reduce to 1 mm near the sharp edges of the window opening. The number of elements varies between 20×10^3 and 60×10^3. Meshing, current density, and intensity of the electric field for case c are shown in Figure 20.13. The induced voltages are obtained by integrating the electric field in the middle plane over the thin dashed lines indicated in Figure 20.14. Integration occurs by the trapezium rule up to 5 mm from sharp edges. In order to deal with the diverging nature of the fields near those edges, field values between 1 and 5 mm from an edge are fitted to the expected $1/\sqrt{r_0^2 - \rho^2}$ behavior—see (20.2) for the circle—and integrated analytically. A similar fit is used for the windows. The voltages can be converted into sensitivity of the sensor, expressed as effective flux capturing area A:

$$V = j\omega\mu_0 A H_0,$$ (20.11)

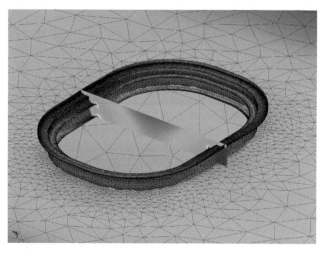

FIGURE 20.13 Meshing, intensity of the electric field and current density near the window, shown upside down. The window mounting cross section is case c in Figure 20.14.

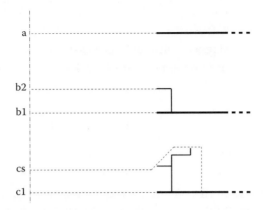

FIGURE 20.14   Three mountings of the window sensor, shown half but with correctly scaled shape. The heavy horizontal lines represent the fuselage extending to the right, thin lines the mounting. Inside is up, outside down. The vertical dash-dot line is the center line of the window. Thin dashes indicate the path of integration to obtain the induced voltage. The upper thin dashes in case c are a good approximation of the actual wire path shown in Figure 20.10.

TABLE 20.2
Values for Effective Flux Capturing
Area A of the Sensor, in Units of cm²

Path	1	2	Sensor
$r = 16.5$ cm	271.9	—	—
Sensor a	240.2	—	—
Sensor b	217.0	192.4	—
Sensor c	202.7	183.3	184.2

The field integration paths are indicated in Figure 20.14.

---

with μ_0 the permeability of free space. Table 20.2 summarizes the values for A. The effective area of the circle agrees within 0.2% with the analytical expression $r_0^2 = 272.3\,\text{cm}^2$. Because of the agreement, there was no need felt to sacrifice the TEM cell and actually make the opening in the bottom. The value of A for case a is the maximum attainable for the window. As the comparison of cases b1 and b2 shows, the presence of a second rim deeper inside reduces the sensitivity appreciably. The path c2 is similar to b2, but it is not shown in Figure 20.14 for the sake of clarity. The actual sensor has about 77% of the sensitivity of case a without flanges. Based on the variation between cases b2 and c2, the accuracy is estimated to be better than 3%.

20.6.3   MAGNETIC FIELD MEASUREMENTS ON THE A320

The measurements were taken with an injected current of 3.11 kA maximum value. Table 20.3 gives the corresponding magnetic field maxima. Those for the window sensors are based on the effective area of Table 20.2 and have been corrected for the signal transfer as discussed in Section 20.6.1. Both recordings 10 and 11 show consistent results. Table 20.3 also includes the magnetic fields determined by the CM method. The H01 field is within 2%, the H02 within 6% of the CM value. The ratio of H02 and H01 is of interest here. It depends sensitively on several factors: the radius of

TABLE 20.3
Magnetic Field at Four Sensors for
an Injection Current $I_{inj} = 3.11$ κA

	H01	H02	H07	H08
Rec. 10	114	466	148	158
Rec. 11	114	464	147	158
CM	119	503	167	167
FDTD	—	—	141	—

the aircraft, the distance between GP and fuselage bottom, and the actual width of the GP. The latter is prone to inaccuracy, since the assumed GP may be effectively widened by currents induced in the reinforcement of the concrete floor. The actual ratio of 4.08 compared reasonably well with the calculated value by CM: 4.23. An infinitely wide GP would result in a ratio of 3.32, a thin wire in a ratio of 11.

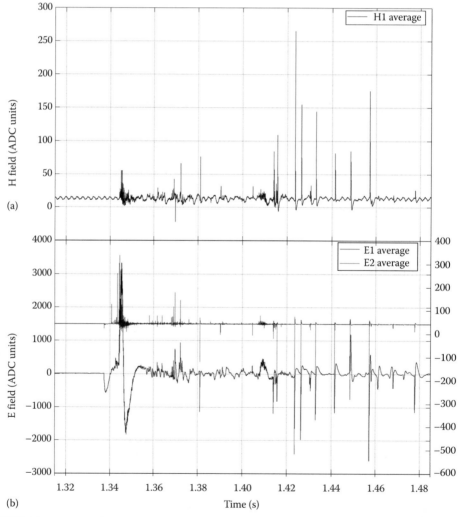

FIGURE 20.15 Record of several strokes through the magnetic field (a) and electric field (b), both in arbitrary units. The lower and upper curves for the electric field are data split for the low and high frequency channels.

The position of the window center is 2.56 m above the ACF bottom. The window sensor, shown in Figure 20.10, has a figure "8" configuration, as discussed in Part I. The effective areas of Table 20.2 should then be multiplied by a factor of 2. The measured H07 field is 12% less than calculated by CM, the H08 field 5%. This agreement is acceptable, in view of the limitations of the CM method, which fully omits the wings and GP thereunder. The measured value for the H07 sensor [20] deviates by about 5% from the FDTD value, calculated for a fuselage diameter of 4.1 m. If the diameter would be scaled to the actual 3.95 m through CM, the field would become 144 A/m, and the agreement would even be better.

20.7   A340 FLIGHT TESTS

After successful completion of the EU FP6 project ILDAS, NLR and Airbus decided to further develop the system in the project ILDAS-2. With a reduced version of the data acquisition system developed for ILDAS, test flights have been carried out early 2011 with an A340 aircraft. During these flights, active thunderstorms were deliberately sought and crossed. Only two sensors were installed in windows, a sensor for the electric field described in [21] and the magnetic field sensor discussed before. The magnetic field sensor had only a single wire at mid-height of the window. Both windows were located halfway between the nose and the leading edge of the wing. Many strikes were recorded. A selection has been presented at the ICOLSE 2011 conference [22]. Figure 20.15 shows a record consisting of a large number of strokes, measured by the magnetic field (Figure 20.15a) and electric field (Figure 20.15b) variations. The time axis is in seconds, and both vertical axis in arbitrary ADC units. In ILDAS-2, it has been decided to uniquely use the window sensors inside the cabin, because of the ease of installation and the comfortable environment. The lack of sensors on the wings and tail airfoils reduces the information for the determination of the lightning attachment points. It is studied now whether this lack can be compensated by a more elaborate inversion process of magnetic field data into attachment points.

20.8   CONCLUSIONS

The window sensor accurately determines the wave shape of the magnetic field outside the aircraft. Details of the window edge and mounting influence the sensitivity, and these have been considered carefully in an MoM approach with fine local mesh. With the aircraft on the ground, the accuracy of the measurements also depends on the current distribution in nearby conductors such as the floor under the GP. Such items are difficult to model. Still, the recorded magnetic field amplitudes are consistent with the other sensors used in the tests on the A320. The window sensor measurements agree within the ILDAS requirement of 10% well with the FDTD value available and reasonably well with a simple CM model for the aircraft. This accuracy shows that the window sensor is well suited for the current pattern recognition on aircraft during lightning strike, the final goal of the ILDAS system. The main benefit of this sensor is the simple mounting, which does not require components extending outside the fuselage.

Applications are not limited to aircraft. Lightning current measurements on tower for television or for windmills often rely on Rogowski coils. Ideally, an R. coil fully encircles the tower at the outside, and its output is then strictly proportional to $\partial I/\partial t$. Being outside the tower, the R. coil itself is exposed to lightning. A large single coil is often inconvenient, and it is replaced by a number of smaller segments, each sensitive to the local magnetic field. The signals from the segments are then added. In such a case, the window sensor may also be used on any available opening in the tower. Other applications can be imagined as well: coaxial structures, such as those that occur in pulsed power setup. If even minute openings in the wall were available, the sensor could be placed there.

ACKNOWLEDGMENT

The author thanks all ILDAS partners [2], in particular, A. de Boer and M. Bardet at NLR (The Netherlands), J. Hardwick at Cobham Technical Services (United Kingdom), I. Revel at EADS Innovation Works (France), and F. Flourens at Airbus Industries (France) for sharing the procedures, measurement and model data, photographs of the June 2009 measurement campaign, and measurement data of the 2011 test flights.

REFERENCES

1. M. Uman and V. Rakov, The interaction of lightning with airborne vehicles, *Progr. Aero. Sci.*, 39, 61–81, October 2003.
2. EU FP6-project ILDAS, National Aerospace Laboratory NLR, Anthony Fokkerweg 2, 1059 CM Amsterdam, The Netherlands. [on-line] http://ildas.nlr.nl/.
3. R. Zwemmer, M. Bardet, A. de Boer, J. Hardwick, K. Hawkins, D. Morgan, M. Latorre et al., In-flight lightning damage assessment system (ILDAS); results of the concept prototype tests, in: *International Conference on Lightning and Static Electricity, ICOLSE*, Pittsfield, MA, 2009.
4. S. Alestra, I. Revel, V. Srithammavanh, M. Bardet, R. Zwemmer, D. Brown, N. Marchand, J. Ramos, and V. Stelmashuk, Developing an in-flight lightning strike damage assessment system, in: *International Conference on Lightning and Static Electricity, ICOLSE*, Paris, France, 2007.
5. V. Stelmashuk, A. van Deursen, and R. Zwemmer, Sensor development for the ILDAS project, in: *EMC Europe Workshop*, Paris, France, June 2007, p. on CD.
6. A. van Deursen and V. Stelmashuk, Sensors for in-flight lightning detection on passengers aircrafts, in: *ESA Workshop on Aerospace EMC*, Florence, Italy, March 30–April 1, 2009.
7. P. Lalande, A. Broc, P. Blanchet, S. Laik, S. Luque, J.-A. Rouquette, H. Poirot, and P. Dimnet, *In-Flight Lightning Measurement System: Design and Validation*. Toulouse, France: Onera, EM-Haz-Onera Rep05, 2003.
8. A. van Deursen and V. Stelmashuk, Inductive sensor for lightning current measurement, fitted in aircraft windows. Part I: Analysis for a circular window, *IEEE J. Sensors*, 11, 199–204, 2011.
9. V. Stelmashuk, A. van Deursen, and M. Webster, Sensors for in-flight lightning detection on aircraft, in: *Proceedings of International Symposium on Electromagnetic Compatibility (EMC Europe '08)*, Hamburg, Germany, September 2008, pp. 269–274.
10. H. Bethe, Theory of diffraction by small holes, *Phys. Rev.*, 66, 163–182, October 1944.
11. J. Jackson, *Classical Electrodynamics*. Chichester, U.K.: Wiley, 1999.
12. F. Ollendorff, *Potentialfelder der Elektrotechnik*. Berlin, Germany: Springer, 1932.
13. H. Kaden, *Wirbelströme und Schirmung in der Nachrichtentechnik*. Berlin, Germany: Springer, 1959.
14. Wolfram Research, IL, US. [on-line] http:www.wolfram.com.
15. EM Software & Systems-S.A. (Pty) Ltd, 32 Techno Avenue, Technopark, Stellenbosch 7600, South Africa. [on-line] http://www.feko.info/.
16. EUROCAE WG-31 and SAE committee AE4L, *ED-84 Aircraft Lightning Environment and Related Test Waveforms Standards*. Paris, France: EUROCAE, 1997.
17. Cobham plc, Brook Road, Wimborne, Dorset, BH21 2BJ, UK. [on-line] http://www.cobham.com/.
18. A. van Deursen, Inductive sensor for lightning current measurement, fitted in aircraft windows. Part II: Measurements on an A320 aircraft, *IEEE J. Sensors*, 11, 205–208, 2011.
19. M. L. Crawford, Generation of standard EM fields using TEM transmission cells, *IEEE Trans. Electromagn. Compat.*, EMC-16, 189–195, November 1974.
20. M. Latorré and I. Revel, ILDAS: Reconstruction d'un courant foudre par méthode inverse utilisant des simulation FDTD, in: *CEM 2010*, Limoges, France, April 7–9, 2010, pp. E2–E4.
21. P. Blanchet, ILDAS-WP3 E sensor and shunt voltage sensor test results, ILDAS Internal Report, 2008.
22. A. de Boer, M. Bardet, C. Escure, G. Peres, V. Srithammavanh, K. Abboud, T. Abboud et al., In-flight lightning damage assessment system (ILDAS): Initial in-flight lightning tests and improvement of the numerical methods, in: *International Conference on Lightning and Static Electricity, ICOLSE*, Oxford, U.K., September 2011.

21 Technologies for Electric Current Sensors

G. Velasco-Quesada, A. Conesa-Roca, and M. Román-Lumbreras

CONTENTS

21.1 INTRODUCTION

When electric charges are the physical carriers of energy or information, any operation related to energy or information processing involves the movement of these charges. The two magnitudes utilized to represent this movement are the difference in electric potential and the electric current.

The difference in electric potential is related to the energy implied in the movement of electric charges between two points and the electric current corresponds to the physical phenomenon caused by the motion of electric charges. The voltage and the current intensity are the magnitudes utilized to measure these two physical properties.

From the definition of electric current, it follows that the volume of moving charge per time unit and the direction of this movement are two important parameters in the current intensity measurement. These parameters are not completely independent and allow the classification of electric currents into two major groups: alternating current (AC) and direct current (DC).

Electric currents are defined as AC when their average value is zero. To characterize these currents, the root-mean-square (RMS) value is often used, and it is determined from the current instantaneous values using the expression for "square root of the mean squared" value calculation.

Electric currents are defined as DC when their average value is nonzero. In this case, two new kinds of current can be differentiated in accordance with their instantaneous value evolution. A DC current is defined as constant when all of its instantaneous values are equal, and it is defined as pulsating when its instantaneous values are different.

It is usual to classify electric current meters according to the basic principles or physical properties on which their operating principle is based. On the basis of the above, next sections are devoted to a short description of these operating principles and the main characteristics of current meters based on these features.

21.2 CURRENT SENSORS BASED ON OHM'S LAW

Ohm's law states that the voltage across a resistor is proportional to the flowing current intensity. The component utilized to exploit this principle is known as current shunt, and, basically, it is a low-resistance precision resistor.

Shunts are low-cost and robust components, they can measure DC and AC simultaneously, but they present some important disadvantages, which are that the circuit under measurement must be interrupted to insert the shunt and it is electrically connected to the measuring circuit, shunt power losses can lead to great heat dissipation, and they are bulky and heavy devices for large current measurement.

Current shunts are usually described by the following:

- *Rated current*: Maximum current through the shunt without damaging it.
- *Output voltage*: Voltage across the shunt produced by the rated current. Typical values are 50, 60, 100, 120, and 150 mV.
- *Resistance accuracy*: Variation margin of the shunt nominal resistance value. Common accuracy values are ±0.1%, ±0.25%, and ±0.5%.
- *Resistance drift*: Shunt's resistance changes, expressed as a percentage or in parts per million, per °C of temperature change.
- *Rated power*: Maximum power that can continuously dissipate through the shunt without damaging it or adversely affecting its resistance. In general, this value can be calculated as two-thirds of the rated current.
- *Power derating*: Indicates how the shunt continuous current value is derated as a function of the ambient temperature. Typically, a shunt can be used at full continuous current if the ambient temperature is 0°C, but only at 40% of full rating with 100°C of ambient temperature.

The frequency response of shunts is related to their self-inductance and the mutual inductance between loops built by sense wires and the main current circuit. Although these parasitic inductances affect the magnitude of the shunt impedance at relatively high frequencies, their effects on the current phase at medium or low frequencies are sufficient to cause significant errors. Another important factor affecting negatively the frequency response of shunts is the skin effect, because this effect increases the shunt resistance for high-frequency currents.

Shunts can be constructed with two, four, or five terminals. Two-terminal shunts (Figure 21.1a) are often utilized in applications for measurements of currents lower than 100 A, such as portable multimeters.

Four-terminal shunts (Figure 21.1b) can be utilized to measure continuous currents up to 100 kA and pulsed currents up to 650 kA of peak value. Two terminals are used to connect the shunt with the main current circuit and the other two terminals are used for connecting the sense wires to the measuring circuit. This approach eliminates the effects of contact resistance between the main current circuit wires and the shunt.

Five-terminal shunts (Figure 21.1c) are used in power metering applications. Two sense wires are devoted to measure the phase current; the third sense wire is dedicated to the phase voltage measurement and the neutral conductor is the second point needed for this voltage measurement.

In many applications, it is possible to incorporate the current sensor on the same printed circuit board (PCB) used for placing the electronic systems. For this use, it is possible to find many shunts specially intended for PCB mounting (Figure 21.1d). These shunts are used to sense currents up to 100–200 A and they are designed to be mounted using through-hole technology (THT) or

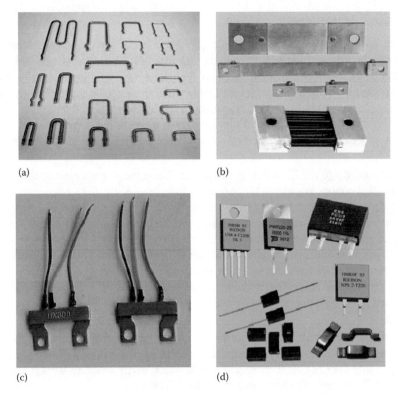

(a)   (b)

(c)   (d)

FIGURE 21.1 Several types of shunt resistors. (a) two-terminal shunts, (b) four-terminal shunts, (c) five-terminal shunts and (d) shunts intended for PCB mounting.

surface-mount technology (SMT). Shunts can also be created using the intrinsic resistance of a PCB copper trace. This is a low-cost realization, but the large temperature dependence of the cooper resistance makes this approach only suitable for low accuracy applications.

Finally, the lack of isolation between the main current circuit and the measurement circuit can be overcome using isolation amplifiers. This is an expensive component that raises the cost of the measurement system and also adversely affects the bandwidth and precision of measurements. For this reason, the use of shunts is not suggested in applications where electrical isolation is required.

21.3 CURRENT SENSORS BASED ON FARADAY'S LAW OF ELECTROMAGNETIC INDUCTION

The operating principle of these sensors must be exposed starting with the Ampere–Maxwell law. Ampere's law relates magnetic flux density with electrical current in stationary situations, but in the case of ACs, Ampere's law needs to be corrected by introducing the effect of the displacement current.

For practical purposes, the contribution of the displacement current is negligible for low-frequency currents when it is compared with the magnetic flux density due to charges' velocity. For these quasi-stationary currents, the relationship between magnetic flux density and electrical current is properly defined by the original form of Ampere's law.

On the other hand, Faraday's law of induction relates the electromotive force induced in a closed electric circuit with the variation of the magnetic flux that crosses the surface formed by this circuit and Lentz's law gives the sign of this relation.

The combination of these three fundamental laws of the electromagnetic theory describes the operation of current transformers (CTs) and Rogowski coils. The following section describes their most important features.

21.3.1 Current Transformers

The construction of CTs is based on a high-permeability toroidal core and two independent windings. In general, the primary winding is a single turn through the central window of the core but more turns can be used to measure small currents. The secondary winding is usually connected to a small sense resistor or burden, although ideally this winding should be short-circuited. The amplitude of the secondary current in ideal CTs is determined by the ratio of the number of secondary turns to the number of primary turns, and they operate as a controlled current source.

By contrast, practical CTs are affected by several factors, which result in measurement errors:

- Voltage drops due to the CT winding resistance. These resistances are responsible for the usually called copper losses.
- Voltage drops due to leakage flux between the two CT windings. Careful construction of CTs guarantees a very small value of leakage flux, making this effect negligible when it is compared with winding resistance.
- Power losses in the magnetic circuit core due to hysteresis and Foucault currents. These losses are modeled by a resistor and they are usually called core losses.
- Magnetizing current needed to produce the magnetic flux for the CT operation. This effect is modeled by an inductor, usually named magnetizing inductance.

CTs are very simple and robust components for AC current measurement with high galvanic isolation between primary and secondary circuits, they do not require external power sources, and the output signal usually does not need additional amplification when adequate burden is utilized.

Commercial CTs are usually characterized by the following:

- *Application*: Protection or metering. Protective CTs are designed to quantify currents in distribution power systems, and its output current is used as input to the protective relays. Measuring CTs are used to adapt the level of measured currents to the span of utilized instruments.
- *Rated frequency*: Range of frequencies in which the CT is designed to operate.
- *Rated primary voltage*: Maximum RMS voltage value of the circuit where the CT primary winding can be connected.
- *Insulation voltage*: Maximum voltage that the CT can withstand between the primary and secondary circuits without permanent damage.
- *Rated primary current*: Input current for continuous and normal operation.
- *Rated continuous thermal current*: Primary current value for continuous operation without thermal overloads, and, unless otherwise specified, it is equal to rated primary current.
- *Rated thermal short-circuit current*: RMS value of the primary current that the CT can withstand for a specific time (usually one second) without damage. This current is specified when the secondary winding is short-circuited and it is indicated as multiples of the rated primary current (80 is a common ratio).
- *Rated dynamic short-circuit current*: Peak value of the primary current that a CT can withstand without thermal, electrical, or mechanical damage when the secondary winding is short-circuited. This current is proportional to the rated thermal short-circuit current, and 2.5 is a common factor.
- *Rated secondary current*: Value of output current due to the rated primary current. Values such as 1, 2, and 5 A are usual for protective CTs.
- *Rated current ratio*: Ratio of primary and secondary rated currents. Normally, the values of the two currents are indicated, such as 1000/5.
- *Accuracy class*: Typical values of the accuracy class for metering CTs are 0.1, 0.2, 0.5, 1, and 3, indicating the maximum percentage of error in the rated current ratio at the rated primary

current. For protective CTs, the accuracy class is usually 5P or 10P. This value indicates the maximum percentage of the composite error (amplitude and phase) in the rated current ratio.

- *Rated burden*: The maximum load that may be connected on the CT secondary, preserving the specified accuracy class. For metering CTs, burden is expressed as impedance, and for protective CTs, it is expressed as the apparent power of the secondary circuit, at the rated secondary current and at a specific power factor value.
- *Rated security factor*: Value of the ratio between the primary current and the rated primary current up to which the CT will operate complying with the accuracy class requirements. For metering CTs, the value of this factor is up to 1.5. For protective CTs, this characteristic is known as the *accuracy limit factor* (ALF) and indicates the maximum number of times that the primary current can exceed the rated primary current. The ALF value can be up to 30, but 10 and 20 are the most typical values.

Protective CTs are used for monitoring power transmission and distribution networks, detecting fault conditions, or for load survey purposes. Using these CTs, it is possible to measure rated primary currents up to 5 kA, rated thermal and dynamic short-circuit currents up to 80 and 200 kA, respectively, and with rated primary voltages from 1 to 765 kV.

Measuring CTs are used in a wide range of applications, which include current monitoring and metering for instrumentation devices, calculating energy consumption in electric power supply billing, and generating error signals in automatic control systems. The currents that can be measured using these devices range from mA to kA, with rated primary voltage lower than 600 V and rated secondary currents between mA and A.

Other possibility for CT classification is according to the architecture used in their construction. In this sense, they can be reduced to four basic types:

- *Window type*: This type has no primary winding, and the primary current wire is passed through the window of the CT (Figure 21.2a).
- *Bushing type*: These CTs are similar to window-type CTs, but they are for use with a fully insulated conductor acting as a primary winding, because they have no insulation in the window (Figure 21.2b).
- *Bar type*: In these CTs, the primary conductor passes through the CT core, forming an integral part of the device (Figure 21.2c).
- *Wound type*: These CTs have an internal primary winding, which usually consists of more than one turn. The CT core and the primary and secondary windings are assembled as an integral structure (Figure 21.2d).

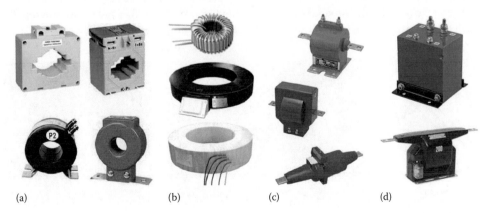

(a) (b) (c) (d)

FIGURE 21.2 Several types of metering current transformers. (a) window type, (b) bushing type, (c) bar type and (d) wound type.

The high-frequency response of CTs is related to the secondary winding resistance and the parasitic capacitances between its turns and layers; for this reason, precise CTs for high-frequency applications use high-permeability cores and careful design for the construction of this winding. Protective CTs and a large variety of measuring CTs are designed for line frequency (50 or 60 Hz), but measuring CTs with bandwidths from 0.1 Hz to 500 MHz can be found in the market.

All window-type and bushing-type CTs shown in Figure 21.2 have in common a solid core. This feature can be a drawback since it implies that the primary current wire must be interrupted for the CT installation. This trouble is solved in split-core CTs because this core construction allows fast and easy CT assembly. Split cores present an undesirable effect over the CT accuracy because the leakage flux increases near the core closure, but if the primary circuit wire passes through the core window as far as possible from the core closure, this inconvenience can be avoided.

It is possible to find CTs intended for PCB mounting and designed for THT and SMT. These CTs can measure currents up to 100 A, and their bandwidth is strongly dependent on the mounting technology. The bandwidth of CTs based on a THT is around 500 kHz, but if an SMT is used, the bandwidth can be extended up to 10 or 15 MHz.

21.3.2 ROGOWSKI COILS

One important drawback of CTs is related to hysteresis and saturation features that the material used for core construction exhibits when the primary current exceeds the value established by the rated security factor. This disadvantage of CTs is not present in Rogowski coils. Rogowski coils are essentially constructed as CTs but no ferromagnetic material is used as a core, and, as a consequence, they do not exhibit saturation features and are linear devices.

The simplest design of a Rogowski coil is a solenoid placed around the primary current carrying conductor, as shown in Figure 21.3a. An important feature of the solenoid construction is that the end of the wire used for the coil is returned to the beginning via the central axis of the solenoid. If the Rogowski coil design does not incorporate this return loop, the sensor could be sensitive to magnetic fields perpendicular to the coil plane. Figure 21.3 also outlines two designs of Rogowski coils, wound on a rigid core (Figure 21.3b) and wound on a flexible and open core (Figure 21.3c).

The Rogowski coil operating principle is defined by Ampere's and Faraday's laws, and the mutual inductance of the coil relates the induced output voltage to the primary current rate of change.

This relation is valid if the coil core shape is not circular or if the primary current conductor is not centered on the coil; nevertheless, a nonuniform density of turns and a variation in the area of the turns along the coil lead to an increase in measurement errors. This effect is greater at the junction of the coil, where a short gap of turns appears. Some manufacturers compensate this gap error by increasing the turn density around the coil junction; however, this effect can be minimized if the primary current wire is installed as far as possible from the coil junction.

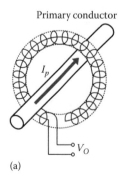
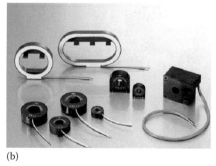
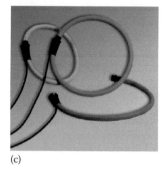

(a)   (b)   (c)

FIGURE 21.3   (a) Rogowski coil construction, (b) wound on rigid core, and (c) wound on flexible core.

In order to obtain an output signal proportional to the measured current, the integration of the coil output voltage is required. Among the possible methods for implementing this operation, electronic integrators are commonly used. The high input impedance of integrators connected as load to Rogowski coils implies that this kind of transducer presents very low insertion impedance when it is compared with CTs.

The following summarizes the most common technical specifications of Rogowski coils:

- *Rated peak input current*: The Rogowski coil has an extremely large measurement range because it does not exhibit saturation effects; normally, the limits are imposed by the integrator saturation. Devices with ranges from few mA up to hundreds of kA can be found on the market.
- *Peak di/dt*: The Maximum primary current change ratio above which the transducer fails to measure correctly. Commercially available devices can provide capability from hundreds of A/s to thousands of kA/s.
- *Maximum peak di/dt*: Absolute maximum rating for primary current change ratio that must not be exceeded because otherwise transducers can be damaged. This specification is in the range of 0.1–100 kA/µs.
- *Maximum RMS di/dt*: Even if the maximum peak *di/dt* is not exceeded, the current transducer can be damaged by primary currents with sufficiently high repetitive *di/dt*. The typical value for this parameter is around one A_{RMS}/µs.
- *Output voltage or sensitivity*: The ratio between the input current and the transducer output voltage signal. It can be specified as the output voltage at rated input current, as the current to voltage ratio value (i.e., 1 mV/A) or as the current value to give 1 V output voltage.
- *Current linearity:* The linearity of Rogowski current transducers typically includes the effects of both the Rogowski coil and integrator circuit, and typical values range from 0.2% to 0.05%.
- *Measurement accuracy*: Accuracy values around 1% are typical on flexible core devices, while values around 0.2% are usual on rigid core Rogowski coils. The transducer accuracy varies slightly according to the position of the primary current wire within the measurement window. This "Positional Accuracy" leads to errors that can reach 2% or 3%.
- *Frequency response or bandwidth*: The Range of primary current frequencies in which the stated accuracy applies. The low-frequency bandwidth is set by the practical limitation of the integrator gain at low frequencies. This frequency is several tens of Hz for most devices, but it can be 0.1 Hz for high-end devices. Besides, the high-frequency bandwidth is related to the parasitic inductance and capacitance of the Rogowski coil and the gain-frequency characteristic of the amplifier used for the integrator implementation. This frequency is up to 5–50 kHz for most devices and can reach values around tens of MHz for high-performance transducers.
- *Phase error*: Output phase error can reach high values when a load is connected to the Rogowski coil. Nevertheless, this error can be quantified and compensated using a phase-shift compensation stage. Since this correction depends on the primary current frequency, it is necessary to define the frequency that has been used to design this compensation stage. The phase error reported by many manufacturers is around 1° at 50–60 Hz.
- *Thermal drift*: It is basically determined by two factors: the thermal drift on integrator characteristics and the variation of the coil cross-sectional area due to thermal expansions/contractions. This parameter value ranges from +0.03%/°C to +0.08%/°C.
- *Isolation voltage*: Rogowski coils provide galvanic isolation between the primary current conductor and the measurement circuit. This voltage for most transducers is compressed between 60 V and 1 kV. However, specialized transducers are available with isolation voltages up to 10 kV.

- *External power requirements*: Many commercial Rogowski current transducers include batteries to power the internal electronic parts, and some transducers can be powered by external AC adaptor or external power supply. Typical power requirements are from 3 to 24 V_{DC}.
- *Battery life*: Estimated time that the battery can power the current transducer. When batteries are used, life times from 50 to 100 h are typical, but this value is reduced from 25 to 50 h when rechargeable batteries are utilized.

Measurement of current transients is the area of applications where Rogowski coils are especially valuable. Some examples of this application are: current monitoring in precision welding systems, arc melting furnaces, or electromagnetic launchers; in short-circuit test of electric generators; and in electrical plants for protection purposes. Another area where these transducers are used due to their excellent linearity is in the measurement of harmonic components on electric currents.

21.4 MAGNETIC FIELD SENSORS

According to the Ampere–Maxwell law, magnetic fields and electrical currents are closely related and, as a direct consequence, magnetic field sensors can be used to measure electrical currents. The capacity of magnetic field sensors to measure static and dynamic magnetic fields makes them an interesting option for measuring DC and AC currents. Current meters using these sensors can be constructed using two basic configurations: open loop and closed loop.

In the open-loop configuration, a magnetic core is used to concentrate the magnetic field onto the magnetic field sensor, as shown in Figure 21.4a. The performance of these sensors is influenced by the properties of the core material and, consequently, core losses due to hysteresis and eddy currents and the possible saturation of the magnetic core can lead to sensor overheating and important measurement errors.

In current sensors based on closed-loop configuration, a secondary winding is incorporated into the magnetic core and the output signal of the magnetic field sensor is used for generating the secondary current, as is illustrated in Figure 21.4b. The secondary current (I_S) generates a magnetic field opposed to the field generated by the primary current (I_P), and, when the secondary current compensates the magnetic flux inside the sensor core ($\Phi = 0$), this current is proportional to the primary current. The ratio between primary and secondary currents is fixed by the number of secondary winding and the ratio between the secondary current and the output voltage is fixed by the value of the used burden (R_B).

The closed-loop configuration reduces the influence of the thermal behavior of the magnetic field sensor on the measurement accuracy and core losses are greatly reduced if the current sensor operates close to zero flux conditions. By contrast, the more complicated construction of this configuration implies larger sizes and costs.

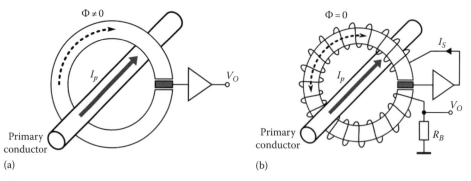

FIGURE 21.4 Current sensors using magnetic field sensing devices in (a) open-loop configuration and (b) closed-loop configuration.

The three technologies more commonly utilized for current measurements are introduced and briefly explained in the following sections.

21.4.1 HALL EFFECT CURRENT SENSORS

When a current circulates through a slab of semiconductor or metal and it is penetrated by a magnetic field, the charge carriers are deflected to one side of the slab and, as a consequence of the difference on the conductive carriers density, a voltage is generated perpendicular to both current and magnetic field directions.

In practice, the behavior of Hall elements also depends on other physical factors such as the pressure and the temperature. The output voltage dependence on the mechanical pressure is not an important factor to be considered by users of these devices because it is considered and compensated by manufacturers. In contrast, temperature adversely affects the Hall elements' accuracy, and they have high values of thermal drift in accuracy specifications.

Another important drawback of these sensors is the presence of offset voltages at the sensor output. This is an output voltage different from zero in the absence of magnetic field. The commercial current sensors include additional electronic circuitry devoted to compensate this voltage and differences on thermal drifts.

The most common technical specifications of Hall current sensors are summarized in the following:

- *Rated current*: Input current for continuous and normal operation. Rated currents ranging from 1 A up to 5 kA can be found in commercial devices.
- *Saturation current*: Input current value for which the output deviates from the estimate output voltage by more than 10%.
- *Output type*: The output signal of these sensors may be a voltage or a current.
- *Rated output*: Output signal value when the input to the sensor is the rated current.
- *Residual output (output offset)*: Output value when the input current is zero.
- *Bandwidth*: Range of primary current frequencies in which the stated accuracy applies. Hall effect transducers operate from DC to several hundreds of kHz.
- *Output linearity*: Error between the value of the sensor output and the estimated output value. It is calculated by the least mean squares method from the output and residual output when the input to the sensor is the rated current.
- *Linearity limits*: Range of the input current value for which the output is within 1% of the estimated output voltage.
- *Output temperature characteristic*: Rate of output change due to temperature variations when the rated current is input. It is evaluated per 1°C variation using the output at 25°C as the reference.
- *Residual output temperature characteristic*: Change of the residual output due to temperature change when the input is the rate current. The change per 1°C is shown as residual output temperature coefficient.
- *Response time*: Output response time when the input is a pulse current. This time is evaluated when the input and output waveforms drop to a specified level of their initial values.
- *External power supply*: Typical power requirements are from 1 to 24 V_{DC}, and it is usual that the bidirectional sensors need a bipolar power supply, with ±5 V_{DC}, ±12 V_{DC}, or ±15 V_{DC} being the typical value.
- *Isolation voltage*: Hall effect sensors provide galvanic isolation between the primary current conductor and the measurement circuit. This voltage is compressed between 100 V and 2.5 kV for most transducers.

The accuracy of these devices is reasonable for open-loop configuration (values between 0.5% and 2% are typical) and high when the closed-loop configuration is implemented (reaching values

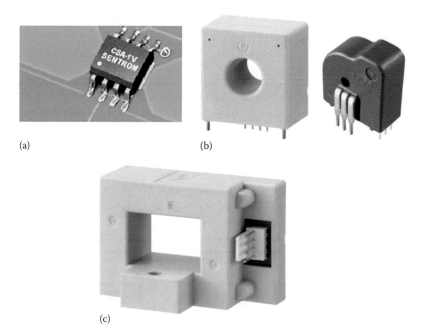

(a) (b)

(c)

FIGURE 21.5 Hall effect current sensors: (a) SMT device (www.melexis.com), (b) THT devices for on-board mount (www.lem.com), and (c) screw-mount device.

around 0.02% and 0.05%). Both types of configurations are commonly used in these transducers, and it is possible to find Hall effect current sensors intended for screw-mount and on-board mount types, as shown in Figure 21.5.

Typical applications for Hall effect sensors are related to DC current measurement. This is the case of electric power conversion systems, motor drives and electric traction, uninterruptible power supplies, electric welding, and electrolyzing equipment. These sensors are also utilized in current clamps used for DC and low-frequency current measurement.

21.4.2 Fluxgate Current Sensor

Fluxgate sensors appeared in the early 1930s and are still used today in many applications that require high precision, because this is one of the most accurate technologies for sensing magnetic fields.

The basic operating principle of these transducers is based on the detection of saturation states in a magnetic circuit. These states are related to the nonlinear relationship between the magnetic field and the magnetic flux density shown by magnetic materials. The three basic designs used for the construction of fluxgate sensors are depicted in Figure 21.6.

The basic fluxgate transducer (Figure 21.6a) uses a toroidal magnetic circuit, which includes an air gap with a saturable inductor as the element for magnetic field detection. This inductor is led to saturation periodically between positive and negative values by the effect of the signal applied to its excitation winding. In the design shown in Figure 21.6b, the sensor uses its own ring core as a magnetic field detector and includes no gap on the magnetic path. An auxiliary winding N_A (called excitation winding) is added to the core, which is used as a saturable inductor for flux detection purposes.

In the absence of the primary current (I_P), the flux through the magnetic field detector is zero. Under these conditions, if an adequate square voltage is applied to the excitation winding, a growing magnetic flux appears in the magnetic circuit until the core saturation state. The primary current effect on the excitation current implies a waveform deformation, and this phenomenon can be utilized to estimate the primary current value.

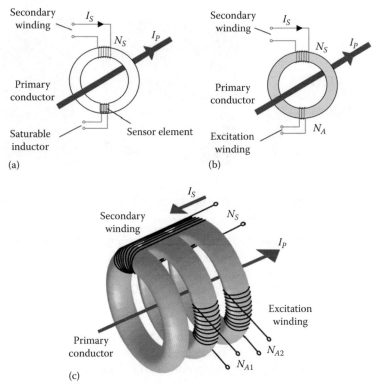

FIGURE 21.6 Fluxgate current sensors: (a) using a saturable inductor as detector, (b) using a closed toroid core as detector, and (c) using multiple cores for voltage disturbances reduction and to extend the device bandwidth.

In open-loop fluxgate sensors, the primary current value is obtained by measuring the excitation current second harmonic component or average value. In closed-loop devices, one of these signals is used as an error signal for generating an additional magnetic field by means of a secondary current (I_S) applied to an additional secondary winding (N_S). When a zero-flux condition is achieved, the secondary current is a replica of the primary current, and the ratio between these two currents is fixed by the number of turns used in the secondary winding.

Fluxgates based on one-core designs present two important drawbacks:

- Noise injection from the excitation winding circuit into the primary current circuit. The solution adopted is the use of a second core with a new excitation auxiliary winding. If these auxiliary windings are identical and they are excited with the same current but in opposite directions, their effect over the primary circuit is canceled.
- The bandwidth of these configurations is limited by the time spent in driving the magnetic core between two saturation states. In order to increase the device bandwidth, a third core is included in the transducer design (Figure 21.6c). The combination of this third core, the secondary winding and the primary current wire operates exclusively as a conventional CT, and this feature allows medium- and high-frequency current measurements.

The fluxgate technology offer several advantages over other technologies based on magnetic field measurement, such as low offset, low offset thermal drift, and excellent over-current recovery because any permanent magnetization on the fluxgate core is reset by subsequent saturation cycles. Furthermore, its large dynamic range allows measurements of small and large currents with the same device and also it presents a large temperature range of accurate operation, limited only by used materials and electronic components.

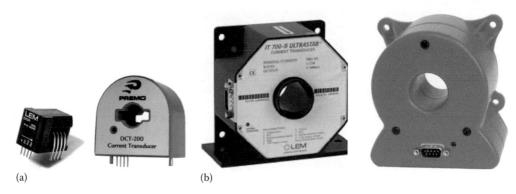

(a)   (b)

FIGURE 21.7   Fluxgate current sensors: (a) THT devices for on-board mount and (b) screw-mount devices (www.grupopremo.com and www.lem.com).

Both open-loop and closed-loop configurations are commonly used for these transducers, and they are intended for screw-mount or on-board mount types (Figure 21.7).

The measuring range of a typical commercial fluxgate is up to 500 A, accuracy 0.2%, linearity 0.1%, response time <1 μs, and bandwidth from 0 to 200 kHz. Moreover, the measuring range of three core fluxgate sensors is up to 1 kA, accuracy 0.0002%, linearity 0.0001%, response time <1 μs, and bandwidth from 0 to 100 kHz.

Finally, these sensors are commonly used in high precision measurement applications due to their high cost and size. Typical applications are in calibration and diagnosis systems, medical equipment, precise energy metering, high-performance laboratory equipment, or as feedback element in high precision systems.

21.4.3   Magnetoresistance Current Sensors

Magnetoresistance is defined as the property of a material to change the value of its electrical resistance when it is submitted to an external magnetic field. Nowadays, it is possible to find commercial current sensors based on this property.

The sensing direction of these devices is in the same plane as the magnetoresistive element, so they cannot be located into narrow air gap in magnetic cores and, consequently, these sensors are commonly coreless.

The magnetoresistive elements are normally configured in a Wheatstone bridge to compensate the external magnetic fields and thermal drifts. Also, these sensors can be found in the market in both open-loop and closed-loop configurations, but they are only intended for on-board mount type.

There are several known physical effects that cause variations in the resistance of certain materials when a magnetic field is present, but only two are currently applied to the electrical current measurement. They are discussed in the following sections.

21.4.3.1   Anisotropic Magnetoresistance Sensors

The anisotropic magnetoresistance (AMR) effect is associated with Hall effect elements. The external magnetic field deflects the trajectory of the charge carriers traveling through the measuring element. This implies an increase in the length of the path followed by the charge carriers and, as a consequence, an increase in the device resistance. In most materials, this effect is less important when compared with the Hall effect; but in anisotropic materials, the application of an external magnetic field may produce resistance variations from 2% to 5%.

One of the major drawbacks of AMR sensors is that a strong external magnetic field can permanently change the direction of the permalloy initial magnetization, introducing a permanent measurement error until a new permalloy remagnetization in the right orientation has been performed.

The bandwidth of these sensors is from DC to several MHz and commonly is limited by the amplifiers used in the signal conditioning circuitry.

FIGURE 21.8 Commercial AMR current sensors (www.fwbell.com, www.sensitec.com, sensing.honeywell.com, and www.diodes.com).

AMR current sensors are available in both open-loop and closed-loop configurations but are always configured in a Wheatstone bridge. In the closed-loop configuration, the Wheatstone bridge output voltage is used as an error signal in order to generate the compensation current and the associated magnetic field. The value of the compensation current is adjusted until its magnetic field cancels the magnetic field associated with the primary current. In this condition, the compensation current is proportional to the primary current.

Commercial AMR current sensors are manufactured by Sensitec, F.W. Bell, Zetex Semiconductors, and Honeywell. They offer sensors up to 1 kA of primary rated current, accuracy lower than 0.3%, linearity errors lower than 0.1%, and bandwidths up to 2 MHz with an isolation voltage of 3.5 kV. Figure 21.8 shows some of these devices.

21.4.3.2 Giant Magnetoresistance Sensors

The giant magnetoresistance (GMR) effect is described as a large change in electrical resistance that occurs when thin stacked layers of ferromagnetic and nonmagnetic materials are exposed to a magnetic field. This effect allows the detection of very small currents due its high sensitivity, and it is a cheap technology because the massive production of small devices is possible using standard semiconductor technology. However, GMR technology has important drawbacks such as high nonlinear behavior and distinct thermal drift. Furthermore, the GMR sensor behavior may be permanently altered after a magnetic shock.

Current sensors based on this technology are constructed using open-loop configuration and, in order to increase their immunity to external magnetic fields and temperature variations, they are configured in a Wheatstone bridge setup. The half-bridge configuration with two active resistances and two shielded ones is commonly used because the manufacturing process of this GMR structures is cheaper if it is compared with the process needed for the full-bridge configuration implementation.

Commercial GMR current sensors can be found in on-board mount types and are designed for SMTs. NVE Corporation is the most important manufacturer of GMR current sensors and offers devices for linear measurement of electric currents in the range of ±80 mA and with frequency response of 100 kHz. Figure 21.9 depicts some of these devices.

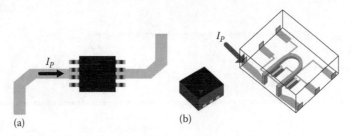

(a)   (b)

FIGURE 21.9 Commercial GMR by NVE Corporation: (a) the AA00X-02 series current sensor and (b) the AAV003-10E device (www.nve.com).

The main application field of GMR current sensors is related to the measurement of low currents using very small size and low-power consumption devices, this is the case of portable equipment and battery powered systems.

21.5   CURRENT SENSORS BASED ON FARADAY EFFECT

This effect describes the interaction between magnetic fields and light and explains that the rotation in the light polarization plane is proportional to the component of the magnetic field in the same direction of light propagation. It is said that the medium that changes the state of light polarization is birefringent, and Faraday discovered that a certain type of birefringence can be induced into a medium by a magnetic field.

Faraday effect can be exploited to design electrical current meters, providing exceptional electrical isolation, and enable measurements of DC and AC currents up to 500 kA, using a negligible amount of energy and space.

Current sensors based on Faraday effect can be constructed using two different configurations usually referred to as extrinsic and intrinsic fiber sensor architectures. They are briefly described in the following sections.

21.5.1   EXTRINSIC FIBER OPTIC SENSORS

These sensors, also named magneto-optic current transducer (MOCT), use a piece of glass wrapping the current carrying conductor, as is shown in Figure 21.10a, and optical fibers are used for light transmission and not for sensing purposes. The sensitivity of these transducers can be increased using multiple light reflections and creating a three-dimensional path with several turns around the current conductor.

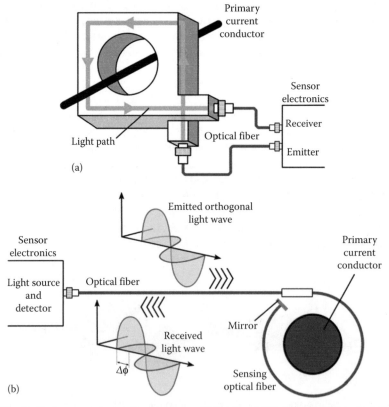

FIGURE 21.10   Optical current sensors types commercialized by ABB: (a) MOCT type and (b) FOCS type.

ABB manufactures these sensors for metering and protective applications in 72.5–800 kV and 50–60 Hz electric energy distribution systems. The MOCT system is suitable for outdoor application and has an accurate metering current range (0.2 class) from less than 1 A to 4 kA using the same device.

21.5.2 INTRINSIC FIBER OPTIC SENSORS

These sensors, also called fiber optic current sensors (FOCS), utilize an optical fiber wrapping the current carrying conductor, as is depicted in Figure 21.10b. The sensitivity of these transducers can be easily increased using multiple fiber turns creating a fiber optic coil around the current conductor. Moreover, the accuracy may be deteriorated due to birefringence induced by bending the fiber optic cable. The use of high birefringent materials with low sensitivity to stress for optical fiber manufacturing makes the intrinsic sensors an interesting alternative to the larger and expensive extrinsic sensors.

In order to reduce the sensitivity to mechanical stress and vibrations, back-and-forth propagation through the sensing optical fiber can be used. This approach exploits the fact that the birefringence induced into the optical fiber by bending stress is reciprocal. This feature is implemented by reflecting with a mirror the light beam at one side of the optical fiber and detecting the differential rotation in the light polarization at the other end.

Some manufacturers of commercial intrinsic fiber optic current sensors are ABB, DynAmp LLC, Ariak Inc., and ALSTOM. They offer sensors in the range of 1 A–600 kA with measurement accuracy and linearity error around 0.1%. These sensors are designed for DC and from 50–60 Hz to 5 kHz applications with isolation voltages of several kV.

As is easily deduced from its specifications, the main fields of application of these sensors are the current metering for protection and measurement of electric energy distribution and production systems and very high DC current metering for control systems in electrochemical industries.

21.6 DO YOU WANT TO LEARN MORE?

This text only describes the current sensor technologies actually applied to commercial devices, but there are still several physical effects or properties applicable to the measurement of magnetic fields or electric currents. Some examples of these are: the tunneling magnetoresistance and the giant magnetoimpedance effects, the colossal magnetoresistance property, devices based on superconducting quantum interference detectors (used for measuring extremely small currents), devices based on magnetically sensitive Complementary Metal-Oxide-Semiconductor (CMOS) transistors and sensors that exploit Lorentz force.

In order to learn more about these future technologies or to increase our knowledge on the current sensors described in this text, we can access various information sources and technical documents.

Two very interesting reviews dedicated to electric current sensors are as follows:

1. S. Ziegler et al., "Current sensing techniques: A review" published in April 2009 (volume 9, number 4) in *IEEE Sensors Journal* (IEEE Sensors Council).
   Doi: 10.1109/JSEN.2009.2013914.
2. P. Ripka, "Electric current sensors: A review" published in November 2010 (volume 21, number 11) in *Measurement Science and Technology Journal* (IOP Publishing).
   Doi: 10.1088/0957-0233/21/11/112001.

Other interesting documents are the application notes and technical reviews published by manufacturers and their research and technical staff. This kind of documents usually includes a technical overview of the operating principle of the described devices and some ideas or hints on how to use these devices in a particular application. These documents can be easily found on the Internet through the manufacturers' websites.

22 Ferrofluids and Their Use in Sensors

B. Andò, S. Baglio, A. Beninato, and V. Marletta

CONTENTS

This chapter gives an overview of ferrofluids and their applications with a special focus on transducers development. The addressed application fields include standard inertial measurements as well as bio-medical systems requiring physical and electrical decoupling between the target biofluid and the electric parts of the device.

22.1 INTRODUCTION

Magnetic fluids, which can be controlled and manipulated by external magnetic fields, are of interest for those applications involving fluid management. By the use of a magnetic field of suitable characteristic, it is possible to completely control the physical properties (e.g., density and viscosity) of the magnetic fluid.

The pioneering activity on magnetic fluid is due to Resler and Rosensweig [1], while the first stable synthesis of a ferrofluid was obtained in 1965 by Papell [2]. Modern ferrofluids are colloids of superparamagnetic nanoparticles [3], such as Fe_3O_4, γ-Fe_2O_3, $CoFe_2O_4$, Co, Fe, or Fe-C, stably dispersed in a carrier liquid, usually oil or water [4]. These nanomaterials manifest simultaneously fluid and magnetic properties.

From a microscopic point of view, several mechanisms rule the interaction of paramagnetic nanoparticles, such as attractive Van der Waals forces, thermal agitation, and anticohesion coatings [1].

The size of the particles makes the difference between two main kinds of magnetic fluids: ferrofluids and magneto-rheological fluids. Magneto-rheological fluids are stable suspensions of magnetically polarizable micron-sized particles dispersed in a carrier fluid. These fluids vary their viscosity with the applied magnetic field and can become hard in the presence of a strong magnetic field. In comparison, ferrofluids retain liquid flowability even with the most intense magnetic field.

From a magnetic point of view, traditional ferrofluids are characterized by a magnetism vector, which follows the applied field without hysteresis, and a small applied field is required to produce saturation.

Switching to the mechanical domain, it can be observed that high magnetization strength implies high magnetic pressure exerted by the fluid. The latter property is strategic for the implementation of transducers adopting ferrofluids as functional or inertial masses. Moreover, ferrofluids adapt to any geometry, e.g., they can be easily moved through micro-channels.

When a ferrofluid mass is subjected to an external magnetic field, its flat surface could become unstable and it shows a behavior ruled by the "Rosensweig effect" [1]. At a certain intensity of the field, peaks appear at the fluid surface, which typically form a static hexagonal pattern at the final stage of the pattern forming process. Such a typical effect is shown in Figure 22.1.

FIGURE 22.1   An example of a ferrofluid pattern. (© B. Andò, A. Ascia, S. Baglio, and A. Beninato, A ferrofluidic inertial sensor exploiting the Rosensweig effect, IEEE Transactions on Instrumentation and Measurements, 59(5), 1471–1476, 2010.)

Magnetic fluids also show interesting patterns, coming from ferrohydrodynamic instabilities, such as lines, labyrinths, and various structures, which could be exploited to produce actuation. In [5], an alternating electric field superimposed over an orthogonal constant magnetic field is used to form labyrinth structures.

Magnetic fluids belong to a multidisciplinary research area involving chemistry for the fluid synthesis, physics for studying their behavior, basic theories, and models, engineering for studying possible applications, and biology and medicine for exploiting their applicability in the biomedical field.

22.2   APPLICATIONS OF MAGNETIC FLUIDS

Ferrofluids have applications in several contexts [6–34]. Sealing for several industrial processes, loudspeakers, inertial dampers, angular position sensors, and computer disk drives are examples of applications where these materials have been widely adopted [7].

For example, in hard disks, they are used to form liquid seals around the spinning drive shafts, while in loudspeakers, ferrofluids are used for coil cooling and as dampers. Ferrofluids have been conveniently applied for the measurement of velocity profile and shear stresses in the fluid/surface interface [8]. The possibility to observe this phenomenon is of primary importance in hydraulic and oceanic engineering due to its role in erosion processes. Furthermore, shear forces generated by liquids or gases flowing over solid surfaces can significantly influence the performance of aircraft, ships, or surface transport vehicles. Shear stress measurement is also needed for a number of aerodynamic studies, fluid dynamic monitoring, diagnostics applications, underwater applications, and biomedical applications. Accurate modeling of these phenomena is then crucial for different fields of research and for industry. In order to develop such models, it is necessary to perform reliable measurements on controlled and reproducible systems. Shear stress measurements are usually accomplished by surface mounted force balance, atomic-force microscope, surface mounted hot wire, surface mounted hot plate. Micro electro-mechanical systems (MEMS) hot film, and oil film interferometry. The main drawback of the aforementioned approaches is the risk of perturbing the original fluid condition. The methodology proposed in [8] tries to fix this problem by using as sensing element a drop of ferrofluid deposited over the surface where the fluid velocity profile has to be estimated. The velocity profile has been estimated by an image processing procedure.

Small dimensions (from few nanometers up to micrometers) of ferrofluid particles make them attractive for developing new devices and methodologies for the biomedical field with diagnostic and therapeutic purposes [9]. Literature is continuously reporting on new applications of ferrofluids in magnetic bio-assay, magnetic separation, drug delivery, hyperthermia treatments, magnetic resonance imaging, and magnetic labeling [10–16].

In magnetic bio-assays, magnetic particles coated with a suitable biocompatible molecule can be used as markers to identify a specific target bio-entity. In case of separation tasks, magnetic fields are used to transport the magnetic labeled entities. Magnetic separation [13–16] is used to separate small amounts of target cells from their environment as in the detachment process or in the removal of unwanted biomaterials from a fluid. Magnetic particles bound with specific drugs are employed in drug delivery techniques [17]. Particles are injected through the circulatory system and delivered by external magnetic fields only in specific locations of the human body to attack tumor tissues [18]. Because of the very small amount of the drug, no side effects have been observed when the magnetic field is turned off, and the drug disperses in the body. In hyperthermia tissue treatments, magnetic particles are inserted in the target tissues and heated by external magnetic fields in order to destroy cancer cells [10]. Cell localization requires remote sensing typically performed by alternating current (AC) susceptometry, superconducting quantum interference device (SQUID) magnetometry, or giant magnetoresistance sensors [19]. A low-cost solution based on fluxgate magnetometers has been proposed in [20].

Practical interest in magnetic fluids derives also from the possibility to implement efficient sensors and actuators [6]. Sensors are mainly based on the fact that, under particular conditions, a ferrofluid volume subjected to a magnetic field can behave like a mass connected to a tunable equivalent spring whose properties can be controlled by the magnetic field amplitude. Actuators exploit the possibility of using ferrofluids to move and control liquids.

The use of magnetic fluids in transducers is now widespread due to valuable properties of these fluids compared to traditional materials. As an example, the idea of using a drop of ferrofluid as the active mass in inertial sensors offers the opportunity to control the device specifications in terms of operating range and responsivity by electrically manipulating the ferrofluid properties (such as its viscosity). Moreover, the absence of mechanical moving parts and solid-inertial masses provides high reliability and robustness against mechanical shocks. Shock resistance and their intrinsic feature to be shapeless allow developing suitable sensors and actuators. Actually, in traditional inertial sensors, a mechanical shock could compromise the device functionality; while by using a mass of ferrofluid, it is possible to recover the system functionality by reaggregating the dispersed particles. Another advantage of ferrofluidic sensors is the electrical insulation between the readout electronics and the liquid medium. This gives the possibility to easily control fluids and to implement measurements in liquids. Moreover, the physical decoupling between the liquid housing (the cheap part of the system) and the electronics enables contexts requiring the substitution of the liquid housing (e.g., due to the use of invasive media or contaminants) such as in biomedical applications.

Various examples of devices exploiting ferrofluids to implement sensors and transducers are available in the literature. In particular, a cantilever beam dipped in magnetic fluid has been proposed to change the operative range of the device simply exploiting external magnetic field actions on the ferrofluid [21]. Another example is represented by an inductive tilt sensor composed of a cylindrical structure housing a mobile magnetic core clamped to two permanent magnets [22]. The magnetic liquid coats the permanent magnet so as to create a cushion, allowing the core sliding consequently to a device tilt. When the device is tilted, the repulsion force between the supporting magnets and two magnets disposed outside the container imposes an equilibrium position at the magnetic core, which is detected by an inductive readout strategy.

Although the conventional micropipette can sample small liquid volumes with a precision ranging in 3%–5%, the development of novel solutions showing high reliability (against shocks), biocompatibility, adaptability (configurable specifications), and flexibility (easy implementation into preexistent pipes) are highly demanded. Moreover, the use of low-cost transducers is becoming mandatory, especially in the contexts of point-of-care.

The research community is focusing toward the development of innovative and cheap solutions to control liquids in channels and micro-channels, especially for biomedical applications (diagnostics, drug delivery, genetic sequencing, lab-on-a-chip, and flow control of bio-fluids) using both new materials and innovative actuation mechanisms. The idea of using a small volume of ferrofluid as the active mass in pumping systems is emerging.

Several kinds of pumps were realized by using magnetic fluids. Examples are micropipettes [23], rotative micropumps [24], alternative micropumps and valves [25,26], electromagnetic micropumps [27], and linear magnetic pumps [28]; all these devices use magnetic fluids to implement pumping operations. In [23], a micropipette exploiting the magnetic force generated by external electromagnets is proposed. The fluid inlet and outlet have been implemented on the same side of the micropipette, thus requiring mechanical actions to move the liquid sample from one tank to the other. In [24], a rotating micropump using a ring tube is presented, where the fluid sampling is accomplished through two ferrofluidic caps created by two external permanent magnets moved by a direct current (DC) motor. In [25], an alternating micropump based on a ferrofluidic plunger and valves is presented. Moving components are controlled by external magnets managed by a DC motor, while valves are realized by channel deformation. In [26], a ferrofluidic plunger is used to implement an alternating micropump, where an external magnet actuated by a DC motor is used to move the plunger in the channel. The ferrofluidic pump, FP3, realized by a glass channel filled with deionized water and three drops of ferrofluid implementing two valves and one plunger is discussed in [29]. The main advantage of this solution resides in the absence of mechanical moving parts, thus avoiding stress and increasing the life time and the reliability of the device. Moreover, such approach allows for the implementation of a pump in a section of a preexistent pipe without actions damaging or modifying the original structure of the channel. This feature is valuable for applications where the pump must be installed occasionally in specific locations (e.g., vascular surgery application), thus increasing the flexibility and the applicability of the methodology proposed. In [30], a novel pumping architecture using one drop of ferrofluid is presented. This architecture exhibits the same advantages of the pump FP3, while it uses a single drop of ferrofluid managed by an array of electromagnets to be clamped on the pipe. This solution boosts advantages of the FP3 device in terms of dimension shrinking and possible implementations in preexistent pipes. To verify the mass movement, an optical system was used [31].

22.3   AN INERTIAL SENSOR BASED ON THE ROSENSWEIG EFFECT

In this section, a novel strategy implementing a suitable noninvasive methodology to sense perturbations imposed on a liquid medium contained in a beaker is presented [32].

The device is based on the Rosensweig effect, which is exploited to generate spikes in a ferrofluid mass injected in a glass plate filled with deionized water. The glass plate resembles the environment under test, while the water is the liquid medium subjected to external stimulus.

A permanent magnet is used to generate a magnetic force, which maintains the ferrofluid drop in a compliant position [7,23].

An external stimulus applied to the system, e.g., an acceleration, is transferred to the ferrofluid mass, thus producing perturbation of spikes around their equilibrium position. Two planar coils in differential configuration, implemented in a separate structure juxtaposed to the glass plate, have been used to monitor the spikes' position. In the following, the developed sensing strategy is illustrated along with some details on the device realization and the characterization of the lab-scale prototype.

With respect to other solutions, the exploitation of the Rosensweig effect allows for reducing the static friction between the magnetic mass and the plate which, as an example, could compromise the device sensitivity.

In a first approximation, the behavioral model of the device can be expressed as follows:

$$\rho_f V_f \ddot{x}_f = F_h + \rho_f V_f u(t) + F_m \tag{22.1}$$

where
  x_f is the position of the ferrofluid along the sensing axis
  V_f is the ferrofluid volume
  ρ_f is the ferrofluid density
  F_h is the hydrodynamic drag force
  $u(t)$ is the external acceleration applied to the device
  F_m is the magnetic force generated along the sensing axis by the permanent magnet

Under particular conditions, it is reasonable to asses that the magnetic force along the sensing axis acts on the ferrofluid mass as an equivalent spring [6,33]. In this case, the magnetic force can be expressed as follows:

$$\vec{F}_m = -K_m(x_f - x_0) \tag{22.2}$$

where
  K_m is the equivalent elastic constant
  x_0 is the equilibrium position of the ferrofluid volume

This force acts as a retaining force, which contrasts external perturbations on the device (and consequently on the ferrofluid volume) and governs the movement of the inertial mass around its equilibrium position.

The elastic constant K_m considering the form of model (22.1) can be approximated with [3,13]

$$K_m = \omega_n^2 \rho_f V_f \tag{22.3}$$

where ω_n is related to the natural frequency of the system.

Figure 22.2 shows real views of the experimental prototype. In particular, Figure 22.2a is the assembled prototype where the glass support filled with deionized water and the ferrofluid pattern can be distinguished. The adopted ferrofluid is the EFH1 by Ferrotec whose main specifications are reported in Table 22.1. The planar coils implemented in printed circuit board (PCB) technology and the permanent magnet are shown in Figure 22.2b.

Figure 22.3 schematizes the whole experimental set-up including the readout electronics and the shaker system used to perturb the device.

The inductive readout strategy exploits the variation of the planar coils inductance due to the ferrofluid motion. Actually, the movement of the magnetic fluid peaks along the sensing axis, x_f, produces a modulation of the output voltage, V_b, of a bridge configuration followed by an instrumentation amplifier with gain G. In the small displacement regime, assuming a linear dependence of the coil inductance value on the ferrofluid mass position, $L = L_0(1 + cx_f)$ (where L_0 is the nominal

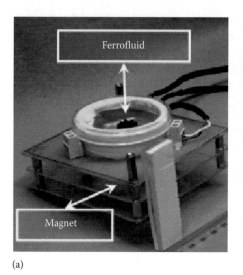
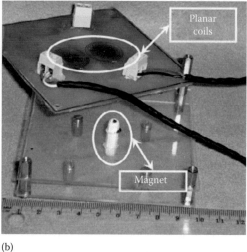

(a)   (b)

FIGURE 22.2 Real views of the sensor prototype. (a) The device assembled and (b) details of the planar coils. (© B. Andò, A. Ascia, S. Baglio, and A. Beninato, A ferrofluidic inertial sensor exploiting the Rosensweig effect, IEEE Transactions on Instrumentation and Measurements, 59(5), 1471–1476, 2010.)

TABLE 22.1

EFH1 Technical Specifications

Medium	Light mineral oil
Saturation magnetization	400 Gauss
Initial susceptibility	1.70
Flash point	92°C
Pour point	−94°C
Volatility (1 h at 50°C)	9%
Density	1100 kg/m³
Viscosity	25 Pa s at 27°C
Surface tension	29 dyn/cm

Source: © B. Andò, A. Ascia, S. Baglio, and A. Beninato, A ferrofluidic inertial sensor exploiting the Rosensweig effect, IEEE Transactions on Instrumentation and Measurements, 59(5), 1471–1476, 2010.

---

value of the inductance and c a sensor constant), the relationship between the mass position and the output voltage, V_o, would be

$$\frac{V_o}{V_d} = -G\frac{(k+1)(1+cx_f)-2k}{2(k+1)} \tag{22.4}$$

where

V_d is the bridge driving voltage

x_f is the position of the ferrofluid along the sensing axis

$k = R_1/R_2$

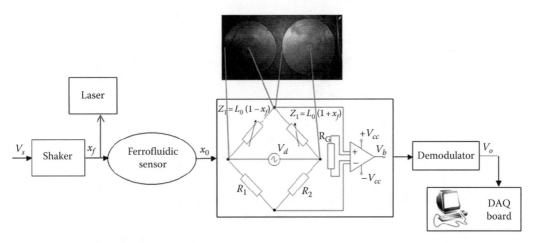

FIGURE 22.3 Schematization of the whole experimental set-up. (© B. Andò, A. Ascia, S. Baglio, and A. Beninato, A ferrofluidic inertial sensor exploiting the Rosensweig effect, IEEE Transactions on Instrumentation and Measurements, 59(5), 1471–1476, 2010.)

In the case of $R_1 = R_2$, Equation 22.4 becomes

$$\frac{V_o}{V_d} = -G \cdot \frac{cx_f}{2} \qquad (22.5)$$

To characterize the behavior of the inertial sensor, the system was solicited by a frequency sweep signal ranging from 4 to 20 Hz. The movement, x, imposed on the inertial sensor is measured independently by using the OADM-12U6430/S35A laser device.

Figure 22.4 shows the response to such stimulus of both the shaker system, measured through the laser system, and the inertial sensor. These results are related to a ferrofluidic volume of 0.2 mL.

The obtained normalized frequency responses of the inertial sensor for two different values of the ferrofluid volume are given in Figure 22.5. In particular, resonant frequencies of 7 and 6 Hz have been obtained in the case of 0.1 and 0.2 mL of ferrofluid, respectively.

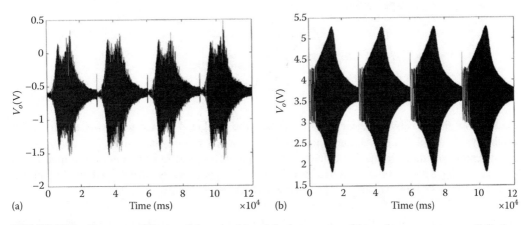

FIGURE 22.4 Response of the inertial sensor (a) and the laser system (b) to a frequency sweep solicitation. (© B. Andò, A. Ascia, S. Baglio, and A. Beninato, A ferrofluidic inertial sensor exploiting the Rosensweig effect, IEEE Transactions on Instrumentation and Measurements, 59(5), 1471–1476, 2010.)

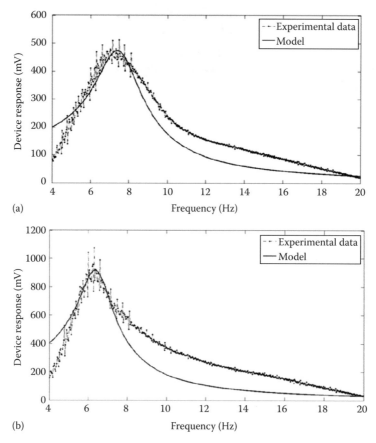

FIGURE 22.5 Frequency response of the Rosensweig inertial sensor. (a) $V_f = 0.1$ mL; (b) $V_f = 0.2$ mL. (© B. Andò, A. Ascia, S. Baglio, and A. Beninato, A ferrofluidic inertial sensor exploiting the Rosensweig effect, IEEE Transactions on Instrumentation and Measurements, 59(5), 1471–1476, 2010.)

Data shown in Figure 22.5 have been used to fit model (22.1) by the Nelder-Mead optimization algorithm, which minimizes the squared difference between the experimental and predicted device responses, R_{exp}, R_{pred}, respectively:

$$J\% = \frac{\sqrt{\sum (R_{pred} - R_{exp})^2}}{\sqrt{\sum (R_{exp})^2}} 100 \qquad (22.6)$$

The suitable fitting between the experimental and the predicted device responses confirms the suitability of model (22.1). Although the model developed represents a simplified form of the real mechanism ruling the device behavior, it can provide useful information on the qualitative behavior of the device. As an example, real information gathered by model (22.1) is the natural frequency of the system, which is well predicted in both the real cases presented.

Figure 22.6 shows the device response as a function of the displacement imposed in the case of a ferrofluid volume of 0.2 mL and a forcing frequency at 6 Hz. In particular, values averaged over several tests have been reported along with the estimated uncertainty.

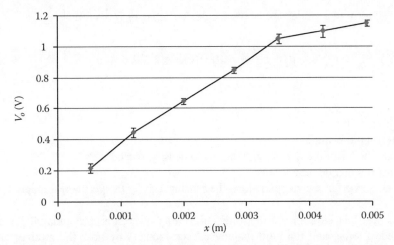

FIGURE 22.6 Response of the inertial sensor as a function of the perturbation imposed at 6 Hz. (© B. Andò, A. Ascia, S. Baglio, and A. Beninato, A ferrofluidic inertial sensor exploiting the Rosensweig effect, IEEE Transactions on Instrumentation and Measurements, 59(5), 1471–1476, 2010.)

Observing the relationship in Figure 22.6 between V_o and x and taking into account the linearity of model (22.1) with the hypothesis (22.2), it is reasonable to assume the consistency of assumption (22.4).

The physical decoupling between the electrical readout and the glass plate, the intrinsic insulation between the electric parts and the liquid medium, the low cost and disposable housing for the liquid media (consisting of a low cost beaker), the reusability of the readout system (the coils and the conditioning electronics, which are the expensive parts of the system), and the adaptability of the proposed sensing strategies to a wide set of real applications, are main advantages of the methodology proposed.

22.4 A FERROFLUIDIC TILT SENSOR

In this section, a tilt sensor, which exploits a ferrofluid oscillating in a resonant regime to prevent the mass adhesion on the housing structure is presented [33,34].

The device consists of a glass pipe filled with deionized water and a drop of ferrofluid. A retainer coil driven by a DC current, I_{DC}, induces a restoring force that attracts the ferrofluidic mass to an equilibrium position along the z axis (the center of the coil). In the case of a null tilt, the equilibrium position is around the center of the coil, while when a tilt is applied the mass of ferrofluid gains a new equilibrium position. The maximum operating range of the device is constrained by the maximum allowable displacement of the ferrofluidic mass roughly corresponding to the length of the retainer coil.

Two sensing coils in differential configuration wounded around the channel implement the readout strategy with the aim of revealing the position of the ferrofluid mass inside the device.

An interesting feature of the device is the use of supplementary coils, which move the ferrofluid mass back and forth at the resonant frequency of the whole system. Such strategy is used to prevent adhesion of the ferrofluidic mass to the glass pipe and consequently to reduce the static friction and to increase the device sensitivity. Actually, the supplementary coils (left and right) are driven at the mechanical resonance frequency of the device by two quadrature AC currents, I_a^L and I_a^R, with the same amplitude.

Another advantage of the novel resonant configuration proposed is the possibility to trap the ferrofluidic mass within one of the two exciter coils in the case of shocks, thus increasing the recovering feature of the device.

A simplified form of a model describing the device behavior is the following:

$$\rho_f V_f \ddot{z} = F_h + (\rho_f - \rho_l])V_f g \sin(\theta) + F_m \qquad (22.7)$$

where
   θ is the applied tilt
   V_f is the ferrofluid volume
   ρ_f and ρ_l are the ferrofluid density and the liquid density, respectively
   F_h is the hydrodynamic drag force
   F_m is the magnetic force generated along the sensing axis by the permanent magnet

Terms on the right side are the hydrodynamic drag force, the gravitational force including the Archimede's force, and the total magnetic force induced by both the retainer coil and the exciter coils.

Figure 22.7 shows real views of the resonant inclinometer prototype developed. The glass pipe has an outer diameter of 8 mm and a length of about 100 mm. A 0.2 mL volume of EFH1 ferrofluid has been injected into the glass pipe [34]. The retainer coil and the sensing coils are wounded on a plastic support and their lengths are $L_r = 0.030$ m and $L_s = 0.015$ m, respectively [34].

(a)

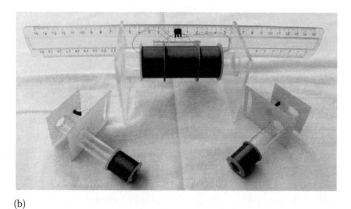

(b)

FIGURE 22.7   (a) The real prototype of the ferrofluid inclinometer. (b) The system components. (© B. Andò, A. Ascia, and S. Baglio, A ferrofluidic inclinometer in the resonant configuration, IEEE Transactions on Instrumentation and Measurements, 59(3), 558–564, 2010.)

In order to preserve the monotonic trend of the readout tool response in the whole displacement range of the ferrofluidic mass, the sensing coils must be juxtaposed to the retainer coil [34].

The exciter coils have been realized on stand-alone plastic supports sliding inside the retainer and sensing coils housing (see Figure 22.7b). This solution allows for the optimal positioning of the exciter coils, which must totally cover the area of the retainer coil to properly drive the ferrofluidic mass. Dedicated electronics was developed for demodulating and filtering signals coming from the two sensing coils [33,34].

In order to test the tilt sensor, a suitable experimental set-up was adopted. It consists of a rotating platform housing the inclinometer and controlled by a high resolution stepper motor to impose the desired tilt on the device.

The frequency of the $I_a^{L,R}$ currents was chosen by estimating the resonance frequency of the device through model (22.1) [34].

Figure 22.8 shows the demodulated output signal, V_{avg}, as a function of the imposed tilt for $I_{DC} = 90$ mA and $I_a^{L,R} = 15$ mA at 0.4 Hz. A tilt approximately ranging between $-20°$ and $20°$ was given to the device with a step of 0.6°. The behavior for a null $I_a^{L,R}$ is also shown for the sake of comparison. In this case, the ferrofluidic mass is not moved around the equilibrium position, and the device collapses to

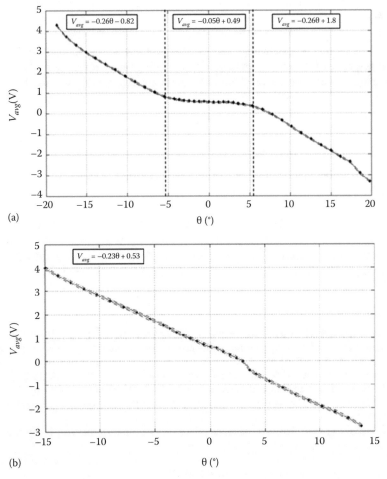

(a)

(b)

FIGURE 22.8 The device response. (a) Nonresonant operation; (b) resonant operation. (© B. Andò, A. Ascia, and S. Baglio, A ferrofluidic inclinometer in the resonant configuration, IEEE Transactions on Instrumentation and Measurements, 59(3), 558–564, 2010.)

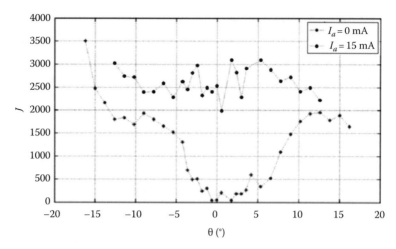

FIGURE 22.9 The J performance index. (© B. Andò, A. Ascia, and S. Baglio, A ferrofluidic inclinometer in the resonant configuration, IEEE Transactions on Instrumentation and Measurements, 59(3), 558–564, 2010.)

the operation mode described in [33]. In this operating condition, the device sensitivity in the interval [−6°–6°] is compromised as evidenced by three different transduction function reported in Figure 22.8a.

Actually, Figure 22.8b evidences a linear trend in the device response, which is related to the resonant operating mode, the mechanism ruling the total force acting on the ferrofluidic mass [34] and the linear response of the readout system in the working range considered.

To estimate the expected device resolution, the following relationship was used:

$$\sigma_\theta = \frac{\sigma_{V\,avg}}{\partial V_{avg}/\partial \theta} \tag{22.8}$$

where

$\sigma_{V\,avg}$ is the standard deviation of several observations of the V_{avg} signal
$\partial V_{avg}/\partial \theta$ represents the sensor responsivity

In order to provide a quantitative estimation of device performances, the figure of merit $J = D\theta/\sigma_\theta$ was used, where $D\theta$ is the operating range for the specific operating conditions. Figure 22.9 shows the behaviors of J for the static and resonant working modes, which confirm the advantages of the resonant operating mode [34].

REFERENCES

1. E. L. Resler, Jr. and R. E. Rosensweig, Magnetocaloric power, *AIAA Journal*, 2(8), 1418, 1964.
2. S. S. Papell, Low viscosity magnetic fluid obtained by the colloidal suspension of magnetic particles, U.S. Patent No. 3, 215, 572, 1965.
3. R. E. Rosensweig, *Ferrohydrodynamics*, Cambridge University Press, New York, 1985.
4. G. S. W. Charles, The preparation of magnetic fluids, in: S. Odenbach (ed.), *Ferrofluids: Magnetically Controllable Fluids and Their Applications*, Lecture Notes in Physics, Springer-Verlag, Berlin, Germany, pp. 3–18, 2002.
5. Yu. I. Dikanskii and O. A. Nechaev, Magnetic fluid structural transformations in electric and magnetic field, *Colloid Journal*, 65(3), 305–330, 2003.
6. B. Andò, A. Ascia, S. Baglio, and N. Pitrone, Magnetic fluids and their use in transducers. *IEEE Magazine on Instrumentation and Measurements*, 9(6), 44–47, 2006.
7. K. Raj et al. Commercial applications of ferrofluids, *Journal of Magnetism and Magnetic Materials*, 85(1–3), 233–245, 1990.

8. B. Andò, S. Baglio, C. Faraci, and C. Trigona, Ferrofluids for a novel approach to the measurement of velocity profiles and shear stresses in boundary layers, *Sensors Conference*, 2009 IEEE Christchurch, New Zealand, pp. 1069–1071.

9. B. Andò, S. Baglio, and A. Beninato, Magnetic fluids for bio-medical application, in: S. C. Mukhopadhyay, and A. Lay-Ekuakille, (eds.), *Advances in Biomedical Sensing, Measurements, Instrumentation and Systems*, Series: Lecture Notes in Electrical Engineering, Vol. 55, Springer-Verlag, Berlin, Germany, pp. 16–28, 2010.

10. Q. A. Pankhurst, J. Connolly, S. K. Jones, and J. Dobson, Applications of magnetic nanoparticles in bio-medicine, *Journal of Physics D: Applied Physics*, 36, 167–181, 2003.

11. B. Y. Kularatne, P. Lorigan, S. Browne, S. K. Suvarna, M. O. Smith, and J. Lawry, Monitoring tumour cells in the peripheral blood of small cell lung cancer patients, *Cytometry*, 50(3), 160–167, 2002.

12. R. S. Molday and D. MacKenzie, Immunospecific ferromagnetic iron–dextran reagents for the labeling and magnetic separation of cells, *Journal of Immunological Methods*, 52(3), 353–367, 1982.

13. S. Morisada, N. Miyata, and K. Iwahori, Immunomagnetic separation of scum-forming bacteria using polyclonal antibody that recognizes mycolic acids, *Journal of Microbiological Methods*, 51(2), 141–148, 2002.

14. C. V. Mura, M. I. Becker, A. Orellana, and D. Wolff, Immunopurification of Golgi vesicles by magnetic sorting, *Journal of Microbiological Methods*, 260(1–2), 263–271, 2002.

15. A. Tibbe, B. de Grooth, J. Greve, P. Liberti, G. Dolan, and L. Terstappen, Optical tracking and detection of immunomagnetically selected and aligned cells, *Nature Biotechnology*, 17, 1210–1213, 1999.

16. R. E. Zigeuner, R. Riesenberg, H. Pohla, A. Hofstetter, and R. Oberneder, Isolation of circulating cancer cells from whole blood by immunomagnetic cell enrichment and unenriched immunocytochemistry in vitro, *The Journal of Urology*, 169(2), 701–705, 2003.

17. C. Scherer and A. M. Figueiredo Neto, Ferrofluids: Properties and applications, *Brazilian Journal of Physics*, 35(3A), 718–727, 2005.

18. C. H. Alexiou, R. Schmid, R. Jurgons, C. H. Bergemann, W. Arnold, and F. G. Parak, Targeted tumor therapy with magnetic drug targeting: Therapeutic efficacy of ferrofluid bound mitoxantrone, in: S. Odenbach (ed.), *Ferrofluids: Magnetically Controllable Fluids and Their Applications*, Springer, Berlin, Germany, p. 233, 2002.

19. M. M. Miller, P. E. Sheehan, R. L. Edelstein, C. R. Tamanaha, L. Zhong, S. Bounnak, L. J. Whitman, and R. J. Colton, A DNA array sensor utilizing magnetic microbeads and magnetoelectronic detection, *Journal of Magnetism and Magnetic Materials*, 225(1–2), 138–144, 2001.

20. B. Andò, A. Ascia, S. Baglio, and A. R. Bulsara, and In V, RTD fluxgate performance in magnetic label-based bioassay: Preliminary result, *EMBC 2006 28th Annual International Conference of the IEEE*, 1, 5060–5063, 2006.

21. G. Q. Hu and W. H. Liao, A feasibility study of a microaccelerometer with magnetorheological fluids, *Proceedings of the 2003 IEEE International Conference on Robotics, Intelligent System and Signal Processing*, Changsha, Hunan, China, October 2003, pp. 825–830.

22. R. Olaru and D. D. Dragoi, Inductive tilt sensor with magnets and magnetic fluid, *Sensors and Actuators A*, 120, 424–428, 2005.

23. N. Greivell and B. Hannaford, The design of a ferrofluidic magnetic pipette, *IEEE Transactions on Biomedical Engineering*, 44(3), 129–135, 1997.

24. A. Hatch, A. E. Kamholz, G. Holman, P. Yager, and K. F. Böhringer, A ferrofluidic magnetic micropump, *Journal of Microelectromechanical System*, 10(2), 215–221, 2001.

25. H. Hartshorne, C. J. Bakhouse, and W. E. Lee, Ferrofluid-based microchip pump and valve, *Sensors and Actuators B*, 99(2–3), 592–600, 2004.

26. C. Yamahata, M. Chastellain, V. K. Parashar, A. Petri, H. Hofmann, and M. A. M. Gijs, Plastic micro-pump with ferrofluidic actuation, *IEEE Journal of Microelectromechanical System*, 14(1), 96–102, 2005.

27. J. Joung, J. Shen, and P. Grodzinski, Micropumps based on alternating high-gradient magnetic fields, *IEEE Transactions on Magnetics*, 36(4), 2012–2014, 2000.

28. G. S. Park and S. H. Park, Design of magnetic fluid linear pump, *IEEE Transactions on Magnetics*, 35(5), 4058–4060, 1999.

29. B. Andò, A. Ascia, S. Baglio, and N. Pitrone, Ferrofluidic pumps: A valuable implementation without moving parts, *IEEE Transactions on Instrumentation and Measurements*, 58(9), 3232–3237, 2009.

30. B. Andò, A. Ascia, S. Baglio, and A. Beninato, The "one drop" ferrofluidic pump with analog control: "FP1_A", *Sensors and Actuators*, 156(1), 251–256, 2009.

31. B. Andò, S. Baglio, and A. Beninato, An IR methodology to assess the behavior of ferrofluidic transducers—Case of study: A contactless driven pump, *Sensors Journal, IEEE*, 11(1), 93–98, November 2010.
32. B. Andò, A. Ascia, S. Baglio, and A. Beninato, A ferrofluidic inertial sensor exploiting the Rosensweig effect, *IEEE Transactions on Instrumentation and Measurements*, 59(5), 1471–1476, 2010.
33. B. Andò, A. Ascia, S. Baglio, and N. Savalli, A novel ferrofluidic inclinometer, *IEEE Transactions on Instrumentation and Measurements*, 56(4), 1114–1123, 2007.
34. B. Andò, A. Ascia, and S. Baglio, A ferrofluidic inclinometer in the resonant configuration, *IEEE Transactions on Instrumentation and Measurements*, 59(3), 558–564, 2010.

Part IV

Sound and Ultrasound Sensors

23 Low-Cost Underwater Acoustic Modem for Short-Range Sensor Networks

Bridget Benson and Ryan Kastner

CONTENTS

23.1   INTRODUCTION

Small, dense, wireless sensor networks are beginning to revolutionize our understanding of the physical world by providing fine resolution sampling of the surrounding environment. The ability to have many small devices streaming real-time data physically distributed near the objects being sensed brings new opportunities to observe and act on the world, which could provide significant benefits to mankind. For example, dense wireless sensor networks have been used in agriculture to improve the quality, yield, and value of crops by tracking soil temperatures and informing farmers of fruit maturity and potential damages from freezing temperatures [1]. They have been deployed in sensitive habitats to monitor the causes for mortality in endangered species [2]. Dense wireless sensor networks have also been used to detect structural damages to bridges and other civil structures to inform authorities of needed repair [3] and have been used to monitor the vibration signatures of industrial equipment in fabrication plants to predict mechanical failures [4].

While wireless sensor-net systems are beginning to be fielded in applications on the ground, underwater sensor nets remain quite limited by comparison [5]. Still, a large portion of ocean research is conducted by placing sensors (that measure current speeds, temperature, salinity, pressure, bioluminescence, chemicals, etc.) into the ocean and later physically retrieving them to download and analyze their collected data. This method does not provide for real-time analysis of data, which is critical for event prediction. Real-time underwater wireless sensor networks that do exist are often sparsely deployed over wide areas. For example, the Deep-Ocean Assessment and Reporting of Tsunami (DART) project consists of 39 stations worldwide acquiring critical data for early detection of tsunamis [6]. The FRONT network consists of about 10 subsurface wirelessly networked sensors spaced about 9 km apart in the inner continental shelf outside Block Island Sound to increase scientific understanding of the coastal ocean [7]. The Seaweb network consists of tens of nodes spaced 2–5 km apart for oceanographic telemetry, underwater vehicle control, and other uses of underwater wireless digital communications [8,9]. Other real-time networks that exist are wired and extremely expensive [10–13].

The existence of small, dense wireless sensor networks on land was made possible by the advent of low-cost radio platforms such as PicoRadio and Mica2 [14,15]. These radio platforms cost a few hundred U.S. dollars, enabling researchers to purchase many nodes with a fixed budget, allowing for dense, short-range deployment. The aquatic counterpart to the terrestrial radio is the underwater acoustic modem. There are a number of acoustic modems currently available, including commercial offerings from companies like Teledyne Benthos, DSPComm, LinkQuest, and Tritech and academic projects, most notably the WHOI MicroModem. Unfortunately, these existing modems' power consumption, ranges, and price points are all designed for sparse, long-range, expensive systems rather than small, dense, and inexpensive sensor-nets [5,16,17]. It is widely recognized that an aquatic counterpart to inexpensive terrestrial radio is required to enable deployment of small dense underwater wireless sensor networks for advanced underwater ecological analyses.

This chapter describes the design of a short-range underwater acoustic modem starting with the most critical component from a cost perspective—the transducer. The transducer is the device that converts electrical energy to/from acoustic energy, which is equivalent to the antenna in radios. The design substitutes a commercial underwater transducer with a home-made underwater transducer using cheap piezoceramic material and builds the rest of the modem's components around the properties of the transducer to extract as much performance as possible. We describe the modem's transducer design, followed by its analog transceiver design and digital transceiver design. We end the chapter by describing real-world tests performed on the complete low-cost modem design, which illustrate that the modem provides bit rates of up to 200 bps for ranges up to 400 m at a components cost of U.S. $350.

This chapter consists of excerpts from the author's PhD thesis [18].

23.2   TRANSDUCER DESIGN

This section describes the design of a low-cost transducer (an electromagnetic device responsible for converting electrical energy to mechanical energy (sound pressure) and vice versa) used as the basis for the design of our low-cost underwater acoustic modem. We first describe the selection of the transducer's piezoceramic based on its type and geometry. We then describe our transducer construction techniques including the selection of wiring and potting compound. We finally describe the calibration procedure used to measure the electromechanical properties of our homemade transducer and present the experimentally determined electromechanical properties that are used to govern the rest of the low-cost modem design.

23.2.1   PIEZOCERAMICS

In 1880, Jacques and Pierre Curie discovered that certain naturally occurring crystalline substances (such as quartz) exhibit an unusual characteristic: when subjected to a mechanical force, the crystals became polarized and when exposed to an electric field the crystals lengthened or shortened according to the polarity and in proportion to the strength of the field. These behaviors were labeled the piezoelectric effect and the inverse piezoelectric effect, respectively [19–21].

In the twentieth century, researchers began to manufacture synthetic materials that exhibit the piezoelectric effect using polycrystalline ceramics or certain synthetic polymers. These materials are relatively inexpensive to manufacture, physically strong, and chemically inert. Common compositions include lead zirconate titanate and barium titanate [19]. The type of the ceramic and its geometry affect the ceramic's piezoelectric properties and are described in more detail in the following subsections.

23.2.1.1   Type

Piezoceramics are categorized into two general groups: hard and soft ceramics. Hard ceramics have low dielectric and mechanical loss and are generally better at producing a signal, whereas soft ceramics have large dielectric losses, low mechanic quality factors, and poor linearity, but are generally better at receiving a signal [19,24]. Either of these ceramics is still capable of producing and receiving signals regardless of which group it is in because of piezoelectric reciprocity. For underwater network communication where one transducer is used for transmitting and receiving for cost effectiveness, a piezoelectric element that is good at doing both is desired. We selected to use a "hard" ceramic due to its low dielectric and mechanical loss and high electromechanical coupling efficiency making it suitable as both a transmitter and receiver.

23.2.1.2   Geometry

After selecting the type of ceramic necessary for the application, the geometry selection is the next important step. The cost of the PZT element can vary greatly and is significantly affected by geometry. Not only are some shapes harder to make, more intricate shapes make poling the ceramic more difficult as well. Element geometry and polarization direction determine the radiated direction of acoustic signals as well as the electromechanical properties of the ceramic element itself, such as resonance frequency, capacitance, generated voltage under load, and displacement. Exactly how they are affected depends on the geometry selected. For underwater communication, ceramics are usually omnidirectional in the horizontal plane to reduce reflection off the surface and bottom [21].

Thus, a ring transducer with radial resonance mode, 26 mm outer diameter, 22 mm inner diameter, and 2.54 mm wall thickness was selected. Specifically, part SMC26D22H13SMQA from Steiner and Martins, Inc. [26] was purchased for approximately $10 per element. For a single radial expanding ceramic ring, the resonance frequency occurs when the circumference approximately equals the operating wavelength [21]. The SMC26D22H13SMQA has an outer diameter of 26 mm (approx. 1 in.) and has a nominal resonance frequency of about 43 kHz.

(a)   (b)   (c)

FIGURE 23.1   (a) Raw PZT, (b) prepotted transducer, and (c) potted transducer.

23.2.2 TRANSDUCER CONSTRUCTION

Although the piezoelectric element is a key component of the transducer, there are other aspects to manufacturing a transducer that are important to its performance. Wiring electrical leads, potting the piezoceramic, and reducing unwanted acoustic radiation should be paid special attention. Figure 23.1 depicts our raw piezoceramic, transducer before potting, and fully potted transducer.

23.2.2.1 Wiring

Using shielded cables to attach to the ceramic will greatly enhance the performance of the transducer. Unshielded wires can act as antennas and pick up much unwanted electromagnetic noise that can bury small signals received by the transducer.

23.2.2.2 Potting

The piezoelectric ceramic needs to be encapsulated in a potting compound to prevent contact with any conductive fluids. Urethanes are the most common material used for potting because of their versatility. The most important design consideration is to find a urethane that is acoustically transparent in the medium that the transducer will be used. Generally, similar density provides similar acoustical properties. A two-part urethane potting compound, EN12, manufactured by Cytec Industries [28] was selected as it has a density identical to that of water, providing for efficient mechanical to acoustical energy coupling.

To pot the ceramic, a tennis ball with a hole cut in it and an ABS pipe cap with holes drilled in it were glued together using silicone and used as our transducer's mold. The ABS cap was only used for mounting. Mixed and uncured urethane was poured into the mold through the holes in the ABS cap. After the urethane was fully cured, the tennis ball and silicone were removed.

The next section describes the procedures used to determine the electromechanical properties of the homemade transducer.

23.2.3 CALIBRATION PROCEDURE

The calibration of a transducer consists of the determination of its electromechanical response as a function of frequency, namely its transmitting voltage response (TVR) and its receiving voltage response (RVR). The TVR is defined as the sound pressure level experienced at 1 m range, generated by the transducer per 1 V of input voltage as a function of frequency. The RVR is a measure of the voltage generated by a plane wave of unit acoustic pressure at the receiver and is a function of frequency. The units of the transmitting response are typically expressed in dB/1 µPa/m, and the units of the receiving response are typically expressed in dB/1 V/µPa. Although numerous calibration procedures exist [31], we use the "comparison method" as it is the simplest transducer calibration procedure wherein the output of the unknown transducer is compared with that of a previously calibrated reference transducer.

In the comparison method, the unknown and reference transducers are placed in a tank of water at known separation (typically 1 m). To obtain the RVR, the reference transducer sends sinusoidal signals of a known duration across the desired frequency range, and the unknown transducer collects the sinusoidal signals over the same duration at each frequency. The collected data represent the combination of the transmitting response of the reference transducer plus the receiving response of the unknown transducer and the effects of attenuation at the separation distance. The RVR may be calculated using the following equations:

$$D = 20 \log_{10} \left(\frac{V_{receiver}}{V_{transmitter}} \right) \tag{23.1}$$

$$\text{RVR} = D_{fromref} + \text{TVR}_{ref} + A \tag{23.2}$$

where
$V_{receiver}$ is the average amplitude voltage of the receiver over the known duration
$V_{transmitter}$ is the average amplitude voltage of the transmitter over the known duration
A is the attenuation of the signal due to the separation distance

To obtain the TVR, the unknown transducer sends sinusoidal signals of reference duration across the desired frequency range and the reference transducer collects the sinusoidal signals over the reference duration at each frequency. The collected data represent the combination of the transmitting response of the unknown transducer plus the receiving response of the reference transducer and the effects of attenuation at the separation distance. The RVR may similarly be calculated as

$$\text{TVR} = D_{toref} - \text{RVR}_{ref} + A \tag{23.3}$$

23.2.4 EXPERIMENTAL MEASUREMENTS

To execute the comparison method, we suspended both the homemade transducer and reference transducer, a spherical ITC 1042 [32], 0.18 m apart in the middle of a 3 m deep, 2 m wide cylindrical test tank filled with salt water. Burst signals of duration 2 ms, across frequencies in 1 kHz increments, were sent from the reference transducer to the homemade transducer and vice versa. Signals were sent and collected via LabView. Adding the TVR and RVR gives a quantity known as the "Figure of Merit," which gives an indication of the transducer's best operating frequencies when acting as both a transmitter and receiver. Figure 23.2 shows the Figure of Merit calculated from the LabView data, indicating an operating frequency range around 40 kHz for transducer T1. The peaks and valleys of the Figure of Merit can be attributed to constructive and destructive interference caused by reflections off the sides of the small calibration tank.

23.2.5 SUMMARY

This section described the design of our low-cost transducer, its ceramic type and geometry, its potting compound and procedure, and its electromechanical properties. The low-cost transducer costs ~$40, is omnidirectional in the horizontal plane, operates in a narrow frequency band around

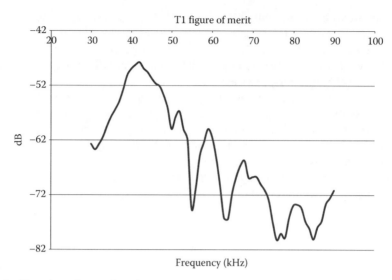

FIGURE 23.2 Transducer figure of merit.

40 kHz, and has a source level of about 140 dB re 1 μPa at 1 m. The next section describes the modem's analog transceiver, which was designed to operate in the transducer's operating frequency range.

23.3 ANALOG TRANSCEIVER DESIGN

The modem's analog transceiver consists of a power amplifier, a power management circuit, an impedance matching circuit, and a preamplifier (Figure 23.3), all of which are described in this section.

23.3.1 POWER AMPLIFIER DESIGN

When designing the power amplifier, we considered the following requirements:

- The amplifier should provide a linear, undistorted output over a relatively wide bandwidth (10–100 kHz) to allow for use with a variety of underwater transducers.
- The amplifier must be power efficient (especially for large output power), as a deployed modem must be powered from batteries.

FIGURE 23.3 Analog transceiver.

An amplifier is said to be linear if it preserves the details of the signal waveform, that is

$$V_0(t) = A V_i(t) \tag{23.4}$$

The amplifier is said to be efficient if it can convert the majority of the DC power of the supply into the signal power delivered to the load. Efficiency is defined as

$$efficiency = \frac{signal\_power\_delivered\_to\_load}{DC\_power\_supplied\_to\_output\_circuit} \tag{23.5}$$

We designed a unique architecture that consists of a Class AB (known for being linear) and a Class D (known for being efficient) amplifier working in parallel to meet our design requirements. The Class AB amplifier provides a highly linear voltage gain of 27 across input voltages and frequencies. The output of the Class AB amplifier is connected to current sense circuitry that in turn controls the secondary amplifier, which is a Class D switching amplifier. The Class D amplifier is inherently nonlinear, but when working in tandem with the Class AB amplifier, it produces a linear output for input voltages greater than 500 mV_{pp} across frequencies. The Class D amplifier provides high power efficiency to the complete amplifier for large power outputs (where the load resistance is below 15 Ω) but must be turned off for lower power outputs where its efficiency drops below that of the Class AB amplifier alone.

23.3.2 Power Management Circuit

The power management circuit is provided to adjust the output power of the transceiver in real-time to match the actual distance between transmitter and receiver. The power management circuit takes the output of the power amplifier and further amplifies it to one of five power levels depending on how far the modem must transmit the signal. This is enabled through five different outputs at the taps on the secondary coil of the transformer. The number of windings on the secondary coil (N) divides the effective resistance of the load by N^2 thus increasing the power. The different transformer taps are connected to the impedance matching network (explained in the next section) by a series of relays. The number of windings (N), voltage output, and transceiver power consumption of each power setting when connected to the homemade transducer and given a 1 V_{pp} input is given in Table 23.1.

23.3.3 Impedance Matching

Impedance matching is the practice of setting the input impedance of an electrical load equal to the output impedance of the signal source to which it is connected in order to maximize power transfer.

TABLE 23.1
Power Management Characteristics

Power Level	No. Windings (N)	V_o (V_{pp})	$P_{consumed}$ (W)
0	0	23	1.2
1	2.5	62	1.8
2	5	122	2.7
3	7.5	180	4.5
4	10	230	6.9

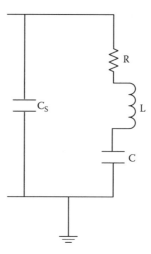

FIGURE 23.4 Electrical equivalent circuit model for a transducer.

Thus, the output impedance of the power amplifier must match the input impedance of the transducer. Matching is obtained when $Z_S = Z_L^*$ (where Z_S is the impedance of the source (or power amplifier) and Z_L^* is the complex conjugate of the impedance of the load [or transducer]).

The transducer's impedance varies across frequencies as it is an active element that can be modeled with the following RLC circuit shown in Figure 23.4 [21]. The static capacitance is the only physical component and is a direct consequence of the type of piezoelectric material and transducer geometry. It can be measured by an RLC meter and for the homemade transducer equals approximately 6.0 nF across frequencies.

To experimentally determine the transducer's electrical impedance, we measured the voltage, current, and phase difference between the voltage and current across frequencies. We then modeled the circuit in Figure 23.4 in PSpice [38] with values of the RLC circuit selected to match the characteristics of the measured values. The RLC values of 750 Ω, 2.3 mH, and 700 pF, respectively, provided good agreement between the measured and simulated values.

As the transducer is mostly capacitive, the impedance matching circuit consists of a single 2.5 mH inductor. This inductor value was experimentally chosen to make the circuit look mostly resistive (0 phase) around the operating frequency (40 kHz).

Although the power coupled to the transducer cannot directly be measured, we can use an equivalent circuit in PSpice to estimate the power delivered to the load. Figure 23.5 shows the majority of the amplifier's output power is coupled to the transducer between 35 and 45 kHz.

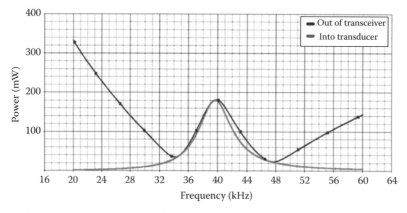

FIGURE 23.5 Estimated power coupled to the transducer.

23.3.4 PREAMPLIFIER DESIGN

When designing the preamplifier for the receiver, we considered the following requirements:

- The preamplifier must amplify signals around the transducer's resonance frequency (40 kHz) and filter out all other frequencies.
- The preamplifier must provide high gain to pick up signals as small as a couple hundred microvolts.
- The design must be easily modifiable to accommodate different transducers with different resonance frequencies and bandwidths.

To meet the aforementioned design requirements of a highly sensitive, high gain, narrow band receiver, the architecture consists of two main components: (1) a 40 dB per decade roll off high-pass filter and (2) an 80 dB per decade rolloff band-pass filter. The combined preamplifier provides an ~80 dB gain around 40 kHz while attenuating low frequencies at a rate of 120 dB per decade and high frequencies at a rate of 80 dB per decade (Figure 23.6). The preamplifier's gain, cut-off, and center frequencies can be easily modified by replacing a few standard resistor and capacitor components.

The current receiver configuration consumes about 240 mW when in standby mode and less than about 275 mW when fully engaged.

23.3.5 SUMMARY

This section described the full design of the analog transceiver including the power amplifier, power management circuit, impedance matching circuit, and preamplifier. The power amplifier is linear in the 10–100 kHz band for inputs greater than 500 mV$_{pp}$ and up to 95% efficient for high power outputs. The impedance matching circuit makes the transducer look resistive between 35 and 45 kHz, coupling the majority of the amplifier's output power to the transducer in that frequency band. The preamplifier provides a flat, high gain for frequencies 38–42 kHz matching the operating frequency of the transducer and allowing for reception of a signal as low as 200 μV. The power management circuit provides five different output power levels that consume 1.2, 1.8, 2.7, 4.5, and 6.9 W, respectively, allowing the modem to adjust the output power in real-time to match it to the actual distance between transmitter and receiver. All components can be easily modified by replacing a few standard components.

The next section describes the design of the digital transceiver.

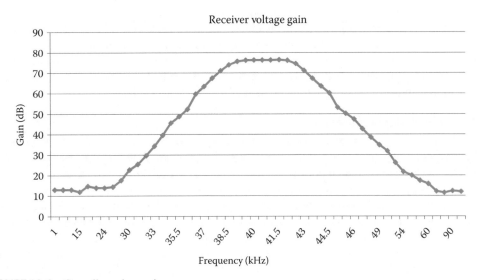

FIGURE 23.6 Overall receiver gain.

23.4 DIGITAL DESIGN

This section describes the design and resource requirements of the digital transceiver of the low-cost acoustic modem. The digital transceiver implements frequency shift keying (FSK), as its proven robustness and simplicity make it an attractive modulation scheme for a low-cost modem design. The design is implemented on a field programmable gate array (FPGA) to serve as a comparison to modem implementations on other digital hardware platforms.

23.4.1 DIGITAL TRANSCEIVER

This subsection provides a brief description of our FPGA implementation of the FSK digital transceiver including the digital down converter, the modulator/demodulator, the symbol synchronizer, and the HW/SW co-design controller. The complete transceiver design is shown in Figure 23.7. Each component was designed in Verilog and initially tested individually in ModelSim [69] to verify its operation prior to system integration.

The digital transceiver design makes use of the parameters given in Table 23.2. The carrier frequency and frequency separation were selected to match the transducer's resonance frequency and narrow bandwidth. Nyquist sampling necessitates that the signal be sampled at 2× the frequency of the highest frequency component of the signal, but in practical applications 4–6× sampling is desired. Thus the given sampling frequency and processing frequency were selected to provide sufficient oversampling of the desired frequency component while being integer multiples of one another. The symbol duration was selected to provide a suitable raw bit rate for low-data rate sensor networking applications.

23.4.1.1 Digital Down Converter

The digital down converter is responsible for converting high resolution signals to lower resolution signals to simplify subsequent processing. It takes the incoming signal *adc_in* and multiplies it with a locally generated 40 kHz signal. The mixed signal then passes through a low pass filter to filter out the high frequency components. Then the signal is downsampled from 192 to 16 kHz to reduce processing power. The low-pass filter is a small 20-tap FIR filter designed using Spiral [70].

23.4.1.2 Modulator/Demodulator

The modulator/demodulator is responsible for translating a bit stream into a waveform and vice versa by shifting the frequency of a continuous carrier to the "mark" or "space" frequency each symbol period. The modulator takes a binary input and selects to generate a sinusoidal wave using a cosine look up table. The demodulator uses the classic "matched" filter structure, which is optimal for FSK detection with white Gaussian noise interference. It works by sending a symbol duration of the received signal through two add-and-shift band-pass filters. An energy detection block is applied to determine the relative amount of energy in each frequency band.

23.4.1.3 Symbol Synchronizer

Symbol synchronization, the ability of the receiver to synchronize to the first symbol of an incoming data stream, is the most critical and complex component in our digital transceiver design. When the modem receiver obtains an input stream, it must be able to find the start of the data sequence to set accurate sampling and decision timing for subsequent demodulation. Without accurate symbol synchronization, higher bit error rates incur, thus reducing the reliability of the wireless network.

Our symbol synchronization approach relies on the transmission of a predefined sequence of symbols, often referred to as a training, or reference sequence. The transmitter sends a packet that begins with the reference sequence and the receiver correlates the received sequence and the known reference sequence in order to locate the start of the packet (and start of the first symbol). When the reference and receiving sequence exactly align with each other, the correlation result reaches a

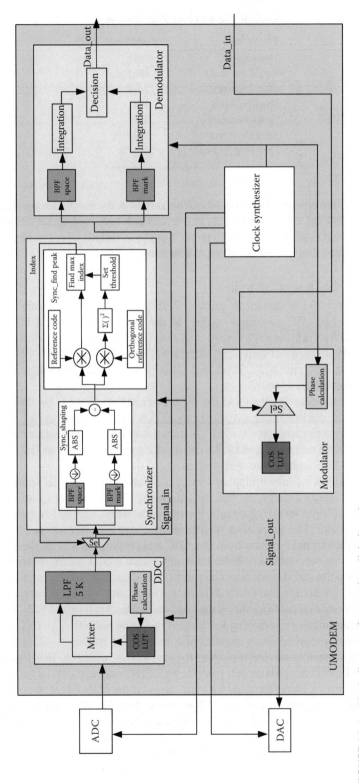

FIGURE 23.7 Block diagram of a complete digital receiver.

TABLE 23.2
Digital Transceiver Parameters

Properties	Assignment
Modulation	FSK
Carrier frequency	40 kHz
Mark frequency	2 kHz
Space frequency	1 kHz
Symbol duration	5 ms
Sampling frequency	192 kHz
Processing frequency	16 kHz

maximum value, and the synchronization point can be located as the maximum point above a predetermined threshold. We use a 15-bit Gold code as our reference sequence and perform a correlation with a 15-bit orthogonal Gold code to set a dynamic threshold. Details of our symbol synchronization implementation and design considerations can be found in [75].

23.4.1.4   HW/SW Co-Design Controller

Xilinx Platform Studio 10.1 is applied to build a HW/SW co-design for accurate control and I/O of the digital transceiver. The co-design consists of the digital transceiver, a UART (Universal Asynchronous Receiver Transmitter) to connect to serial sensors or to a computer serial port for debugging, an interrupt controller to process interrupts received by the UART or the transceiver, logic to configure the on board ADC, DAC, and clock generator, and MicroBlaze, an embedded microprocessor to control the system.

The MicroBlaze processor interfaces to the digital transceiver through two fast simplex links (FSLs), point-to-point, unidirectional asynchronous FIFOs that can perform fast communication between any two design elements on the FPGA that implement the FSL interface. The MicroBlaze interfaces to the interrupt controller and UART core over a peripheral local bus (PLB), based on the IBM standard 64-bit PLB architecture specification.

Upon start-up, the MicroBlaze initializes communication with the digital transceiver through sending a command signal through the FSL bus, signaling the transceiver to turn on. When the transceiver is ready to begin receiving signals, it sends an interrupt back to MicroBlaze to indicate initialization is complete. The transceiver then begins the down conversion and synchronization process, processing the signal received from the ADC and looking for a peak above the threshold to indicate a packet has been received. If the transceiver finds a peak above the threshold, it finds the synchronization point and demodulates the packet. The demodulated bits are stored in the FSL FIFO. When the full packet has been demodulated, the transceiver sends an interrupt indicating a packet has been received and the MicroBlaze may retrieve the packet from the FSL. The transceiver then returns to synchronization, searching for the next incoming packet.

After initialization, the MicroBlaze remains idle, waiting for interrupts either from the transceiver or UART. If it receives an interrupt from the transceiver indicating that a packet has been demodulated, the MicroBlaze reads the bits from the FSL FIFO and sends the bits over the UART to be printed on a computer's HyperTerminal for verification. If the MicroBlaze receives an interrupt from the UART, indicating that the user would like to send data, the MicroBlaze sends a command to the transceiver to send the bitstream the MicroBlaze places in the FSL. The transceiver then modulates the data from the FSL and sends the modulated waveform to the DAC for transmission. The MicroBlaze then returns to waiting for interrupts from the transceiver or the UART and the transceiver returns to synchronization, searching for the next incoming packet. This control flow is depicted in Figure 23.8.

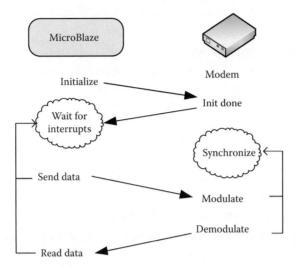

FIGURE 23.8   Digital transceiver control flow.

23.4.2   RESOURCE REQUIREMENTS

Table 23.3 shows the FPGA hardware resources occupied for each component of the digital transceiver. The resources reported for "Total" include the complete digital transceiver and HW/SW co-design controller. These resources were mapped for a Spartan 3 XC3S4000, the smallest device in the established Spartan-3 family that can fit the design.

We obtained a power estimate of the complete FSK modem design on various FPGA devices by entering the resource values of the total modem into the Xilinx XPower Estimator 9.1.03 and the Altera Cyclone IV PowerPlay Early Power Estimator. We acknowledge that the resource values determined for one device are not an exact estimate of the resource values needed for another device, but they do provide us with a reasonable first-order resource (and power) estimate. Also note that the estimates reported for the Xilinx devices are more accurate than those reported for the Altera device as the current design makes use of a Xilinx MicroBlaze core, which would have to be replaced with the Altera NIOS processor, should the design be implemented on an Altera chip. The devices reported are in device families known for their low power consumption (the Xilinx Spartan 6 and the Altera Cyclone IV being some of the newest FPGA device families on the market). The particular devices reported are the smallest devices in their family that fit the total modem design. The letters "Q," "D," and "T" in Table 23.4 stand for "quiescent," "dynamic," and "total" power, respectively. The last column of Table 23.4 reports the size of device family's CMOS technology.

TABLE 23.3
Digital Design Resource Usage

	Occupied Slices	LUTs	BRAMs
Modulator	95	184	9
DDC	284	541	9
Demodulator	1,025	1,980	1
Synchronizer	12,000	22,101	2
Total	16,706	29,076	55
(% Spartan 3)	(60%)	(51%)	(57%)

TABLE 23.4

FPGA Power Consumption

Family	Device	Q Pwr (W)	D Pwr (W)	T Pwr (W)	Tech (nm)
Spartan 3	XC3S4000	0.274	0.105	0.379	90
Spartan 6	XC6SLX150T	0.212	0.021	0.255	45
Cyclone IV	EP4CE30	0.087	0.06	0.147	60

TABLE 23.5

Digital Transceiver Design Comparison

Modem	Mod	Category	Platform	Total Pwr (W)	Cost ($)
[50]	FSK	FixedDSP	TMS320C5416	0.180	45
[50]	PSK	FP DSP	TMS320C6713	2.0	25
[51]	DSSS	FP DSP	TMS320C6713	1.6	25
[54]	FSK	MCU	Blackfin 533	0.280	25
Ours	FSK	FPGA	XC6SLX150T	0.233	14
			EP4CE30	0.147	40

From Table 23.4, we observe that the quiescent power contributes significantly to the total power of the design. Thus, it is important to select a device that offers low quiescent power. Altera has invested considerable resources in reducing static power in their products as evidenced by our design's low quiescent power on the Cyclone IV. Furthermore, the newer technologies (Spartan 6 and Cyclone IV) consume less power than the older technology, suggesting that with continued technological advancements, FPGA power consumption will continue to decrease.

Table 23.5 compares the total digital hardware platform processing power and cost of our modem design with existing research underwater modem digital designs that report their total processing power. As previously described, these research modem designs use various modulation schemes (FSK, PSK, or DSSS) and are implemented on various hardware platforms.

From Table 23.5, we notice that the designs implemented on floating point DSPs consume considerably more power than any of the other designs. However, these designs use more complex modulation schemes and thus require more resources. The FSK design in [50] on the fixed point DSP and the FSK design in [54] on a microcontroller provide comparable power consumption to our FSK design on an FPGA.

The cost reported for the DSP, microcontroller, and FPGA devices were all found on electronic distributer websites. These estimates suggest that the FPGA design also provides a comparable cost to other digital hardware platforms.

23.4.3 SUMMARY

This section described the design of the low cost modem's FSK FPGA-based digital transceiver including the modulator, digital down converter, symbol synchronizer, demodulator, clock generator, and HW/SW co-design. The design operates at 200 bps in a narrow band around a 40 kHz carrier (to match the operating frequency of the transducer and the design of the analog transceiver) and provides comparable cost and power consumption to other frequency shift keying based modem designs. The next section describes the system tests used to evaluate the functionality and performance of the complete modem design.

23.5   SYSTEM TESTS

In order to verify the operation of the low-cost modem, we first tested the analog components and digital components separately and then tested the full, integrated system in a real-world underwater environment at various ranges. This section describes the results of the integrated tests and summarizes the performance of our modem.

23.5.1   Integrated Tests

After verifying the correct operation of the analog and digital hardware separately, we conducted integrated system tests of the complete modem design in a tank, pool, and lake. We measured the multipath and bit error rate for each test. The multipath measurements and test results are described in the following subsections.

23.5.1.1   Multipath Measurements

Underwater, there exist multiple paths from the transmitter to receiver or multipath. Two fundamental mechanisms of multipath formation are reflection at the boundaries (bottom, surface, and any objects in the water) and ray bending (where rays of sound bend toward regions of lower propagation speed). The amount of multipath seen at the receiver depends on the locations of the transmitter and receiver and the geometric and physical properties of the environment.

The extent of the multipath at a receiver can be characterized by the multipath delay spread. Given the amplitude delay profile, $A_c(\tau)$, with effective signal length, M, the mean delay, $\overline{\tau}$, and the rms delay spread, τ_{rms}, are given as [80]

$$\overline{\tau} = \frac{\sum_{n=1}^{M} \tau A_c(\tau)^2}{\sum_{n=1}^{M} A_c(\tau)^2} \tag{23.6}$$

$$\tau_{rms} = \sqrt{\frac{\sum_{n=1}^{M} (\tau - \overline{\tau})^2 A_c(\tau)^2}{\sum_{n=1}^{M} A_c(\tau)^2}} \tag{23.7}$$

For frequency shift keying, multipath will cause intersymbol interference when the multipath delay spread is larger than the symbol duration. Intersymbol interference is a form of distortion of a signal in which one symbol interferes with subsequent symbols. This is an unwanted phenomenon as the previous symbols have similar effect as noise, thus making the communication less reliable [81]. Therefore, because the modem has a symbol duration of 5 ms, the delay spread of the channel must be less than 5 ms to ensure reliable communication.

23.5.1.2   Westlake Tests

After completing tests in a tank and a pool, we conducted full integrated system tests in Westlake, a freshwater lake in Westlake Village, CA, at 5, 50, 95, and 380 m distances. The transmitter was located on a dock, and the receiver was located on a boat. Five packets consisting of the reference code followed by 1000 randomized bits were sent from transmitter to receiver using all power levels. Tests at the 95 and 380 m distance were incomplete due to the inability to remain near a private dock for a length of time. The calculated delay spreads for the 5, 50, and 95 m distance tests were 2, 1.5, and 2.66 ms, respectively. Multipath at the 380 m distance was not measured.

The integrated system test results in terms of distance, multipath delay spread, bit error rate, and signal to noise ratio (SNR) are summarized in Figure 23.9. As anticipated, the modem performed well (having a bit error rate of <5%) in environments with a multipath spread less than 5 ms. The results also suggest that higher SNR will only improve performance for environments with low multipath.

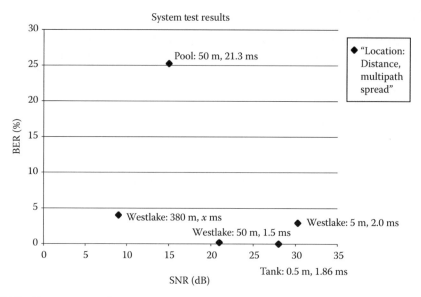

FIGURE 23.9 System test results.

TABLE 23.6
Commercial Underwater Acoustic Modem Comparison

Company	Modem	Freq (kHz)	Tx Pwr (W)	Range (km)	Rx Pwr (W)	Mod	Bit Rate (bps)	Cost ($)
Aquatec	AQUAModem	8–16	20	10	0.6	DSSS	300–2,000	>7,600
DSPComm	AquaComm	16–30	Varied	3	Varied	DSSS/OFDM	480	6,600
TriTech	MicronModem	20–24	7.92	0.5	0.72	DSSS	40	3,500
WHOI	MicroModem	25	<50	1–10	0.23/2	FSK/PSK	80/5,400	8,100/9,400
Benthos	ATM885	16–21	28–84	2–6	0.7	FSK/PSK	140–15,360	7,200–11,000
EvoLogics	S2CM48/78	48–78	2.5–80	1	0.5	S2C	15,000	12,500
LinkQuest	UWM2000H	NS	1.5	0.8	NS	Proprietary	9,600	7,000
Ours	Low-cost	40	1.3–7.0	400	0.42	FSK	200	350[a]

[a] Component cost only.

TABLE 23.7
Research Underwater Acoustic Modem Comparison

Modem	Platform	Mod	Bit Rate (bps)	Range (km)	BER	References
USC	MCU	FSK	NS	NS	10^{-5} (coded)	[52]
UCI	Tmote	FSK	12	5	10%	[83]
uConn	DSP	Varied	Varied	Varied	NS	[84]
AquaModem	DSP	DSSS	133	440	1%	[46]
Kookmin	MCU	NS	5000	30	NS	[53]
AquaNode	MCU	FSK	300	400	NS	[54]
Ours	FPGA	FSK	200	400	4%	[18]

23.5.2 Summary

This section described the integrated system tests used to evaluate the functionality and performance of the complete modem design. These tests prove that a short-range underwater acoustic modem can be designed from low-cost components. The tests indicate that the modem can support data rates of 200 bps for ranges up to ~400 m with the power characteristics given in Table 23.1. Tables 23.6 and 23.7 show how the modem compares with existing commercial and research modems, respectively.

23.6 CONCLUSION

This chapter described the design and initial testing of a functional low-cost underwater acoustic modem prototype for short-range underwater sensor networks. The modem can support data rates of 200 bps for ranges up to ~400 m in environments with less than 5 ms multipath delay spread.

REFERENCES

1. R. Beckwith, D. Teibel, and P. Bowen, Unwired wine: Sensor networks in vineyards, *Proceedings of IEEE Sensors*, 2, 561–564, 2004.
2. A. Mainwaring, J. Polastre, R. Szewczyk, and D. Culler, Wireless sensor networks for habitat monitoring, *Proceedings of the ACM Workshop on Sensor Networks and Applications*, New York, September 2002.
3. J. Lynch, K. Law, E. Straser, A. Kiremidjian, and T. Kenny, The development of a wireless modular health monitoring system for civil structures, *Proceedings of the MCEER Mitigation of Earthquake Disaster by Advanced Technologies Workshop*, Las Vegas, NV, November 2000.
4. N. Ramanathan, M. Yarvis, J. Chhabra, N. Kushalnagar, L. Krishnamurthy, and D. Estrin, A stream-oriented power management protocol for low duty cycle sensor network applications, *Proceedings of the IEEE Workshop on Embedded Networked Sensors*, Sydney, New South Wales, Australia, May 2005.
5. J. Heidemann, W. Ye, J. Wills, A. Syed, and Y. Li, Research challenges and applications for underwater sensor networking, *Proceedings of IEEE Wireless Communications and Networking Conference (WCNC)*, Istanbul, Turkey, pp. 263–270, 2006.
6. H. Milburn, A. Nakamura, and F. I. Gonzalez, Deep Ocean Assessment and Reporting of Tsunamis (DART): Real-time tsunami reporting from the deep ocean, NOAA online report, http://www.ndbc.noaa.gov/Dart/milburn_1996.shtml, 1996.
7. D. Codiga, J. A. Rice, and P. S. Bogden, Real-time wireless delivery of subsurface coastal circulation measurements from distributed instruments using networked acoustic modems, *Proceedings of MTS/IEEE Oceans 2000*, Piscataway, NJ, 2000.
8. J. G. Proakis, E. M. Sozer, J. A. Rice, and M. Stojanovic, Shallow water acoustic networks, *IEEE Communications Magazine*, 3, 114–119, 2001.
9. E. M. Sozer, M. Stojanovic, and J. G. Proakis, Underwater acoustic networks, *IEEE Journal of Oceanic Engineering*, 25, 72–83, 2000.
10. Ocean Research Interactive Observatory networks Website (ORION), http://www.orionprogram.org.
11. Laboratory for the Ocean Observatory Knowledge Integration Grid (LOOKING) Website, http://lookingtosea.ucsd.edu.
12. Monterey Accelerated Research System (MARS) Website, http://www.mbari.org/mars/.
13. NEPTURE Website, http://www.neptune.washington.edu.
14. D. Estrin, L. Girod, G. Pottie, and M. Srivastava, Instrumenting the world with wireless sensor networks, *IEEE International Conference on Acoustics, Speech, and Signal Processing*, Salt Lake City, UT, 2001.
15. J. M. Rabaey, M. J. Ammer, J. L. da Silva, D. Patel, and S. Roundy, PicoRadio supports ad-hoc ultra-low power wireless networking, *IEEE Computer*, 33, 42–48, 2000.
16. I. F. Akyildiz, D. Pompili, and T. Melodia, Challenges for efficient communication in underwater acoustic sensor networks, *ACM Sigbed Review*, 1(2), 3–8, July 2004.
17. R. Jurdak, C. V. Lopes, and P. Baldi, Battery lifetime estimation and optimization for underwater sensor networks, in: *Sensor Network Operations*, S. Phoha, T. F. LaPorta, and C. Griffin, Eds. IEEE Press, New York, 2004.
18. B. Benson, Design of a low-cost underwater acoustic modem for short-range sensor networks. PhD Thesis. Department of Computer Science and Engineering. University of California, San Diego, CA, 2010.

19. *Sensor Design Fundamentals: Piezoelectric Transducer Design for Marine Use*, Airmar Technology Corporation, 2000.
20. Piezo Theory, APC International Ltd., http://www.americanpiezo.com/piezo_theory/index.html.
21. C. H. Sherman and J. L. Butler, *Transducers and Arrays for Underwater Sound*, Springer, New York, 2007.
22. Revision of DOD-STD-1376A, Ad Hoc Subcommittee Report on Piezoceramics, April 1986.
23. MIL-STD-1376B (Notice 1), Military Standard Piezoelectric ceramic material and measurements guidelines for sonar transducers, July 1999.
24. Channel Industries Inc., Piezoelectric Ceramics Catalog, online document, www.channelindustries.com, Jun 2009.
25. Morgan ElectroCeramics, http://www.morganelectroceramics.com.
26. Steminc, Steiner & Martins, Inc., http://www.steminc.com/.
27. O. B. Wilson, *An Introduction to Theory and Design of Sonar Transducers*, Peninsula Publishing, Monterey, CA, 1985.
28. Cytec Industries, http://www.cytec.com/.
29. Urethanes, Silicones, and Epoxies Seminar, *Deep Sea Power and Light Facilities*, San Diego, CA, 92123-1817, May 2009.
30. How-To Guide for Mold Making and Casting, Smooth-On, www.smooth-on.com.
31. R. Urick, *Principles of Underwater Sound for Engineers*, McGraw-Hill Book Company, New York, 1967.
32. ITC-1042, Deep water omnidirectional transducer, http://www.itc-transducers.com.
33. Marport Deep Sea Technology, http://www.marport.com/.
34. H. Krauss, C. Bostian, and F. Raab, *Solid State Radio Engineering*, John Wiley & Sons, Hoboken, NJ, 1980.
35. J. Honda and J. Adams, Application Note AN-1071, International Rectifier, Class D Audio Amplifier Basics, www.irf.com/technical-info/appnotes/an-1071.pdf.
36. Electronics Tutorial about Amplifiers, Electronics-Tutorials, http://www.electronics-tutorials.ws/amplifier/amp_1.html.
37. C. Verhoeven, A. van Staveren, G. Monna, M. Kouwenhoven, and E. Yildiz, *Structured Electronic Design: Negative Feedback Amplifiers*, Kluwer Academic, Boston, MA, 2003.
38. Cadence PSpice A/D and Advanced Analysis http://www.cadence.com/products/orcad/pspice_simulation/Pages/default.aspx.
39. L. Freitag, M. Grund, S. Singh, J. Partan, P. Koski, and K. Ball, The WHOI micro-modem: An acoustic communications and navigation system for multiple platforms, *MTS/IEEE OCEANS*, Washington, DC, pp. 1086–1092, 2005.
40. D. B. Kilfoyle and A. B. Baggeroer, The state of the art in underwater acoustic telemetry, *IEEE Journal of Oceanic Engineering*, 25(1), 4–27, January 2000.
41. I. F. Akyildiz, D. Pompili, and T. Melodia, Underwater acoustic sensor networks: Research challenges, *Ad Hoc Networks (Elsevier)*, 3(3), 257–279, May 2005.
42. L. Liu, S. Zhou, and J.-H. Cui, Prospects and problems of wireless communication for underwater sensor networks, *Wireless Communications and Mobile Computing*, Crete Island, Greece, pp. 977–994, 2008.
43. A. Phadke, *Handbook of Electrical Engineering Calculations, Technology and Engineering*, Marcel Dekker, New York, 1999.
44. M. Stojanovic, J. Catipovic, and J. G. Proakis, Phase coherent digital communications for underwater acoustic channels, *IEEE Journal of Oceanic Engineering*, 19(1), 100–111, 1994.
45. L. Freitag, M. Stojanovic, S. Singh, and M. Johnson, Analysis of channel effects on direct-sequence and frequency-hopped spread spectrum acoustic communications, *IEEE Journal of Oceanic Engineering*, 26(4), 586–593, October 2001.
46. R. A. Iltis, H. Lee, R. Kastner, D. Doonan, T. Fu, R. Moore, and M. Chin, An underwater acoustic telemetry modem for eco-sensing, *Proceedings of MTS/IEEE Oceans*, Washington, DC, September 2005.
47. B. Li, S. Zhou, J. Huang, and P. Willett, Scalable OFDM design for underwater acoustic communications, *Proceedings of International Conference on ASSP*, Las Vegas, NV, March 3–April 4, 2008.
48. J. A. C. Bingham, Multicarrier modulation for data transmission: An idea whose time has come, *IEEE Communications Magazine*, 28, 5–14, May 1990.
49. Z. Wang and G. B. Giannakis, Wireless multicarrier communications: Where Fourier meets Shannon, *IEEE Signal Processing Magazine*, 17(3), 29–48, May 2000.
50. Micro-Modem Overview, Woods Hole Oceanographic Institution, http://acomms.whoi.edu/umodem/.

51. D. Doonan, AquaModem Electronics Engineer, Personal Communication, May 2006.
52. J. Wills, W. Ye, and J. Heidemann, Low-power acoustic modem for dense underwater sensor networks, *Proceedings of ACM International Workshop on Underwater Networks*, Los Angeles, CA, 2006.
53. J. Namgung, N. Yun, S. Park, C. Kim, J. Jeon, and S. Park, Adaptive MAC protocol and acoustic modem for underwater sensor networks, Demo Presentation, *Proceedings of ACM International Workshop on Underwater Networks*, Berkeley, CA, November 2009.
54. I. Vasilescu, C. Detweiler, and D. Rus, AquaNodes: An underwater sensor network, *Proceedings of ACM International Workshop on Underwater Networks*, Montreal, Quebec, Canada, September 2007.
55. Analog Devices' New Blackfin Family Offers the Fastest and Most Power-Efficient Processors for Their Class, Radio Locman, 2003. http://www.radiolocman.com/news/new.html?di = 475.
56. M. Yovits, *Advances in Computers*, Academic Press, New York, pp. 105–107, 1993.
57. A. Amara, F. Amiel, and T. Ea, FPGA vs. ASIC for low power applications, *Microelectronics Journal*, 37, 669–677, 2006.
58. E. Rocha, *Implementation Tradeoffs of Digital FIR filters, Military Embedded Systems*, Open Systems Publishing, 2007.
59. R. Kastner, A. Kaplan, and M. Sarrafzadeh, *Synthesis Techniques and Optimizations for Reconfigurable Systems*, Kluwer Academic, Boston, MA, 2004.
60. W. Mangione-Smith, B. Hutchings, D. Andrews, A. DeHon, C. Ebeling, R. Hartenstein, O. Mencer et al., Seeking solutions in configurable computing, *Computer*, 30(12), 38–43, December 1997.
61. A. DeHon and J. Wawrzynek, Reconfigurable computing: What, why, and implications for design automation, *Proceedings 1999 Design Automation Conference, IEEE*, New Orleans, LA, pp. 610–615, 1999.
62. K. Bondalapati and V. K. Prasanna, Reconfigurable computing systems, *Proceedings of the IEEE*, 90, 1201–1217, 2002.
63. K. Compton and S. Hauck, Reconfigurable computing: a survey of systems and software, *ACM Computing Surveys*, 34, 171–210, 2002.
64. P. Schaumont, I. Verbauwhede, K. Keutzer, and M. Sarrafzadeh, A quick safari through the reconfiguration jungle, *Proceedings of the 38th Design Automation Conference (IEEE Cat. No.01CH37232)*, Las Vegas, NV, ACM, pp. 172–177, 2001.
65. M. LaPedus, FPGAs can outperform DSPs, study says, DSP DesignLine, 11/13/2006.
66. BDTI, FPGAs vs. DSPs: A look at the unanswered questions, DSP Design Line 1/11/2007.
67. Z. Yan, J. Huang, and C. He, Implementation of an OFDM underwater acoustic communication system on an underwater vehicle multiprocessor structure, *Frontiers of Electrical and Electronic Engineering in China*, 2(2), 151–155, April 2007.
68. E. M. Sozer and M. Stojanovic, Reconfigurable acoustic modem for underwater sensor networks, *Proceedings of ACM International Workshop on Underwater Networks*, Los Angeles, CA, 2006.
69. ModelSim SE 6.4a. http://model.com/content/modelsim-downloads.
70. Spiral. http://spiral.net/hardware/filter.html.
71. S. Bregni, *Synchronization of Digital Telecommunications Networks*, Wiley, Chichester, U.K., 2002.
72. E. M. Sozer and M. Stojanovic, Underwater acoustic networks, *IEEE Journal of Oceanic Engineering*, 25(1), January 2000.
73. C. Chien, *Digital Radio System on a Chip: A System Approach*, Springer, 2001.
74. B. Watson, FSK: Signals and semodulation, W. J. Communications, July 2004 http://www.wj.com/pdf/technotes/FSK_signals_demod.pdf.
75. Y. Li, B. Benson, X. Zhang, and R. Kastner, Hardware implementation of symbol synchronization for underwater FSK, *IEEE International Conference on Sensor Networks, Ubiquitous, and Trustworthy Computing*, Newport Beach, CA, 2010.
76. Y. Li, B. Benson, L. Chen, and R. Kastner. Determining the suitability of FPGAs for a low-cost, low-power, underwater acoustic modem, *Future Control and Automation*, 173, 509–517, October 2010.
77. DINI Group, DNMEG_ADDA, http://www.dinigroup.com/index.php?product = DNMEG_ADDA.
78. M-Audio Ltd, http://www.m-audio.com.
79. Xilinx Chipscope Pro, http://www.xilinx.com/tools/cspro.htm.
80. H. Li, D. Liu, J. Li, and P. Stoica, Channel order and RMS delay spread estimation with application to AC power line communications, *Digital Signal Processing*, 13, 284–300, 2003.
81. W. Dally and J. Poulton, *Digital Systems Engineering*, Cambridge University Press, Cambridge, U.K., pp. 280–285, 1998.

82. B. Benson, Y. Li, B. Faunce, K. Domond, D. Kimball, C. Schurgers, and R. Kastner, Design of a low-cost, underwater acoustic modem for short-range sensor networks, *IEEE Oceans Conference*, Seattle, WA, 2010.
83. R. Jurdak, C. V. Lopes, and P. Baldi, Software acoustic modems for short range mote-based underwater sensor networks, *Proceedings of IEEE Oceans Asia Pacific*, Singapore, May 2006.
84. H. Yan, S. Zhou, Z. Shi, and B. Li, A DSP implementation of OFDM acoustic modem, *Proceedings of ACM International Workshop on Underwater Networks*, Montreal, Quebec, Canada, September 2007.

24 Integrating Ultrasonic Standing Wave Particle Manipulation into Vibrational Spectroscopy Sensing Applications

Stefan Radel, Johannes Schnöller, and Bernhard Lendl

CONTENTS

24.1 INTRODUCTION

24.1.1 MOTIVATION

Vibrational spectroscopy is a group of optical measurement techniques increasingly popular in process analytical chemistry because of the ability to directly provide molecular-specific information about a given sample under investigation. Spectra can be recorded from solids, liquids, and gases, providing qualitative as well as quantitative information on the chemical composition of the sample. In the context of real-time process analytical chemistry, advances in vibrational spectroscopy are of interest as great part of the desired (bio)chemical information can be extracted from near infrared, mid-infrared, or Raman spectra.

The growing use of bioprocesses as a manufacturing route for, e.g., antibiotics and other medical compounds, enforces the development of reliable, automated sensors for bioprocess monitoring and control. Such sensors are a key for optimal system performance [1,2] as continued analysis is needed in order to control the monitored bioprocess. State-of-the-art sensor systems provide information on physical parameters (p, T) but only a few chemical parameters. Among these, pH, oxygen, and carbon dioxide concentrations in liquid and gas phase may be mentioned. Fast response times of at least one order of magnitude faster compared to the generation time of the observed microorganism are necessary [3,4]. Real-time information on the chemical composition and on the physiological status of the employed microorganism would be of high diagnostic value; however, due to experimental difficulties, such sensors are not available so far. Moreover, the development of new sensor designs is triggered by the rapid progress in biotechnology.

In most cases, it is necessary to apply some kind of handling technique to control the location of the living sample when investigating biological cells. Common approaches based on optical and electrical techniques are used for the positioning of microparticles. Optical tweezers exploit gradient forces exerted on dielectric particles in highly focused light beams [5]. Dielectrophoresis utilizes forces brought about by non-uniform electric fields, inducing a dipole moment in uncharged particles or cells [1,3].

24.1.2 ULTRASONIC PARTICLE MANIPULATION

Although a rather old phenomenon [6–8], the manipulation of small particles by the use of an ultrasonic standing wave (USW) has attracted broad notice for application purposes during the past decades. The technique exploits the so-called radiation forces acting on μm particles when a suspension is irradiated by a USW in the low MHz range. The general source of these forces is the spatial gradient of the sound waves' acoustic pressure [6,9,10].

The formation of a USW in a resonator is shown in Figure 24.1. A wave is emitted by a sound source, the transducer, in the direction toward a reflector (Figure 24.1A). The incoming and reflected waves are both progressing; however, their superposition (Figure 24.1B) is stationary, i.e., the wave does not change its location over time (and therefore is called a *standing* wave). The envelope (Figure 24.1C) describes the maximum of displacement or sound pressure of the ultrasonic field for a given location.

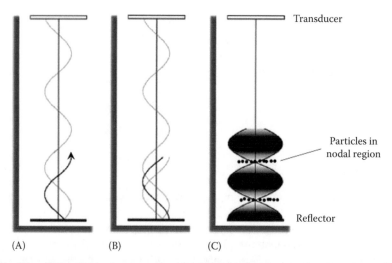

(A)   (B)   (C)

FIGURE 24.1 So-called radiation forces exerted on small (compared to the sound wavelength) particles by a USW: A transducer emits a wave, which is reflected at the reflector (A). A USW is generated by the superposition of the incident wave and the reflection (B). Commonly a USW is depicted as amplitude distribution (envelope), which is spatially fixed (C). Suspended particles are driven into the nodes, i.e., regions with vanishing acoustic pressure. (With kind permission from Springer Science+Business Media: *Elektrotechnik und Informationstechnik*, Ultrasonic particle manipulation exploited in on-line infrared spectroscopy of (cell) suspensions, 125(3), 2008, 76–81, Radel, S., Schnöller, J., Lendl, B., Gröschl, M., and Benes, E.)

The radiation forces direct solid particles like cells toward the pressure nodes of the USW, i.e., regions of vanishing sound pressure. Particle diameter typically ranges from 1 to 50 μm. If the transducer emits a plane wave, the pressure nodes will be planes as well. The line pattern of Figure 24.1C represents these pressure nodal planes of the standing wave field; parallel planes of concentrated particles are half a wavelength apart.

USWs have been investigated for filtration purposes in different set-ups [11,12]. Exploitation in industrial environments was especially successful in bio-tech applications; utilizing the "virtual" mesh of the ultrasonic field, the BioSep (Applikon, the Netherlands) is a commercially available cell retention system for continuous filtration of high-density cultures [7,10,13,14]. The special nature of the filtration mechanism has some advantages, no fouling or clogging occurs and the design allows for in-place hot-steam sterilization. The technology is used, among others, in pharmacological fermentations, separation efficiency reaches 99% and above.

In recent years, USW particle manipulation in small cavities and microfluidic devices has become popular [15]. Among the reported applications were microparticle manipulation [16–18], separation and filtration of suspensions [19–21], and particle sorting by size [22–24]. USW microfluidic devices have also been combined with other manipulation techniques like dielectrophoresis [22,25,26] or micromechanical devices [25].

An advantage of using a USW in comparison with the aforementioned handling approaches is the simultaneous impact on the sonicated volume. Therefore, all particles in this region are manipulated, previous knowledge of their position is not necessary [9]. Typically, it takes less than a second until a spatial distribution, like in Figure 24.1C, is reached.

Much interest was received by application of USWs for the micromanipulation of biological material; applicability has been shown for hybridomas, animal and plant cells, red blood cells, and DNA [27–31]. Reports throughout the time showed the exertion of acoustic radiation forces on various cell lines to be gentle and not affecting viability as long as the cells were kept in the pressure nodes of a USW in the MHz range [32,33]. This has been holding true for microfluidic systems as well [32,34].

The application of USW particle manipulation has long been intensively investigated for reliable sample concentration to enhance sensor applications [23,26,35], including exploitation for diagnostic purposes in medical environments [36,37]. Moreover, it was possible to selectively manipulate cells of a certain type on basis of their viability [38–40]. In this context, the application of a USW for cell washing, i.e., the exchange of the host liquid of a cell suspension should be mentioned [38].

A cell system often used in studies about ultrasonic particle manipulation is *Saccharomyces cerevisiae*, baker's yeast [41–43]. This microorganism is of advantage as a biological model for availability and safety while being important in many fields of biotechnological environments, e.g., in the pharmaceutical industry. The spherical shape of the yeast cell (Ø 4–10 μm) is advantageous in respect to the applicability of theoretical results, usually assuming spherical particles [43,44].

24.1.3 FOURIER TRANSFORM INFRARED ATTENUATED TOTAL REFLECTION

Fourier transform infrared (FT-IR) spectroscopy is a well-developed method in chemical analysis. The incident IR radiation excites parts of the molecules in the sample; therefore, a certain amount of light energy at a given light wavelength is converted into vibrational energy, hence absorbed. Generally speaking, the resonance frequencies of the resulting vibrations depend on the type of the chemical bond and the masses of the involved atoms or groups, i.e., the chemical environment. The acquisition of an IR absorption spectrum can be conducted in minutes to seconds delivering specific molecular information about the sample in the optical pathway [45]. Constantly new devices and concepts for advanced chemical analysis have been developed during recent years [46].

Many influences are to be considered when examining at an IR spectrum; any energy losses along the light path influence the intensity at a given wavenumber measured. Therefore, absorption spectra are given in reference to the "background," i.e., a spectrum recorded prior to introducing the sample into the beam, hence including all influences except the sample's. The absorption A is then calculated by the Beer–Lambert–Bouguer law:

$$A = \log \left(\frac{I_0}{I} \right) = c * \varepsilon * l$$

where
  I_0 denotes the intensity of the incoming light
  I is the intensity when the sample is additionally present in the light path

The absorbance A is proportional to the concentration c and the molar absorptivity ε of an analyte and the light path length l, respectively.

The result of a measurement is therefore the order of magnitude in the decrease of light intensity due to the presence of the sample at a given wavenumber. IR spectra are given in dimensionless absorbance units (AU) versus wavenumber (cm^{-1}), i.e., the reciprocal of the light wavelength.

Using this technique, the presence of practically any molecule can be determined if present at sufficiently high concentrations. For mid-IR spectroscopy, concentration sensitivities for measurements in aqueous phase are in the order of few hundred mg/L. This region in the electromagnetic spectrum (approximately 4000–400 cm^{-1}) is as well called the "fingerprint" region, because most biological substances display specific absorption patterns. Beyond that, dependent measurands like the pH have shown to be assessable by mid-IR spectroscopy [47].

FT-IR spectroscopy in combination with attenuated total reflection (ATR) sensing elements is a currently developing, very promising means for process and bio-process monitoring. The ATR technique exploits the occurrence of total reflection of light at the interface of two media with different optical densities. FT-IR ATR spectroscopy is a surface sensitive technique; the detection range is only some μm.

FT-IR ATR spectroscopy exploits the light "skin" (evanescent field), that occurs when light is totally reflected at the interface between two media of different refractive indices. If the incident beam is approaching the interface from the material with higher refractive index ($n_1 > n_2$; see Figure 24.2A), total reflection is observed when the angle of incidence is larger than the angle of total reflection, which is in turn determined by the ratio of the refractive indices.

Any substance covering the interface influences the incoming light at certain wavenumbers and thus specific information about its chemical composition can be obtained from the absorption spectrum. Due to the exponential decay within the evanescent field, the closer the sample is located to the ATR surface the higher its contribution to the recorded spectrum will be, whereas almost no absorption takes place at greater distances. Therefore, additional measures are necessary to bring a sufficient amount of sample, e.g., suspended particles into this region (see Figure 24.2B).

However, the limited detection range is especially advantageous when measuring aqueous samples in the information rich mid-IR spectral range. Due to the strong infrared absorption of water, the optical path must be short (<10 μm) to keep the detection limit low in common FT-IR spectrometers using thermal radiation emitters. Short optical path lengths like this are realized by the ATR technique without putting geometrical constraints on the sample volume [48].

24.1.4 RAMAN SPECTROSCOPY

In Raman spectroscopy, a sample is irradiated with a focused laser beam and the rare inelastically scattered photons are recorded. The measured intensities at different wavenumbers are termed Raman spectrum providing information on vibrational transitions characteristic for the chemical composition of the sample under study—similar to infrared absorption spectroscopy from a phenomenological point of view.

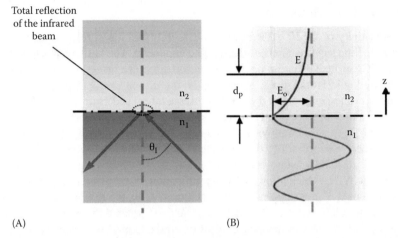

(A) (B)

FIGURE 24.2 (A) Total reflection of an IR beam at the boundary to a medium with lower refractive index $n_2 < n_1$. At the point of total reflection, the exponentially decreasing evanescent field is observed in medium 2. (B) E_0 denotes the electrical field amplitude of the electromagnetic field at the interface, and E is the exponentially decreasing electrical field amplitude of the evanescent field. d_p is the penetration depth, i.e., the distance from the interface at which the amplitude has decreased to the 1/eth part of E_0. (From Radel, S., Schnöller, J., Gröschl, M., Benes, E., and Lendl, B., On chemical and ultrasonic strategies to improve a portable FT-IR ATR process analyzer for online fermentation monitoring, *IEEE Sensors J.*, 10(10), 1615–1622, 2010. With permission. Copyright 2010 IEEE.)

Raman microspectroscopy is a powerful technique to deliver spatially resolved information about chemical composition of materials in many fields, including the study of biological structures. The spatial resolution of Raman microspectroscopy in the low-micrometer scale and its ability to probe samples under in vivo conditions allow for new insights into living single cells without the need for fixatives, markers, or stains [49,50].

In the case of monitoring reacting suspensions, Raman spectroscopy holds great promise due to possible non-invasive measurement strategies. This turns out to be difficult for standard on-line Raman spectroscopy because Raman photons from the solid matter need to be discriminated from Raman signals originating from the pure liquid phase. This problem is of special relevance in the case of low concentration of suspended particles.

Therefore, means to concentrate samples in the light path are applied on a standard basis. Acoustic levitation in air for monitoring containerless chemical reactions has been used traditionally in Raman microspectroscopy [51], recently the investigation of red blood cells and microorganisms with this technique was reported [52]. In another study, the combination of a microfluidic system with optical trapping and Raman microspectroscopic detection for red blood cells has been described [53]. This approach delivered the possibility of monitoring chemical processes in real-time with good control over the addition of reagents employing two different lasers.

The combination of a USW for particle manipulation and confocal Raman microspectroscopy is a novel approach to increase selectivity and sensitivity of on-line Raman measurements of suspensions. Applying a USW to a suspension can provide an experimentally simpler approach to immobilize and manipulate microparticles, resulting in higher sensitivity due to the local enrichment. Moreover, particles may be deliberately concentrated or removed from the Raman measurement spot, thus allowing one to selectively measure the liquid and solid phases, respectively.

Furthermore, the strategy of trapping particles in a flow cell allows for exposure to a variety of reactants and thus the execution of a given set of chemical reactions while being continuously monitored by Raman microspectroscopy. This approach as well overcomes drawbacks related to levitation in air, such as solvent evaporation.

24.1.5 STOPPED FLOW TECHNIQUE

The surface sensitivity of FT-IR ATR opens the possibility to measure the spectra of suspended particles (cells) and the supernatant (i.e., the liquid component). The basic idea is to keep cells from the horizontal ATR surface by pumping the suspension through a detection cell while the spectrum of the suspending liquid is measured. Subsequently, the throughput is stopped, the cells settle on the surface in the evanescent field of the ATR. Subsequently, a spectrum is taken from sediment.

Off-line measurements of whole cells have shown that the physiological state of the cells is accessible by mid-IR spectra employing chemometric techniques; feasibility of the stopped flow technique was recently successfully demonstrated with an *Escherichia coli* fermentation for the production of an intracellular biopolymer [54]. During the first step (Figure 24.3a), the cells were kept away from the ATR, which allowed the determination of dissolved analytes, e.g., the glucose level of the supernatant. Upon stopping the flow (Figure 24.3b), the *E. coli* settled on the ATR surface and thus, coming into reach of the evanescent field, dominated the recorded infrared spectra. Based on this approach, the intracellular content of product could be determined. In the case of *E. coli*, the cleanness and thus the sensitivity of the ATR element could be maintained by periodic rinsing with NaHCO$_3$ (Figure 24.3c).

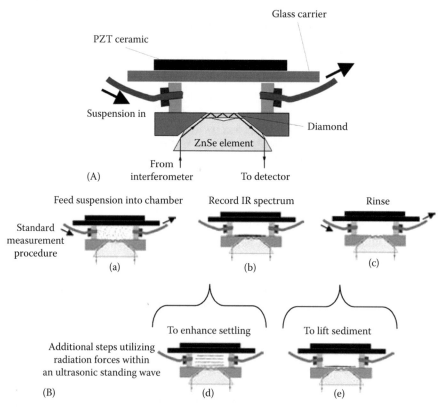

FIGURE 24.3   (A) Flow cell comprising the ATR element at the bottom and the PZT-sandwich transducer at the top. (B) Stopped flow technique to specifically measure the IR absorbance of suspended particles: the suspension is pumped into the detection volume (a). When the flow is switched off, particles settle onto the ATR surface and the spectrum is recorded (b). After the measurement, the cell is rinsed (c). A USW was applied to accelerate the measurement time by agglomerating the yeasts prior to the settling (d) and to improve the cleaning by actively lifting the sediment from the ATR prior to the rinse (e). (With kind permission from Springer Science+Business Media: *Elektrotechnik und Informationstechnik*, Ultrasonic particle manipulation exploited in on-line infrared spectroscopy of (cell) suspensions, 125(3), 2008, 76–81, Radel, S., Schnöller, J., Lendl, B., Gröschl, M., and Benes, E.)

There is great promise that monitoring systems based on the stopped technique can be applied on-line in biotech environments, if a sensor delivering robust and reliable measurements during long fermentation times can be designed.

24.2 TASKS

In the context of enhancing vibrational spectroscopy sensing applications by exploitation of a USW, the results of four attempts will be presented in the following.

- We have used the agglomeration of particles in suspension at predictable positions to increase the Raman signal.
- Special beads have been trapped, i.e., immobilized versus streaming liquid to enable a certain chemical protocol modifying their surface, enabling surface enhanced Raman (SER) spectroscopy.
- A USW was used to lift sediment material in a stopped flow set-up to enhance the concluding rinse and thus help prevent the formation of a biofilm.
- Application of a USW prior to measurement accelerated the settling process in a stopped flow cell, therefore increasing the time resolution of the measurement.

To excite the ultrasonic plane wave field, 1 mm piezoelectric sheets of lead zirconate titanate (PZT) equipped with silver electrodes at both surfaces were used. The respective sheet was glued to the wall or lid of a cavity containing the suspension under investigation. Further experimental particulars can be found in [16,55–57].

24.2.1 AGGLOMERATING

Raman scattering is brought about by interaction of light emitted by the instrument and the molecules of the sample. The signal is inherently weak, therefore the measurement of a Raman spectrum benefits from a high spatial particle concentration in the light path when the sample in question is a suspension. Exploiting the forces exerted within a USW, we used model suspensions to investigate the applicability of the ultrasonic particle manipulation to be employed for Raman spectroscopy to provide molecular information about the dispersed particles. Measurements were performed on suspensions of different polymer microparticles: dextran, polyvinyl alcohol, and melamine resin-based beads, with and without functionalization, were detected and successfully discriminated based on their characteristic Raman spectrum [16]. In all measurements, the ultrasonic field did not significantly influence the structure of the Raman spectra.

An application of an on-line sensor could be the assessment of the physiological status of microorganisms in fermentations based on the Raman spectrum [58]. To examine the feasibility of the technique for biotech applications, we used the yeast suspension. Another example would be the distinction between polymorphs of a given substance during crystallization processes [39,59]. For this case, the ophylline crystals in a saturated solution (9.3 mg/mL) were employed as a model.

A small resonator was used to control the spatial distribution of suspended particles relative to the light path of the Raman microscope, i.e., agglomerates of particles were deliberately positioned within and out of the focus of the instrument.

The set-up comprised an aluminum spacer between two glass sheets. For the excitement of the USW, a PZT ceramics was bonded to one sidewall of the spacer. The other sidewall acted as an ultrasound reflector to build up the USW. The device was placed under the Raman microscope with the glass sheets perpendicular to the light path (Figure 24.4, top). Thus, the pressure nodal planes where oriented parallel to the incident light beam, allowing to control their locations relative to the light path by slightly changing the excitation frequency. Illustrating the influence of the radiation

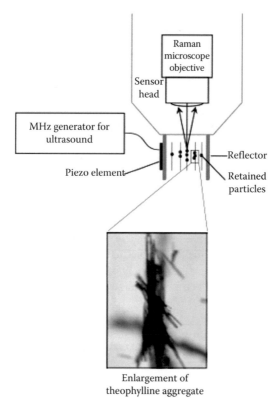

Enlargement of
theophylline aggregate

FIGURE 24.4 Top: Raman microscope with light path into flow cell. The ultrasonic standing wave (USW) is generated by the superposition of the ultrasonic wave emitted by the piezoelectric element and its reflection. Bottom: Picture of theophylline crystals agglomerated by the USW. (With kind permission from Springer Science+Business Media: *Elektrotechnik und Informationstechnik*, Raman spectroscopy of particles in suspension concentrated by an ultrasonic standing wave, 125(3), 2008, 82–85, Radel, S., Schnöller, J., Dominguez, A., Lendl, B., Gröschl, M., and Benes, E.)

forces on the spatial distribution of particles, a light micrograph of the agglomeration of theophylline crystals is shown in the enlargement at the bottom of Figure 24.4.

24.2.1.1 Yeast Cells

The gray line in Figure 24.5 shows the Raman spectrum when the scattered light was coming from the homogeneous yeast suspension. When the USW was applied, the Raman spectrum resulting from the agglomeration of yeast cells in the nodal plane is depicted as black line with dots in Figure 24.5. Significant features of yeast could be identified around wave numbers 2850, 1660, and 1437 cm^{-1}, which arise from the symmetric CH_2 stretching vibration, the amide I band and the amide III band, respectively. For reasons of comparison, Figure 24.5 includes the Raman spectrum of dried yeast on a quartz glass plate (black line), where the scatter intensity was found to be almost twice as high as for the measurement of the agglomerate; however, the feature structure was conclusively similar. Agglomerates of yeast cells showed a slightly lower level of scatter intensity when compared to dried material (Figure 24.5, black and black with dots, respectively).

24.2.1.2 Theophylline Crystals

The results of investigations of dissolved theophylline as well as suspended crystals are depicted in Figure 24.6. The aim was to compare the Raman signal of homogeneously suspended theophylline crystals (black line) with measurements of agglomerates brought about by the USW (black with dots).

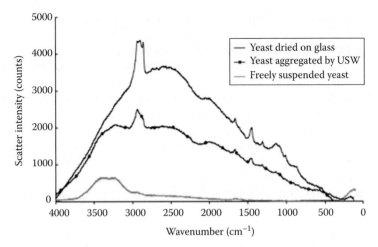

FIGURE 24.5 Comparison of Raman spectra of yeast freely suspended in water (gray), yeast cells agglomerated in the nodal plane of the ultrasonic field (black with dots), and as reference dried yeast cells on quartz (black). (With kind permission from Springer Science+Business Media: *Elektrotechnik und Informationstechnik*, Raman spectroscopy of particles in suspension concentrated by an ultrasonic standing wave, 125(3), 2008, 82–85, Radel, S., Schnöller, J., Dominguez, A., Lendl, B., Gröschl, M., and Benes, E.)

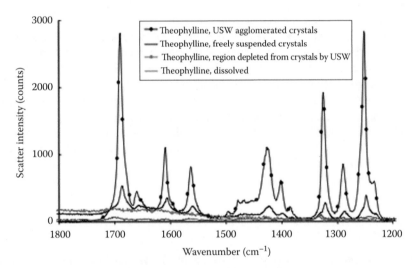

FIGURE 24.6 Raman spectra of theophylline solution (gray) and freely suspended theophylline crystals (black) in comparison to theophylline crystals agglomerated by ultrasound (black with dots) and the theophylline solution in a region where the crystals were depleted by the ultrasonic standing wave (gray with dots). (With kind permission from Springer Science+Business Media: *Elektrotechnik und Informationstechnik*, Raman spectroscopy of particles in suspension concentrated by an ultrasonic standing wave, 125(3), 2008, 82–85, Radel, S., Schnöller, J., Dominguez, A., Lendl, B., Gröschl, M., and Benes, E.)

Figure 24.6 shows a significant increase (three- to sixfold) of scatter intensity when the ultrasonic field was applied. Moreover, the data suggest better resolution, e.g., between wavenumbers 1600 and 1700 cm^{-1} when the theophylline crystals were concentrated by the USW.

In contrast, no significant differences were found for regions where no particles are present (in the displacement nodes). The Raman spectrum of a crystal-free theophylline solution (Figure 24.6, gray line) was not different from a measurement taken with the optical focus positioned within this depleted region (Figure 24.6, gray with dots).

When comparing sediment theophylline crystals (data not shown, see [55]), which can be expected to be packed tightly with agglomerates brought about by the ultrasonic radiation forces, a slight increase of scatter intensity was measured when the USW was present.

The presented results of Raman measurements indicate that a USW and the involved agglomeration of particles in suspension is a suitable means to achieve higher Raman scatter intensities. Different types of particles were used in this study to cover a broad range of potential applications of this combination of Raman spectroscopy and ultrasonic particle handling.

The beneficial effects on the measurement of Raman spectra of the two components of a suspension were the consequence of the vast increase of the particle concentration in the nodal planes of the USW, in comparison to the homogeneous distribution of particles when no ultrasonic field was present. Moreover, it was shown for theophylline crystals that the very low concentration of particles between the nodal planes enables one to specifically take measurements of host liquid's composition.

24.2.2 Trapping

The transversal primary radiation force, which is exerted perpendicular to the sound propagation direction, enables one to keep particle agglomerates versus a streaming liquid. This was exploited to gain full control over the media surrounding trapped particles, which allowed us to perform chemical reactions on these particles and at the same time monitor according to spectral changes. The potential of this ultrasonic trapping method to monitor on-bead chemical reactions was investigated with an example of automated generation of a SER-active layer on ultrasonic-trapped beads [16]. The use of the USW for trapping was especially valuable when synthesizing SER-active beads online. The process included steps of preparation and subsequent recording of SER spectra of different analytes during which the beads were retained in the focus of the Raman microscope. The repeatability in the recorded SER intensities was in the order of 4%–5%.

This methodology for SER measurement circumvents many problems related to the difficulty of producing reproducible SER substrates. For example, degradation of substrates and memory effects as a consequence of adhesion of nanoparticles to the walls of the cell are a common problem in flow systems. Here, the use of a fresh, unused SER-active substrate for each measurement helps avoid both degradation and memory effects. The demand of additional steps when discarding beads from the flow cell was reported [60]; in the case of using a USW, it is a question of simply switching off the field.

24.2.3 Lifting

In a way, an inverse task was accomplished by the use of a USW in a flow cell for stopped flow ATR spectroscopy. Originally, the exploitation was triggered by the observation of detrimental biofilm load of the horizontal ATR of a stopped flow set-up when measuring glucose and ethanol in baker's yeast fermentation [61].

24.2.3.1 Biofilm Formation

At the beginning of a fed-batch fermentation, the ATR element was manually cleaned (wiped), and a background spectrum of the water filled flow cell was recorded. Subsequently, infrared measurements of the broth were performed hourly: sample was drawn from the bioreactor and pumped through the flow cell. Then the flow was stopped for 10 min to allow settling of the cells, and the sample spectrum was measured. The following cleaning cycle consisted of a rinse with a 2% $NaHCO_3$ cleaning solution introduced by the flow system for 15 min. The cleaning step was finished by rinsing the cell with distilled water and followed by the recording of a new background spectrum.

In parallel to the online measurements, samples were drawn and kept frozen for subsequent offline validation. After un-freezing these samples, the filling-settling-cleaning procedure was

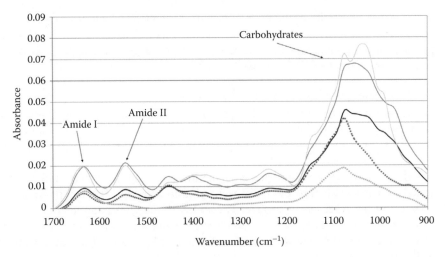

FIGURE 24.7 Infrared spectra measured during fermentation at the beginning (black) and after 10 h and 20 h (gray and light gray), respectively. Continuous lines refer to recordings with a carefully cleaned surface, while dashed lines represent data obtained, when a biofilm was coating the ATR. (With kind permission from Springer Science+Business Media: *Elektrotechnik und Informationstechnik*, Raman spectroscopy of particles in suspension concentrated by an ultrasonic standing wave, 125(3), 2008, 82–85, Radel, S., Schnöller, J., Dominguez, A., Lendl, B., Gröschl, M., and Benes, E.)

repeated; however, the absorbance spectrum was now assessed in comparison to a background acquired with a clean (wiped) ATR element.

Spectra acquired during the fermentation at the beginning (black) and after 10 h (dark gray, dashed line) and 20 h (light gray, dashed line), respectively, are shown in Figure 24.7. The data achieved during online monitoring suggested conspicuously a significant decrease in the carbohydrate region (1200–900 cm^{-1}) over time. Moreover, the amide I feature (1635 cm^{-1}) had not increased, and the absorption at amide II (1545 cm^{-1}) was nearly extinguished after 20 h. On top of that, the reliability of the spectra was impaired, e.g., most features between 1050 and 950 cm^{-1} were missing in the results at 10 and 20 h, respectively.

The hypothesis that this behavior was brought about by a contamination on the ATR surface was confirmed by the offline validation (Figure 24.7, dark and light gray continuous lines, colors reflect the time the samples were taken as mentioned earlier). When the background was recorded with a carefully cleaned ATR, the expected increase of carbohydrates during the fermentation was detected and the amide features were pronounced and distinct. In the results measured after 10 and 20 h, respectively, the region between 1050 and 950 cm^{-1} was reasonable alike the spectrum acquired at the beginning of the experiment.

This comparison between the online and the offline recordings led to the conclusion that the cleaning cycle employed was not able to prevent the formation of a biofilm on the ATR during a long lasting experiment like this. The corresponding influence was a decrease in signal strength and spectral reliability.

Together, these severely increase the danger of data misinterpretation, especially when spectra are compared after different numbers of measurement cycles, i.e., at different levels of biofilm. For a (bio)process control based on such monitoring, measures to increase the long-term stability of the measurement system are therefore necessary.

The stopped flow technique was reported to be less successful in the case of yeast fermentations before, formation of biofilms was observed when performing similar cycles for on-line measurement [62,63]. Formation of biofilm is a widely known problem in medically, biochemically, and industrially used sensors and filters [64], various means have been developed to avoid or remove [65].

Due to the surface sensitivity of the ATR technique, formation of biofilm is a serious issue for online fermentation monitoring as it impairs the selectivity as well as the sensitivity of a measurement, especially during prolonged fermentation times. A thorough cleaning protocol is therefore needed to prevent or reduce biofilm formation to a minimum.

24.2.3.2 Investigating Biofilm Prevention by a USW

An obvious measure to remove a biofilm is the application of liquid cleaning agents [66]. The effects of acids, surfactants, and oxidizing agents on biofilm removal were tested. The various cleaning agents substituting the originally used were employed at concentrations of 2% to avoid damage of tubes and fittings of the experimental setup.

Additionally, the flow cell was equipped with an ultrasonic transducer at the top, the radiation forces exerted on particles within the USW were used to actively lift the cells to ensure prolonged cleanness of the ATR sensor surface. The ability of the ultrasound field to lift deposited material was assessed by switching the USW on and off (duty cycle 50%) for 100 s, while the flow cell was flushed with distilled water. This was followed by another 100 s of rinsing with the USW turned off.

Ultrasound of a few tens of kHz was exploited before to clean the surface of an ATR sensor [67,68]; however, this is a completely different approach; ultrasound of this frequency utilizes the effect of cavitation like in ultrasonic cleaning and disruptor devices.

For the investigation of a residual biofilm adhered to the ATR after the described filling-settling-cleaning cycle, the model broth containing inactive yeast cells was used. The "sample" spectrum was recorded when the flow cell was filled with distilled water again, the spectrum of a carefully wiped ATR was used as the background. Therefore, the observed absorbance spectrum resulted from the biofilm on the ATR surface that was not eliminated by the cleaning procedure.

Figure 24.8A shows four infrared spectra of the same yeast suspension, one recorded 30 min after the other. A decrease of signal strength and resolution was obvious during the experiment (the black line refers to the first measurement, followed by lines in dark gray, gray, and light gray). This is explained by the fact that a biofilm build-up during the on-going experiment was incorporated with the background spectra recorded over time. As a result, the evanescent field effectively contributing to the measured spectrum was decreased. Figure 24.8B shows the evaluation of this biofilm only: A significant increase of absorbance for a "cleaned" ATR over the whole range of observed wavenumbers was detected, the recording suggested significant absorbance by the suspensions content, i.e., at wavenumbers corresponding to amide I and II at 1635 and 1545 cm^{-1} and carbohydrate features (1200–900 cm^{-1}), respectively.

When the USW procedure was applied as described, i.e., the USW was switched on and off during the rinse, the measured spectra (Figure 24.8C) showed only very small alterations over the duration of the experiment, suggesting a much cleaner surface. This was confirmed to be the case by the biofilm spectra (Figure 24.8D). A mild coating for the amide I and II features was found, but almost no deposition of carbohydrates. The presence of amide features suggests protein on the ATR, which might be less successfully lifted by the USW due to smaller particle size.

24.2.3.3 Various Cleaning Agents and Quantification

A range of different chemical strategies and the application of a USW was studied for their cleaning performance in [69] to find recommendations for automated cleaning procedures. The target of putting a number on the spectral results was accomplished by defining two parameters reflecting the respective cleaning performance over a range of wavenumbers. The cleaning performance is expressed in relation to a "standard" biofilm resulting from rinsing with water only.

Results are shown in Figure 24.9: The left most column group depicts the biofilm itself, thus defining 100%. The second column group shows the instrument's stability, i.e., the absorbance when distilled water was used instead of a sample. The deviations were found to be within 1%.

The results for the "pH effectors" in the following three column groups were not very effective. The application of $NaHCO_3$/HAc and HNO_3 showed only nominal cleaning performance, certain artifacts of remaining NaOH/NaCl in the flow cell show up as additional signal for both protein (260%) and carbohydrate (170%).

Among the surfactants, the spectra obtained when using a commercial spot cleaner showed low cleaning performance and artifacts. Sodium dodecyl sulfate (SDS) decreased the protein apparently to 55% and the carbohydrates to 39%; however, good results—and a dependable evaluation—could only be achieved after 30 min of rinsing.

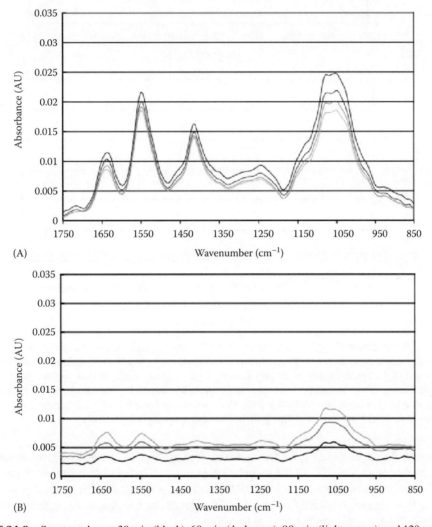

FIGURE 24.8 Spectra taken at 30 min (black), 60 min (dark gray), 90 min (lighter gray), and 120 min (lightest gray). (A) Result when the basic procedure (Figure 24.3a,b,c) was used. (B) Background spectra suggesting a biofilm causing the decrease in resolution and sensitivity of the measurements shown left (in respective gray values).

(*continued*)

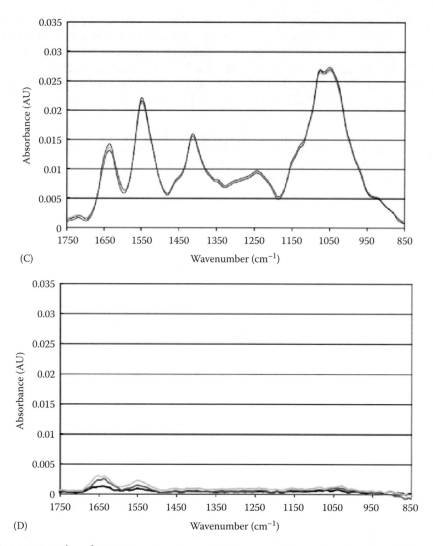

FIGURE 24.8 (continued) (C) Result when a USW was applied during the rinse (Figure 24.3e). (D) Background spectra when the USW was applied. The biofilm was significantly reduced by this additional lift during the rinse. Spectra taken at 30 min (black), 60 min (dark gray), 90 min (lighter gray), and 120 min (lightest gray). (From Radel, S., Schnöller, J., Gröschl, M., Benes, E., and Lendl, B., On chemical and ultrasonic strategies to improve a portable FT-IR ATR process analyzer for online fermentation monitoring, *IEEE Sensors J.*, 10(10), 1615–1622, 2010. With permission. Copyright 2010 IEEE.)

The results for the oxidizing agents were mixed: The application of NaOCl delivered very good cleaning performance, only 5% of the carbohydrate level was found after treatment; however, the application caused some damage to tubings and tightenings of the set-up. The residual biofilm when employing H_2O_2 was slightly better for protein (39%) than for carbohydrates (61%); to be completely removed a reduction agent ($NaNO_2$) had to be added.

The ultrasonic field delivered a removal in the carbohydrate region down to 10% of the original load, while 29% of the protein was left.

In conclusion, the presented results are satisfactory; the stopped flow principle with a horizontal ATR element was successfully applied on yeast suspensions. The adverse biofilm formation could be decreased by employing selected cleaning agents. The application of a USW in the ATR flow cell was advantageous and did not pose additional limitations.

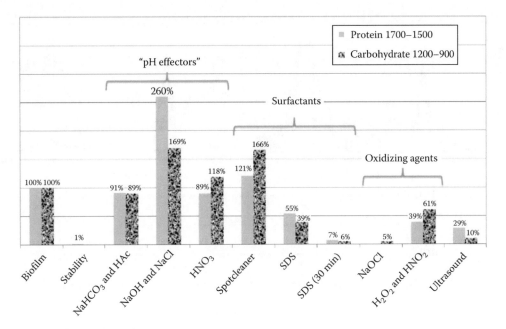

FIGURE 24.9   Protein and carbohydrate residuals after application of the various cleaning strategies. (From Radel, S., Schnöller, J., Gröschl, M., Benes, E., and Lendl, B., On chemical and ultrasonic strategies to improve a portable FT-IR ATR process analyzer for online fermentation monitoring, *IEEE Sensors J.*, 10(10), 1615–1622, 2010. With permission. Copyright 2010 IEEE.)

24.2.4   ACCELERATE SETTLING

Operation of the mentioned commercially available acoustic filters is based on the fact that small particles, e.g., microorganisms in suspension, settle quicker when aggregated, than freely dispersed cells. The reason for this is that Stokes friction forces, which are exerted at the surface of a body, are decreasing in relation to gravity with the surface to volume ratio. The technique is as well called ultrasound enhanced settling [7,70] as the arrangement in a "super-particle," which sediments much faster due to increased size, is brought about by a USW.

When developing a sensor for fermentation, monitoring time resolution is an important issue. In order to increase the settling rate and therefore decrease the measurement time, agglomeration of the yeast cells was induced by having the USW present when the flow cell was filled with suspension (see Figure 24.3D).

The settling was observed by consecutive measurements ($\Delta t = 5$ s) of the infrared absorption. Here, Figure 24.10A and B was taken when the yeast suspension was fed into the vessel cell and the flow was stopped. In the absence of the USW (Figure 24.10A), the suspended yeast cells settled on the ATR surface, the slowness of the process is indicated by small distances in vertical direction between the spectra. When the ultrasonic field was present during the filling and switched off subsequently (Figure 24.10B), larger distances suggest accelerated sedimentation.

These data were further processed to compare sedimentation by gravity and sedimentation enhanced by means of a USW in regard to the settling rate. Three absorbance values in the carbohydrate region were averaged and used as measure for the settling rate versus time. The result is shown in Figure 24.10C. The freely dispersed yeast cells sediment gradually over the observed period of 185 s. The application of a USW changed the sedimentation tremendously. During the first 15 s, the USW prevented the cells—dots indicate that sound was on—to settle at all. It took further 15 s for the settling agglomerates to be picked up by the ATR followed by a strong increase in absorbance within 70 s. The acceleration due to ultrasound induced aggregation was therefore increasing the settling rate by a factor of more than two.

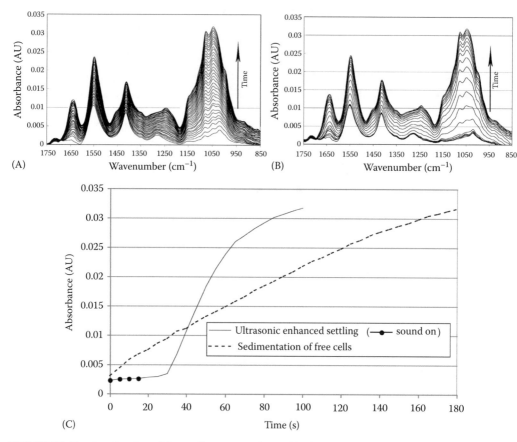

FIGURE 24.10 Acceleration of the settling process by application of a USW (Figure 24.3d). (A) Spectra of settling yeast cells recorded every 5 s. (B) Same with USW applied; increase of distance between the lines indicates faster sedimentation. (C) Increase of absorbance over time calculated from aforementioned graphs. Free settling was nearly finished after 185 s (dotted line), whereas the same absorption (0.03 AU) was reached in only 75 s. (From Radel, S., Schnöller, J., Gröschl, M., Benes, E., and Lendl, B., On chemical and ultrasonic strategies to improve a portable FT-IR ATR process analyzer for online fermentation monitoring, *IEEE Sensors J.*, 10(10), 1615–1622, 2010. With permission. Copyright 2010 IEEE.)

24.3 SUMMARY AND FUTURE DIRECTIONS

The ability to aggregate particles in aqueous suspension when collecting their Raman spectra has been demonstrated. The exploitation of radiation forces increased the signal strength concentrating the particles to be measured in the optical focus of the instrument. Measurements were comparable to sediments of crystals and yeast cells when the USW was applied to the suspension. It has as well been shown that chemical reactions carried out on ultrasonically trapped beads can be monitored. This was found to be especially valuable when dealing with the protocol steps connected to surface-enhanced Raman spectroscopy (SERS) on beads. Due to the flexibility of the proposed system, a broad variety of assays can be performed with application in many fields, such as solid-phase synthesis and bead-based bioassays. The robust but gentle handling also allows for prolonged studies on beads as well as living cells. In a future parallel analysis, employing trapped beads carrying different functionalities can be envisioned.

In connection with ATR spectroscopy, the presented results are promising regarding positive influences of a USW on biofilm creation. Prolonged cleanness of a horizontal ATR sensor surface as used for stopped flow principle was found when recording infrared spectra of a yeast suspension. The results received with a USW employed in the flow cell during the rinse significantly decreased residuals on the ATR, suggesting a significant improvement in terms of long-term stability. Depending on the substance—indicated by the wavenumbers for protein and carbohydrates—a 3–10 times lower absorbance value by the biofilm was detected. The imbalance between protein and carbohydrates in the residual can be interpreted as a result of differences in the size distribution of the biofilms constituents.

The investigations regarding the removal strategies of a "standard" biofilm included various chemical cleaning solutions. Experiments employing such methods to clean the surface of the ATR delivered mediocre results, yet with some investigated agents good cleaning performance could be achieved. However, some practical drawbacks were found. Among them, the attack of the flow system by NaOCl has to be mentioned, which limits the range of applicable materials. Another was the requirement of an in-acceptable long-term rinsing step with water to remove residuals of SDS. The necessary cleaning efficiency for application in on-line (bio)process monitoring was somewhat achieved by the sequential use of H_2O_2 and HNO_2. Obviously, a next step could be to investigate the cleaning performance of chemical measures and USW combined.

These novel attempts to integrate a USW in different vibrational spectroscopy sensing applications have been reasonably successful. While being advantageous in various respects, no adverse effects like, e.g., cross-sensitivity were observed. Both ultrasonic and vibrational spectroscopy techniques are at a stage of development where crucial improvements can be expected in near future, while at the same time, mature enough to be endorsed in industrial environments.

An important step to proliferate this is the establishment of in-line sensing approaches. Due to sterility and practicability issues such are the preferred option for process monitoring, for instance inside a bio-reactor. However, as no recalibration or verification of sensor response is possible, such sensors are especially demanding in regard to long term robustness and calibration stability.

In-line ATR sensors connected to the spectrometer by mid-IR fiber optics or optical conduits were exploited in bioprocess monitoring before. A variety of small molecules like sugars, alcohols, organic and amino acids, as well as phosphate were successfully determined in the fermentation broth [64,71].

However, only chemical information of the supernatant could be assessed with these in-line configurations. From a bio-process control point of view, it is very desirable to access chemical information about the culture as well.

The ability of a USW to deposit particles on a surface was investigated with functionalized surfaces [72] and by optical means [73]. This recently triggered our research to combine a USW with an in-line ATR fiber probe aiming at the independent measurement of mid-IR spectra of the supernatant and the suspended cells in purposely populating or de-populating the ATR's evanescent field of cells by a USW.

Most recently, we managed to build up a USW in close proximity of the tip of a fiber optic in-line ATR probe and consequently showed to be able to control the precise location of particle aggregates in respect to the evanescent field by adjusting the ultrasound frequency [74].

ACKNOWLEDGMENT

The Austrian Science Fund (FWF) supported this work in Project Nos. 13686, 15531, and L416-N17 (translational research program).

REFERENCES

1. M. Gröschl, W. Burger, B. Handl, and O. Doblhoff-Dier, Ultrasonic separation of suspended particles—Part III: Application in biotechnology, *Acta Acustica united with Acustica*, 84 (5), 815–822, January 1998.
2. A. Kundt and O. Lehmann, Longitudinal vibrations and acoustic figures in cylindrical columns of liquids, *Annnual Review of Physical Chemistry*, 153, 1–12, 1874.

3. M. Evander, L. Johansson, T. Lilliehorn, J. Piskur, M. Lindvall, S. Johansson, M. Almqvist, T. Laurell, and J. Nilsson, Noninvasive acoustic cell trapping in a microfluidic perfusion system for online bioassays, *Analytical Chemistry*, 79 (7), 2984–2991, April 2007.

4. T. M. P. Keijzer, F. Trampler, A. Oudshoorn, O. Doblhoff-Dier, and H. van der Berg, *Integrating Acoustic Perfusion in Mammalian Cell Culture*, Cell Culture Engineering VII, United Engineering Foundation, Inc., Snowmass Village, CO, July 25, 2002.

5. F. Trampler, S. A. Sonderhoff, P. W. S. Pui, D. G. Kilburn, and J. M. Piret, Acoustic cell filter for high density perfusion culture of hybridoma cells, *Nature Bio/Technology*, 12, 281–284, January 1994.

6. E. Benes, M. Gröschl, H. Nowotny, F. Trampler, T. Keijzer, H. Böhm, S. Radel, L. Gherardini, J. J. Hawkes, R. König, and C. Delouvroy, Ultrasonic separation of suspended particles, Presented at the *IEEE Ultrasonics Symposium*, Atlanta, Georgia, 2001, 1, 649–659.

7. M. C. F. Dalm, M. Jensen, T. M. P. Keijzer, W. M. J. van Grunsven, A. Oudshoorn, J. Tramper, and D. E. Martens, Stable hybridoma cultivation in a pilot-scale acoustic perfusion system: Long-term process performance and effect of recirculation rate, *Biotechnology and Bioengineering*, 91, 894–900, July 2005.

8. O. Palme, G. Comanescu, I. Stoineva, S. Radel, E. Benes, D. Develter, V. Wray, and S. Lang, Sophorolipids from *Candida bombicola*: Cell separation by ultrasonic particle manipulation, *European Journal of Lipid Science and Technology*, 112 (6), 663–673, May 2010.

9. J. J. Hawkes, D. Barrow, and W. T. Coakley, Microparticle manipulation in millimetre scale ultrasonic standing wave chambers, *Ultrasonics*, 36, 925–931, 1998.

10. M. Ruedas-Rama, A. Dominguez-Vidal, S. Radel, and B. Lendl, Ultrasonic trapping of microparticles in suspension and reaction monitoring using Raman microspectroscopy, *Analytical Chemistry*, 79 (20), 7853–7857, January 2007.

11. J. F. Spengler and W. T. Coakley, Ultrasonic trap to monitor morphology and stability of developing microparticle aggregates, *Langmuir*, 19 (9), 3635–3642, April 2003.

12. L. Gherardini, C. Cousins, J. J. Hawkes, J. Spengler, S. Radel, H. Lawler, B. Devcic-Kuhar, M. Gröschl, W. T. Coakley, and A. J. McLoughlin, A new immobilisation method to arrange particles in a gel matrix by ultrasound standing waves, *Ultrasound in Medicine & Biology*, 31 (2), 261–272, February 2005.

13. J. J. Hawkes and W. T. Coakley, A continuous flow ultrasonic cell-filtering method, *Enzyme and Microbial Technology*, 19, 57–62, 1996.

14. J. J. Hawkes, D. Barrow, J. Cefai, and W. T. Coakley, A laminar flow expansion chamber facilitating downstream manipulation of particles concentrated using an ultrasonic standing wave, *Ultrasonics*, 36, 901–903, 1998.

15. J. J. Hawkes and W. T. Coakley, Force field particle filters, combining ultrasound standing waves and laminar flow, *Sensors and Actuators B*, 75, 213–222, 2001.

16. S. Ravula, D. Branch, C. James, R. Townsend, M. Hill, G. Kaduchak, M. Ward, and I. Brener, A microfluidic system combining acoustic and dielectrophoretic particle preconcentration and focusing, *Sensors and Actuators B: Chemical*, 130 (2), 645–652, 2008.

17. M. Wiklund and H. M. Hertz, Ultrasonic enhancement of bead-based bioaffinity assays, *Lab on a Chip*, 6 (10), 1279, 2006.

18. T. Laurell, T. Laurell, F. Petersson, F. Petersson, A. Nilsson, and A. Nilsson, Chip integrated strategies for acoustic separation and manipulation of cells and particles, *Chemical Society Reviews*, 36 (3), 492, 2007.

19. A. Neild, S. Oberti, F. Beyeler, J. Dual, and B. J. Nelson, A micro-particle positioning technique combining an ultrasonic manipulator and a microgripper, *Journal of Micromechanics and Microengineering*, 16 (8), 1562–1570, June 2006.

20. M. Hill, J. J. Hawkes, N. Harris, and M. McDonnell, Resonant ultrasonic particle manipulators and their applications in sensor systems, *Sensors, Proceedings of IEEE*, Vienna, Austria, pp. 794–797, 2004.

21. P. W. S. Pui, F. Trampler, S. A. Sonderhoff, M. Gröschl, D. G. Kilburn, and J. M. Piret, Batch and semicontinuous aggregation and sedimentation of hybridoma cells by acoustic resonance fields, *Biotechnology Progress*, 11, 146–152, 1995.

22. H. Bierau, A. Perani, M. Al-Rubeai, and A. N. Emery, A comparison of intensive cell culture bioreactors operating with hybridomas modified for inhibited apoptotic response, *Journal of Biotechnology*, 62, 195, 1998.

23. H. Böhm, P. Anthony, M. R. Davey, L. G. Briarty, J. B. Power, K. C. Lowe, E. Benes, and M. Gröschl, Viability of plant cell suspensions exposed to homogeneous ultrasonic fields of different energy density and wave type, *Ultrasonics*, 38, 629–632, January 2000.

24. C. M. Cousins, P. Holownia, J. J. Hawkes, C. P. Price, P. Keay, and W. T. Coakley, Clarification of plasma from whole human blood using ultrasound, *Ultrasonics*, 38, 654–656, 2000.

25. K. Yasuda, M. Kiyama, S.-I. Umemura, and K. Tekeda, Deoxyribonucleic acid concentration using acoustic radiation force, *JASA*, 99 (2), 1248–1251, 1995.
26. J. Hultström, O. Manneberg, K. Dopf, H. Hertz, H. Brismar, and M. Wiklund, Proliferation and viability of adherent cells manipulated by standing-wave ultrasound in a microfluidic chip, *Ultrasound in Medicine and Biology*, 33 (1), 145–151, January 2007.
27. S. Radel, A. McLoughlin, L. Gherardini, O. Doblhoff-Dier, and E. Benes, Viability of yeast cells in well controlled propagating and standing ultrasonic plane waves, *Ultrasonics*, 38 (1), 633–637, 2000.
28. S. Radel, L. Gherardini, A. J. McLoughlin, O. Doblhoff-Dier, and E. Benes, Breakdown of immobilisation/separation and morphology changes of yeast suspended in water-rich ethanol mixtures exposed to ultrasonic plane standing waves, *Bioseparation*, 9 (6), 369–377, 2001.
29. W. T. Coakley, Ultrasonic separations in analytical biotechnology, *Trends in Biotechnology*, 15 (12), 506–511, 1997.
30. M. A. Sobanski, S. J. Gray, M. Cafferkey, R. W. Ellis, R. A. Barnes, and W. T. Coakley, Meningitis antigen detection: Interpretation of agglutination by ultrasound-enhanced latex immunoassay, *British Journal of Biomedical Science*, 56 (4), 239–246, 1999.
31. M. Wiklund, S. Nilsson, and H. M. Hertz, Ultrasonic trapping in capillaries for trace-amount biomedical analysis, *Journal of Applied Physics*, 90 (1), 421, 2001.
32. J. J. Hawkes, R. W. Barber, D. R. Emerson, and W. T. Coakley, Continuous cell washing and mixing driven by an ultrasound standing wave within a microfluidic channel, *Lab on a Chip*, 4 (5), 446–452, 2004.
33. H. Wikström, P. Marsac, and L. S. Taylor, In-line monitoring of hydrate formation during wet granulation using Raman spectroscopy, *Journal of Pharmaceutical Sciences*, 94(1), 209–219, 2005.
34. T. Gaida, O. Doblhoff-Dier, K. Strutzenberger, H. Katinger, W. Burger, M. Gröschl, B. Handl, and E. Benes, Selective retention of viable cells in ultrasonic resonance field devices, *Biotechnology Progress*, 12 (1), 73–76, 1996.
35. J. J. Hawkes, M. S. Limaye, and W. T. Coakley, Filtration of bacteria and yeast by ultrasound enhanced sedimentation, *Journal of Applied Microbiology*, 82, 39–47, 1997.
36. J. Spengler, M. Jekel, K. Christensen, R. Adrian, J. Hawkes, and W. Coakley, Observation of yeast cell movement and aggregation in a small-scale MHz-ultrasonic standing wave field, *Bioseparation*, 9 (6), 329–341, 2000.
37. M. Gröschl, Ultrasonic separation of suspended particles—Part I: Fundamentals, *Acta Acustica united with Acustica*, 84 (3), 432–447, January 1998.
38. A. Doinikov, Acoustic radiation force on a spherical particle in a viscous heat-conducting fluid. III. Force on a liquid drop, *Journal of the Acoustical Society of America*, 101(2), 731–740, 1997.
39. J. Chalmers and P. Griffiths, *Handbook of Vibrational Spectroscopy*, John Wiley & Sons Inc., New York, January 2002.
40. N. Harrick, *Internal Reflection Spectroscopy*, John Wiley & Sons Inc., New York, 1967.
41. J. Schenk, I. W. Marison, and U. von Stockar, pH prediction and control in bioprocesses using mid-infrared spectroscopy, *Biotechnology and Bioengineering*, 100 (1), 82–93, 2008.
42. J. E. Bertie and Z. Lan, Infrared intensities of liquids XX: The intensity of the OH stretching band of liquid water revisited, and the best current values of the optical constants of $H_2O(l)$ at 25°C between 15,000 and 1 cm -1, *Applied Spectroscopy*, 50 (8), 1047–1057, August 1996.
43. J. R. Baena and B. Lendl, Raman spectroscopy in chemical bioanalysis, *Current Opinion in Chemical Biology*, 8 (5), 534–539, October 2004.
44. R. Swain and M. M. Stevens, Raman microspectroscopy for non-invasive biochemical analysis of single cells, *Biochemical Society Transactions*, 35 (3), 544–549, 2007.
45. N. Leopold, M. Haberkorn, T. Laurell, J. Nilsson, J. R. Baena, J. Frank, and B. Lendl, On-line monitoring of airborne chemistry in levitated nanodroplets: In situ synthesis and application of SERS-active Ag-sols for trace analysis by FT-Raman spectroscopy, *Analytical Chemistry*, 75 (9), 2166–2171, 2003.
46. L. Puskar, R. Tuckermann, T. Frosch, J. Popp, V. Ly, D. McNaughton, and B. R. Wood, Raman acoustic levitation spectroscopy of red blood cells and Plasmodium falciparum trophozoites, *Lab on a Chip*, 7 (9), 1125–1131, 2007.
47. K. Ramser, J. Enger, M. Goksör, D. Hanstorp, K. Logg, and M. Käll, A microfluidic system enabling Raman measurements of the oxygenation cycle in single optically trapped red blood cells, *Lab on a Chip*, 5 (4), 431–436, 2005.
48. G. Jarute, A. Kainz, G. Schroll, J. R. Baena, and B. Lendl, On-line determination of the intracellular poly(β-hydroxybutyric acid) content in transformed Escherichia coli and glucose during PHB production using stopped-flow attenuated total reflection FT-IR spectrometry, *Analytical Chemistry*, 76 (21), 6353–6358, November 2004.

49. S. Radel, J. Schnöller, A. Dominguez, B. Lendl, M. Gröschl, and E. Benes, Raman spectroscopy of particles in suspension concentrated by an ultrasonic standing wave, *Elektrotechnik und Informationstechnik*, 125 (3), 82–85, March 2008.

50. S. Radel, J. Schnöller, B. Lendl, M. Gröschl, and E. Benes, Ultrasonic particle manipulation exploited in on-line infrared spectroscopy of (cell) suspensions, *Elektrotechnik und Informationstechnik*, 125 (3), 76–81, March 2008.

51. S. Radel, J. Schnöller, E. Benes, and B. Lendl, *Proceedings of IEEE Sensors*, Vienna, Austria, 2004, pp. 757–759.

52. K. C. Schuster, I. Reese, E. Urlaub, J. R. Gapes, and B. Lendl, Multidimensional information on the chemical composition of single bacterial cells by confocal Raman microspectroscopy, *Analytical Chemistry*, 72 (22), 5529–5534, November 2000.

53. A. Llinàs and J. M. Goodman, Polymorph control: past, present and future, *Drug Discovery Today*, 13 (5), 198–210, March 2008.

54. M. J. Ayora Cañada, A. Ruiz Medina, J. Frank, and B. Lendl, Bead injection for surface enhanced Raman spectroscopy: Automated on-line monitoring of substrate generation and application in quantitative analysis, *Analyst*, 127 (10), 1365–1369, August 2002.

55. J. Schnöller and B. Lendl, *Proceedings of IEEE Sensors*, Vienna, Austria, 2004, pp. 742–745.

56. X. Chen and P. S. Stewart, Biofilm removal caused by chemical treatments, *Water Research*, 34 (17), 4229–4233, December 2000.

57. G. Mazarevica, J. Diewok, J. R. Baena, E. Rosenberg, and B. Lendl, On-line fermentation monitoring by mid-infrared spectroscopy, *Applied Spectroscopy*, 58 (7), 804–810, July 2004.

58. D. J. Pollard, R. Buccino, N. C. Connors, T. F. Kirschner, R. C. Olewinski, K. Saini, and P. M. Salmon, Real-time analyte monitoring of a fungal fermentation, at pilot scale, using in situ mid-infrared spectroscopy, *Bioprocess and Biosystems Engineering*, 24 (1), 13–24, January 2001.

59. B. Meyer, Approaches to prevention, removal and killing of biofilms, *International Biodeterioration and Biodegradation*, 51 (4), 249–253, January 2003.

60. K. Merritt, V. Hitchins, and S. Brown, Safety and cleaning of medical materials and devices, *Journal of Biomedical Materials Research B*, 53 (2), 131–136, January 2000.

61. U. Wolf and U. Wolf, Application of infrared ATR spectroscopy to in situ reaction monitoring, *Catalysis Today*, 49 (4), 411–418, March 1999.

62. U. Wolf, Device for continuous spectroscopic analysis according to the principle of attenuated total reflection, Patent DE 4333560 AI, 1995.

63. S. Radel, J. Schnöller, M. Gröschl, E. Benes, and B. Lendl, On chemical and ultrasonic strategies to improve a portable FT-IR ATR process analyzer for online fermentation monitoring, *IEEE Sensors Journal*, 10 (10), 1615–1622, 2010.

64. S. Radel, Ultrasonically enhanced settling: The effects of ultrasonic plane wave fields on suspensions of the yeast *Saccharomyces cerevisiae*, PhD thesis, Vienna University of Technology, Vienna, Austria, 157, March 2003.

65. D. L. Doak and J. A. Phillips, In situ monitoring of an Escherichia coli fermentation using a diamond composition ATR probe and mid-infrared spectroscopy, *Biotechnology Progress*, 15 (3), 529–539, June 1999.

66. J. J. Hawkes, M. J. Long, W. T. Coakley, and M. B. McDonnell, Ultrasonic deposition of cells on a surface, *Biosensors and Bioelectronics*, 19 (9), 1021–1028, April 2004.

67. P. Glynne-Jones, R. J. Boltryk, M. Hill, F. Zhang, L. Dong, J. Wilkinson, T. Melvin, N. R. Harris, and T. Brown, Flexible acoustic particle manipulation device with integrated optical waveguide for enhanced microbead assays, *Analytical Sciences*, 25 (2), 285–291, 2009.

68. S. Radel, M. Brandstetter, and B. Lendl, Observation of particles manipulated by ultrasound in close proximity to a cone-shaped infrared spectroscopy probe, *Ultrasonics*, 50 (2), 240–246, February 2010.

25 Wideband Ultrasonic Transmitter and Sensor Array for In-Air Applications

Juan Ramon Gonzalez, Mohamed Saad,
and Chris J. Bleakley

CONTENTS

25.1 INTRODUCTION

Ultrasonic technology has been increasing in importance in recent years. It has been shown to be effective in a range of applications, including range estimation (Toda and Dahl 2007), indoor local positioning systems (LPSs) (Hazas and Hopper 2006), and nondestructive testing (Dobie et al. 2011). Demand for improved performance in these applications has created new requirements that are difficult to fulfill using commonly available ultrasonic transducers. Most commonly available ultrasonic transducers have a narrowband response, which makes them unsuited to wideband applications. In recent years, wideband ultrasonic transducers have been developed using polyvinylidene fluoride (PVDF) materials (Bloomfield et al. 2000, Fiorillo 1992). However, these transducers require a high supply voltage and only achieve short-range signaling, making them unsuitable for mobile and low power applications. Other researchers have proposed the use of electromechanical film (EMFi) transducers, but these are not commercially available (Barna et al. 2007).

In this chapter, we describe our research on wideband piezoelectric transducers and microelectromechanical systems (MEMS) sensors for ultrasonic in-air applications. The transducers and sensors described are compact, low cost, and have low power consumption (Gonzalez and Bleakley 2011). These characteristics enable their use in mass market mobile applications. Furthermore, their wide bandwidth enables the introduction of frequency hopped spread spectrum (FHSS) modulation, significantly improving signaling robustness and increasing the update rate (Gonzalez and Bleakley 2009). The small size of the sensors allows construction of compact uniform circular arrays (UCAs) to improve sensitivity and provide accurate angle of arrival (AoA) estimation. We describe application of the transducers and sensors in a prototype LPS. The LPS achieves sub-millimeter ranging accuracy (Saad et al. 2011) and high accuracy AoA estimation using a 7 mm diameter UCA of sensors (Gonzalez and Bleakley 2009).

In Sections 25.2 through 25.4, we review previous work by other authors on in-air ultrasonic transducers; we describe the wideband transducers and sensor arrays investigated in our work and evaluate their performance. In Section 25.5, we describe use of the transducers and sensor array in an indoor ultrasonic LPS. Section 25.6 concludes the chapter.

25.2 ULTRASONIC TRANSDUCER TECHNOLOGY

There are four main categories of ultrasonic in-air transducer—piezoelectric, PVDF, EMFi, and capacitive MEMS ultrasonic transducers (CMUTs). These are reviewed in the context of wideband mobile applications in the following subsections.

25.2.1 PIEZOELECTRIC TRANSDUCERS

Piezoelectric ultrasonic transducers are typically zirconate titanate polymer or composite materials (Gururaja et al. 1985). Typically, piezoceramic transducers have high electrical-to-mechanical efficiency, narrow bandwidth, and high impedance and offer a range of options for characteristic modification in production. The technology is very cost effective for large-scale production. However, due to their narrow bandwidth (usually 1–3 kHz), they are not suitable for broadband applications. However, they do offer simple electronic polarization and have low supply voltage requirements. Hence, they have been widely used in narrowband indoor location systems, such as in Cricket (Priyantha et al. 2000) and Bats (Harter et al. 2002).

Piezoelectric transducers rely on a mechanical phenomenon called piezoelectricity, a linear phenomenon that converts the mechanical tension in the material to an induced voltage, and vice versa. This phenomenon makes the piezoelectric material vibrate according to the applied voltage. There are two main resonance modes (Gururaja et al. 1985): thickness and planar. Thickness mode resonance, as shown in Figure 25.1a, produces longitudinal waves, giving acoustic signals at low frequency, usually less than 1 MHz. Planar mode resonance, as shown in Figure 25.1b, produces radial vibrations, giving acoustic signals at high frequency, usually greater than 1 MHz.

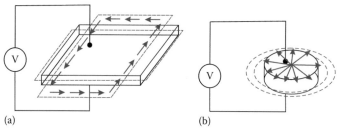

FIGURE 25.1 Resonant modes in piezoelectric transducers: (a) Thickness mode and (b) planar mode. (From Gonzalez, J.R. and Bleakley, C.J., Low cost, wide band ultrasonic transmitter and receiver for array signal processing applications, *IEEE Sens. J.*, 11(5), 1284–1292, 2011. With permission. Copyright 2011 IEEE.)

There are other spurious resonances modes, such as lateral mode resonances, which are out of the scope of this chapter, but are explained in Gururaja et al. (1985).

Commonly used procedures for increasing bandwidth (Kossoff 1966) are based on mechanically damping the piezoelectric element with a well-matched medium, which reduces the element's sensitivity, or on using an impedance matching layer of thickness λ/4 and impedance equal to the geometric mean of the transducer-load impedance. Other work on bandwidth modification for piezoelectric transducers has focused on finding new piezoelectric materials that provide better properties, such as broader bandwidth and matched acoustic impedance (Cochran et al. 1997, 1999, Smith 1989). The results show slight improvements but not enough to call the transducers broadband. Some researchers (Takeuchi et al. 2002) have investigated increasing transducer bandwidth by modifying the materials in order to obtain a double resonance peak for harmonic imaging, increasing the bandwidth and resolution in imaging applications. These procedures are difficult to replicate and do not provide a significant bandwidth improvement. In contrast to these approaches, the method described herein does not modify the internal structure of the transducer but adds passive components that are easy to replicate and provides a wide bandwidth response.

Piezoceramic transducers are usually made from polycrystal materials. The piezoelectric properties of these crystals are obtained by applying a high voltage to them (usually several kV/mm) during the production process. This orients the dipoles in the direction necessary to provide the desired piezoelectric properties. A number of publications have described research on piezoelectric-composite materials for low frequency applications (<40 kHz) (Klicker et al. 1981, Newnham et al. 1978, Skinner et al. 1978). Higher frequencies are used for medical diagnostic and nondestructive evaluation (1–10 MHz).

One of the greatest problems in using piezoceramic transducers in air is that the transducer and medium impedance are badly matched. The transducer impedance determines the Q factor, which determines the bandwidth of the transducer. This mismatch means that the coupling of the acoustic energy at the transducer-load interface is very poor. This limits the maximum achievable sound pressure level (SPL), decreasing signaling range for a given signal-to-noise ratio (SNR). Also, the high Q (narrow bandwidth) causes a slow pulse-rise time and prolonged ring-down, reducing resolution in ranging applications.

25.2.2 PVDF TRANSDUCERS

PVDF was patented by Ford and Hanford (1948). Since then, there has been a great deal of development work on ultrasonic devices that take advantage of the inherent properties of piezoelectric polymers, i.e., their relatively good acoustic impedance match to water and tissue; their flexible form; availability in large sheets; broadband acoustic performance; and ability to be dissolved and coated onto various substrates.

A number of papers have developed theory for building transducers based on PVDF materials (Brown 2000, Fiorillo 1992, Toda 2002). In addition, several papers have studied the acoustic and electric characteristics of PVDF transducers (Bloomfield et al. 2000, Lan et al. 1999).

Because of the difficulties in creating an efficient PVDF transducer, there are few applications that make use of PVDF materials. Recently, the commercial availability of PVDF cylindrical transducers, such as the US40KT-01 transducer from MSI (MSI 2001), has increased the number of applications that exploit this material. For example, the authors Jimenez and Seco (2005) used the MSI transducer to improve object position and contour estimation in outdoor environments. Villadangos et al. (2007) used the transducer to improve coverage in an indoor location system. However, the mechanical setup required in a PVDF transducer, due to the PVDF film length variation, makes it difficult to use in custom acoustic applications. Also, the high polarization voltage needed means that they are not ideal for low power applications. The maximum range that transducers based on PVDF can achieve is comparable to the range provided by piezoelectric transducers, having a typical mean SPL of 110 dB (MSI 2001).

25.2.3 EMFi Transducers

EMFi is a low-cost thin film of microporous polypropylene foam with high resistivity and permanent charge due to being polarized by the corona method. The resultant inner air voids act as dipoles, which make it particularly sensitive to forces normal to its surface. When glued to a rigid substrate and excited by an external voltage, EMFi can be used as an actuator, operating in thickness mode without the influence of the substrate geometry (Ealo et al. 2006, 2008). EMFi has been used to build acoustic transducers, such as physiological sensors (Alametsa et al. 2006), keyboards (Sorvoja et al. 2005), and force position sensors (Evreinov and Raisamo 2005). The usable frequency range of EMFi film for in-air applications begins at audible frequencies and extends up to its measured resonance frequency of 300 kHz.

The supply voltage of around 110 Vpp (Ealo et al. 2006) makes EMFi materials only suitable for fixed devices for which power consumption is not a major issue. However, even in these applications, power can be an issue if a high density of transducers is needed. These high voltages are not acceptable in mobile devices (MDs), which require low power consumption for long battery life.

There are no commercially available EMFi-based transducers. EMFi films are available, so building an EMFi transducer requires transducer prototyping using a film (Jimenez et al. 2007, Karki and Lekkala 2008).

25.2.4 CMUTs

Recently, MEMS, or more specifically, CMUT, have emerged as an alternative technology offering advantages such as wide bandwidth, ease of fabricating large arrays, and potential for integration. Sensors, actuators, and signal processing components can be integrated into miniaturized smart systems, capable of performing tasks that previously needed the use of a range of fabrication processes (Sarihan et al. 2008, Schuenemann et al. 2000). MEMS technology has demonstrated its economic strength in batch fabrication of large volumes of more or less identical devices (Tummala and Madisetti 1999).

CMUTs are fabricated using standard silicon integrated circuit fabrication technology (Schubring and Fujita 2007). This technology makes it possible to create large arrays using simple photolithography. Two-dimensional CMUT arrays with as many as 128 × 128 elements have already been successfully fabricated and characterized (Cheng et al. 2000). Individual electrical connections to transducer elements are provided by through-wafer interconnects. Another feature, inherent to CMUT technology, is wide bandwidth. A wideband transducer does not simply increase resolution, it also enables the design of new applications and tools. CMUTs are promising for high frequency applications, such as intravascular ultrasound imaging, in which high-frequency operation using miniature probes is vital. CMUTs operating at frequencies as high as 60 MHz have been fabricated and tested successfully.

MEMS ultrasonic transducers are commercially available, e.g., SPM0204 from Knowles Acoustics (Knowles Acoustics). They provide good sensitivity as well as a nearly flat response between 10 and 70 kHz. Their size, 4.72 mm × 3.76 mm × 1.15 mm, makes them useful for compact ultrasonic applications.

25.3 WIDEBAND TRANSDUCERS

Because of their low circuit complexity and low supply voltage, piezoelectric transducers are a very good option for low power applications. MEMS sensors are small in size, making them suitable for array processing applications. Also, their wide bandwidth, high sensitivity, and low supply voltage make them a good choice for implementation in MDs. In this section, we consider use of piezoelectric transducers and MEMS sensors as wideband ultrasonic transmitters and receivers for mobile applications.

25.3.1 Piezoelectric Bandwidth Modification

A piezoelectric transducer can be modeled using an equivalent electronic representation. Several models have been proposed (Church and Pincock 1985, Mason 1935, Ymada et al. 1999). A simple but efficient representation, i.e., one that captures the device's performance with acceptable error, is the Mason circuit (Mason 1935) shown in Figure 25.2. The series and parallel resonance frequencies of the circuit, for both transmitter and receiver operation, are given by

$$f_s = \frac{1}{2\pi\sqrt{L_s C_s}} \tag{25.1}$$

$$f_p = \frac{1}{2\pi\sqrt{L_s(C_s C_p / (C_s + C_p))}} \tag{25.2}$$

The piezoelectric resonant frequency is either the series resonance frequency f_s, the minimum impedance frequency f_m, or the smallest of f_s and f_m, called f_{rr}. Usually, the three of these are similar, so f_m is taken to be the resonant frequency, f_r, because it can be easily calculated.

The acoustic power transmitted by the transducer is proportional to the power dissipated in the resistor. Using the Laplace transform, the current through the resistor R_s is defined as follows:

$$I_M = \frac{V_{in}}{L_s s + (1/C_s s) + R_s} \tag{25.3}$$

$$\frac{I_M}{V_{in}} = \frac{(1/L_s)s}{s^2 + (R_s / L_s)s + 1/L_s C_s} \tag{25.4}$$

By defining w_n and ξ as

$$w_n^2 = \frac{1}{L_s C_s} \qquad \xi = \frac{R_s}{2}\sqrt{\frac{C_s}{L_s}} \tag{25.5}$$

Equation 25.4 can be expressed as

$$Y(s) = \frac{(1/L_s)s}{s^2 + 2\xi\omega_n + \omega_n^2} \tag{25.6}$$

which has the same structure as a bandpass filter, with ω_n as the filter's resonant frequency and ξ as a factor related to the filter's bandwidth.

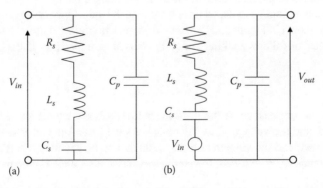

(a) (b)

FIGURE 25.2 Piezoelectric electric equivalent circuit. (a) Transmitter and (b) receiver. (From Gonzalez, J.R. and Bleakley, C.J., Low cost, wide band ultrasonic transmitter and receiver for array signal processing applications, *IEEE Sens. J.*, 11(5), 1284–1292, 2011. With permission. Copyright 2011 IEEE.)

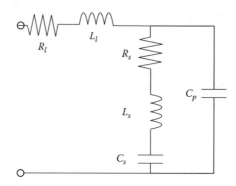

FIGURE 25.3 Compensated piezoelectric electric equivalent circuit. (From Gonzalez, J.R. and Bleakley, C.J., Low cost, wide band ultrasonic transmitter and receiver for array signal processing applications, *IEEE Sens. J.*, 11(5), 1284–1292, 2011. With permission. Copyright 2011 IEEE.)

If an inductance and a resistor are added to the circuit, as shown in Figure 25.3, a second peak appears in the frequency response. If this second peak is correctly chosen, the transducer's bandwidth is increased, but its sensitivity is slightly reduced. The Laplace response $Y_1(s)$ is then

$$\left(R_s + L_s s + \frac{1}{C_s s} \right) \Big\| \frac{1}{C_p s} = X_s \| X_p \tag{25.7}$$

$$X_s \| X_p = \frac{L_s s + R_s + (1/C_s s)}{C_p L_s s^2 + C_p R_s s + (C_s + C_p)/C_s} \tag{25.8}$$

$$Y_1(s) = \frac{X_s \| X_p}{X_s \| X_p + R_L + L_L s} \cdot \frac{1}{R_s + L_s s + (1/C_s s)} \tag{25.9}$$

where
‖ is the parallel equivalent impedance
X_s is the equivalent series impedance of R_s, L_s, and C_s
X_p is the equivalent impedance of C_p

Figure 25.4 shows the effect of inductance and resistor values on the compensated frequency response. The thick black line represents the unmodified frequency response. The thin lines represent the modified frequency responses. As can be seen, the inductance controls the frequency of the secondary peak, and the resistance controls the amplitude of the peaks.

This procedure can be applied to the receiver by adding the inductance and resistor in parallel to the equivalent circuit as shown in Figure 25.5. This provides results similar to those obtained for the transmitter. C_p is calculated as follows, making use of the transducer impedance Z at a given frequency f:

$$C_p = \frac{1}{|Z| 2\pi f} \tag{25.10}$$

In order to obtain the impedance of the transducer for various frequencies, the circuit shown in Figure 25.6 is used, and the voltage V_m is calculated. As the impedance of R is known, the current can be easily obtained, and the voltage in the transducer can be calculated as $V_{tx} = V_{in} - V_m$. Once the parallel capacitance C_p is obtained, the equivalent series resistance is calculated as

$$R_s = \frac{|Z|}{\sqrt{1 - (|Z| C_p 2\pi f_m)^2}} \tag{25.11}$$

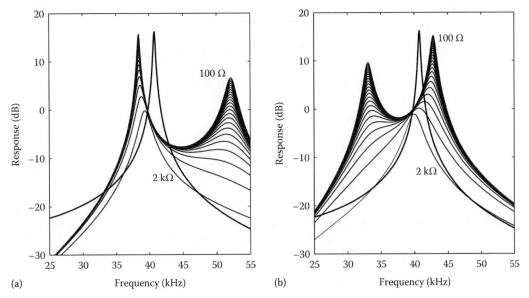

FIGURE 25.4 Frequency response of compensated (thin) and non-compensated (thick) transducer for $R = 100$-2k: (a) $Lc = 5$ mH and (b) $Lc = 10$ mH. (From Gonzalez, J.R. and Bleakley, C.J., Low cost, wide band ultrasonic transmitter and receiver for array signal processing applications, *IEEE Sens. J.*, 11(5), 1284–1292, 2011. With permission. Copyright 2011 IEEE.)

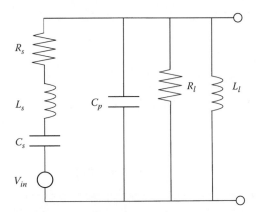

FIGURE 25.5 Compensated receiver piezoelectric electric equivalent circuit. (From Gonzalez, J.R. and Bleakley, C.J., Low cost, wide band ultrasonic transmitter and receiver for array signal processing applications, *IEEE Sens. J.*, 11(5), 1284–1292, 2011. With permission. Copyright 2011 IEEE.)

Use is made of C_p and the impedance at the resonant frequency to determine R_s. The resonant frequency is the one with maximum impedance, which means that voltage is maximum:

$$f_m = \frac{1}{2\pi\sqrt{C_s L_s}} \tag{25.12}$$

L_s and C_s can be calculated using Equation 25.12, which relates them to the resonant frequency. Parametric plots are provided in Figure 25.7 for various values of C_s and L_s and frequencies between 25 and 60 kHz.

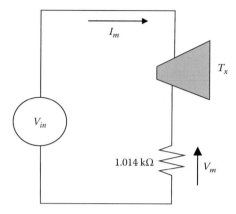

FIGURE 25.6 Transducer test circuit. (From Gonzalez, J.R. and Bleakley, C.J., Low cost, wide band ultrasonic transmitter and receiver for array signal processing applications, *IEEE Sens. J.*, 11(5), 1284–1292, 2011. With permission. Copyright 2011 IEEE.)

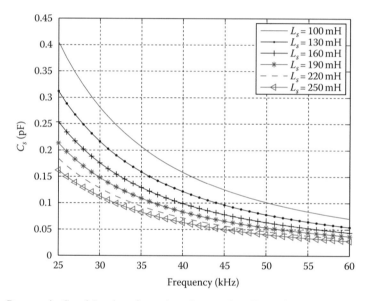

FIGURE 25.7 Parametric C and L values for various frequencies. (From Gonzalez, J.R. and Bleakley, C.J., Low cost, wide band ultrasonic transmitter and receiver for array signal processing applications, *IEEE Sens. J.*, 11(5), 1284–1292, 2011. With permission. Copyright 2011 IEEE.)

25.3.2 CMUT Sensor Array

Array signal processing has been used in many applications. Beamforming techniques have been used to improve system performance (Veen and Buckley 1988) and to extract additional information from an impinging signal (Krim and Viberg 1996, Trees 2002). In (Niculescu and Nath 2003, Rong and Sichitiu 2006), location is estimated in wireless sensor networks by computing the AoA of the exchanged signals. A triangular shaped ultrasonic receiver array was proposed and tested in Cricket Compass (Priyantha et al. 2001). The array is large (21 mm) and suffers from phase ambiguities.

AoA estimators are based on estimating the phase difference between a signal arriving at two different sensors. The maximum unambiguous phase difference between two sensors is $\pm\pi$. If D is the separation between two sensors, then we must have $D \leq \pi c/2\pi f = \lambda/2$ in order to provide a phase difference less than or equal to $\pm\pi$. As can be seen in Table 25.1, for low frequency ultrasonic signals, i.e., <45 kHz, the maximum separation between sensors is 3.81 mm. The distance decreases

TABLE 25.1

Maximum Distance between Sensors for c = 343 m/s

Frequency (kHz)	Wavelength (mm)	Maximum Distance (mm)
25	13.72	6.86
35	9.80	4.90
45	7.62	3.81
65	5.27	2.64
95	3.61	1.80

Source: From Gonzalez, J.R. and Bleakley, C.J., Low cost, wide band ultrasonic transmitter and receiver for array signal processing applications, *IEEE Sens. J.*, 11(5), 1284–1292, 2011. With permission. Copyright 2011 IEEE.

to 1.8 mm for a 95 kHz signal. Commonly, ultrasonic receivers have a diameter greater than that which would allow construction of an ultrasonic array. If this maximum sensor separation requirement is not met, the beamforming or AoA estimation process becomes ambiguous. The system is unable to correctly estimate the phase delays and multiple solutions are possible rather than a single solution. Using CMUT sensors allows smaller arrays, which can support array signal processing without ambiguities.

If the signal arrives at the array from direction (ϕ,θ), where ϕ is the source elevation, and θ is the source azimuth then the received signal at the center of the array can be described as

$$x_{ref}(t) = a(t)\exp(j\omega_0 t) \tag{25.13}$$

where $a(t)$ is the complex-envelope of the signal. The signal received at sensor m is defined as

$$x_m(t) = a(t)\exp(j\omega_0 t)\exp(j\Theta_m) \tag{25.14}$$

where Θ_m is the phase delay between the signal at sensor m and the central sensor. Let us assume that a set of N_s signals arrive at an array from direction $[\phi_n, \theta_n], n = 1,\ldots,N_s$. The received signal at each sensor at an instant n is called a snapshot and is defined as (Krim and Viberg 1996)

$$\underline{X}_n = \underline{\underline{A}}\,\underline{S} + \underline{w}_n \tag{25.15}$$

$$\underline{S} = \begin{bmatrix} S_1(n) \\ S_2(n) \\ \vdots \\ S_{NS}(n) \end{bmatrix} \tag{25.16}$$

$$\underline{\underline{A}} = \left[\underline{a}_1(\varsigma_1,\phi_1),\ \underline{a}_2(\varsigma_2,\phi_2),\ldots,\ \underline{a}_{NS}(\varsigma_{NS},\phi_{NS}) \right] \tag{25.17}$$

where

$S_i(n)$ is the complex envelope of each of the N_{NS} signals at time n

$\underline{\underline{A}}$ is a matrix containing the AoA of each signal

$\underline{w}_n$ is independent white Gaussian noise at each sensor

The snapshot is basically the sum of contributions of a set of sources plus noise. Most AoA estimators operate on the covariance matrix of the received snapshots. The covariance matrix R is ideally obtained by calculating the correlations between the sensors and placing the results in an $M \times M$ matrix, where M is the number of sensors in the array:

$$\underline{\underline{R}} = E\left[X_n \cdot X_n^H \right] \tag{25.18}$$

In practical applications, the number of observations is limited to the number of snapshots N_{snap}. In this case, the covariance matrix is estimated according to

$$\underline{\underline{\hat{R}}} = \frac{1}{N} \sum_{q=0}^{N_{snap}-1} X_{n-q} X_{n-q}^H \tag{25.19}$$

25.4 EVALUATION

This section describes characterization of four modified transmitters and a receiver array.

25.4.1 METHOD

The transducers chosen for modification were Prowave models: 250ST180 (25 kHz), 328ET250 (32 kHz), 400ET180 (40 kHz), and 400EP900 (50 kHz) (Prowave). The first step in the modification process was obtaining the equivalent circuit parameters. These parameters were calculated based on the voltage and impedance of the transducer at a range of frequencies. The circuit shown in Figure 25.6 was built in order to obtain these measurements. The test circuit was connected to a signal generator, applying a 15 Vpp sinusoidal wave signal to the transducer with a resistor $R = 1.014$ kΩ in series. Frequency response characterization was performed by placing the transmitter and a SPM0204 receiver face-to-face and sending sinusoidal pulses between them from 10 to 50 kHz in steps of 100 Hz. For each frequency, the mean amplitude was calculated and used as a data point in the measured frequency response. The results were corrected to allow for the receiver response, the frequency response of which is given in the datasheet (Knowles Acoustics).

The design process had an optimization criterion of maximizing the bandwidth at −15 dB. The transducer parameters were applied to the optimization procedure in order to find the inductance value that provides two resonance peaks equally separated from the original narrow peak. Once the inductance value was chosen, so as to provide a modified frequency response centered on the original resonance, an iterative procedure was applied to search for the resistance providing maximum bandwidth at −15 dB, ideally with the smallest resonance peak variation, in order to ensure small variations in the frequency response within the usable bandwidth. The circuit was constructed and the frequency response of the bandwidth modified transducer was measured.

The receiver array was constructed using eight Knowles Acoustics SPM0204 sensors fixed to a Printed Circuit Board, configured as shown in Figure 25.8. The separation between sensors is 4.71 mm, allowing a maximum frequency of 40 kHz without ambiguity. The voltage supply necessary for the sensors is 3.3 V, and the current consumption varies between 0.1 and 0.25 mA, as stated in the datasheet (Knowles Acoustics). A signal conditioning and amplification circuit was implemented to boost the signals to levels suitable for processing. The mutual coupling and mismatch between channels in the array were measured.

Finally, the SNR achieved between the transmitter and receiver was measured for a variety of ranges. A 14 Vpp sinusoid at 40 kHz was applied to the transducer. The receiver was supplied by a 3.6 V battery. The mean Vpp was calculated for the background noise and the signal.

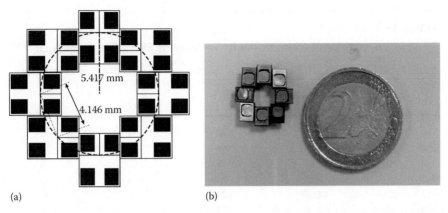

(a)   (b)

FIGURE 25.8   Antenna array: (a) footprint and (b) photograph. (From Gonzalez, J.R. and Bleakley, C.J., Low cost, wide band ultrasonic transmitter and receiver for array signal processing applications, *IEEE Sens. J.*, 11(5), 1284–1292, 2011. With permission. Copyright 2011 IEEE.)

25.4.2   RESULTS

Tables 25.2 and 25.3 show the results obtained when measuring voltage and impedance for the transducers using the circuit presented in Figure 25.6, with $R = 1.014$ kΩ. Table 25.4 presents the equivalent circuit component values for all transducers. Figure 25.9 shows the estimated compensated frequency response for the transducers modeled with the component values listed in Table 25.4. The effect of the compensation process on all of the transducers responses can be clearly seen. The circuits provide significant improvements in bandwidth, achieving a bandwidth of 10–15 kHz, depending on the transducer. Figure 25.10 shows the results of the modifications for the four real transducers. It can be seen that the measurements match the estimates very closely. The second peak is visible in all cases.

TABLE 25.2
Transducer Voltages for Various Excitation Frequencies

Freq (kHz)	Vtx (25 kHz) (Vpp)	Vtx (32 kHz) (Vpp)	Vtx (40 kHz) (Vpp)	Vtx (48 kHz) (Vpp)
20.0	6.70	7.00	7.20	7.00
25.0	5.20	6.70	6.44	6.40
27.5	8.68	6.30	6.30	6.20
30.0	7.60	5.80	6.00	5.90
32.5	6.45	4.80	5.55	5.60
35.0	5.95	9.02	5.00	5.20
40.0	4.00	6.00	3.20	4.50
42.5	6.80	5.50	8.40	4.40
45.0	6.80	5.20	5.70	4.20
47.5	5.70	4.80	5.20	3.90
50.0	5.20	4.60	4.80	3.60
55.0	4.50	4.20	4.40	8.10
60.0	4.05	3.60	3.85	7.00

Source: From Gonzalez, J.R. and Bleakley, C.J., Low cost, wide band ultrasonic transmitter and receiver for array signal processing applications, *IEEE Sens. J.*, 11(5), 1284–1292, 2011. With permission. Copyright 2011 IEEE.

TABLE 25.3
Transducer Impedance Measurements for Various Excitation Frequencies

Frequency (kHz)	Z (25 kHz) (kΩ)	Z (32 kHz) (kΩ)	Z (40 kHz) (kΩ)	Z (48 kHz) (kΩ)
20.0	2.06	2.37	2.26	2.37
25.0	1.10	2.06	1.83	1.80
27.5	6.67	1.73	1.73	1.65
30.0	3.21	1.40	1.52	1.46
32.5	1.84	0.94	1.24	1.29
35.0	1.49	9.13	1.01	1.10
40.0	0.68	1.52	0.48	0.83
42.5	1.02	1.24	5.32	0.80
45.0	2.16	1.10	1.34	0.73
47.5	1.35	0.94	1.10	0.65
50.0	1.10	0.86	0.94	0.57
55.0	0.83	0.73	0.80	4.06
60.0	0.69	0.57	0.63	2.37

Source: From Gonzalez, J.R. and Bleakley, C.J., Low cost, wide band ultrasonic transmitter and receiver for array signal processing applications, *IEEE Sens. J.*, 11(5), 1284–1292, 2011. With permission. Copyright 2011 IEEE.

TABLE 25.4
Transducer Equivalent Circuit Component Values

Tx Model	$C_p(nF)$	$R_s(k\Omega)$	$C_s(nF)$	$L_s(mH)$
250STI80	3.84	2.65	0.25	160
328ET250	4.65	2.08	0.15	160
400ET180	3.05	1.95	0.10	160
400EP900	3.36	2.77	0.07	160

Source: From Gonzalez, J.R. and Bleakley, C.J., Low cost, wide band ultrasonic transmitter and receiver for array signal processing applications, *IEEE Sens. J.*, 11(5), 1284–1292, 2011. With permission. Copyright 2011 IEEE.

Table 25.5 gives the voltage received in all channels when only one sensor in the array is receiving signals. With the information in this table, the mutual coupling matrix was computed (Krim and Viberg 1996, Veen and Buckley 1988). The results are presented in Equation 25.20. The contribution to one channel from the others is very small, allowing for precise array signal processing.

$$G\mid_{40\,\text{kHz}} = \begin{bmatrix} 1 & 0.04 & 0.02 & 0.03 & 0.02 & 0.03 \\ 0.03 & 1 & 0.06 & 0.01 & 0.01 & 0.03 \\ 0.04 & 0.05 & 1 & 0.05 & 0.01 & 0.08 \\ 0.08 & 0.05 & 0.08 & 1 & 0.08 & 0.08 \\ 0.03 & 0.02 & 0.03 & 0.08 & 1 & 0.02 \\ 0.02 & 0.03 & 0.01 & 0.03 & 0.03 & 1 \end{bmatrix} \tag{25.20}$$

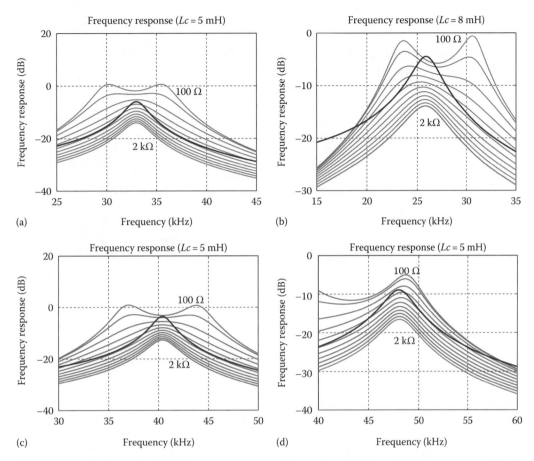

FIGURE 25.9 Estimated modified frequency response for transducer, RL from 100 Ω to 2 kΩ, (a) 250ST180; (b) 328ET250; (c) 400ET180; and (d) 400EP900. (From Gonzalez, J.R. and Bleakley, C.J., Low cost, wide band ultrasonic transmitter and receiver for array signal processing applications, *IEEE Sens. J.*, 11(5), 1284–1292, 2011. With permission. Copyright 2011 IEEE.)

Table 25.6 shows the mismatch between channels. The first column is the amplitude of the received signal at each channel. The second column is the deviation in amplitude of each channel with respect to the reference channel *CHr* in %. The third column is the deviation in the continuous component with respect the reference channel, in volts. As can be seen, the mismatch is small enough so as to be negligible.

Table 25.7 shows the SNR values obtained in the tested room between the transmitter and receiver. Clearly, the transmitter–receiver pair provides very good SNR at the ranges measured.

25.5 APPLICATION

An LPS determines the 3D location, and possibly the orientation, of an MD in a fixed frame of reference. Over the years, many LPS technologies have been proposed and investigated. Ultrasonic LPSs are distinguished by their ability to estimate location with a high degree of resolution at low cost. The high resolution achieved is primarily due to the low propagation velocity of sound in air. Over the years, a number of ultrasonic LPSs have been developed, notably, Constellation (Foxlin et al. 1998), Cricket (Priyantha et al. 2000), Bats (Harter et al. 2002), Whisper (Vallidis 2002), and Dolphin (Hazas and Hopper 2006). While results are promising, the final accuracy obtained from these

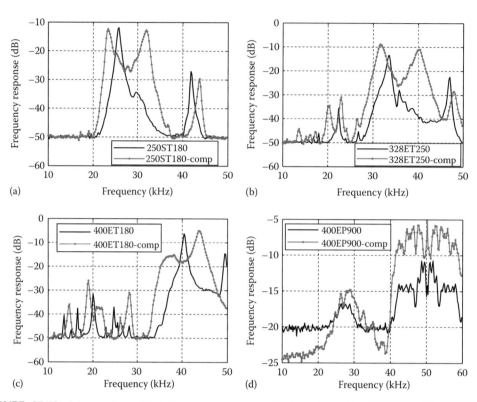

FIGURE 25.10 Measured modified frequency response for transducer (a) 250ST180; (b) 328ET250; (c) 400ET180; and (d) 400EP900. (From Gonzalez, J.R. and Bleakley, C.J., Low cost, wide band ultrasonic transmitter and receiver for array signal processing applications, *IEEE Sens. J.*, 11(5), 1284–1292, 2011. With permission. Copyright 2011 IEEE.)

TABLE 25.5
Receiver Channel Mutual Coupling Index

	CHr (mV)	CH1 (mV)	CH2 (mV)	CH3 (mV)	CH4 (mV)	CH5 (mV)	CH6 (mV)
CHr	51.0	1.04	1.34	1.14	1.26	0.95	0.92
CH1	1.96	49.0	2.18	1.19	1.50	1.11	1.25
CH2	0.83	0.63	18.5	1.05	0.24	0.22	0.56
CH3	1.04	0.87	1.24	24.3	1.16	0.25	2.05
CH4	1.05	1.21	0.83	1.25	15.8	1.30	1.20
CH5	0.95	1.18	0.94	0.99	3.01	37.6	0.93
CH6	1.22	1.17	1.29	0.64	1.29	1.52	50.9

Source: From Gonzalez, J.R. and Bleakley, C.J., Low cost, wide band ultrasonic transmitter and receiver for array signal processing applications, *IEEE Sens. J.*, 11(5), 1284–1292, 2011. With permission. Copyright 2011 IEEE.

systems is dependent on temperature, humidity, multipath, ambient noise, and air flow. Temperature can be accounted for numerically with the aid of a temperature sensor, while errors due to variations in humidity and air flow can generally be assumed to be negligible in indoor environments. Multipath and ambient noise, however, have a significant impact on performance in real-world indoor environments. This is particularly problematic when conventional, narrowband ultrasonic transducers

TABLE 25.6

Continuous Component and Amplitude Mismatch between Channels

Channel	Amplitude (V)	Amp. Mismatch (%)	Continuous C. (V)
CHr	3.67	0.00	0.00
CH1	3.75	2.40	0.01
CH2	3.66	0.22	0.00
CH3	3.50	4.52	0.01
CH4	3.75	2.06	0.01
CH5	3.69	0.67	0.01
CH6	3.59	2.04	0.02

Source: From Gonzalez, J.R. and Bleakley, C.J., Low cost, wide band ultrasonic transmitter and receiver for array signal processing applications, *IEEE Sens. J.*, 11(5), 1284–1292, 2011. With permission. Copyright 2011 IEEE.

TABLE 25.7

SNR Values in Tested Room

Distance (m)	SNR (dB)
1.12	33.40
1.67	30.45
2.60	25.45
3.57	20.12

Source: From Gonzalez, J.R. and Bleakley, C.J., Low cost, wide band ultrasonic transmitter and receiver for array signal processing applications, *IEEE Sens. J.*, 11(5), 1284–1292, 2011. With permission. Copyright 2011 IEEE.

are employed. Since the signals are impulsive, the direct path signal is difficult to distinguish from reflected signals, particularly under conditions of ambient noise. To alleviate this problem, a number of researchers have proposed the use of broadband ultrasonic transducers (e.g., Hazas and Hopper 2006, Villadangos et al. 2007). However, the broadband ultrasonic transducers used were bulky, costly, and have high power consumption. Clearly, these features are not desirable for mass market mobile applications.

We have developed a prototype indoor ultrasonic LPS based on the transducers and sensors described in Sections 25.3 and 25.4. In the following sub-sections, we describe the LPS, the signal processing algorithms used, and the performance achieved by the overall system.

25.5.1 Local Positioning System

The LPS prototype consisted of three fixed beacon devices, each with a single ultrasonic transmitter. The beacons were deployed on tripods in the corners of the location space. Based on the ultrasonic signals received from the beacons and a wired timing reference signal, the beacon-MD range and MD

FIGURE 25.11   Photograph of the prototype LPS.

orientation were determined. The MD consisted of the ultrasonic sensors and an associated amplifier circuit. A digital signal processing (DSP) board from Sundance (model 361A) was used for signal generation and signal acquisition. The board included a C6416 DSP from Texas Instruments with two daughter boards, a SMT377 with eight independent digital-to-analog converters and a SMT317 with an eight-channel analog-to-digital converter. The beacon and MD units were connected to the DSP board via coaxial connections. Signal processing was performed off-line in MATLAB® on a PC. Based on the options in the DSP board, a sampling frequency of 117.5 kHz was selected for the experiments. Sound velocity was assumed to be constant during the experiments. Variations in humidity and air flow were not accounted for. A photograph of the prototype system is provided in Figure 25.11.

25.5.2   SIGNAL PROCESSING ALGORITHM

Each beacon continuously transmits an FHSS signal. The carrier frequency of the signals hops between a set of frequencies within the available bandwidth according to a pseudorandom hopping sequence. Each beacon has a unique hopping sequence to avoid collisions between signals and to allow beacon identification based on the received signal. The transmitted signal $x_k(n)$ for beacon k can be described as

$$x_k(n) = \sin(2\pi f_k n/F_s + \phi_k) \tag{25.21}$$

where
   f_k is the carrier frequency, which is a function of time according to the pseudorandom sequence of user k
   F_s is the sampling frequency
   ϕ_k is the phase offset

Range estimation is performed by calculating the cross-correlation $r(l)$ of the signal $y_0(n)$ received at sensor 0 with the known signal transmitted by beacon k:

$$r_k(l) = \sum_{n=0}^{N-1} x_k(n) y_0(n-l) \tag{25.22}$$

where N is the length of the window. The delay of the earliest peak p_k in the cross-correlation is taken as the time of flight of the direct path signal from the transmitter to the receiver. The earliest peak is defined as the earliest peak with an amplitude greater than 0.7 of the highest peak. The estimated transmitter–receiver distance $\bar{d}_k$ is then

$$\bar{d}_k = \frac{p_k c}{F_s} \tag{25.23}$$

where c is the speed of sound.

This coarse range estimate is refined by adding a phase adjustment term. The phase adjustment term improves the resolution of ranging from sample-level to sub-sample. The phase angle for the signal from beacon k at hop m, $\phi_{m,k}$, can be calculated as

$$\phi_{m,k} = \text{ang}(X_{m,k}(\omega_m) Y_{m,0}^*(\omega_m)) \tag{25.24}$$

where * is the complex conjugate operator and the ang (·) operator returns the phase angle of a complex number, and $X_{m,k}(\omega_m)$ and $Y_{m,0}(\omega_m)$ are the discrete Fourier transforms (DFTs) of the mth hop of the signal transmitted by beacon k delayed by p_k samples and the received signal at sensor 0, respectively. The estimated phase adjustment is the phase shift expressed as a distance averaged over multiple hops:

$$\delta_k = \frac{c}{M} \sum_{m=0}^{M-1} \frac{\phi_{m,k}}{\omega_m} \tag{25.25}$$

where

ω_m is the carrier frequency of hop m

M is the number of hops

Since the phase adjustment is limited in the range $\lambda/2$, phase adjustments are calculated for a range of candidate values of integer delay, e.g. $p_k 8$. The candidate integer delay giving the minimum variance V in the phase adjustment at each hop is selected as the final integer delay $\hat{p}_k$ together with its associated mean phase adjustment $\hat{\Delta}_k$.

$$V = \frac{1}{M} \sum_{m=0}^{M-1} (\delta_k - \delta_{m,k})^2 \tag{25.26}$$

where $\delta_{m,k}$ is the phase adjustment for beacon k at hop m.

The final range estimate is then

$$\widehat{\widehat{d}}_k = \frac{\hat{p}_k c}{F_s} + \hat{\delta}_k \tag{25.27}$$

Given range estimates for at least three beacons, the 3D location of the MD can be determined by trilateration (Foy 1976).

The AoA of the signal from beacon k is estimated based on difference in the times of arrival (ToA) of the signals at the sensors in the array. The phase angle of the signals at sensors i and j for hop m from beacon k is calculated as

$$\theta_{m,k,i,j} = \text{ang}\left(Y_{m,i}(\omega_m) Y_{m,j}^*(\omega_m)\right) \tag{25.28}$$

where

* is the complex conjugate operator

$Y_{m,i}(\omega_m)$ is the DFT of the received signal for hop m at sensor i

The phase angle can be converted to a time difference of arrival (TDoA) by accounting for the frequency of the carrier and the TDoAs can be averaged over multiple hops:

$$\Theta_{k,i,j} = \frac{1}{M} \sum_{m=0}^{M-1} \frac{\theta_{m,k,i,j}}{\omega_m} \qquad (25.29)$$

Given these inter-sensor TDoAs, the AoA of the signal at the UCA can be determined using the multiple signal classification (MUSIC) algorithm (Mathews and Zoltowski 1994). Given AoA estimates to three or more beacons, the 3D orientation of the MD can be determined by a system of geometrical equations (Gonzalez and Bleakley 2009).

25.5.3 EXPERIMENTAL RESULTS

Experiments were conducted to assess the performance of the LPS in estimation of beacon-MD range. The beacons were fitted with Prowave 250ST180 transducers with bandwidth expansion from 2 to 10 kHz. The MD was fitted with three SPM0204 sensors with known separations. The LPS was deployed in a normal office measuring 3.5 m × 2.8 m × 2.7 m. The distance between the sensors was measured using the ultrasonic LPS at three different beacon-MD ranges, namely 2.1, 2.3, and 2.5 m. The cumulative error in the estimated ranges is shown in Figure 25.12. The error is less than 0.5 mm in 90% of cases.

Experiments were conducted to assess the performance of the LPS in estimating MD orientation. The beacons were fitted with Prowave 400ST160 transducers with bandwidth expansion from 2 to 15 kHz. The MD was fitted with an eight element UCA of SPM0204 ultrasonic sensors, as shown in Figure 25.8. The LPS was deployed in a normal office measuring 2 m × 4 m × 2 m. The MD was placed in the center of the location space and the orientation of the device estimated using the ultrasonic signals and compared with manual measurements. The mean error in orientation estimation was calculated over 10 orientation estimates. Each orientation estimate was derived from AoA estimates calculated over 15 hops. The mean errors in the estimated pitch, roll and yaw of the MD are shown in Figures 25.13 through 25.15. The average error for pitch and roll is around 3.5°, while for yaw the average is around 1.5°.

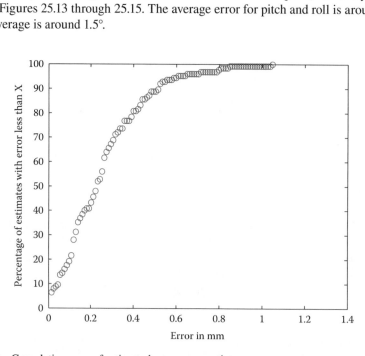

FIGURE 25.12 Cumulative error of estimated sensor separation.

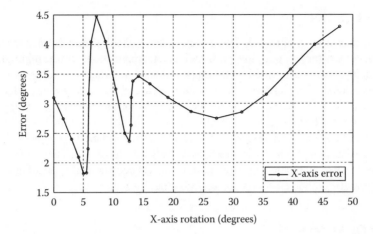

FIGURE 25.13 Orientation estimation accuracy—Pitch.

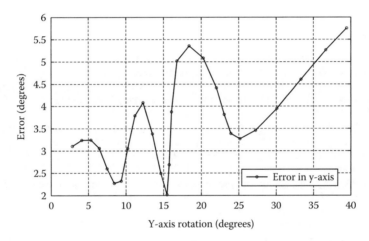

FIGURE 25.14 Orientation estimation accuracy—Roll.

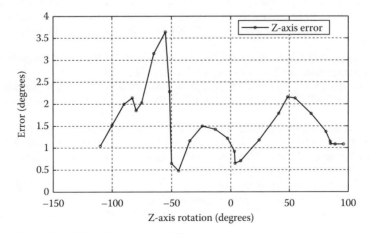

FIGURE 25.15 Orientation estimation accuracy—Yaw.

25.6 CONCLUSION

In this chapter, a review of ultrasonic in-air transducer technologies was presented. The review focused on transducers for low cost, low power wideband applications. It was highlighted that conventional technologies are not optimal for this application. A modification process was presented for ultrasonic transmitters, which allows a significant increase in piezoelectric transducer bandwidth by adding two passive components to the circuit—an inductor and a resistor. The theory used was explained and validated by experimental results. The chapter described how MEMS sensors can be used to construct ultrasonic receiver arrays. The results of array characterization are provided. The use of these transducers and sensors in a ultrasonic LPS was described. These results demonstrate the potential of these technologies for exploitation in novel commercial applications, such as body tracking, 3D human–computer interaction, and robot navigation.

ACKNOWLEDGMENTS

This work was funded by Science Foundation Ireland (SFI) and the Higher Education Authority (HEA) of Ireland.

REFERENCES

Alametsa, J., Rauhala, E., Huupponen, E. et al. 2006. Automatic detection of spiking events in EMFi sheet during sleep. *Medical Engineering and Physics* 28(3):267–275.

Barna, L., Koivuluoma, M., Hasu, M., Tuppurainen, J., and Varri, A. 2007. The use of electromechanical film (EMFi) sensors in building a robust touch sensitive tablet-like interface. *IEEE Sensors Journal* 7(1):74–80.

Bloomfield, P.E., Lo, W.J., and Lewin, P.A. 2000. Experimental study of the acoustical properties of polymers utilized to construct PVDF ultrasonic transducers and the acousto-electric properties of PVDF and P(VDF/TrFE) films. *IEEE Transactions on Ultrasonics, Ferroelectrics and Frequency Control* 47(6):1397–1405.

Brown, L.F. 2000. Design considerations for piezoelectric polymer ultrasound transducers. *IEEE Transactions on Ultrasonics, Ferroelectrics and Frequency Control* 47(6):1377–1396.

Cheng, C.H., Khuri, E., and Yakub, B. 2000. An efficient electrical addressing method using through-wafer vias for two-dimensional ultrasonic arrays. *Proceedings of the IEEE Ultrasonics Symposium*, San Juan, Puerto Rico, Vol. 2, pp. 1179–1182.

Church, D. and Pincock, D. 1985. Predicting the electrical equivalent of piezoceramic transducers for small acoustic transmitters. *IEEE Transactions on Sonics and Ultrasonics* 32(1):61–64.

Cochran, A., Hayward, G., and Murray, V. 1999. Multilayer piezocomposite ultrasonic transducers operating below 50 kHz. *Proceedings of the IEEE Ultrasonics Symposium*, Caesars Tahoe, NV, Vol. 2, pp. 953–956.

Cochran, A., Reynolds, P., and Hayward, G. 1997. Multilayer piezocomposite transducers for applications of low frequency ultrasound. *Proceedings of the IEEE Ultrasonics Symposium*, Toronto, Canada, Vol. 2, pp. 1013–1016.

Dobie, G., Summan, R., Pierce, S.G., Galbraith, W., and Hayward, G. 2011. A noncontact ultrasonic platform for structural inspection. *IEEE Sensors Journal* 11(10):2458–2468.

Ealo, J.L., Jimenez, A.R., Seco, F. et al. 2006. Broadband omnidirectional ultrasonic transducer for air ultrasound based on EMFi. *Proceedings of the IEEE Ultrasonics Symposium*, Vancouver, Canada, pp. 812–815.

Ealo, J.L., Seco, F., and Jimenez, A.R. 2008. Broadband EMFi based transducers for ultrasonic air applications. *IEEE Transactions on Ultrasonics, Ferroelectrics and Frequency Control* 55(4):919–929.

Evreinov, G. and Raisamo, R. 2005. One-directional position-sensitive force transducer based on EMFi. *Sensors and Actuators A: Physical* 123:204–209.

Fiorillo, A.S. 1992. Design and characterization of a PVDF ultrasonic range sensor. *IEEE Transactions on Ultrasonics, Ferroelectrics and Frequency Control* 39(6):688–692.

Ford, T.A. and Hanford, W.E. 1948. Polyvinylidene fluoride and process for obtaining the same, U.S. Patent 2,435,537.

Foxlin, E., Harrington, M., and Pfeifer, G. 1998. Constellation: A wide-range wireless motion-tracking system for augmented reality and virtual set applications. *Proceedings of the International Conference on Computer Graphics and Interactive Techniques (SIGGRAPH)*, Orlando, FL, pp. 371–378.

Foy, W.H. 1976. Position-location solutions by Taylor-series estimation. *IEEE Transactions on Aerospace Electronic Systems* 12(2):187–194.

Gonzalez, J.R. and Bleakley, C.J. 2009. High precision robust broadband ultrasonic location and orientation estimation. *IEEE Journal of Selected Topics in Signal Processing* 3(5):832–844.

Gonzalez, J.R. and Bleakley, C.J. 2011. Low cost, wideband ultrasonic transmitter and receiver for array signal processing applications. *IEEE Sensors Journal* 11(5):1284–1292.

Gururaja, T.R., Schulze, W.A., Cross, L.E. et al. 1985. Piezoelectric composite materials for ultrasonic transducer applications. I: Resonant modes of vibration of PZT rod-polymer composites. *IEEE Transactions on Sonics and Ultrasonics* 32(4):481–498.

Harter, A., Hopper, A., Steggles, P., Ward, A., and Webster, P. 2002. The anatomy of a context-aware application. *Wireless Networks* 8(2):187–197.

Hazas, M. and Hopper, A. 2006. Broadband ultrasonic location systems for improved indoor positioning. *IEEE Transactions on Mobile Computing* 5(5):536–547.

Jimenez, A., Hernandez, A., Urena, J. et al. 2007. Piezopolymeric transducer for ultrasonic sensorial systems. *IEEE International Symposium on Industrial Electronics (ISIE)*, Vigo, Spain, pp. 1458–1463.

Jimenez, A.R. and Seco, F. 2005. Precise localisation of archaeological findings with a new ultrasonic 3D positioning sensor. *Sensors and Actuators A: Physical* 123:224–233.

Karki, S. and Lekkala, J. 2008. Film-type transducer materials PVDF and EMFi in the measurement of heart and respiration rates. *IEEE International Conference on Engineering in Medicine and Biology Society (EMBS)*, Vancouver, Canada, pp. 530–533.

Klicker, K.A., Biggers, J.V., and Newham, R.E. 1981. Composites of PZT and epoxy for hydrostatic transducer applications. *Journal of American Ceramic Society* 64(1):5–9.

Knowles Acoustics, SPM0204 MEMS Sensor Datasheet. http://www.knowles.com (accessed on Jan 9, 2013).

Kossoff, G. 1966. The effects of backing and matching on the performance of piezoelectric ceramic transducers. *IEEE Transactions on Sonics and Ultrasonics* 13(1):20–30.

Krim, H. and Viberg, M. 1996. Two decades of array signal processing research: The parametric approach. *IEEE Signal Processing Magazine* 13(4):67–94.

Lan, J., Boucher, S.G., and Tancrell, R.H. 1999. Investigation of broadband characteristics of PVDF ultrasonic transducers by finite element modeling and experiments. *Proceedings of the IEEE Ultrasonics Symposium*, Vol. 2, pp. 1109–1112.

Mason, W.P. 1935. An electromechanical representation of a piezoelectric crystal used as a transducer. *Proceedings of the IRE* 23(10):1252–1263.

Mathews, C.P. and Zoltowski, M.D. 1994. Eigenstructure techniques for 2-D angle estimation with uniform circular arrays. *IEEE Transactions on Signal Processing* 42(9):2395–2407.

MSI. 2001. PVDF 40 kHz Transducer Datasheet. (http://www.meas-spec.com/product/t-product.aspx?id=2490 (accessed on Jan 9, 2013).

Newnham, R.E., Skinner, D.P., Cross, L.E. et al. 1978. Connectivity and piezoelectric-pyroelectric composites. *Materials Research Bulletin* 13(5):525–536.

Niculescu, D. and Nath, B. 2003. Ad hoc positioning system (APS) using AOA. *Proceedings of the IEEE Conference on Computer Communications (INFOCOM)*, San Francisco, CA, pp. 1734–1743.

Priyantha, N.B., Chakraborty, A., and Balakrishnan, H. 2000 The cricket location-support system. *Proceedings of the International Conference on Mobile Computing and Networking*, Boston, pp. 32–43.

Priyantha, N.B., Miu, A.K.L., Balakrishnan, H., and Teller, S. 2001. The cricket compass for context-aware mobile applications. *Proceedings of the International Conference on Mobile Computing and Networking (MobiCom)*, Rome, Italy, pp. 1–14.

Prowave. Piezoelectric Ultrasonic Transducers Datasheets. http://www.prowave.com.tw/ (accessed on Jan 9, 2013).

Rong, P. and Sichitiu, M. 2006. Angle of arrival localization for wireless sensor networks. *Proceedings of the IEEE Sensor Ad Hoc Communications and Networks (SECON)*, Reston, VA, pp. 374–382.

Saad, M.M., Bleakley, C.J., and Dobson, S. 2011. Robust high accuracy ultrasonic range measurement system. *IEEE Transactions on Instrumentation and Measurement* 60(10):3334–3341.

Sarihan, V., Wen J., Li, G. et al. 2008. Designing small footprint, low-cost, high-reliability packages for performance sensitive MEMS sensors. *Proceedings of the Electronic Components and Technology Conference (ECTC 2008)*, Lake Buena Vista, FL, pp. 817–818.

Schubring, A. and Fujita, Y. 2007. Ceramic package solutions for MEMS sensors. *IEEE International Electronic Manufacturing Technology. Symposium (IEMT)* San Jose, CA, pp. 268.

Schuenemann, M., Jam, K.A., Grosser, V. et al. 2000. MEMS modular packaging and interfaces. *Proceedings of the Electronic Components and Technology Conference*, Las Vegas, NV, pp. 681–688.

Skinner, D.P., Newnham, R.E., and Cross, L.E. 1978. Flexible composite transducers. *Materials Research Bulletin* 13(6):599–607.

Smith, W.A. 1989. The role of piezocomposites in ultrasonic transducers. *Proceedings of the IEEE Ultrasonics Symposium*, Montreal, Canada, pp. 755–766.

Sorvoja, H., Kokko, V.M., Myllyla, R., and Miettinen, J. 2005. Use of EMFi as a blood pressure pulse transducer. *IEEE Transactions on Instrumentation and Measurement* 54(6):2505–2512.

Takeuchi, S., Al Zaabi, M.R.A., Sato, T., and Kawashima, N. 2002. Study on ultrasound transducer with double peak type frequency characteristics for sub-harmonic imaging. *Proceedings of the IEEE Ultrasonics Symposium*, Vol. 2, pp. 1101–1105.

Toda, M. 2002. Cylindrical PVDF film transmitters and receivers for air ultrasound. *IEEE Transactions on Ultrasonics, Ferroelectrics and Frequency Control* 49(5):626–634.

Toda, M. and Dahl, J. 2007. PVDF corrugated transducer for ultrasonic ranging sensor. *Sensors and Actuators A: Physical* 134(2):427–435.

Trees, H.L. Van. 2002. *Optimum Array Processing: Part IV of Detection, Estimation and Modulation Theory.* John Wiley & Sons, New York.

Tummala, R.R. and Madisetti, V.K. 1999. System on chip or system on package. *IEEE Design & Test of Computers* 16(2):48–56.

Vallidis, N.M. 2002. *WHISPER: A Spread Spectrum Approach to Occlusion in Acoustic Tracking.* PhD thesis, University of North Carolina at Chapel Hill.

Veen, B.D. Van and Buckley, K.M. 1988. Beamforming: A versatile approach to spatial filtering. *IEEE Signal Processing Magazine* 5(2):4–24.

Villadangos, J.M., Urena, J., Mazo, M. et al. 2007. Improvement of cover area in ultrasonic local positioning system using cylindrical PVDF transducer. *Proceedings of the IEEE International Symposium on Industrial Electronics (ISIE)*, Vigo, Spain, pp. 1473–1477.

Ymada, K., Sakamura, J.I., and Nakamura, K. 1999. Equivalent network analysis of piezoelectrically-graded broadband ultrasound transducers. *Proceedings of the IEEE Ultrasonics Symposium,* Vol. 2, pp. 1119–1124.

26 Sensing Applications Using Photoacoustic Spectroscopy

Ellen L. Holthoff and Paul M. Pellegrino

CONTENTS

26.1 INTRODUCTION

In recent years, photoacoustic spectroscopy (PAS) has emerged as an attractive and powerful technique well suited for sensing applications. The development of high power radiation sources and more sophisticated electronics, including sensitive microphones and digital lock-in amplifiers, have allowed for significant advances in PAS. Recent research suggests that PAS is capable of trace gas detection at parts per trillion (ppt) levels.[1,2] Furthermore, photoacoustic (PA) detection of infrared absorption spectra using modern tunable lasers offers several advantages, including simultaneous detection and discrimination of numerous molecules of interest.

Successful applications of PAS in gases and condensed matter have made this a notable technique and it is now studied and employed by scientists and engineers in a variety of disciplines. Therefore, a substantial body of literature on PAS exists today. The following discussion summarizes PAS and the experimental components and arrangements for PA detection, as well as its use in the past 15 years for sensing applications. PA sensing, specifically laser-based PA sensing schemes, of gas, liquid, and solid samples is reviewed. It is not the intention of the authors to discuss PA theory in great detail, as this has been presented in numerous publications on the topic. Instead, we hope to provide the reader with a general understanding of PAS as it applies to sensing.

26.2 FUNDAMENTALS OF PHOTOACOUSTICS

26.2.1 PHOTOTHERMAL PHENOMENA

One should always begin discussion on PA or optoacoustic spectroscopy with a more general discussion of the phenomena of spectroscopy, namely photothermal spectroscopy. Photothermal spectroscopy encompasses a group of highly sensitive methods that can be used to detect trace levels of optical absorption and subsequent thermal perturbations of the sample in gas, liquid, or solid phases. The underlying principle that connects these various spectroscopic methods is the measurement of physical changes (i.e., temperature, density, or pressure) as a result of a photo-induced change in the thermal state of the sample. In general, most scientists classify photothermal methods as indirect methods for detection of trace optical absorbance because the transmission of the light used to excite the sample is not measured directly. Upon closer inspection, a counter position asserting that these techniques may be a more direct measure of optical absorption could be appropriate due to its sole dependence on optical absorption and its immunity to optical scattering and reflections. Examples of photothermal techniques include photothermal interferometry (PTI), photothermal lensing (PTL), photothermal deflection (PTD), and PAS. All photothermal processes consist of several linked steps that result in a change of the state of the sample. In general, the sample undergoes an optical excitation, which can take various forms of radiation, including laser radiation. This radiation is absorbed by the sample, placing it in an excited state (i.e., increased internal energy). Some portion of this energy decays from the excited state in a non-radiative fashion. This increase in local energy results in a temperature change in the sample or the coupling fluid (e.g., air). The increase in temperature can result in a density change; and, if it occurs at a faster rate than the sample or coupling fluid can expand or contract, the temperature change will result in a pressure change. Figure 26.1 pictorially shows the process associated with photothermal phenomena. As mentioned earlier, all photothermal methods attempt to key in on the changes in the thermal state of the sample

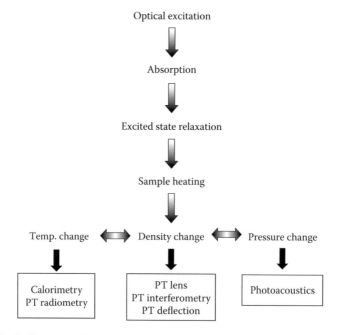

FIGURE 26.1 The basic process for signal generation with photothermal spectroscopy. Following absorption of radiation of the appropriate wavelength, the sample undergoes non-radiative excited-state relaxation resulting temperature, density, and pressure changes. (From Bialkowski, S.E.: *Photothermal Spectroscopy Methods for Chemical Analysis.* Vol. 134. 1996. Copyright Wiley-VCH Verlag GmbH & Co. KGaA. Reproduced with permission.)

by measuring the index of refraction change as with PTI, PTL, and PTD; temperature change as with photothermal calorimetry and photothermal radiometry; or pressure change as with PAS.[3]

26.2.2 PHOTOACOUSTIC SPECTROSCOPY

In order to generate acoustic waves in a sample, periodic heating and cooling of the sample is required to produce pressure fluctuations. This is accomplished using modulated or pulsed excitation sources.[4–6] The pressure waves detected in PAS are generated directly by the absorbed fraction of the modulated or pulsed excitation beam. Therefore, the signal generated from a PA experiment is directly proportional to the absorbed incident power. However, depending on the type of excitation source (i.e., modulated or pulsed), the relationship between the generated acoustic signal and the absorbed power at a given wavelength will differ.[7] The theory of PA signal generation and detection has been extensively outlined in the literature[3,8–11] and will not be discussed here.

The oldest application of PAs dates back to Alexander Graham Bell's photophone circa 1880.[12] Viengerov's study of absorption in gases in the late 1930s represents the first example of PAS.[13] The first studies involving lasers were performed by Kerr and Atwood,[14] but the technique found more popularity when Kreuzer detected methane and ammonia by laser excitation at parts per billion (ppb) and sub-ppb levels, respectively.[15]

Early work by Kreuzer[8,15–18] demonstrated the analytic power of PAS in gases, but the technique can also be applied readily to liquids and solids. PAS in these phases can be accomplished using both a direct or indirect coupling method. Direct coupling is the most straightforward and, as the name implies, the acoustic wave generated in the sample is detected by a transducer in direct contact with the solid or liquid sample. Since the acoustic wave generated in the sample never crosses a high impedance interface, it is easily detected by the transducer. In comparison, indirect coupling methods are not as straightforward. In fact, this method was demonstrated at the onset of PAs by Bell but was not rediscovered until Parker noted increased signal contributions from his cell windows in his gaseous PAS experiments.[19] One possible explanation for its elusiveness may have been the fact that the original acoustic wave is not the origin of the main signal in an indirect PAS method. Usually, there is a large acoustical impedance at the sample–fluid interface such that most acoustic energy will be reflected back into the sample. The indirect PA detection of liquids or solids relies on the gas-coupling method and is explained clearly by Rosencwaig as the gas–piston model.[10] In this model, the periodic heating of the sample surface occurs within a diffusion length of the surface, and this thermal wave is responsible for the subsequent heating of the layer of gas directly above the surface (diffusion length in the gas). The periodic expansion in this gas layer produces an acoustic wave that can be detected using standard gas-phase microphones.

26.2.3 EXPERIMENTAL ARRANGEMENTS FOR PA DETECTION

In comparison to other photothermal techniques, which measure the changes in refractive index or temperature using combinations of probe sources and detectors, PAS measures the pressure wave produced by sample heating. Although the basic experimental embodiment of PAS can take many forms, several key elements are constants. This typical setup has been discussed in the literature.[6,11,20–22] Figure 26.2 shows the main components of a PAS apparatus. An excitation source, usually a laser or filtered lamp, is either modulated or pulsed and directed at a sample cell. The resultant pressure wave, which is created due to sample heating, is detected by a pressure transducer, in most cases a microphone, of the appropriate frequency response. The signal generated by the microphone is proportional to the amplitude of the pressure wave, but other information is contained in the phase and delay of the wave as well. This information is captured either by lock-in amplification or use of a gated accumulation, such as a boxcar amplifier. Sample cells can take the form of a simple sealed tube to a complex resonant chamber or multi-pass cell (see Section 26.2.3.2). In the example shown in Figure 26.2, a personal computer is used to read and record the voltage

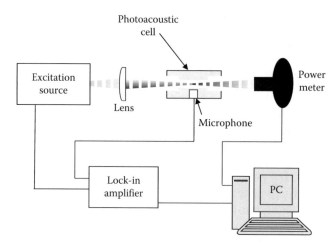

FIGURE 26.2 Simplified diagram of a typical PA sensor system with microphone detection. Electronic connections are indicated by a solid line.

outputs from the lock-in amplifier, and a power meter is employed to measure the transmitted laser power, which allows for normalization of the PA signal for any residual drift associated with the excitation source. Different adaptations of this basic scheme have been presented in the literature. The main features are summarized in the following subsections.

26.2.3.1 Radiation Sources

The first step in a PA experiment is the introduction of light. The main criterion for the starting point of PAS is that the molecules of interest need to be excited either electronically, vibrationally, or rotationally. In the broadest sense, this light can be at many different wavelengths and even span large differences in imparted energy (e.g., ultraviolet through the infrared). Once excited, numerous pathways exist for the energy to dissipate, but a preponderance of the light energy is removed through a non-radiative pathway or through sample heating, which is the basis for all photothermal phenomena, including PAS. Light sources can take various forms, depending on the application at hand. A selection of the common examples currently used for PA sensing is briefly outlined in the following paragraphs.

There are two main categories of light sources used for PAS; broadband sources, such as lamps, and narrow-band laser sources. The general category of lamps includes the following: arc lamps, filament lamps, and glow bars. These sources were some of the first sources used to study the phenomenon of PAS[13] and have several advantages and disadvantages. The broadband output of these sources can be significant (e.g., ultraviolet to infrared) and in most cases can potentially cover all regions of interest for optical excitation and subsequent PA examinations. These sources are generally inexpensive and, depending on the overall wattage required, can be somewhat compact in size. Unfortunately, lamp sources also have low spectral brightness, require spectral selection through the use of filters or monochromators, and are usually restricted to low source modulation frequency and optical efficiencies.

Although lamp-based PAS is still common, modern PAS research has been mainly performed using laser sources. The remainder of this section will focus on these sources and their various formats in more recent PA examinations. As with lamps, lasers have numerous advantages but also present some disadvantages. Some of the key advantages of these sources are their large spectral brightness, collimated output, ease of modulation, and narrow spectral linewidth. Disadvantages include expense and limited tunability; however, these are not a generic quality of every laser architecture. As mentioned previously, PAS can be performed as a pulsed or modulated measurement with respect to the light excitation. That allowance is seen vividly in the sources used for PA experimentation. Early work in detection of gases used a variety of these sources, including pulsed and continuous wave (CW) dye laser sources[23,24] and CW laser sources, such as grating tunable CO and CO_2[17] and helium–neon (He–Ne)[15] lasers. These sources usually provided reasonable or even high power levels, some limited tunability, and were based in the near-infrared or infrared wavelength regions. All of these features allowed for some level of spectroscopic studies to be performed. Several PA studies involving liquid and solid samples, which require more laser irradiation, used solid state sources such as neodymium-doped yttrium aluminum garnet (Nd:YAG) lasers. The harmonics or pumping of an optical parametric oscillator enable tuning of these sources. Semiconductor lasers based on a direct bandgap transition with various feedback mechanisms (e.g., distributed Bragg reflectors, distributed feedback reflectors [DFB]) were used for numerous studies, especially in regard to the study of atmospheric or small molecular gas targets that could be identified with tuning ranges from fractions to single-integer cm^{-1} (see Table 26.1). This type of tuning was easily accomplished by current and/or temperature tuning of the laser diode. Occasionally, efforts used other laser sources such as lead-salt diode lasers, which are centered in the infrared and theoretically can be tuned through mode hopping over a larger spectral band; however, these sources were plagued by cryogenic cooling requirements and low output powers (e.g., 0.1 mW typical).[25]

In 1994, the introduction of the quantum cascade laser (QCL) changed the prospects of laser PAs and, in general, infrared spectroscopy. In that year, Bell Laboratories first demonstrated the QCL as a new infrared laser source.[26] Since that time, continuing and aggressive evolution have occurred. PA sensing capability employing QCLs was identified early on, and demonstrations by Paldus et al. using these sources can be seen as early as 1999.[27] Although QCLs took years to evolve into their current state, work continued on PA studies using these sources throughout this development cycle.[28–37] Furthermore, these sources, operating in low duty cycles, have demonstrated that PAS based on lock-in amplification can still be performed and indeed shows great promise.

26.2.3.2 Acoustic Resonators

An essential element of a PA sensor is the cell, which serves as a container for the sample as well as the detector. Therefore, optimum design of a PA cell is necessary to facilitate signal generation and detection. To date, a variety of cell configurations have been reported for solid, liquid, and gas samples, including cells operated at acoustically resonant and non-resonant modes, single- and multipass cells, and cells for intracavity operation.[1,11,38–45]

As is often the case for trace gas sensing, the detection sensitivity is limited by the ratio of signal to noise (SNR). High sensitivity can be achieved in non-resonant gas cells; however, noise sources (e.g., amplifier noise and external acoustic noise) show a characteristic $1/f$ frequency dependence resulting in a small SNR. Furthermore, light absorption at the cell windows and walls results in a background signal, which is difficult to separate from the PA signal generated by the gas sample itself.[39] An improvement in the SNR of the cell can be achieved with cell design modifications. Stray reflections can be reduced using Brewster windows,[46] and acoustic baffles[47] and "windowless"[48] or open cells[49] will minimize the effect of window heating noise. Meyer and Sigrist[50] modified the "windowless" PA cell by adding a buffer volume on either side of the central cylinder. In this design, both the laser beam and the gas sample flow enter and exit the cell at nodal positions of the cell's operating first radial mode. With this configuration, the authors minimized flow noise and allowed for gas flow rates of up to 1 L/min without a decrease in the SNR.

Applying higher modulation frequencies and acoustic amplification of the PA signal will also result in an improved SNR. When the modulation or pulse frequency is the same as an acoustic resonance frequency of the PA cell, the resonant eigenmodes (i.e., acoustic modes) of the cell can be excited, resulting in an amplification of the signal.[51–54] The resonance frequencies are dependent on the shape and size of the PA cell. The most commonly employed resonator is the cylinder. In order to achieve signal amplification, resonant PA cells operating on longitudinal, azimuthal, or radial resonances have been developed. The theoretical determination of the corresponding eigenfrequencies of these modes has been discussed in detail elsewhere.[55] Furthermore, Helmholtz resonators are widely used. A general description of these acoustic resonators has been discussed elsewhere.[39]

TABLE 26.1

Examples of Photoacoustic Studies on Gases

Laser	Laser Power	Spectral Region	Analyte	Detector	Detection Limit	References
Nd:YAG	—	355 nm, 532 nm	H_2	Mic	200 ppm	[94]
CO_2	—	9–11 μm	CO_2 isotopes	Mic	ppb	[95]
EC-Diode	2 mW	1125 nm	H_2O vapor	Mic	13 ppm	[96]
Diode	1 mW	1.67 μm	C_6H_6	Mic	70 ppb	[97]
					100 ppb	
			$C_6H_5CH_3$		160 ppb	
			$C_6H_4(CH_3)_2$			
Diode	1 mW	1.67 μm	$C_6H_5CH_3$	Mic	1.1 ppm	[98]
					0.35 ppm	
			C_6H_6			
QCL	Few mW	9.4 μm	C_2H_6O	Mic	1 ppm	[99]
CO_2	1.4 W	10.55 μm	SF_6	Mic	3.5 ppb	[100]
CO_2	6 W	10.58 μm	SF_6	Mic	65 ppt	[59]
CO_2	2 W	9.22 μm	NH_3	Mic	220 ppt	[101]
Argon	35 mW	Visible	NO_2	Mic	50–130 ppb	[102]
Q-switched Nd:YAG	≤40 mW	532 nm	NO_2	Mic	50 ppb	[103]
Diode	2 mW	1.65 μm	CH_4	Mic	0.3 ppm	[104]
DFB diode	38 mW	1.53 μm	NH_3	TF	0.65 ppm	[105]
DFB ICL	3.4 mW	3.53 μm	H_2CO	TF	0.6 ppm	[105]
DFB QCL	8 mW	5.3 μm	NO	Mic	500 ppb	[106]
QCL	19 mW	4.55 μm	N_2O	TF	4 ppb	[30]
Diode	20–30 mW	1531.7 nm	NH_3	Mic	120 ppb	[107]
Diode w/EDFA	750 mW	1532 nm	NH_3	Mic	3–6 ppb	[108]
Diode	18 mW	1651 nm	CH_4, H_2O	Piezo	80 ppb	[109]
	22 mW	1368.6 nm	Vapor		24 ppb	
	16 mW	1737.9 nm	HCl		30 ppb	
DFB diode	mW	1370 nm	H_2O	Mic	40 ppb	[110]
		1740 nm				
			HCl		60 ppb	
DFB diode	30 mW	760 nm	O_2	CL	20 ppm	[111]
VCSEL	0.5 mW				5 ppt	
Diode-pumped Nd:YAG	~70 mW	2.76–2.91 μm	N_2O	Mic	~313 ppb	[112]
QCL	25 mW	7.9 μm	CH_4	Mic	3 ppb	[113]
DFB QCL	2 mW	6.2 μm	NO_2	Mic	80–100 ppb	[32]
	5 mW	8 μm	N_2O			
DFB laser	3.4 mW	2–2.5 μm	NH_3	Mic	Sub-ppm	[114]
Q-switched Nd:YAG	126 mW	2 μm	CO_2	Ultrasonic sensor	$\alpha_{min} = 3.3 \times 10^{-9}$	[112]
DFB laser	8 mW	1648–1652 nm	CH_4	Mic	Sub-ppm	[115]
Diode	40 mW	1574.5 nm	H_2S	Mic	0.5 ppm	[116]
QCL	5.3 mW	8.41 μm	Freon 134a	TF	$\alpha_{min} = 2.0 \times 10^{-8}$	[117]
CO_2	1 W	9–11 μm	O_3	Mic	5 ppb	[118]
DFB QCL	8 mW	8 μm	CH_4	Mic	34 ppb	[119]

TABLE 26.1 (continued)
Examples of Photoacoustic Studies on Gases

Laser	Laser Power	Spectral Region	Analyte	Detector	Detection Limit	References
DFB diode	30 mW,	1572 nm	CO_2	CL	1.9 ppm	[120]
DFB QCL	1 mW	4.3 μm	CO_2	Mic	0.023% vol	[121]
DFB diode	6.2 mW	2.0 μm	CO_2, NH_3	TF	18 ppm, 3 ppm	[122]
Diode w/EDFA	1.17–1.89 W	1.53 μm	C_2H_2	Mic, CL	440 ppb, 14.5 ppb	[123]
DFB laser	16 mW	1738.9 nm	HCl	Mic	Sub-ppm	[124]
DFB diode	8 mW	1.65 μm	CH_4	TF	$\alpha_{min} = 1.0 \times 10^{-8}$	[125]
DFB diode	1.5 mW	2.7 μm	CO_2	Mic	30 ppb	[126]
DFB diode	14 mW	1.53 μm	C_2H_2	CL	$\alpha_{min} = 1.2 \times 10^{-7}$	[127]
CO_2	10 W	9.22 μm	NH_3	Mic	—	[128]
Nd:YAG	—	266 nm	O_3	Mic	10 ppb	[129]
DFB QCL	4 mW	5.6 μm	CH_2O	Mic	150 ppb	[130]
DFB diode	8 mW	1.396 μm	H_2O vapor	TF	$\alpha_{min} = 1.68 \times 10^{-8}$	[131]
DFB diode	15 mW	1.62 μm	C_2H_4	TF	0.3–4 ppm	[132]
TEDFL w/EDFA	500 mW	1531.7 nm	NH_3	Mic	3 ppb	[133]
EC-QCL	~250 mW	5.26 μm	NO	Mic, TF	60 ppb, 15 ppb	[134]
DFB diode	46 mW	1.53 μm	C_2H_2	TF	$\alpha_{min} = 3.3 \times 10^{-9}$	[135]
Diode	3 mW	1450 nm	H_2O vapor	Mic	—	[136]
DFB diode	25 mW	1.53 μm	NH_3	TF	60 ppb	[137]
DFB diode	20 mW	1371 nm	H_2O vapor	Mic	80 ppm	[138]
EC-QCL	100 mW	5.26 μm	NO	TF	15 ppb	[139]

Mic, microphone; ICL, interband cascade laser; EDFA, erbium-doped fiber amplifier; VCSEL, vertical-cavity surface-emitting; TEDFL, tunable erbium-doped fiber laser; α_{min} units are $cm^{-1} W \sqrt{Hz}$; CL, cantilever.

Cells have also been designed for multipass or intracavity operation.[38,56,57] In the following paragraphs, some selected examples of PA cell designs are discussed in more detail.

The sensitivity of a PA sensor is strongly dependent on the geometry of the PA cell, and the pressure distribution in the cell must be understood for optimization. Various PA cell modeling approaches have been investigated for the systematic optimization of PA cells. Bijnen et al.[1] examined a PA cell configuration similar to the design introduced by Meyer and Sigrist.[50] The authors applied acoustic transmission line theory[58] pertaining to experimental results from a cylindrical resonant PA cell excited in the first longitudinal mode. Numerous approaches to optimize a PA cell for trace gas detection were investigated, resulting in suggested parameters for the construction of a small and sensitive resonant cell with a fast response time. The criteria considered for the geometry of the cell were dependent on various background and noise sources (e.g., window absorption signal, AR window reflection, absorption of radiation at resonator wall, chopper noise, gas sample flow noise, and laboratory noise), which lower the detection limit.

In conjunction with research to examine performance and design issues associated with microelectromechanical systems (MEMS) scale PA cells, we fabricated and tested a miniature non-MEMS (macro) resonant cell.[59] To date, limited research has been done to demonstrate the feasibility of a miniaturized PA sensor. The basic design for the macro PA cell is a modified version of the cell studied by Bijnen et al.[1] Our design is a quarter scale down from the "Bijnen" cell. Experimental results suggested that miniaturization of a PA cell is viable without a significant loss in signal, and no adverse effects of the size scaling were visualized in the optics or acoustics of the macro cell. We achieved a detection limit of 65 ppt for sulfur hexafluoride (SF_6) using this macro cell.

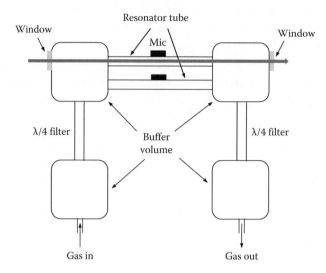

FIGURE 26.3 Optimized differential PA cell geometry with two resonator tubes and λ/4 filters. (From Miklós, A. et al., *AIP Conf. Proc.*, 463, 126, 1999; Pellegrino, P. et al., *Proc. SPIE*, 5416, 42, 2004. With permission.)

Miklós et al. introduced a differential PA cell specifically designed for fast time response, low acoustic and electric noise characteristics, and high sensitivity.[60] The authors developed a fully symmetrical design in order to reduce flow noise and electromagnetic disturbances. Figure 26.3 presents this optimized differential PA cell. The gas sample flows through both tubes, which produces about the same flow noise in both resonators; however, the laser light passes through only one of them, thus generating the PA signal in only one tube. The PA signal from each resonator is amplified with a differential amplifier, thus all coherent noise components in the two tubes are effectively suppressed. In our PA sensor platforms for trace gas detection (see Section 26.3.1), we have used modified versions of the differential PA cell developed by Miklós et al. In one configuration, we fabricated a MEMS-scale version of the "Miklós" cell. Initial examination of the scaling principles associated with PAS with respect to MEMS dimensions indicated that PA signals would remain at similar sensitivities or even surpass those commonly found in macro-scale devices.[61–66] Figure 26.4 includes a photograph of the internal structure of the MEMS cell and the complete PA cell package. The resonators have square cross sections with l_{res} = 8.5 mm and r_{res} = 0.465 mm. Another cell design consisted of two open resonators having square cross sections (l_{res} = 10 mm), each with r_{res} = 0.432 mm. To further suppress gas flow noise, the cell had a convoluted, split sample inlet/outlet

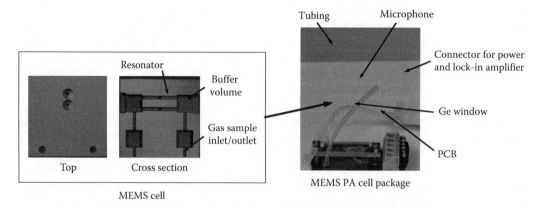

FIGURE 26.4 Photograph of the internal structure of the MEMS-scale cell and complete PA cell package.

design.[37] For each of these MEMS-scale PA cell designs, resonator and buffer volume dimensions were determined based on the criteria presented for the "Bijnen" cell.

PA studies on liquid and solid samples utilizing modified cell designs have been reported. For example, Jalink and Bicanic[67] developed a PA heat pipe cell for use with liquid samples having low vapor pressures, and Schmid et al.[68] utilized a conventional 1 cm glass cuvette for construction of a PA sensor for opaque dyes. Sanchez et al.[69] attached an aluminum (Al) absorber to polymer thick films. An expanded He–Ne laser beam was directed to the Al surface to ensure that the incident light beam generated a surface sample heating, resulting in sample optical opaqueness. The sample-absorber system closed one of the openings of a cylindrical cell cavity, and a glass window closed the other cavity opening.

Finally, is it important to note the condition and quality of the PA cell surface. These characteristics influence the background signal due to scattering and molecule adsorption. Various cell materials and surface treatments and coatings have been investigated.[70]

26.2.3.3 Detectors

The acoustic waves generated in a PA cell as a result of the absorption of radiation by a sample are detected by a pressure sensor. The appropriate choice depends on the application (e.g., sensitivity requirements, ease of operation, and ruggedness). A selection of examples is briefly outlined in the following paragraphs.

The most widely used PA sensor scheme employs commercial microphones as pressure sensors. Typically, a lock-in amplifier is used to detect a small voltage produced by the microphone as a result of sample absorption of radiation. Microphone types include miniature electret microphones (e.g., Knowles, Sennheiser, and Intricon Tibbetts) and condenser microphones (e.g., Brüel and Kjaer). These devices are easy to use, sensitive enough for PA studies of solids, liquids, and gases, and the responsivity only weakly depends on frequency. In most cases, the detection threshold of a PA system is neither determined by microphone responsivity nor electrical noise, but instead by other noise sources (e.g., external noise, window heating, absorption of desorbing molecules from the cell walls, etc.).

Recently, alternative methods of transduction have been employed. Kauppinen et al. replaced a capacitive microphone with a cantilever-type pressure sensor made out of silicon.[71–73] In this configuration, the sensor in the cantilever microphone is a flexible bar. The typical dimensions for width and length are a few millimeters, with a thickness of 5–10 μm. The cantilever is separated on three sides from a thicker frame with a narrow gap (3–5 μm) and moves like a flexible door due to the pressure variations in the surrounding gas (Figure 26.5). The fabrication and characterization of the cantilever sensor is described in detail elsewhere.[74] As the pressure changes, the cantilever bends, but does not stretch. A laser interferometer was used to measure the displacement of the cantilever. Although cantilevers exhibit superior sensitivity, PA cells containing these devices and the associated interferometric detection system are much more expensive and fragile, compared to a cell equipped with a conventional capacitive microphone.

Quartz enhanced PAS (QEPAS) is another PA detection approach in which the microphone is replaced with a quartz crystal tuning fork (TF). Kosterev et al.[75,76] suggested inverting the typical resonant PA approach in which the absorbed energy is accumulated in the gas and instead

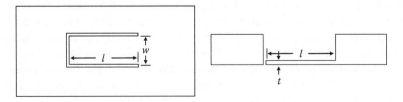

FIGURE 26.5 Dimensions of a cantilever-type pressure sensor. (Kauppinen, J. et al., *Microchem. J.*, 76(1–2), 151, 2004. With permission.)

accumulates the energy in a sharply resonant acoustic transducer. This approach removes limitations imposed on the PA cell by acoustic resonance conditions because the resonant frequency is determined by the TF. Therefore, the cell is optional in QEPAS and is utilized only to separate the gas sample from the environment and control its pressure. This approach allows for gas samples that are 1 mm^3 in volume. Crystal quartz was chosen as a suitable material because it is a common low-loss piezoelectric material and is mass produced and inexpensive. Furthermore, quartz TFs have become common devices for atomic force and optical near-field microscopy and therefore are well characterized.[77] Only the symmetric vibration of a TF (i.e., the two prongs bend in opposite directions) is piezoelectrically active. Therefore, efficient excitation of this vibration is achieved when the excitation beam passes through the gap between the TF prongs. The pressure wave generated when the optical radiation interacts with a gas excites a resonant vibration of the TF. This event is converted into an electric signal due to the piezoelectric effect. This electric signal is proportional to the concentration of the gas and is measured by a transimpedance amplifier. The initial feasibility experiments performed by Kosterev et al.[75] utilized a quartz-watch TF.

For PA applications such as studies on liquid and solid samples, the use of conventional microphones was reported to be inefficient by Hordvik and Schlossberg[78] and Farrow et al.[79] Both groups were concerned with acoustic impedance mismatching between the solid–gas or liquid–gas interface, resulting in most of the acoustic energy being reflected or absorbed back into the sample rather than transferred across the boundary. The authors demonstrated improvements in sensitivity with the use of piezoelectric transducers in contact with solid and liquid samples, respectively. Piezoelectric elements used in this manner offer the advantage of good impedance matching. Similar to a conventional microphone detection scheme, a lock-in amplifier is used to detect the voltage change produced by the piezoelectric sensor. This direct coupling method is simple, and there have been numerous studies employing piezoelectric elements in contact with liquids or solids for PA detection.[5,78,80–82] More recent reports describing PA detection of solid samples utilize conventional microphones[83–86] and the indirect coupling method described by Rosencwaig[9,10]; however, piezoelectric transducers are still widely used for liquid studies.[68,87–90]

26.3 PHOTOACOUSTIC SENSING APPLICATIONS

There are numerous publications with thorough discussions on the use of PAS for various applications.[11,21,57,91,92] In this offering, we are focused on conventional PA sensing, including detection in gas, liquid, and solid media. We aim to summarize the state of the art in detection by laser PA techniques. Because PA sensing has evolved tremendously since the initial investigations, we have not attempted to include every important reference.

26.3.1 GASEOUS SAMPLES

Investigations of gaseous species continue to be the most common application for PAS. In Table 26.1, numerous examples of PA studies on gases are given. The table lists the laser source with the corresponding wavelength or wavelength range, the laser power, the analyte studied, the type of detector, and the minimum detectable concentration. In the past, many investigations were performed with CO_2 lasers[57]; however, more recently, as evidenced by this truncated list, DFB lasers, including diodes and QCLs, have become a popular choice for PA studies due to their smaller size, room temperature operation, and the availability of a variety of wavelengths. For example, Grossel et al. employed a room temperature QCL for the PA detection of methane (CH_4).[93] A Helmholtz resonator was again used for the PA cell in these experiments. The laser emitted between 1276 and 1283 cm^{-1} (7.8 μm region). Two CH_4 absorption features appear in this wavelength region at 1276.84 and 1277.47 cm^{-1}. The authors achieved a CH_4 detection limit of 17 ppb. The ability of the sensor to detect CH_4 and

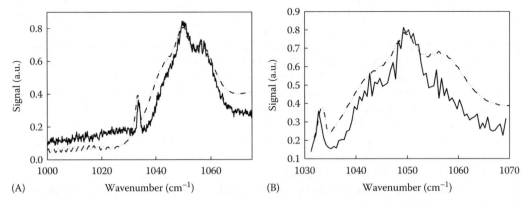

FIGURE 26.6 Measured (A) pulsed and (B) CW modulated laser PA spectra (—) of DMMP. Data derived from our own PA measurements are compared to FTIR reference spectra (– ·· –).

N_2O in ambient air was also demonstrated, and the laser PA spectrum recorded for these species was in excellent agreement with the calculated spectrum using the parameters of the HITRAN database.

Recent advances in QCL technology (see Section 26.2.3.1) have allowed for continuous wavelength tuning ranges of ≥ 200 cm⁻¹. These broad tuning ranges permit laser PA absorbance spectra of various analytes to be collected. Figure 26.6 shows the laser PA spectrum for the standard nerve agent stimulant, dimethyl methyl phosphonate (DMMP). The spectrum in Figure 26.6A was collected as a pulsed external cavity (EC)-QCL was continuously tuned from 990 to 1075 cm⁻¹ (10.10–9.30 μm), in 1 cm⁻¹ increments. The spectrum in Figure 26.6B was collected as a CW modulated EC-QCL was continuously tuned from 1032 to 1070 cm⁻¹ (9.69–9.34 μm). Due to the mode hopping[140] nature of this source, we reduced the number of measured wavelengths, which was possible due to the broad absorption features of DMMP in this wavelength region. The absorbance spectrum for DMMP, which was recorded using a Fourier transform infrared (FTIR), is also provided in Figure 26.6A and B for comparison to the PA data. There is good agreement between the laser PA data and the FTIR spectroscopy absorbance spectrum. A MEMS-scale differential PA cell (described in Section 26.2.3.2) was employed for spectroscopic and sensitivity data collection. The cell had two resonator tubes, both housing a commercial microphone. Minimal detectable DMMP concentrations of 54 and 20 ppb were achieved with the pulsed QCL and CW modulated QCL-based PA sensor systems, respectively.

We used a similar PA sensing system to study Freon 116, a propellant analog and a component commonly found in refrigerants. Figure 26.7 is the spectrum of Freon 116 collected as a pulsed

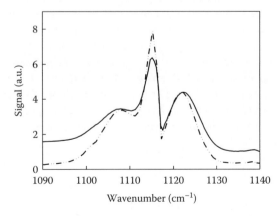

FIGURE 26.7 Measured laser PA spectrum (—) of Freon 116. Data derived from our own PA measurements are compared to FTIR reference spectrum (– ·· –).

EC-QCL was continuously tuned from 1050 to 1240 cm^{-1} (9.52–8.06 µm), in 1 cm^{-1} increments. The absorbance spectrum for this species, recorded using an FTIR, is also provided in Figure 26.7 for comparison to the PA data. There is excellent agreement between the *PA and FTIR data*. A MEMS-scale differential PA cell was employed for spectroscopic and sensitivity data collection. We achieved a minimal detectable Freon 116 concentration of 19 ppb.

Although the microphone continues to be the most commonly used detector for gaseous sensors, quartz crystal TFs and microcantilevers have also emerged as desired transducers for trace vapor detection. Specifically, the combination of tunable QCLs employed with QEPAS has become an attractive sensing scheme. Lewicki and coworkers developed a mid-IR QEPAS sensor device for tri-acetone triperoxide (TATP) detection.[141] The authors employed two different EC-QCLs (pulsed and CW), having different center wavelengths and tuning ranges. These tuning ranges allowed for these absorption features to be clearly seen in the laser PA spectra. The authors report a TATP detection limit of 1 ppm using this QEPAS sensor.

26.3.2 Liquid Samples

Investigating liquids using the PA technique is attractive as it allows optical absorption measurements to be made for optically opaque samples. This capability of PAS, along with its insensitivity to scattered light, makes this technique a very attractive spectroscopic tool for the investigation of liquids. For instance, in industrial applications where ultraviolet-visible online process monitoring is hampered by light scattering and opacity of samples, PAS allows for the measurements of both high and low absorptions and no need for sample preparations. Schmid et al. described the construction and characterization of a PA sensor for optical absorption measurements in transparent and opaque liquids.[68] A Q-switched frequency-doubled Nd:YAG laser was employed for excitation. The pulsed source had an emission wavelength of 532 nm. A glass cuvette with piezoelectric transducers placed on two sides was used to contain the samples for PA data collection. The samples were aqueous solutions of a black textile dye with concentrations in the range of 20 mg/L to 1.34 g/L, aqueous solutions of red textile dye in the range of 100 mg/L to 50 g/L, and suspensions of TiO$_2$ in water. The PA sensor allowed for the determination of absorption coefficients ranging from 0.1 to 1000 cm^{-1}.

Furthermore, PAS is attractive for industrial or environmental applications for the identification of contaminants in solution. For example, Hodgson et al. developed a laser PA measurement technique for the detection of oil contamination in water for the continuous monitoring of hydrocarbons in return process water from oil production installations.[142] Two pulsed diode lasers were employed, having emission wavelengths of 0.904 and 1.55 µm. A quartz cuvette equipped with a piezoelectric transducer (PZT-5A) ceramic disk was used for PA data collection. The authors studied crude oil emulsions and hydrocarbons in solution in the range of 0–900 mg/L and demonstrated that dispersed and dissolved hydrocarbon components give an additive PA response, whereas with most optical instrumentation, the components must be measured separately. Hernández-Valle et al. presented a PA technique for the trace analysis of pesticides in water.[143] The light from a 460 nm pulsed dye laser was directed into a doubler crystal in which 230 nm light was generated. Different concentrations of the pesticides atrazine and methyl parathion were prepared. Samples were poured into quartz cells, and two piezoelectric sensors were used to monitor PA signals. The authors reported the ability to monitor concentrations as low as tenths of ng/L.

26.3.3 Solid Samples

Similar to liquid samples, the initial PA studies on solid samples demonstrated that optical absorption measurements could be obtained for optically opaque materials, therefore making this technique an attractive spectroscopic tool for the investigation of solid materials. Although this topic is discussed extensively in the literature, specifically with regard to applications of PA FT-IR spectroscopy, the use of lasers in PA experiments on solids has not been widely reported. Wen and Michaelian used an EC-QCL for the measurement of the PA spectrum of acetyl polystyrene beads.[85] There have been several reports discussing the use of these tunable infrared lasers, such as QC and CO_2, to obtain PA spectra of gases; however, the application of these sources in PA experiments on solids are minimal. In this work, the pulsed EC-QCL was tunable from 990 to 1075 cm^{-1} (10.10–9.30 µm), and the results demonstrated better peak definition in the PA spectrum compared to the FTIR spectrum, allowing identification of bands at ~1005 and 1030 cm^{-1} for the solid material.

Numerous PA studies on explosive materials have been reported due to an increased interest in the identification and quantification of these substances. The detection of these compounds in the solid form is attractive because the majority of explosives have extremely low vapor pressures, making vapor detection difficult. Chaudhary et al. presented the low-limit PA detection of solid 1,3,5-Trinitroperhydro-1,3,5-triazine (RDX) and 2,4,6-Trinitrotoluene (TNT) using a CO_2 laser.[144] The source was grating-tunable and could be precisely tuned in the 9.25–10.74 µm spectral region. The authors successfully demonstrated for the first time the use of a CO_2 laser-based PA technique to record the high resolution PA absorption spectra of RDX and TNT in solid form at room temperature. Detection limits of 16.5 and 10.0 ppb were reported for TNT and RDX, respectively. Van Neste et al. demonstrated the detection of trace amounts of RDX adsorbed on a quartz crystal TF using illumination from a QCL.[145] The authors used three pulsed EC-QCLs having an overall optical tuning range from 7.83 to 10.93 µm when used in combination. The RDX signatures recorded using the tunable sources had excellent agreement with published infrared absorption spectra.

Furthermore, PAS of solid samples is often applied to depth profiling of layered samples and two- and three-dimensional tomographic imaging for biomedical applications.[146–148] Viator et al. developed an in vivo PA probe that employed a Nd:YAG (neodymium, yttrium, Al, and garnet) laser operating at 532 nm to generate acoustic pulses in skin.[149] A piezoelectric element was used to detect the acoustic waves arising from thermoelastic expansion, which were analyzed for epidermal melanin content using a PA melanin index (PAMI). Melanin content was compared to results obtained using visible reflectance spectroscopy (VRS), which is currently employed to estimate epidermal melanin content. The authors reported a good correlation between PAMI and VRS measurements, and the 200 µm active area of the PA method allowed for pinpoint measurements.

26.4   CONCLUSIONS AND OUTLOOK

The versatility of laser-based PAS for sensing applications has been demonstrated by numerous state-of-the-art examples. Detection limits for gaseous and condensed media are often in the ppb or sub-ppb range, with some reports of ppt level sensing capabilities.

The development of continuously tunable sources, such as QCLs, allows for PA absorption spectra to be collected. Data collected with a simple hearing aid microphone have been presented for numerous chemical species and are in excellent agreement with absorption spectra collected for the same analytes using an FTIR spectrometer. Additionally, these broadly tunable sources allow for the simultaneous detection of several molecules of interest as well as increased molecular discrimination, which assists in overcoming the absorption interference problem often associated with PAS.

Although numerous reports have verified the sensitivity of PA sensors at trace levels, the total system size represents a large logistics burden in terms of bulk, cost, and power consumption. The future of PAS for sensing applications includes the continued development of laser sources with respect to broad continuous tunability and decreased system sizes and power requirements. In addition, the successful demonstration of PA sensing platforms by employing miniaturized PA cells is an important step toward the development of man-portable sensor systems. We expect continued success from PA-based sensing applications for the detection of a diverse range of chemical and biological agents and for use in environmental monitoring.

REFERENCES

1. F. G. C. Bijnen, J. Reuss, and F. J. M. Harren, Geometrical optimization of a longitudinal resonant photoacoustic cell for sensitive and fast trace gas detection. *Rev. Sci. Instrum.* **1996**, *67* (8), 2914–2923.
2. M. Nagele and M. W. Sigrist, Mobile laser spectrometer with novel resonant multipass photoacoustic cell for trace-gas sensing. *Appl. Phys. B.* **2001**, *70*, 895–901.
3. S. E. Bialkowski, *Photothermal Spectroscopy Methods for Chemical Analysis.* John Wiley & Sons, New York, 1996, Vol. 134.
4. J. B. Kinney and R. H. Staley, Applications of photoacoustic spectroscopy. *Ann. Rev. Mater. Sci.* **1982**, *12*, 295–321.
5. C. K. N. Patel and A. C. Tam, Pulsed optoacoustic spectroscopy of condensed matter. *Rev. Mod. Phys.* **1981**, *53* (3), 517–550.
6. A. C. Tam, Applications of photoacoustic sensing techniques. *Rev. Mod. Phys.* **1986**, *58* (2), 381–431.
7. A. Miklos and P. Hess, Modulated and pulsed photoacoustics in trace gas analysis. *Anal. Chem.* **2000**, *72*, 30A–37A.
8. L. B. Kreuzer, The physics of signal generation and detection. In: *Optoacoustic Spectroscopy and Detection*, Y.-H. Pao, Ed. Academic Press, New York, **1977**, 1–25.
9. A. Rosencwaig, Theoretical aspects of photoacoustic spectroscopy. *J. Appl. Phys.* **1978**, *49* (5), 2905–2910.
10. A. Rosencwaig and A. Gersho, Theory of photoacoustic effect with solids. *J. Appl. Phys.* **1976**, *47* (1), 64–69.
11. A. C. Tam, Photoacoustics: Spectroscopy and other applications. In: *Ultrasensitive Laser Spectroscopy*, D. S. Klinger, Ed. Academic Press, New York, **1983**, 1–108.
12. A. G. Bell, The photophone. *Science* **1880**, *1* (11), 130–134.
13. M. L. Viengerov, *Dolk. Akad. Nauk. SSSR* **1938**, *19*, 687.
14. E. L. Kerr and J. G. Atwood, Laser illuminated absorptivity spectrophone—A method for measurement of weak absorptivity in gases at laser wavelengths. *Appl. Optics* **1968**, *7* (5), 915–922.
15. L. B. Kreuzer, Ultralow gas concentration infrared absorption spectroscopy. *J. Appl. Phys.* **1971**, *42* (7), 2934–2943.
16. L. B. Kreuzer, Laser optoacoustic spectroscopy—New technique of gas-analysis. *Anal. Chem.* **1974**, *46* (2), A237–A244.
17. L. B. Kreuzer, N. D. Kenyon, and C. K. N. Patel, Air-pollution—Sensitive detection of 10 pollutant gases by carbon-monoxide and carbon-dioxide lasers. *Science* **1972**, *177* (4046), 347–349.
18. L. B. Kreuzer and C. K. N. Patel, Nitric oxide air pollution—Detection by optoacoustic spectroscopy. *Science* **1971**, *173* (3991), 45–47.
19. J. G. Parker, Optical-absorption in glass—Investigation using an acoustic technique. *Appl. Optics* **1973**, *12* (12), 2974–2977.
20. G. A. West, J. J. Barrett, D. R. Siebert, and K. V. Reddy, Photo-acoustic spectroscopy. *Rev. Sci. Instrum.* **1983**, *54* (7), 797–817.
21. A. Rosencwaig, *Photoacoustics and Photoacoustic Spectroscopy.* John Wiley & Sons, New York, 1980.
22. M. W. Sigrist, Laser generation of acoustic-waves in liquids and gases. *J. Appl. Phys.* **1986**, *60* (7), R83–R121.
23. P. C. Claspy, C. Ha, and Y. H. Pao, Optoacoustic detection of NO_2 using a pulsed dye-laser. *Appl. Optics* **1977**, *16* (11), 2972–2973.
24. A. M. Angus, E. E. Marinero, and M. J. Colles, Opto-acoustic spectroscopy with a visible cw dye laser. *Opt. Commun.* **1975**, *14* (2), 223–225.
25. T. H. Vansteenkiste, F. R. Faxvog, and D. M. Roessler, Photoacoustic measurement of carbon-monoxide using a semiconductor diode-laser. *Appl. Spectros.* **1981**, *35* (2), 194–196.
26. J. Faist, F. Capasso, D. L. Sivco, C. Sirtori, A. L. Hutchinson, and A. Y. Cho, Quantum cascade laser. *Science* **1994**, *264* (5158), 553–556.
27. B. A. Paldus, T. G. Spence, R. N. Zare, J. Oomens, F. J. M. Harren, D. H. Parker, C. Gmachl, F. Capasso, D. L. Sivco, A. L. Baillargeon, A. L. Hutchinson, and A. Y. Cho, Photoacoustic spectroscopy using quantum-cascade lasers. *Opt. Lett.* **1999**, *24* (3), 178–180.
28. M. G. da Silva, H. Vargas, A. Miklos, and P. Hess, Photoacoustic detection of ozone using a quantum cascade laser. *Appl. Phys. B Lasers Opt.* **2004**, *78* (6), 677–680.
29. D. Hofstetter, M. Beck, J. Faist, M. Nagele, and M. W. Sigrist, Photoacoustic spectroscopy with quantum cascade distributed-feedback lasers. *Opt. Lett.* **2001**, *26* (12), 887–889.
30. A. A. Kosterev, Y. A. Bakhirkin, and F. K. Tittel, Ultrasensitive gas detection by quartz-enhanced photoacoustic spectroscopy in the fundamental molecular absorption bands region. *Appl. Phys. B Lasers Opt.* **2005**, *80* (1), 133–138.

31. R. Lewicki, G. Wysocki, A. A. Kosterev, and F. K. Tittel, QEPAS based detection of broadband absorbing molecules using a widely tunable, cw quantum cascade laser at 8.4 mu m. *Opt. Express* **2007**, *15* (12), 7357–7366.
32. J. P. Lima, H. Vargas, A. Miklos, M. Angelmahr, and P. Hess, Photoacoustic detection of NO_2 and N_2O using quantum cascade lasers. *Appl. Phys. B-Lasers Opt.* **2006**, *85* (2–3), 279–284.
33. A. Mukherjee, M. Prasanna, M. Lane, R. Go, I. Dunayevskiy, A. Tsekoun, and C. K. N. Patel, Optically multiplexed multi-gas detection using quantum cascade laser photoacoustic spectroscopy. *Appl. Optics* **2008**, *47* (27), 4884–4887.
34. M. Taslakov, M. Simeonov, M. Froidevaux, and H. van den Bergh, Open-path ozone detection by quantum-cascade laser. *Appl. Phys. B-Lasers Opt.* **2006**, *82* (3), 501–506.
35. D. Weidmann, A. Kosterev, C. Roller, R. F. Curl, M. P. Fraser, and F. K. Tittel, Monitoring of ethylene by a pulsed quantum cascade laser. *Appl. Optics* **2004**, *43* (6), 3329–3334.
36. E. Holthoff, J. Bender, P. Pellegrino, A. Fisher, and N. Stoffel, Photoacoustic spectroscopy for trace vapor detection and molecular discrimination. *Proc. SPIE Int. Soc. Opt. Eng.* **2010**, *7665*, 766510 (7 pp.)–766510 (7 pp.).
37. E. L. Holthoff, D. A. Heaps, and P. M. Pellegrino, Development of a MEMS-scale photoacoustic chemical sensor using a quantum cascade laser. *IEEE Sens. J.* **2010**, *10* (3), 572–577.
38. F. J. M. Harren, F. G. C. Bijnen, J. Reuss, L. Voesenek, and C. Blom, Sensitive intracavity photoacoustic measurements with a CO2 wave-guide laser. *Appl. Phys. B-Photophys. Laser Chem.* **1990**, *50* (2), 137–144.
39. A. Miklos, P. Hess, and Z. Bozoki, Application of acoustic resonators in photoacoustic trace gas analysis and metrology. *Rev. Sci. Instrum.* **2001**, *72* (4), 1937–1955.
40. J. P. Monchalin, L. Bertrand, G. Rousset, and F. Lepoutre, Photoacoustic-spectroscopy of thick powdered or porous samples at low-frequency. *J. Appl. Phys.* **1984**, *56* (1), 190–210.
41. S. Oda and T. Sawada, Laser-induced photoacoustic detector for high-performance liquid-chromatography. *Anal. Chem.* **1981**, *53* (3), 471–474.
42. A. Rosencwaig, Photoacoustic spectroscopy of solids. *Opt. Commun.* **1973**, *7* (4), 305–308.
43. L. G. Rosengren, Optimal optoacoustic detector design. *Appl. Optics* **1975**, *14* (8), 1960–1976.
44. K. Veeken, N. Dam, and J. Reuss, A multipass transverse photoacoustic cell. *Infrared Phys.* **1985**, *25* (5), 683–696.
45. R. J. W. Hodgson, Regularization techniques applied to depth profiling with photoacoustic spectroscopy. *J. Appl. Phys.* **1994**, *76* (11), 7524–7529.
46. C. K. N. Patel, and R. J. Kerl, New optoacoustic cell with impronved performance. *Appl. Phys. Lett.* **1977**, *30* (11), 578–579.
47. C. F. Dewey, Jr., Design of optoacoustic system. In: *Optoacoustic Spectroscopy and Detection*, Y.-H. Pao, Ed. Academic Press, New York, 1977, 47–77.
48. R. Gerlach and N. M. Amer, Brewster window and windowless resonant spectrophones for intra-cavity operation. *Appl. Phys.* **1980**, *23* (3), 319–326.
49. A. Miklos and A. Lorincz, Windowless resonant acoustic chamber for laser-photoacoustic applications. *Appl. Phys. B.* **1989**, *48* (3), 213–218.
50. P. L. Meyer and M. W. Sigrist, Atmospheric-pollution monitoring using CO2-laser photoacoustic-spectroscopy and other techniques. *Rev. Sci. Instrum.* **1990**, *61* (7), 1779–1807.
51. C. F. Dewey, R. D. Kamm, and C. E. Hackett, Acoustic amplifier for detection of atmospheric pollutants. *Appl. Phys. Lett.* **1973**, *23*, 633–635.
52. R. D. Kamm, Detection of weakly absorbing gases using a resonant optoacoustic method. *J. Appl. Phys.* **1976**, *47*, 3550–3558.
53. A. Karbach and P. Hess, High precision acoustic spectroscopy by laser excitation of resonator modes. *J. Chem. Phys.* **1985**, *83*, 1075–1084.
54. A. Miklos, P. Hess, and Z. Bozoki, Application of acoustic resonators in photoacoustic t race gas analysis and metrology. *Rev. Sci. Instrum.* **2001**, *72*, 1937–1955.
55. P. Hess, Resonant photoacoustic spectroscopy. In: *Topics in Current Chemistry*. Springer-Verlag, Berlin, Germany, 1983, Vol. 111, pp. 1–32.
56. J. M. Rey, D. Marinov, D. E. Vogler, and M. W. Sigrist, Investigation and optimisation of a multipass resonant photoacoustic cell at high absorption levels. *Appl. Phys. B Lasers Opt.* **2005**, *80* (2), 261–266.
57. M. W. Sigrist, Air monitoring by laser photoacoustic spectroscopy. In: *Air Monitoring by Spectroscopic Techniques*, M. W. Sigrist, Ed. Wiley Interscience, New York, 1994, Vol. 127, 163–238.
58. S. Bernegger and M. W. Sigrist, Co-laser photoacoustic-spectroscopy of gases and vapors for trace gas-analysis. *Infrared Phys.* **1990**, *30* (5), 375–429.
59. P. M. Pellegrino and R. G. Polcawich, Advancement of a MEMS photoacoustic chemical sensor. *Proc. SPIE Int. Soc. Opt. Eng.* **2003**, *5085*, 52–63.

60. A. Miklós, P. Hess, A. Mohacsi, J. Sneide, S. Kamm, and S. Schafer, Improved photoacoustic detector for monitoring polar molecules such as ammonia with a 1.53 mu m DFB diode laser. *AIP Conf. Proc.* **1999**, (463), 126–128.

61. S. L. Firebaugh, K. F. Jensen, and M. A. Schmidt, Miniaturization and integration of photoacoustic detection with a microfabricated chemical reactor system. *JMEMS* **2001**, *10*, 232–237.

62. S. L. Firebaugh, K. F. Jensen, and M. A. Schmidt, Miniaturization and integration of photoacoustic detection. *J. Appl. Phys.* **2002**, *92*, 1555–1563.

63. D. A. Heaps, and P. Pellegrino, Examination of a quantum cascade laser source for a MEMS-scale photoacoustic chemical sensor. *Proc. SPIE* **2006**, *6218*, 621805-1–621805-9.

64. D. A. Heaps and P. Pellegrino, Investigations of intraband quantum cascade laser source for a MEMS-scale photoacoustic sensor. *Proc. SPIE* **2007**, *6554*, 65540F-1–65540F-9.

65. P. Pellegrino and R. Polcawich, Advancement of a MEMS photoacoustic chemical sensor. *Proc. SPIE* **2003**, *5085*, 52–63.

66. P. Pellegrino, R. Polcawich, and S. L. Firebaugh, Miniature photoacoustic chemical sensor using micorelectromechanical structures. *Proc. SPIE* **2004**, *5416*, 42–53.

67. H. Jalink and D. Bicanic, Concept, design, and use of the photoacoustic heat pipe cell. *Appl. Phys. Lett.* **1989**, *55* (15), 1507–1509.

68. T. Schmid, U. Panne, R. Niessner, and C. Haisch, Optical absorbance measurements of opaque liquids by pulsed laser photoacoustic spectroscopy. *Anal. Chem.* **2009**, *81* (6), 2403–2409.

69. R. R. Sanchez, J. B. Rieumont, S. L. Cardoso, M. G. da Silva, M. S. Sthel, M. S. O. Massunaga, C. N. Gatts, and H. Vargas, Photoacoustic monitoring of internal plastification in poly(3-hydroxybutyrate-co-3-hydroxyvalerate) copolymers: Measurements of thermal parameters. *J. Braz. Chem. Soc.* **1999**, *10* (2), 97–103.

70. S. M. Beck, Cell coatings to minimize sample (NH3 and N2H4) adsorption for low-level photoacoustic detection. *Appl. Optics* **1985**, *24* (12), 1761–1763.

71. J. Kauppinen, K. Wilcken, I. Kauppinen, and V. Koskinen, High sensitivity in gas analysis with photoacoustic detection. *Microchem J.* **2004**, *76* (1–2), 151–159.

72. V. Koskinen, J. Fonsen, J. Kauppinen, and I. Kauppinen, Extremely sensitive trace gas analysis with modern photoacoustic spectroscopy. *Vib. Spectrosc.* **2006**, *42* (2), 239–242.

73. V. Koskinen, J. Fonsen, K. Roth, and J. Kauppinen, Progress in cantilever enhanced photoacoustic spectroscopy. *Vib. Spectrosc.* **2008**, *48* (1), 16–21.

74. P. Sievila, V. P. Rytkonen, O. Hahtela, N. Chekurov, J. Kauppinen, and I. Tittonen, Fabrication and characterization of an ultrasensitive acousto-optical cantilever. *J. Micromech. Microeng.* **2007**, *17* (5), 852–859.

75. A. A. Kosterev, Y. A. Bakhirkin, R. F. Curl, and F. K. Tittel, Quartz-enhanced photoacoustic spectroscopy. *Opt. Lett.* **2002**, *27* (21), 1902–1904.

76. A. A. Kosterev, F. K. Tittel, D. V. Serebryakov, A. L. Malinovsky, and I. V. Morozov, Applications of quartz tuning forks in spectroscopic gas sensing. *Rev. Sci. Instrum.* **2005**, *76* (4), 043105-1–043105-9.

77. D. V. Serebryakov, A. P. Cherkun, B. A. Loginov, and V. S. Letokhov, Tuning-fork-based fast highly sensitive surface-contact sensor for atomic force microscopy/near-field scanning optical microscopy. *Rev. Sci. Instrum.* **2002**, *73* (4), 1795–1802.

78. A. Hordvik and H. Schlossberg, Photoacoustic technique for determining optical-absorption coefficients in solids. *Appl. Optics* **1977**, *16* (1), 101–107.

79. M. M. Farrow, R. K. Burnham, M. Auzanneau, S. L. Olsen, N. Purdie, and E. M. Eyring, Piezoelectric detection of photoacoustic signals. *Appl. Optics* **1978**, *17* (7), 1093–1098.

80. J. A. Burt, Response of a fluid-filled piezoceramic cylinder to pressure generated by an axial laser-pulse. *J. Acoust. Soc. Am.* **1979**, *65* (5), 1164–1170.

81. D. C. Emmony, M. Siegrist, and F. K. Kneubuhl, Laser-induced shock-waves in liquids. *Appl. Phys. Lett.* **1976**, *29* (9), 547–549.

82. R. M. White, Generation of elastic waves by transient surface heating. *J. Appl. Phys.* **1963**, *34* (12), 3559–3567.

83. S. S. Goncalves, M. G. Da Silva, M. S. Sthel, S. L. Cardoso, R. R. Sanchez, J. B. Rieumont, and H. Vargas, Determination of thermal and sorption properties of poly-3-hydroxy octanoate using photothermal methods. *Phys. Status Solidi A Appl. Res.* **2001**, *187* (1), 289–295.

84. G. Giubileo and A. Puiu, Photoacoustic spectroscopy of standard explosives in the MIR region. *Nucl. Instr. Meth. Phys. Res. A.* **2010**, *623* (2), 771–777.

85. Q. Wen and K. H. Michaelian, Mid-infrared photoacoustic spectroscopy of solids using an external-cavity quantum-cascade laser. *Opt. Lett.* **2008**, *33* (16), 1875–1877.

86. M. D. Rabasovic, M. G. Nikolic, M. D. Dramicanin, M. Franko, and D. D. Markushev, Low-cost, portable photoacoustic setup for solid samples. *Meas. Sci. Technol.* **2009**, *20* (9), 1–6.

87. H. K. Park, D. Kim, C. P. Grigoropoulos, and A. C. Tam, Pressure generation and measurement in the rapid vaporization of water on a pulsed-laser-heated surface. *J. Appl. Phys.* **1996**, *80* (7), 4072–4081.

88. S. Kaneko, S. Yotoriyama, H. Koda, and S. Tobita, Excited-state proton transfer to solvent from phenol and cyanophenols in water. *J. Phys. Chem. A* **2009**, *113* (13), 3021–3028.

89. A. A. Samokhin, V. I. Vovchenko, N. N. Il'ichev, and P. V. Shapkin, Explosive boiling in water exposed to q-switched erbium laser pulses. *Laser Phys.* **2009**, *19* (5), 1187–1191.

90. H. Kim, S. J. Yu, and S. H. Cho, Detection of pressure waves in water by using optical techniques. *J. Korean Phys. Soc.* **2008**, *53* (4), 1906–1909.

91. K. H. Michaelian, *Photoacoustic Infrared Spectroscopy*. Wiley-Interscience, Hoboken, NJ, 2003.

92. A. Mandelis and P. Hess, *Progress in Photothermal and Photoacoustic Science and Technology: Life and Earth Sciences.* SPIE—The International Society for Optical Engineering, Bellingham, WA, 1997, Vol. 3.

93. V. Zeninari, A. Grossel, L. Joly, T. Decarpenterie, B. Grouiez, B. Bonno, and B. Parvitte, Photoacoustic spectroscopy for trace gas detection with cryogenic and room-temperature continuous-wave quantum cascade lasers. *Cent. Eur. J. Phys.* **2010**, *8* (2), 194–201.

94. J. K. S. Wan, M. S. Ioffe, and M. C. Depew, A novel acoustic sensing system for on-line hydrogen measurements. *Sens. Actuator B Chem.* **1996**, *32* (3), 233–237.

95. I. G. Calasso, V. Funtov, and M. W. Sigrist, Analysis of isotopic CO2 mixtures by laser photoacoustic spectroscopy. *Appl. Optics* **1997**, *36* (15), 3212–3216.

96. Z. Bozoki, J. Sneider, Z. Gingl, A. Mohacsi, M. Szakall, Z. Bor, and G. Szabo, A high-sensitivity, near-infrared tunable-diode-laser-based photoacoustic water-vapour-detection system for automated operation. *Meas. Sci. Technol.* **1999**, *10* (11), 999–1003.

97. A. Beenen and R. Niessner, Development of a photoacoustic trace gas sensor based on fiber-optically coupled NIR laser diodes. *Appl. Spectrosc.* **1999**, *53* (9), 1040–1044.

98. A. Mohacsi, Z. Bozoki, and R. Niessner, Direct diffusion sampling-based photoacoustic cell for in situ and on-line monitoring of benzene and toluene concentrations in water. *Sens. Actuator B Chem.* **2001**, *79* (2–3), 127–131.

99. S. Barbieri, J. P. Pellaux, E. Studemann, and D. Rosset, Gas detection with quantum cascade lasers: An adapted photoacoustic sensor based on Helmholtz resonance. *Rev. Sci. Instrum.* **2002**, *73* (6), 2458–2461.

100. M. A. Gondal, M. A. Baig, and M. H. Shwehdi, Laser sensor for detection of SF6 leaks in high power insulated switchgear systems. *IEEE Trns. Dielectr. Electr. Insul.* **2002**, *9* (3), 421–427.

101. M. B. Pushkarsky, M. E. Webber, O. Baghdassarian, L. R. Narasimhan, and C. K. N. Patel, Laser-based photoacoustic ammonia sensors for industrial applications. *Appl. Phys. B Lasers Opt.* **2002**, *75* (2–3), 391–396.

102. G. Santiago, V. Slezak, and A. L. Peuriot, Resonant photoacoustic gas sensing by PC-based audio detection. *Appl. Phys. B-Lasers Opt.* **2003**, *77* (4), 463–465.

103. V. Slezak, J. Codnia, A. L. Peuriot, and G. Santiago, Resonant photoacoustic detection of NO2 traces with a Q-switched green laser. *Rev. Sci. Instrum.* **2003**, *74* (1), 516–518.

104. V. Zeninari, B. Parvitte, D. Courtois, V. A. Kapitanov, and Y. N. Ponomarev, Methane detection on the sub-ppm level with a near-infrared diode laser photoacoustic sensor. *Infrared Phys. Technol.* **2003**, *44* (4), 253–261.

105. M. Horstjann, Y. A. Bakhirkin, A. A. Kosterev, R. F. Curl, F. K. Tittel, C. M. Wong, C. J. Hill, and R. Q. Yang, Formaldehyde sensor using interband cascade laser based quartz-enhanced photoacoustic spectroscopy. *Appl. Phys. B-Lasers Opt.* **2004**, *79* (7), 799–803.

106. A. Elia, P. M. Lugara, and C. Giancaspro, Photoacoustic detection of nitric oxide by use of a quantum-cascade laser. *Opt. Lett.* **2005**, *30* (9), 988–990.

107. M. E. Webber, T. S. MacDonald, M. B. Pushkarsky, C. K. N. Patel, Y. J. Zhao, N. Marcillac, and F. M. Mitloehner, Agricultural ammonia sensor using diode lasers and photoacoustic spectroscopy. *Meas. Sci. Technol.* **2005**, *16* (8), 1547–1553.

108. J. P. Besson, S. Schilt, E. Rochat, and L. Thevenaz, Ammonia trace measurements at ppb level based on near-IR photoacoustic spectroscopy. *Appl. Phys. B Lasers Opt.* **2006**, *85* (2–3), 323–328.

109. J. P. Besson, S. Schilt, and L. Thevenaz, Sub-ppm multi-gas photoacoustic sensor. *Spectrochim. Acta A Mol. Biomol. Spectrosc.* **2006**, *63* (5), 899–904.

110. J. P. Besson, S. Schilt, F. Sauser, E. Rochat, P. Hamel, F. Sandoz, M. Nikles, and L. Thevenaz, Multi-hydrogenated compounds monitoring in optical fibre manufacturing process by photoacoustic spectroscopy. *Appl. Phys. B Lasers Opt.* **2006**, *85* (2–3), 343–348.

111. H. Cattaneo, T. Laurila, and R. Hernberg, Photoacoustic detection of oxygen using cantilever enhanced technique. *Appl. Phys. B Lasers Opt.* **2006**, *85* (2–3), 337–341.

112. M. G. da Silva, A. Miklos, A. Falkenroth, and P. Hess, Photoacoustic measurement of N_2O concentrations in ambient air with a pulsed optical parametric oscillator. *Appl. Phys. B Lasers Opt.* **2006**, *82* (2), 329–336.

113. A. Grossel, V. Zeninari, L. Joly, B. Parvitte, D. Courtois, and G. Durry, New improvements in methane detection using a Helmholtz resonant photoacoustic laser sensor: A comparison between near-IR diode lasers and mid-IR quantum cascade lasers. *Spectrochim. Acta A Mol. Biomol. Spectrosc.* **2006**, *63* (5), 1021–1028.

114. M. Mattiello, M. Nikles, S. Schilt, L. Thevenaz, A. Salhi, D. Barat, A. Vicet, Y. Rouillard, R. Werner, and J. Koeth, Novel Helmholtz-based photoacoustic sensor for trace gas detection at ppm level using GaInAsSb/GaAlAsSb DFB lasers. *Spectrochim. Acta A Mol. Biomol. Spectrosc.* **2006**, *63* (5), 952–958.

115. S. Schilt, J. P. Besson, and L. Thevenaz, Near-infrared laser photoacoustic detection of methane: The impact of molecular relaxation. *Appl. Phys. B Lasers Opt.* **2006**, *82* (2), 319–328.

116. A. Varga, Z. Bozoki, M. Szakall, and G. Szabo, Photoacoustic system for on-line process monitoring of hydrogen sulfide (H2S) concentration in natural gas streams. *Appl. Phys. B Lasers Opt.* **2006**, *85* (2–3), 315–321.

117. M. D. Wojcik, M. C. Phillips, B. D. Cannon, and M. S. Taubman, Gas-phase photoacoustic sensor at 8.41 mu m using quartz tuning forks and amplitude-modulated quantum cascade lasers. *Appl. Phys. B Lasers Opt.* **2006**, *85* (2–3), 307–313.

118. M. A. Gondal and Z. H. Yamani, Highly sensitive electronically modulated photoacoustic spectrometer for ozone detection. *Appl. Optics* **2007**, *46* (29), 7083–7090.

119. A. Grossel, V. Zeninari, B. Parvitte, L. Joly, and D. Courtois, Optimization of a compact photoacoustic quantum cascade laser spectrometer for atmospheric flux measurements: application to the detection of methane and nitrous oxide. *Appl. Phys. B Lasers Opt.* **2007**, *88* (3), 483–492.

120. V. Koskinen, J. Fonsen, K. Roth, and J. Kauppinen, Cantilever enhanced photoacoustic detection of carbon dioxide using a tunable diode laser source. *Appl. Phys. B Lasers Opt.* **2007**, *86* (3), 451–454.

121. B. Lendl, W. Ritter, M. Harasek, R. Niessner, and C. Haisch. Photoacoustic monitoring of CO2 in biogas matrix using a quantum cascade laser, *IEEE Sensors*, Daegu, South Korea, October 22–25, 2006, IEEE, Daegu, South Korea, 2007, pp 338–341.

122. R. Lewicki, G. Wysocki, A. A. Kosterev, and F. K. Tittel, Carbon dioxide and ammonia detection using 2 mu m diode laser based quartz-enhanced photoacoustic spectroscopy. *Appl. Phys. B Lasers Opt.* **2007**, *87* (1), 157–162.

123. R. E. Lindley, A. M. Parkes, K. A. Keen, E. D. McNaghten, and A. J. Orr-Ewing, A sensitivity comparison of three photoacoustic cells containing a single microphone, a differential dual microphone or a cantilever pressure sensor. *Appl. Phys. B Lasers Opt.* **2007**, *86* (4), 707–713.

124. J. P. Besson, S. Schilt, and L. Thevenaz, Molecular relaxation effects in hydrogen chloride photoacoustic detection. *Appl. Phys. B Lasers Opt.* **2008**, *90* (2), 191–196.

125. A. A. Kosterev, Y. A. Bakhirkin, F. K. Tittel, S. McWhorter, and B. Ashcraft, QEPAS methane sensor performance for humidified gases. *Appl. Phys. B Lasers Opt.* **2008**, *92* (1), 103–109.

126. M. Wolff, M. Germer, H. G. Groninga, and H. Harde, Photoacoustic CO_2 sensor based on a DFB diode laser at 2.7 mu m. *Eur. Phys. J. Spec. Top.* **2008**, *153*, 409–413.

127. B. D. Adamson, J. E. Sader, and E. J. Bieske, Photoacoustic detection of gases using microcantilevers. *J. Appl. Phys.* **2009**, *106* (11), 114510-1–114510-4.

128. G. Giubileo, A. Puiu, F. Dell'Unto, M. Tomasi, and A. Fagnani, High resolution laser-based detection of ammonia. *Laser Phys.* **2009**, *19* (2), 245–251.

129. M. A. Gondal, A. Dastageer, and Z. H. Yamani, Laser-induced photoacoustic detection of ozone at 266 nm using resonant cells of different configuration. *J. Environ. Sci. Health A Toxic/Hazard. Subst. Environ. Eng.* **2009**, *44* (13), 1457–1464.

130. A. Elia, C. Di Franco, V. Spagnolo, P. M. Lugara, and G. Scamarcio, Quantum cascade laser-based photoacoustic sensor for trace detection of formaldehyde gas. *Sensors* **2009**, *9* (4), 2697–2705.

131. K. Liu, J. Li, L. Wang, T. Tan, W. Zhang, X. Gao, W. Chen, and F. K. Tittel, Trace gas sensor based on quartz tuning fork enhanced laser photoacoustic spectroscopy. *Appl. Phys. B Lasers Opt.* **2009**, *94* (3), 527–533.

132. S. Schilt, A. A. Kosterev, and F. K. Tittel, Performance evaluation of a near infrared QEPAS based ethylene sensor. *Appl. Phys. B Lasers Opt.* **2009**, *95* (4), 813–824.

133. P. Yong, Z. Wang, L. Liang, and Y. Qingxu, Tunable fiber laser and fiber amplifier based photoacoustic spectrometer for trace gas detection. *Spectrochim. Acta A Mol. Biomol. Spectrosc.* **2009**, *74* (4), 924–927.

134. C. Di Franco, A. Elia, V. Spagnolo, P. M. Lugara, and G. Scamarcio, Advanced optoacoustic sensor designs for environmental applications. *Proc. SPIE Int. Soc. Opt. Eng.* **2010**, *7808*, 78081A (8pp.).
135. L. Dong, A. A. Kosterev, D. Thomazy, and F. K. Tittel, QEPAS spectrophones: Design, optimization, and performance. *Appl. Phys. B Lasers Opt.* **2010**, *100* (3), 627–635.
136. J. M. Rey, C. Romer, M. Gianella, and M. W. Sigrist, Near-infrared resonant photoacoustic gas measurement using simultaneous dual-frequency excitation. *Appl. Phys. B Lasers Opt.* **2010**, *100* (1), 189–194.
137. D. V. Serebryakov, I. V. Morozov, A. A. Kosterev, and V. S. Letokhov, Laser microphotoacoustic sensor of ammonia traces in the atmosphere. *Quantum Electron.* **2010**, *40* (2), 167–172.
138. Z. Bozoki, A. Szabo, A. Mohacsi, and G. Szabo, A fully opened photoacoustic resonator based system for fast response gas concentration measurements. *Sens. Actuator B Chem.* **2010**, *147* (1), 206–212.
139. V. Spagnolo, A. A. Kosterev, L. Dong, R. Lewicki, and F. K. Tittel, NO trace gas sensor based on quartz-enhanced photoacoustic spectroscopy and external cavity quantum cascade laser. *Appl. Phys. B Lasers Opt.* **2010**, *100* (1), 125–130.
140. T. A. Heumier and J. L. Carlsten, App Note 8: Mode hopping in semiconductor lasers. ILX Lightwave Corporation, Bozeman, MT, 2005, 1–12.
141. C. Bauer, U. Willer, R. Lewicki, A. Pohlkotter, A. Kosterev, D. Kosynkin, F. K. Tittel, and W. Schade, A mid-infrared QEPAS sensor device for TATP detection. *J. Phys., Conf. Ser.* **2009**, *157*, 012002 (6 pp.).
142. P. Hodgson, K. M. Quan, H. A. Mackenzie, S. S. Freeborn, J. Hannigan, E. M. Johnston, F. Greig, and T. D. Binnie, Application of pulsed-laser photoacoustic sensors in monitoring oil contamination in water. *Sens. Actuator B Chem.* **1995**, *29* (1–3), 339–344.
143. F. Hernandez-Valle, M. Navarrete, E. V. Mejia, and M. Villagran-Muniz, Trace analysis of pesticides in water using pulsed photoacoustic technique. *Eur. Phys. J. Spec. Top.* **2008**, *153*, 507–510.
144. A. K. Chaudhary, G. C. Bhar, and S. Das, Low-limit photo-acoustic detection of solid RDX and TNT explosives with carbon dioxide laser. *J. Appl. Spectrosc.* **2006,** *73* (1), 123–129.
145. C. W. Van Neste, M. E. Morales-Rodriguez, L. R. Senesac, S. M. Mahajan, and T. Thundat, Quartz crystal tuning fork photoacoustic point sensing. *Sens. Actuator B Chem.* **2010**, *150* (1), 402–405.
146. J. E. De Albuquerque, D. T. Balogh, and R. M. Faria, Quantitative depth profile study of polyaniline films by photothermal spectroscopies. *Appl. Phys. A Mater. Sci. Process.* **2007**, *86* (3), 395–401.
147. M. H. Xu and L. H. V. Wang, Photoacoustic imaging in biomedicine. *Rev. Sci. Instrum.* **2006**, *77* (4), 041101-1–041101-22.
148. K. Uchiyama, K. Yoshida, X. Z. Wu, and T. Hobo, Open-ended photoacoustic cells: Application to two-layer samples using pulse laser-induced photoacoustics. *Anal. Chem.* **1998**, *70* (3), 651–657.
149. J. A. Viator, J. Komadina, L. O. Svaasand, G. Aguilar, B. Choi, and J. S. Nelson, A comparative study of photoacoustic and reflectance methods for determination of epidermal melanin content. *J. Invest. Dermatol.* **2004**, *122* (6), 1432–1439.

Part V

Piezoresistive, Wireless, and Electrical Sensors

27 Piezoresistive Fibrous Sensor for On-Line Structural Health Monitoring of Composites

Saad Nauman, Irina Cristian, François Boussu, and Vladan Koncar

CONTENTS

27.1 INTRODUCTION

The use of high-performance composite materials in aerospace, automotive, marine, and civil engineering applications accelerated rapidly in the past decades. The main attraction of fiber reinforced composites is their light weight and high specific strength and stiffness that can be optimized for specific loading conditions. Good quality and reliability are basic requirements for advanced composite structures, which are often used under harsh environments. In-service health monitoring of the fiber reinforced composite structures, using non-destructive evaluation (NDE), can help in better understanding the specific deformation modes and also in keeping the structure operating reliably and safely.

Different approaches that can be used for structural health monitoring (SHM), including ultrasonic scanning, acoustic emission, shearography, stimulated infrared thermography, fiber brag grating sensors, vibration testing, etc. have been discussed in detail in the literature (Chung, 2002; Buyukozturk and Yu, 2003; Balageas, 2006; Farrar and Worden, 2007; Black, 2009). The classical NDE techniques hardly address this concern because of difficulties in making in situ implementation. Today, design engineers lay special emphasis on the integration of sensors during the manufacturing process, which enables them to perform in situ health monitoring of the composite parts, reduce cost, and improve the accuracy of measurements.

For textile composites, one possible solution is to use intelligent textile materials and structures, which provide real possibility for on-line and in situ monitoring of structural integrity. Such intelligent materials are made by coating or treating textile yarns, filaments, or fabrics with nanoparticles or conductive and semi conductive polymers, giving them special properties.

A review of piezoresistive sensing approaches already being applied to measure strain in textile composites shows that several sensing mechanisms exist (Scilingo et al., 2003; Dharap et al., 2004; Fiedler et al., 2004; Lorussi et al., 2005). These approaches may be categorized on the basis of manufacturing technology as nanotube networks, use of carbon tows for self-sensing, and semiconductive coatings.

Nanotubes have been investigated in detail for use as sensing mechanisms, both for smart textile applications and for SHM of composites (Kang et al., 2006; Lee et al., 2006; Hecht et al., 2007; Li and Chou, 2008; Thostenson and Chow, 2008; Zhao et al., 2010). Significant challenges still exist in their development, for example, the efficient growth of macroscopic-length carbon nanotubes controlled growth of nanotubes on desired substrates, durability of nanotube-based sensors and actuators, and effective dispersion in polymer matrices and their orientation. Therefore, there is a need to develop both experimental and analytical techniques to bridge the nano and macro scales toward optimization so as to use nanotube networks as sensors inside macroscale (fabric) or mesoscale (tow) composites (Li et al., 2008).

Carbon fiber reinforced composites offer a unique possibility of using carbon tows as sensing network because of their conductivity (Kaddour et al., 1994; Abry et al., 2001; Kupke et al., 2001; Muto et al., 2001; De Baere et al., 2010). However, such an approach can only be used for conductive fiber-based composites. Moreover, before applying such an approach for SHM, it is imperative to understand the deformation mechanism of the reinforcement. Any anomaly in the deformation mechanism can threaten the sensing mechanism's validity and efficacy.

As for semiconductive coatings, to date they have only been used for design of active components of intelligent textile structures, such as silicon flexible skins with regular textiles (Katragadda and Xu, 2008), flexible fibrous transistors (Lee and Subramanian, 2005; Fortunato et al., 2008; Guerra et al., 2009; Kagan, 2011), and other smart textile applications to manufacture consumer products and to detect the physiological condition of the wearer (Baurley, 2003; Koncar et al., 2004; Wijesiriwardana, 2006; Kim et al., 2008). But their use as embedded sensors for realization of smart reinforcements needs to be further investigated.

It has been suggested in our previous research work that the use of the intelligent textile approach in order to realize fibrous sensors compatible with SHM and composite technology is a very promising solution for real-time in situ health monitoring of composite parts (Cristian et al., 2011; Nauman et al., 2011a,b, 2012). In the case of high-performance textile composites, these intelligent textile materials can be integrated during the manufacturing phase of the reinforcement or during the lay-up process. These materials perform dual function inside a composite as after integration in the reinforcement they not only act as a part of structural material but also have actuating, sensing, and microprocessing capabilities.

In this chapter, a new approach of on-line SHM using fibrous sensors inserted inside textile composite reinforcements is presented. A novel flexible piezoresistive fibrous sensor has been developed and optimized for in situ structural deformation sensing in textile composites. The morphological and electromechanical properties of the fibrous sensors have been analyzed using scanning electron microscopy, tomography, and yarn tensile strength tester.

The sensors were inserted as weft in woven reinforcement during the weaving process. The reinforcement was then impregnated in epoxy resin and was later subjected to quasi-static tensile loading and bending. An appropriate data acquisition module has been developed and used for data acquisition and its further treatment.

It was found that the sensor was able to detect deformations in the composite structure until rupture since it was inserted together with reinforcing tows. The results obtained for carbon composite specimens under standard testing conditions have validated the in situ monitoring concept using our fibrous textile sensors. Embedding such an intelligent piezoresistive sensor inside the reinforcement during weaving process is the most convenient and cost effective way of insertion of a sensor for SHM.

27.2 SENSOR ARCHITECTURE

27.2.1 General Requirements

Our sensors have been designed to be embedded inside the textile reinforcement during the weaving process and they have to present all the characteristics of a traditional textile material: flexibility, lightweight, and capability of adopting the geometry of the reinforcement. The sensors must be sensitive enough to measure in situ strains inside the composite part. Sensitivity is important as the targeted application usually undergoes very low strains (<2.5%) and even such low strains and/or vibrations during the life time of composite parts are critical. Often, they are used in areas where structural integrity cannot be compromised (aircraft wings, bodies, etc.).

27.2.2 Coating Solution: Sensing Structure and Sensing Principle

Conductive polymers are some of the developments that seem to respond to specific properties of textile materials like flexibility and deformability. Two sub-classes of conductive polymers can be identified: intrinsically conductive polymers and conductive polymer composites. Inherently, conducting polymers are suitable for applications in many domains of intelligent textiles, but they present some substantial disadvantages like infusibility or insolubility in common organic solvents, weak mechanical properties, and poor processability. Composite conductive polymers are obtained by blending (generally by melt mixing) an insulating polymer matrix (thermoplastic or thermosetting plastic) with conductive fillers like carbon black, carbon fibers or nanotubes, metallic particles, or conductive polymers. The presence of filler particles in the matrix may have a negative impact on the mechanical properties of the final composite. Instead of this, the development in the field of composite conductive polymers therefore seems to be a promising approach for intelligent textiles owing to their simplicity of preparation and to their low cost.

As coating solution for our sensors, a conductive polymer composite based on dispersion of carbon black particles (Printex® L6) in polymer (Evoprene® 007) solution, using chloroform as a solvent was chosen (Cochrane et al., 2007).

Conduction in such composites charged with conductive fillers depends on various phenomena, depending upon the filler geometry and their distribution in the polymer.

The transition between the insulating and conductive states occurs at a particular volume concentration of the conductive filler. This critical concentration is termed "percolation threshold." Around the percolation threshold, a sudden and rapid drop in resistivity is observed. This is because of the formation of conductive networks of fillers. At the percolation threshold, these networks are just enough to allow electrons to flow through. When filler particles are in direct contact, the electrical conduction is explained by metallic conduction and hopping. In metallic conduction, the band structure of the material is overlapping, which allows the electrons to flow from one site to another without energy input. When there is gap or barrier (a polymer layer in this case) between the filler particles, the electrons need to jump from one site to another. This hopping can be either short range hopping (sites energetically distant and geographically close) or variable range hopping (sites energetically close and geographically distant). Due to its dual nature, as predicted by quantum mechanics, the electron is able to hop even though its kinetic energy is less than the potential energy of the barrier. The Heisenberg uncertainty principle suggests that an electron has a nonzero probability of moving from one side of any physical barrier to another. When an electron wave meets a potential barrier (polymer film), the wave does not instantly go to zero. Instead, it starts to decay exponentially within the potential barrier.

If the wave does not reach zero by the time it reaches the other end of the barrier then there is a finite probability that it will be found on the other side of the barrier, implying that the wave effectively "tunnels" through the non-conductive barrier (Peratech, 2011). When these composites are subjected to tensile loading, the filler particles displace relative to one another. This results in an average increase in interparticle distance, breakage of certain percolation networks, and an increase in potential energy of barriers for electron hopping. The combined effect of all of these phenomena is a net increase in the resistivity of the composite. This property can be used for detecting tensile loading in any composite part. When compression is applied on a polymer composite filled with conductive particles, the interparticle distance is reduced. This translates into an increase in conductivity as more electrons can flow through filler particles physically in contact and tunnel through the barriers, which under compression become "thinner" and the probability that decaying electron wave crosses over to the other end is higher. This property can be used for compression sensing.

Based on previous research (Cochrane et al., 2007), a 35% carbon black solution was chosen for the coating of fibrous sensors.

27.2.3 FIBROUS SUBSTRATE

In order to characterize the sensitivity and adherence of the coating on different substrates, the carbon black solution was coated on different yarns (71 tex cotton spun yarns; 48.2 tex polyethylene monofilament, and 25 tex polyamide monofilament). Visual inspection of the surfaces of coated yarns shows that the coating is more uniform for synthetic monofilaments compared to cotton yarns. The cotton yarns absorb the conductive solution, which penetrates inside the pores and interstices much like a dye. This particular phenomenon could be a source of nonhomogeneity in sensor electrical and mechanical properties, as the spun yarn is nonuniform as compared to filaments, the coating, and thus the resistivity achieved could be nonuniform. Moreover, the greatest inconvenience with coated cotton spun yarns is their low sensitivity during the initial tensile loading phase. The electrical resistance values were measured on 12 coated samples of each variant using a multimeter. The resistivities were then calculated using the yarn/filament fineness, yarn/filament lengths, and these measured electrical resistance values. Coated polyethylene filaments show relatively lower dispersion of resistivity as compared to coated polyamide filaments (Cristian et al., 2011).

In order to carry out tensile tests on coated yarns and monofilaments, an material testing system (MTS) 1/2 tester was used. Samples underwent quasi-static tensile loading at a constant test speed of 5 mm/min. For the purpose of electrical resistance variation measurement during the tensile testing, a simple voltage divider circuit and Keithley® KUSB-3100 data acquisition module were employed. Initial electrical resistance of the coating on cotton yarns is much lower than on monofilaments, but since the cotton spun yarns are inherently irregular, the coatings obtained are not homogenous, and the results for different coated yarns vary widely in their response to tensile loading. Due to the particular fineness of the polyamide monofilament, it was found that slight non-homogeneity in coating on the surface can result in breakdown of the conductive paths. As a result, the behavior of polyamide is highly inconsistent. Polyethylene monofilaments provide a reasonably good compromise as the substrate. The coatings on polyethylene are easy to achieve due to good substrate/conductive solution interfacial properties. The polyethylene coatings are reproducible as the curves for all the samples are nearly identical as opposed to polyamide and cotton (Cristian et al., 2011). This is because of the fact that the coating achieved on polyethylene monofilaments is relatively homogeneous. In view of all of these advantages, polyethylene monofilaments were chosen for sensor development. In order to reduce the initial electrical resistance, two ply polyethylene filaments were used. These were coated with the polymer solution as described earlier.

27.2.4 Sensor Final Characteristics

The two ends of the coated polyethylene filaments were additionally coated with silver paint in order to reduce contact resistance, and fine copper wire was attached to the two ends with the help of this paint. In this way, secure connections were realized. Details of its connections at the two ends can also be observed in Figure 27.1.

Prior to insertion in the carbon fiber reinforcement, the sensor should be coated with an insulating layer. This is because of the fact that the carbon multifilament tows used for weaving of composite reinforcement are conductive and may disturb the functioning of the piezoresistive fibrous sensor. Natural rubber-based latex supplied by Vosschemie® was chosen as the final coating layer to insulate the sensor from surrounding carbon tows. The liquid latex is prevulcanized with sulfur and contains cross-linking points. Transversal and longitudinal sections of the sensor are given in Figure 27.2.

Sensor structural and geometrical parameters along with initial electrical resistance are shown in Table 27.1.

The resistance has been measured between two electrodes (crocodile clips, 40 cm apart), if being the total electrical resistance of the conductive layer. The contact resistance between the sensor and the crocodile clips of electrodes can be considered as negligible. The electrical resistivity of the carbon black coating is supposed to be homogeneous as the conductive charges are uniformly distributed over the surface of the substrate.

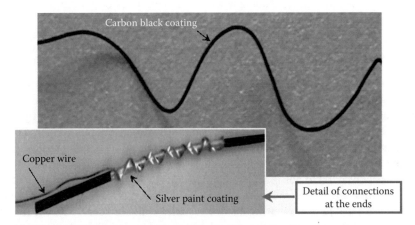

FIGURE 27.1 Carbon black coated sensor with silver coated connections.

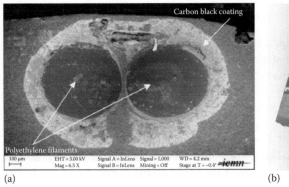
(a)

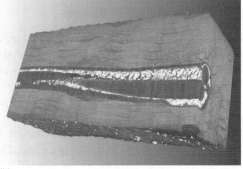
(b)

FIGURE 27.2 (a) Transversal section (SEM); (b) longitudinal section (tomography) of the sensor.

TABLE 27.1
Sensor Characteristics

No.	Parameters	UM	Value
1	Linear density of substrate (polyethylene filament)	tex	48.23
2	Diameter of the filament	mm	0.70
3	Average width of the sensor cross section	mm	1.68
4	Average thickness of the sensor cross section	mm	1.26
5	Aspect ratio of the sensor (width/thickness)	—	1.33
6	Initial resistance of the sensors	kΩ	43.3

27.3   OUT OF COMPOSITE SENSOR TESTING

In order to verify that the sensor is sensitive at very low strains (<2.5%), tensile tests have been carried out, prior to insertion of the sensor in the composite structure. The sensors were tested on MTS 1/2 tester, under quasi-static tensile loading at a constant test speed of 5 mm/min. These tests also lead to the calculation of gauge factor and Young modulus values for the sensor.

27.3.1   Data Acquisition Module

The same Keithley KUSB-3100 data acquisition module was employed for the purpose of voltage variation during tensile testing. This time, a special set-up containing a Wheatstone bridge and an amplifier was used to measure unknown variable resistance of the sensor as a function of output voltage. The simple voltage divider circuit is not adequate for the measurement of resistance change in the case of sensors developed here. These piezoresistive sensors produce a very small percentage change in resistance in response to physical phenomena such as strain. Moreover, the output signal has considerable noise. Generally, a bridge measures resistance indirectly by comparison with a similar resistance. Wheatstone bridges offer an attractive solution for sensor applications as they are capable of measuring small resistance changes accurately. Figure 27.3 shows schematic diagram of the data acquisition module developed and used for data acquisition and its further treatment.

27.3.2   Tensile Testing Results

The resistance variation data thus obtained for different test results were treated for noise reduction using a low-pass filter. The resultant stress–strain-normalized resistance relationship curve up to 2.75% elongation of the out of composite sensor (before insertion in the reinforcement) is shown in Figure 27.4a and b.

It may be noticed in Figure 27.4a that the stress versus strain curve has the same shape as the normalized resistance ($\Delta R/R$) versus strain curve. This validates the electromechanical properties of our fibrous sensor for strains ranging from 0% to 2.75%. In Figure 27.4b, the hysteresis results of the sensor for 10 cycles are given. For this test, the sensor underwent 0.5% extension at a constant cross head displacement rate of 5 mm/min, followed by compression in each cycle. The hysteresis test also shows that the sensor is capable of following the extension and compression patterns in each cycle. It can also be noticed that the hysteresis is high for the first cycle, which reduces gradually and for the 10th cycle the sensor exhibits almost linear behavior. This loss in

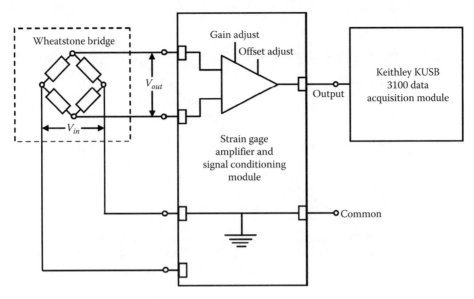

FIGURE 27.3 Schematic of an instrumentation amplifier connected to a Wheatstone bridge.

hysteresis with increasing number of cycles can be attributed to permanent breakage of some of the percolation networks. It should be recalled that these sensors have been optimized so as to have volume concentration of carbon nanoparticles corresponding to the percolation threshold. At the percolation threshold, the conductivity in the nanoparticle filled composites is due to particle–particle contact, electron hopping, and tunnel effect. As the sensor undergoes repeated loading and unloading, some of the percolation networks responsible for conduction due to physical contact break down completely. This results in sensor conductivity depending more and more on the tunnel effect with increasing number of cycles. This causes the sensor behavior to become less hysterical and more linear.

After this test, the gauge factor (K) and Young modulus (E) values of the sensor have been calculated: K = 12.3; E = 520.8 MPa.

The linear relationship between relative electrical resistance change ($\Delta R/R$) and the strain of the strain gauge can be given as follows:

$$\frac{\Delta R}{R} = K \cdot \varepsilon \tag{27.1}$$

where
  ΔR is the change in electrical resistance caused by elongation or contraction (Ω)
  R is the initial resistance of the sensor (Ω)
  K is the constant of proportionality (gauge factor). Its value depends on the type of material
  ε is strain in the sensor

Within the elastic limit, the stress–strain relationship can be given as follows:

$$\varepsilon = \frac{1}{E} \cdot \sigma \tag{27.2}$$

where
  E is Young's modulus
  σ is stress (N/mm^2)

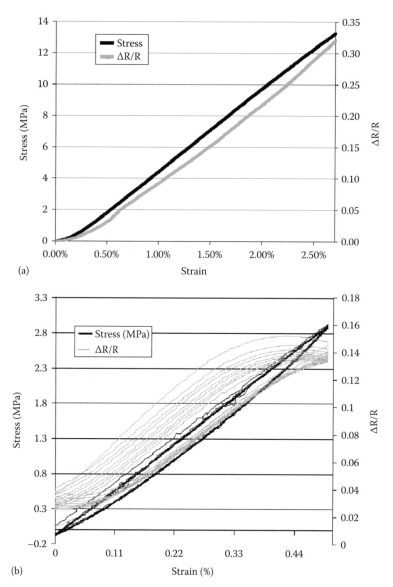

(a)

(b)

FIGURE 27.4 Normalized resistance and stress against strain for the sensor outside the composite. (a) Tensile test up to 2.75% elongation and (b) hysteresis 10 cycles at 0.5% extension.

Combining Equations 27.2 and 27.1, the following relationship is obtained:

$$\frac{\Delta R}{R} = \frac{K}{E} \cdot \sigma \tag{27.3}$$

27.4   COMPOSITE MANUFACTURING AND SENSOR PLACEMENT

The three-dimensional (3D) interlock woven fabric used as reinforcement was woven using 200 Tex carbon tows in warp and weft having 6 K filaments in the cross section. These multifilament tows were provided by Hercules Inc. The resin employed for impregnation of woven reinforcement was epoxy EPOLAM 5015. As 3D weave structure, an orthogonal interlock having 13 weft layers and

TABLE 27.2
Reinforcement Specifications

			Samples for	
No.	Parameter	UM	Tensile Test	Bending Test
1	Linear density of warp tow	tex	200	200
2	Linear density of weft tow	tex	200	200
3	Average thickness of reinforcement	mm	6.5	8.5
4	Warp tows density	tows/cm	24	24
5	Weft tows density	tows/cm	169	169
6	Areal weight	g/m^2	3908	4699

12 warp layers with layer-to-layer binding was chosen. The reinforcement was woven on a modified weaving loom—ARM PATRONIC®. Technical specifications of the reinforcement including its thickness are given in Table 27.2.

The TexGen® generated geometry of the reinforcements together with sensor trajectories is shown in Figure 27.5a and b.

The placement of a sensor in the reinforcement was decided so that the sensor was inserted in the middle of the structure (related to thickness) for the samples intended for traction test (Figure 27.5a) and at the top and bottom of the reinforcement for the samples for bending test (Figure 27.5b). These sensors were integrated at the top and bottom faces in order to allow simultaneous mapping of compression and traction at the top and bottom of the reinforcement when they undergo bending. The sensors were inserted during the weaving process as weft yarns and they follow the same trajectory as the carbon weft yarns inside the reinforcement. The sensors were inserted in addition to the carbon tows in their respective layers so that they do not affect the structural properties of the reinforcement.

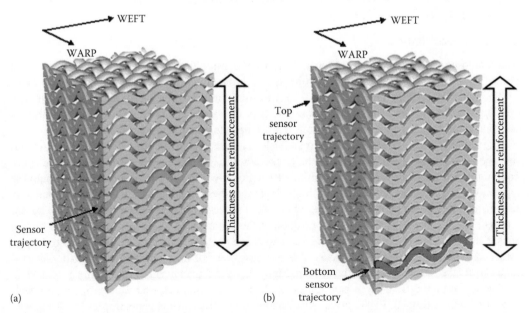

FIGURE 27.5 TexGen® generated geometry of woven reinforcement with sensors inserted as weft tows. (a) Samples for traction test and (b) samples for bending test.

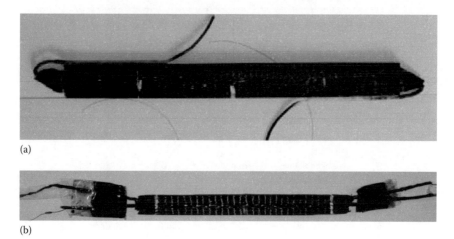

(a)

(b)

FIGURE 27.6 Carbon composite specimen with protruding sensor connections for (a) tensile test and (b) bending test.

After weaving, the reinforcement was carefully removed from the loom. The 13-layer reinforcement with integrated sensors was directly impregnated in resin using vacuum assisted resin transfer molding (VARTM) technology. The connections of sensors, which remain outside the reinforcement at the two ends, were carefully isolated from the rest of the mould so as to protect them and to prevent their resin impregnation. This was achieved by creating two vacuum sub-moulds inside the larger mould.

The impregnated composite was cut into slabs of 25 cm × 2.5 cm for the tensile test and 16.5 cm × 1.5 cm for bending test, according to the norms ISO 527-4, 1997 and NF EN ISO 14125: 1998, respectively. Each of the carbon composite specimens for traction test had one sensor in the middle of a slub, and the samples for three-point bending test had two sensors for compression and traction detection, as shown in Figure 27.6a and b.

27.5 PERFORMANCE EVALUATION OF COMPOSITE WITH INTEGRATED SENSOR

27.5.1 TENSILE TESTING

The composite specimens were tested on an Instron 8500 tester. Tensile strength tests were performed according to ISO 527-4, 1997 in the weft direction, i.e., the direction parallel to the inserted sensor.

The same Wheatstone bridge (Figure 27.3) was used for resistance variation measurement. The configuration of the testing equipment was also kept the same. The composite structural part was tested at a constant test speed of 5 mm/min. The composite underwent traction until rupture.

The resultant stress–strain-resistance relationship curve is shown in Figure 27.7. It can be observed that the normalized resistance follows the stress–strain curve. The stress–strain-resistance curve can be divided into four regions: the initial stiff region—where the composite exhibits toughness against the applied load represented by high slope; the tows straightening region; the second stiff region, and the zone of rupture. The rupture occurred at the strain of 0.52%, after which the tensile strength tester came back to its initial position at the same speed (5 mm/min). Since the fibrous sensor has not been broken, the normalized resistance ($\Delta R/R$) decreased until zero as the tester returned to its initial position. However, this decrease was not linear because the sensor was still intact while the resin–senor interface was partially damaged, which caused its nonlinear behavior.

Due to the high difference in yarn densities (24 warp yarns/cm vs. 169 weft yarn/cm), the weft tows are highly crimped. In the initial stiff region, micro cracks start appearing as the composite specimen undergoes traction, but the interface at resin and multifilament tows is still intact. That is why the composite exhibits rigid behavior. In Figure 27.7, it can be observed that after the initial

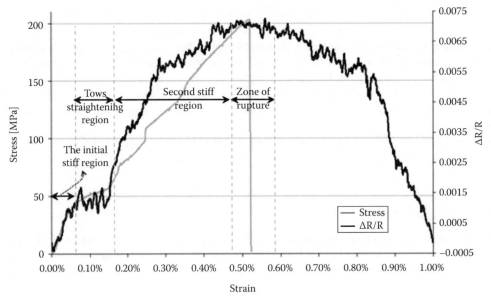

FIGURE 27.7 Normalized resistance and stress against strain for the sensor inside the composite.

stiff region, the highly crimped tows tend to straighten due to increasing tensile load in the second region. In this region, the micro cracks give way to relative slippage of highly crimped tows in their sockets, i.e., the resin–tow interface is relatively weakened. It can also be remarked that the sensor resistance follows the stress–strain curve; but in the second region, the electrical resistance curve is noisier as compared to other regions of the curve, which might signify the slippage of tows as well as the sensor in their sockets. This second region is followed by the third region called the second stiff region where the tows are locked in their sockets. In this region, the tows resist the applied load and exhibit stiff behavior as they regain some of their initial stiffness after the straightening of tows in the second region. The electrical resistance varies almost linearly with the applied load, in this region. The third region is followed by the zone of rupture of the composite in which the electrical resistance, having attained the highest value, starts dropping down. The normalized resistance starts dropping after the rupture. The fact that the sensor resistance reach its initial value after the rupture signifies that the sensor, owing to its elastic properties, is not destroyed with the composite. This fact was confirmed by a tomographical image of the samples, which underwent traction, shown in Figure 27.8. The sensor path near the zone of rupture can be observed.

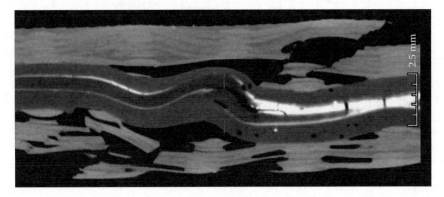

FIGURE 27.8 Tomographical images of a sensor inside a tested sample near the zone of rupture (longitudinal section).

The insulating medium on the sensor surface needs to have good adherence with the epoxy resin and carbon fiber reinforcements. In Figure 27.8, it can be observed that even in the fracture zone, the embedded sensor has not completely debonded from bulk of the composite. This is a proof of very strong latex–epoxy adhesion. A kink in the sensor can be observed, which is caused by the relaxation of the sensor as it tries to regain its original dimensions after the tensile loading damages the composite sample. Small cracks do appear in the conductive layer. However, the latex layer, which acts as insulating medium between the bulk of the composite and the conductive carbon layer, remains intact.

The insulation coating around the sensor renders it thick as well, which is undesirable for high-performance composite materials as thick insulation coatings might adversely affect the mechanical properties.

27.5.2 BENDING TESTING

The three-point bending tests on composite samples were carried out on an Instron 1185 tester, according to the same norm (NF EN ISO 14125: 1998). The top and bottom sensors were connected in a Wheatstone bridge configuration to the data acquisition module in order to record resistance variation. The same acquisition module Keithley KUSB-3100 was employed for the purpose of data acquisition during bending tests. Special data amplification and linearization modules were attached to the data acquisition module in order to reduce noise and to amplify the sensor signal. Schematic representation of the data acquisition setup is shown in Figure 27.9.

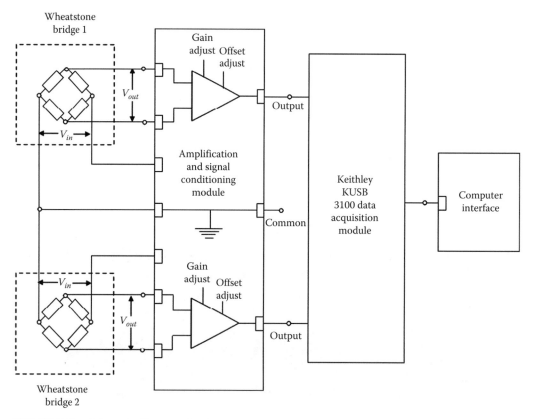

FIGURE 27.9 Schematic diagram of an instrumentation amplifier and a data acquisition module connected to sensors in a Wheatstone bridge configuration.

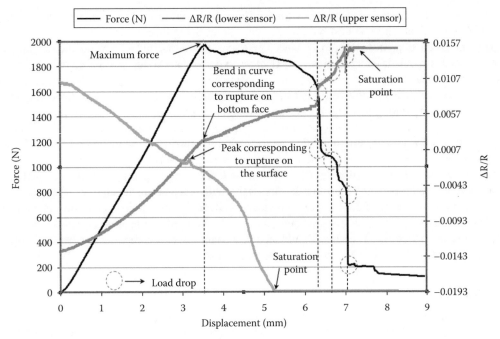

FIGURE 27.10   Force-displacement plot against normalized resistance variation for the two sensors inside the 3D carbon composite specimen tested until fracture at a constant displacement rate of 3.5 mm/min.

The three point bending test was performed until complete fracture of the specimen, at a constant displacement rate of 3.5 mm/min. The resultant force displacement curve plotted against variation in normalized resistance for the two sensors is shown in Figure 27.10.

The analysis of test results plotted in Figure 27.10 shows that the sensors are able to follow the loading and onset of damage in the composite. A small peak in the upper sensor plot is followed by the maximum load in the force-displacement plot. This load coincides with a bend in the lower sensor plot. It might be because the maximum force does not necessarily correspond to rupture in all the composite layers. The maximum compression loading may have been achieved before the maximum tensile loading. The slope of the upper and lower sensor plots generally changes after the rupture of the composite. The upper sensor plot soon afterwards saturates, whereas the lower sensor plot continues to follow the tensile loading of the lower composite part. Each load drop in the force-displacement curve coincides with sharp vertical rise in the lower sensor curve until the composite sample fails completely and the saturation point of upper sensor output is achieved as well.

A crimp in the sensor does not have negative influence over its damage detection capability. This is because of the fact that the sensor is constrained by the tows inside the reinforcement. Moreover, when the resin impregnation takes place, further sensor fixation is affected. Since the sensor adopts the geometry of reinforcement tows, it is expected to deform much like the surrounding tows inside the composite. Therefore, a sensor crimp does not adversely effect its sensitivity and damage detection ability. This conjecture is supported by the plots given in Figure 27.10.

27.6   CONCLUSIONS

Filament substrate-based nanoparticle coated sensors have been shown to detect different mechanical properties of composites, such as their deformation under tensile and flexural loading. These sensors can also be used to register stress–strain history prior to and after the initiation of damage. The propagation of cracks and ultimate fracture events are also detected by the sensors inserted strategically inside the composite. This has been shown by the integration of these sensors as weft tows in specially selected weft positions for tensile and flexural tests. These sensors are not only flexible, lightweight, and low cost but are also capable of being inserted inside the reinforcement during the production phase. Moreover, since these sensors are inserted as weft during weaving on the modified loom, their integration is cost effective, and no separate step is required to be added in the composite manufacturing process. Strategic location of these sensors inside the reinforcement and their sensitivity is imperative for extracting maximum information related to deforming phenomena occurring during loading. For tensile loading, the sensor has been inserted in the middle layer of the reinforcement, whereas for bending, two sensors have been inserted in top and bottom layers. The results presented in this article show that these sensors can be integrated inside the 3D interlock reinforcement to form an "intelligent textile neural network" capable of generating useful information about the composite "health." Nevertheless, the interpretation of information gathered from sensor signals needs to be further developed through data analysis techniques. This task also calls for extensive testing of different reinforcements under different loading conditions. In addition to that, the study of impact of sensor integration inside the reinforcement on composite strength should also be carried out. The study should include a comparison of composite mechanical properties prior to integration and after the integration of a sensor.

REFERENCES

Abry J.K., Choi Y.K., Chateauminois A., Dalloz B., Giraud G., and Salvia M. (2001) In-situ monitoring of damage in CFRP laminates by means of AC and DC measurements. *Composites Science and Technology*, 61: 855–864.

Balageas D. (2006) Introduction to structural health monitoring. In: Balageas D., Fritzen C.P., and Guemes A. (eds.) *Structural Health Monitoring*. London, U.K.: ISTE Ltd., pp. 13–44.

Baurley S.L. (2003) Smart textiles for future intelligent consumer products. In: *Proceedings of the IEE Eurowearable*, Birmingham, U.K., pp. 73–75.

Black S. (2009) Structural health monitoring: Composites get smart. Available at: www.compositesworld.com/articles/structural-health-monitoring-composites-get-smart. (accessed on October 19, 2011.)

Buyukozturk O. and Yu T.Y. (2003) Structural health monitoring and seismic impact assessment. In: *Proceedings of the 4th National Conference on Earthquake Engineering*, Istanbul, Turkey. Available at: web.mit.edu/istgroup/ist/documents/2003_Paper_SHM%20and%20Seismic%20Impact%20Assessment_Istanbul.pdf (accessed on October 19, 2011.)

Chung D.D.L. (2002) Composites get smart. *Materials Today*, 5: 30–35.

Cochrane C., Koncar V., Lewandowski M., and Dufour C. (2007) Design and development of a flexible strain sensor for textile structures based on a conductive polymer composite. *Sensors and Actuators A*, 7: 473–492.

Cristian I., Nauman S., Cochrane C., and Koncar V. (2011) Electro-conductive sensors and heating elements based on conductive polymer composites in woven structures. In: Savvas V. (ed.) *Advances in Modern Woven Fabrics Technology*, Croatia: InTech—Open access, pp. 3–23.

De Baere I., van Paepegem W., and Degrieck J. (2010) Electrical resistance measurement for in situ monitoring of fatigue of carbon fabric composites. *International Journal of Fatigue*, 32: 197–207.

Dharap P., Zhiling L., Nagarajaiah S., and Barrera E.V. (2004) Nanotube film based on single-wall carbon nanotubes for strain sensing. *Nanotechnology*, 15: 379–382.

Farrar C.R. and Worden K. (2007) An introduction to structural health monitoring. *Philosophical Transactions of the Royal Society A*, 365: 303–315.

Fiedler B., Gojny F.H., Wichmann M.H.G. Bauhofer W., and Schulte K. (2004). Can carbon nanotubes be used to sense damage in composites? *Annales de Chimie*, 29: 81–94.

Fortunato E., Correia N., Barquinha P., Pereira L., Goncalves G., and Martins R. (2008) High-performance flexible hybrid field-effect transistors based on cellulose fiber paper. *IEEE Electron Device Letters*, 29: 988–990.

Guerra E.M., Silva G.R., and Mulato M. (2009) Extended gate field effect transistor using V_2O_5 xerogel sensing membrane by sol–gel method. *Solid State Sciences*, 11: 456–460.

Hecht D.S., Liangbing H., and George G. (2007) Electronic properties of carbon nanotube/fabric composites. *Current Applied Physics*, 7: 60–63.

Kaddour A.S., Al-Salehi F.A.R., Al-Hassani S.T.S., and Hinton M.J. (1994) Electrical resistance measurement technique for detecting failure in CFRP materials at high strain rates. *Composites Science and Technology*, 51: 377–385.

Kagan C. (2011) Flexible transistors. Available at: www.technologyreview.com/InfoTech/12259/?a = f (accessed on November 3, 2011.)

Kang I.P., Schulz M.J., Kim J.H., and Shanov V. (2006) A carbon nanotube strain sensor for structural health monitoring. *Smart Materials and Structures*, 15: 737–748.

Katragadda R.B. and Xu Y. (2008) A novel intelligent textile technology based on silicon flexible skins. *Sensors and Actuators A: Physical*, 143: 169–174.

Kim S., Leonhardt S., Zimmermann N., Kranen P., Kensche D., Muller E., and Quix C. (2008) Influence of contact pressure and moisture on the signal quality of a newly developed textile ECG sensor shirt. In: *Proceedings of the 5th International Summer School and Symposium on Medical Devices and Biosensors*, The Chinese University of Hong Kong, HKSAR, China, pp. 256–259.

Koncar V., Kim B., Nebor E.B., and Joppin X. (2004) FICC (floatable intelligent and communicative clothing) project—*The Eighth International Symposium on Wearable Computers*, DoubleTree Crystal City, Arlington, VA.

Kupke M., Schulte K., and Schüler R. (2001) Non-destructive testing of FRP by D.C. and A.C. electrical methods. *Composites Science and Technology*, 61: 837–847.

Lee S.S., Lee J.H., Park I.K., Song S.J., and Choi M.Y. (2006) Structural health monitoring based on electrical impedance of a carbon nanotube neuron. *Key Engineering Materials*, 321–323: 140–145.

Lee J.B. and Subramanian V. (2005) Weave patterned organic transistors on fiber for E-textiles. *IEEE Transactions on Electron Devices*, 62: 269.

Li C. and Chou T.W. (2008) Modeling of damage sensing in fiber composites using carbon nanotube networks. *Composites Science and Technology*, 68: 3373–3379.

Li C., Thostenson E. T., and Chou T. W. (2008). Sensors and actuators based on carbon nanotubes and their composites: A review. *Composites Science and Technology*, 68: 1227–1249.

Lorussi F., Scilingo E.P., Tesconi M., Tognetti A., and De Dossi D. (2005) Strain sensing fabric for hand posture and gesture monitoring. *IEEE Transactions on Information Technology in Biomedicine*, 9: 372–381.

Muto N, Arai Y., Shin S.G., Matsubara H., Yanagida H., Sugita M., and Nakatsuji T. (2001) Hybrid composites with self-diagnosing function for preventing fatal fracture. *Composites Science and Technology*, 61: 875–883.

Nauman S., Cristian I., and Koncar V. (2011a) Simultaneous application of fibrous piezoresistive sensors for compression and traction detection in glass laminate composites. *Sensors (Basel)*, 11: 9478–9498.

Nauman S., Cristian I., and Koncar V. (2012) Intelligent carbon fibre composite based on 3D—Interlock woven reinforcement. *Textile Research Journal*, 82(9): 931–944.

Nauman S., Lapeyronnie P., Cristian I., Boussu F., and Koncar V. (2011b) On line measurement of structural deformations in composites. *IEEE Sensors Journal* 11(6): 1329–1336.

Peratech (2011) QTC science. Available online: http://www.peratech.com/qtcscience.php. (accessed on November 3, 2011.)

Scilingo E. P., Lorussi F., Mazzoldi A., and De Rossi D. (2003) Strain-sensing fabrics for wearable kinaesthetic-like systems. *IEEE Sensors Journal*, 3: 460–467.

Thostenson E.T. and Chou T.W. (2008) Carbon nanotube-based health monitoring of mechanically fastened composite joints. *Composites Science and Technology*, 68: 2557–2561.

Wijesiriwardana R. (2006) Inductive fiber-meshed strain and displacement transducers for respiratory measuring systems and motion capturing systems. *IEEE Sensors Journal*, 6: 571–579.

Zhao H., Zhang Y., Bradford P.D., Zhou Q., Jia Q., Yuan F.G., and Zhu Y. (2010) Carbon nanotube yarn strain sensors. *Nanotechnology*, 21: 1–5.

28 Structural Health Monitoring Based on Piezoelectric Transducers

Analysis and Design Based on the Electromechanical Impedance

Fabricio G. Baptista, Jozue Vieira Filho, and Daniel J. Inman

CONTENTS

28.1 INTRODUCTION

Piezoelectric transducers are widely used in many applications. In recent years, these transducers have been used in a modern and promising application: structural health monitoring (SHM) of various types of structures.

SHM systems are designated to detect, in real-time or in an appropriate time, structural damage. There are both scientific and economic motivations for the use of an SHM system. From the scientific point of view, the monitoring and detection of structural damage mean to achieve a high level of safety. From the economic point of view, systems with this capability allow a significant reduction in maintenance costs.

Currently, the aviation industry is one of most focused fields of application. Although the design and criteria for certification of an aircraft already guarantee a high level of security, an SHM system could significantly reduce the repair and maintenance costs, which represent, according to [1], 27% of the cost of its life cycle. The direct costs related to the repair could be reduced by detecting damage in an early stage. In addition, the indirect costs could be reduced by a lower frequency at which the aircraft would be shut down for maintenance.

Several methods have been developed for damage detection in SHM, such as wave propagation, vibration-based methods, acoustic emission, comparative vacuum, and electromechanical impedance (EMI). Details and references for these and other methods can be found in the literature review done by [2]. Among several methods, EMI technique has been noted as one of the most promising methods for developing SHM systems on a wide variety of structures. This technique uses low-cost, small, and lightweight piezoelectric transducers, being minimally invasive with negligible effects on the mechanical properties of the structure. The most widely used piezoelectric sensors are the Pb-lead zirconate titanate (PZT) ceramics and macro-fiber composite (MFC), which have thickness on the order of a few tenths of millimeters. These characteristics make the EMI technique especially attractive for the monitoring of aircraft structures, which is one of the most prominent application fields nowadays. According to [3], the EMI technique is the only PZT-based technique that has characteristics for the development of a real-time and in-situ SHM system in aircrafts.

This chapter presents a brief literature review regarding the analysis of the EMI technique in aluminum structures, focusing on the impedance measurement, the correct design of PZT transducers, and the appropriate frequency range selection for optimal performance in SHM systems.

28.2 ELECTROMECHANICAL IMPEDANCE TECHNIQUE

28.2.1 BASIC CONCEPT

The basic principle of the EMI technique is based on the piezoelectric effect, which is the property that piezoelectric transducers have to convert mechanical energy into electrical energy (direct effect) and electrical energy into mechanical energy (reverse effect). When a piezoelectric transducer is bonded to the structure to be monitored, there is an interaction between the mechanical impedance of the host structure and the electrical impedance of the transducer. Therefore, changes in the mechanical impedance of the host structure caused by some damage, such as cracks or corrosion, can be evaluated simply by measuring the electrical impedance of the transducer in a suitable frequency range [4–6].

A transducer bonded to a structure or other propagation media can be represented by an equivalent electromechanical model. In the literature, there are many suggestions of equivalent electromechanical models for piezoelectric transducers; a historical review was presented in [7]. Many of these models are based on the Mason's one-dimensional (1-D) model [8]. The Mason's 1-D model can be modified to analyze PZT transducers applied in SHM systems based on the EMI technique. In the EMI technique, it is common to use thin PZT patches with thickness ranging from 0.1 to 0.3 mm. These patches are bonded to the structure of interest using a high-strength adhesive to ensure a good electromechanical coupling, such as superglue and epoxy. Then, the transducer is excited by an alternating voltage as shown in Figure 28.1.

In Figure 28.1, the PZT transducer is square with side ℓ and thickness t, and it is bonded in a host structure with cross-section area A_S. An external sinusoidal signal with amplitude U_m is used to excite the PZT through the top and bottom electrodes, both with surface area A_E, and the response is a current with intensity I. If the PZT patch has small thickness, a wave propagating at velocity v_a

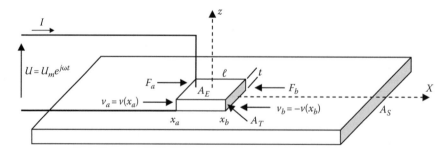

FIGURE 28.1 Basic principle of the electromechanical impedance technique; a square PZT patch bonded to the structure to be monitored. (From Baptista, F. and Vieira Filho, J., Optimal frequency range selection for PZT transducers in impedance based SHM systems, *IEEE Sensors Journal*, 10, 1298, 2010. With permission. Copyright 2010 IEEE.)

in the host structure reaches the patch side with coordinate x_a and surface A_T causing the force F_a. Similarly, in the side of coordinate x_b, there is a force F_b due to the incoming wave propagating at velocity v_b. If the applied voltage U_m is low in the order of a few volts and hence the resultant electric field is also low, the piezoelectric effect is predominantly linear and the non-linearities can be neglected. Furthermore, considering the class $6\ mm$ for PZT ceramics [9], and for 1-D assumption, where the main propagation direction is considered along the length direction (perpendicular to the cross-section area A_S) of the host structure, as shown in Figure 28.1, the electrical impedance of the PZT transducer, Z_E, can be determined by

$$Z_E = \frac{U}{I} = \frac{1}{j\omega C_0} \left\| jZ_T \left(\frac{s_{11}}{d_{31}\ell} \right)^2 \left[\frac{1}{2} \tan\left(\frac{k\ell}{2} \right) - \frac{1}{\sin(k\ell)} + \frac{Z_S}{j2Z_T} \right] \right. \tag{28.1}$$

where
  the symbol ‖ represents a parallel association
  d_{31} is the piezoelectric constant
  s_{11} is the compliance at constant electric field
  C_0 is the static capacitance of the transducer
  ω is the angular frequency
  Z_T is the mechanical impedance of the transducer
  k is the wave number
  $j = \sqrt{-1}$
  Z_S is the mechanical impedance of the host structure [10]

According to Equation 28.1, there is a relationship between the mechanical impedance Z_S of the host structure and the electrical impedance Z_E of the PZT transducer. Therefore, structural damage can be identified by analyzing the variations in the mechanical impedance of the structure through the electrical impedance of the transducer, which is easier to measure.

28.2.2 DAMAGE DETECTION

Typically, the characterization (identification and quantification) of damage is performed through metric indices by comparing two impedance signatures, where one of these is acquired when the structure is considered healthy and used as reference, commonly called the baseline. Various indices have been proposed in the literature, but the most widely used are the root mean square deviation (RMSD) and the correlation coefficient deviation metric (CCDM).

The RMSD index is based on the Euclidean norm [11], which is more sensitive to variations in amplitude of the electrical impedance signature. On the other hand, the CCDM index [12] is less sensitive to variations in amplitude of the electrical impedance, but is more sensitive to the frequency shifts in the impedance signature caused by damage.

Therefore, the basic idea in the EMI technique is to detect structural damage by comparing the baseline signature and the updated signature using these indices. The electrical impedance signatures from PZT transducers should be acquired and stored using appropriate measurement systems. The next section presents the main electrical impedance measurement methods used in SHM systems based on the EMI technique.

28.3 ELECTRICAL IMPEDANCE MEASUREMENT

Although the EMI technique is simple and uses small and low-cost transducers, the measurement of the electrical impedance, which is the basic stage of the technique, is usually performed by commercial impedance analyzers such as the 4192A and 4294A from Hewlett Packard/Agilent, for example. Besides the high costs, these instruments are slow, making it difficult to use the technique in real-world applications.

Many researchers have proposed alternative systems to measure the electrical impedance in SHM applications. One of the pioneering methods was presented in [13,14], which uses a spectrum analyzer and a simple and inexpensive auxiliary circuit.

The impedance analyzer proposed in [15] uses only a resistor as an auxiliary circuit, and the signal response acquisition is performed by a data acquisition (DAQ) device, controlled by the software LabVIEW from National Instruments. It also seems that it is more accurate than the one suggested in [13,14]. However, the excitation signal is provided by an external function generator that needs an additional general purpose interface bus (GPIB) card. Besides increasing the cost, these devices make the system less versatile.

Currently, the interest for low-cost, low-power, and wireless impedance analyzers has increased. The system proposed in [16] is based on a digital signal processor (DSP) and uses a new algorithm that eliminates the use of analog-to-digital converters and digital-to-analog converters. However, the range and the frequency resolution are limited by the small memory space of DSP. As a consequence, a conventional and expensive impedance analyzer is still necessary to determine the frequency range sensitive to structural damage. Recently, analog devices developed a miniaturized high precision impedance converter integrated in a single chip (AD5933). This chip has been used in SHM to develop compact and low-cost measurement systems. These new systems support wireless communication and several sensors through analog multiplexer and can process data locally [17–19].

An accurate, versatile, and easy to assemble measurement system was proposed in [20]. The system diagram is shown in Figure 28.2. The basic principle of operation is based on the frequency response function (FRF) obtained by the discrete Fourier transform in regards to the excitation and response signals from the PZT transducer. Using the FRF and considering the detailed parameters of the analyzed circuit, it is possible to obtain the accurate PZT impedance. The hardware of the system consists of a low-cost DAQ device and a personal microcomputer laptop. The excitation signal is generated by software, which allows the choice of a wide variety of signals. The results presented in this chapter were obtained with the chirp signal.

In Figure 28.2, the resistor R_S should be chosen according to the maximum voltage desired in the PZT patch. It is recommended that this voltage may not be very high because the PZT material tends to lose the linearity for intense electric fields [21]. For example, using a chirp signal with 5 V of amplitude, it was verified that a precision resistor of 10 kΩ ensures a voltage below 1 V for a typical PZT transducer at frequencies above 10 kHz.

For example, Figure 28.3 shows the comparison between two electrical impedance signatures for a PZT patch bonded to an aluminum beam obtained using the measurement system proposed

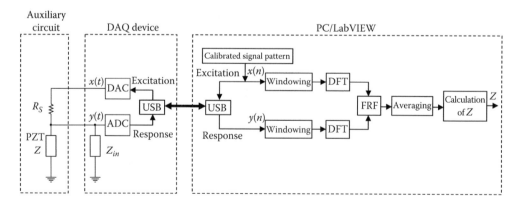

FIGURE 28.2 An alternative system for the measurement of the electrical impedance of PZT patches. (From Baptista, F. and Vieira Filho, J., A new impedance measurement system for PZT-based structural health monitoring, *IEEE Trans. Instrum. Meas.*, 58, 3603, 2009. With permission. Copyright 2010 IEEE.)

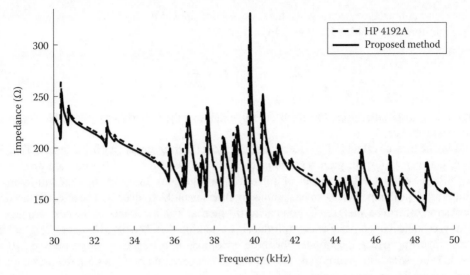

FIGURE 28.3   Comparison between the electrical impedance signatures of a PZT patch obtained using a conventional impedance analyzer and the alternative system. (From Baptista, F. and Vieira Filho, Piezoelectric transducers applied in structural health monitoring: Data acquisition and virtual instrumentation for electro-mechanical impedance technique, *Advances in Piezoelectric Transducers*, F. Ebrahimi, (Ed), Chapter 6, 121, 2011, InTech Open Access.)

in [20] and a conventional impedance analyzer 4192A from Hewlett Packard for a frequency range from 30 to 50 kHz. It can be verified that the shapes of the electrical impedance signatures obtained using the alternative method and the conventional impedance analyzer are very similar. Besides the excellent accuracy, this system provides fast measurements, versatility, and low-cost compared to the conventional impedance analyzers. The methodology can be easily integrated into a system based on microcontroller or DSP together with a wireless communication system allowing the EMI technique to be used in real-time monitoring.

In addition to an appropriate measurement system, the EMI technique faces some problems for its practical application, such as the transducer loading effect, the design of piezoelectric patches, and the correct frequency range selection most sensitive to structural damage. These problems are discussed in the following sections.

28.4   TRANSDUCER LOADING EFFECT

Although many studies indicate that the EMI technique is efficient [22], some practical considerations on the performance of the technique in real applications, especially in large structures, are still desirable.

A practical issue that has not been considered in the application of the EMI technique is the loading of the PZT transducer due to the propagation media, i.e., the structure to be monitored. This effect is a well-known issue in the literature and has been investigated, for example, for piezoelectric transducers loaded by the backing, liquid media [10], and electrodes.

The transducer loading effect in the EMI technique was analyzed in [23]. According to the expression in (28.1), the electrical impedance of the transducer depends on the mechanical imped-ance of the host structure. The electrical impedance can vary only between two values depending on the size of the structure. The minimum electrical impedance is due to the transducer itself, i.e., there is no host structure. So, as a consequence $Z_S = 0$ and from (28.1) we have

$$Z_{E,min} = \frac{1}{j\omega C_0} \left\| jZ_T \left(\frac{s_{11}}{d_{31}\ell} \right)^2 \left[\frac{1}{2} \tan\left(\frac{k\ell}{2} \right) - \frac{1}{\sin(k\ell)} \right] \right. \quad (28.2)$$

The maximum impedance occurs when the cross section area of the host structure is very large and can be determined by making $Z_S \to \infty$ in Equation 28.1:

$$Z_{E,\max} = \lim_{Z_S \to \infty} Z_E = \frac{1}{j\omega C_0} \qquad (28.3)$$

Therefore, for large structures, the electrical impedance of the transducer tends to be closer to its capacitive reactance.

In [23], the loading effect in PTZ piezoceramics from Piezo Systems was investigated. The properties of a square PSI-5H4E piezoceramic patch of 20 mm × 20 mm × 0.267 mm are presented in Table 28.1. Substituting all the values of Table 28.1 in Equations 28.2 and 28.3, the minimum and maximum electrical impedance can be obtained in function of the frequency. The effect of the size of the host structure can be analyzed by comparing the mechanical impedance of the host structure, Z_S, with the mechanical impedance of the transducer, Z_T, making $Z_S = 2Z_T$, $5Z_T$, $10Z_T$, and $20Z_T$ in (28.1). The electrical impedance magnitudes for these conditions in a frequency range between 10 kHz and 30 kHz are shown in Figure 28.4. The inferior and superior dashed lines are the minimum and maximum impedance, respectively.

TABLE 28.1

Properties of the PZT Ceramic Used in the Experimental Tests

Symbol	Quantity	Value
S_{11}	Compliance	$16.1\ 10^{-12}\ \text{m}^2/\text{N}$
d_{31}	Piezoelectric constant	$-320\ 10^{-12}\ \text{m/V}$
ρ_T	Mass density	$7800\ \text{kg/m}^3$
$\varepsilon_{33}/\varepsilon_0$	Relative permittivity	3800
D	Thickness	0.267 mm

The symbol ε_0 is the permittivity in vacuum.

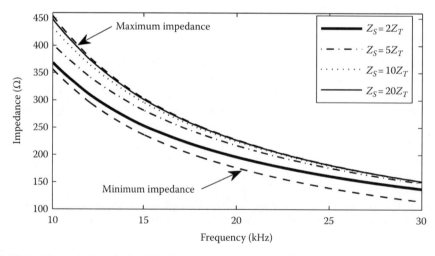

FIGURE 28.4 Theoretical analysis of the transducer loading effect. (From Baptista, F. and Vieira Filho, J., Transducer loading effect on the performance of PZT-based SHM systems, *IEEE Trans. Ultrason. Ferroelectr. Freq. Control.* 57, 935, 2010. With permission. Copyright 2010 IEEE.)

These theoretical results show that the electrical impedance of the transducer increases directed to the upper limit as the mechanical impedance of the host structure, which is proportional to its cross-section area, increases. This is the PZT transducer loading effect due to the size of the host structure. Therefore, if the host structure has a large cross-section area and the transducer operates near to its upper limit, any change in mechanical impedance Z_S of the structure due to some damage causes a very small variation in electrical impedance Z_E of the transducer. Consequently, the values of the metric indexes, such as RMSD and CCDM, will be low and the damage will be difficult to be detected. Therefore, the sensitivity of the SHM system for damage detection should exhibit a considerable decreasing trend as the ratio Z_S/Z_T increases.

The transducer loading effect was also analyzed experimentally in [23]. Experiments were carried out on structures with different cross-section areas and mechanical impedances. Two sets of specimens were used. First, four aluminum beams with fixed length and thickness of 500 and 2 mm, respectively, and widths of 30, 60, 120, and 240 mm were tested. Then, tests were performed on two aluminum plates with fixed length and width of 500 and 300 mm, respectively, and thicknesses of 2 and 16 mm. A square PSI-5H4E piezoceramic patch of 20 mm × 20 mm × 0.267 mm was bonded at distance of 10 mm from the end of each specimen using "Superglue." To simulate structural damages, a steel screw-nut of 4 mm diameter, 2 mm thickness, and 2 g was bonded to each structure at distances of 1, 10, 20, 30, and 40 cm from the patch. The electrical impedance of the transducers was measured using the system developed in [20]. For each specimen, the impedance in healthy condition (without the screw-nut) was compared with the impedance in damaged conditions by employing the calculation of the RMSD and CCDM indices. All measures were taken with the structures in free–free set, i.e., with the two ends suspended by elastics and at room temperature.

Figure 28.5 shows the RMSD indices calculated using the real and imaginary part of the electrical impedance. According to Figure 28.5, the sensitivity of the system to detect structural damage significantly decreased as the cross-section area of the host structure increased by the change of its width or thickness. With some exceptions, the magnitude of the RMSD index decreased as the width of the beam and the thickness of the plate increased, independently of the distance of the transducer from the damage or of the part of the impedance used to calculate the index. Similar results were obtained for the CCDM index [23].

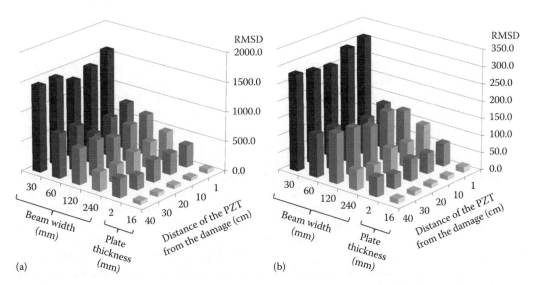

FIGURE 28.5 Decrease observed in the RMSD index due to the transducer loading effect using (a) the real part and (b) the imaginary part of the electrical impedance. (From Baptista, F. and Vieira Filho, J., Transducer loading effect on the performance of PZT-based SHM systems, *IEEE Trans. Ultrason. Ferroelectr. Freq. Control.* 57, 938, 2010. With permission. Copyright 2009 IEEE.)

In general, the experimental results confirm that the sensitivity of the transducer for damage detection significantly decreases as the mechanical impedance of the host structure, which is proportional to its cross-section area, becomes very large compared to the mechanical impedance of the transducer. For structures with high Z_S/Z_T ratio, the values of the metric indices tend to be very low, approximating to the indices obtained for the structure in healthy condition; in this situation, it is difficult to establish a threshold and to diagnose the structure as healthy or damaged. Therefore, there is a significant decline in the sensitivity.

The analysis of the transducer loading effect can be applied for the correct design of PZT patches, as investigated in the next section.

28.5 DESIGN OF PZT PATCHES

Although the feasibility of the EMI technique has been examined for lab-sized and complex structures [22,24], there are few results reported in the literature regarding the correct size of the piezoelectric transducer in relation to the host structure. As mentioned before, two piezoelectric transducers are commonly used in the EMI technique: PZT ceramic patches and MFC patches. The aim of this section is to present a methodology for the correct sizing of PZT patches, which was investigated in [25].

According to the literature, the PZT transducers must be small enough not to be intrusive in the host structure [26]. Generally, it is recommended that the PZT patches have size ranging from 5 to 15 mm and thickness from 0.1 to 0.3 mm [27]. It is also important to consider the relationship between the size of the transducer and the bond layer. The issues related to bonding layer are expected to be more significant as the size of the transducer increases.

It is well known that, if the host structure is thin enough to act as a plate, the electromechanical coupling with the PZT transducer is most efficient when its size matches a half-wavelength in the propagation medium. The half-wavelength matching between the transducer and the propagation medium was applied to the EMI technique in [28], where the authors analyzed the influence of the size and bonding position of PZT patches on the longitudinal mode impedance responses for damage detection in truss structures. However, despite the good results, this methodology is appropriate for monitoring specific natural frequencies of the host structure and requires the transducer bonded in positions close to the nodes of specific mode shapes corresponding to these natural frequencies. These parameters are usually obtained by finite element analysis.

In the EMI technique, the impedance is usually analyzed in a reasonably wide frequency range, which contains several natural frequencies, and it may be necessary to bond the transducer in positions far from specific nodes, where damage is most likely to occur (i.e., hot spots). Moreover, the sensitivity of the transducer for damage detection relative to the size of the host structure has not been considered in the literature. In the previous section, it was shown that the sensitivity of PZT transducers significantly decreases as the ratio between the mechanical impedance of the host structure and the mechanical impedance of the transducer increases due to the loading effect. Therefore, it is important to consider the transducer loading effect on its correct sizing.

The correct sizing of the PZT patches for optimal sensitivity for damage detection was theoretically and experimentally investigated in [25] based on the transducer loading effect. In the proposed methodology, the derivative of the electrical impedance of the transducer in relation to the mechanical impedance of the host structure is used as a parameter for determining the sensitivity of the transducer for damage detection. In addition, the influence of the intrinsic capacitance of the PZT patches on the sensitivity for damage detection and a parallel arrangement of various patches were analyzed. The feasibility of the methodology was verified through experiments on specimens of different sizes.

As discussed in Section 28.2, the basic principle of the EMI technique is to identify small variations in the mechanical impedance of the host structure caused by damage through the measurement of the electrical impedance of the transducer. Thus, the sensitivity of the PZT transducer for damage detection can be estimated by analyzing the derivative of the modulus of the electrical impedance Z_E in Equation 28.1 in relation to the mechanical impedance Z_S of the host structure, as follows:

$$\Delta Z_E = \frac{\partial}{\partial Z_S}\left| \frac{1}{j\omega C_0}\left\| jZ_T\left(\frac{s_{11}}{d_{31}\cdot\ell}\right)^2\left[\frac{1}{2}\tan\left(\frac{k\cdot\ell}{2}\right) - \frac{1}{\sin(k\cdot\ell)} + \frac{Z_S}{j2Z_T}\right]\right\| \right| \tag{28.4}$$

Therefore, in order to obtain good sensitivity, the PZT patches should be designed so that Equation 28.4 provides high values, i.e., the electrical impedance Z_E must have a large variation for small changes in the mechanical impedance Z_S of the host structure.

From the values shown in Table 28.1, the electrical impedance of the PZT patches can be analytically determined. Since the goal is to find the appropriate size of the patch according to the size of the host structure, it is useful to examine the variation in the electrical impedance as a function of the Z_S/Z_T ratio. This analysis is applicable because the mechanical impedances Z_T and Z_S are directly related to the cross-sectional area, i.e., the sizes of the patch and the host structure, respectively.

As an example, the magnitude of the electrical impedance and its derivative for a PZT patch with side (ℓ) of 10 mm and thickness (t) of 0.267 mm operating at a frequency of 10 kHz is shown in Figure 28.6. The electrical impedance is normalized between the values 0 and 1, which correspond to the minimum and the maximum impedance, respectively. Observing Figure 28.6, the electrical impedance has a maximum variation when the Z_S/Z_T ratio is about 6. In this condition, the transducer

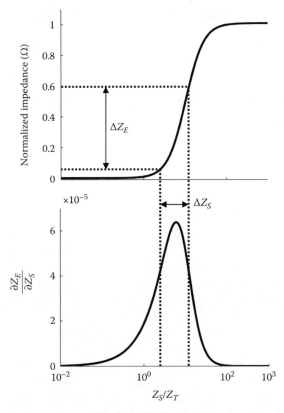

FIGURE 28.6 The electrical impedance and its derivative in relation to the mechanical impedance of the host structure for a PZT patch with size (ℓ) of 10 mm operating at 10 kHz. (From Baptista, F., Vieira Filho, J. and Inman, D., Sizing PZT transducers in impedance-based structural health monitoring, *IEEE Sensors Journal*, 11, 1407, 2011. With permission. Copyright 2009 IEEE.)

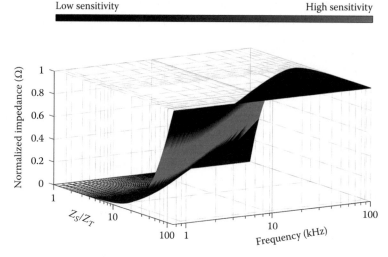

FIGURE 28.7 The electrical impedance of a PZT patch with size (ℓ) of 10 mm as a function of the frequency and the Z_S/Z_T ratio. The light areas indicate good sensitivity for detecting damage. (From Baptista, F., Vieira Filho, J. and Inman, D., Sizing PZT transducers in impedance-based structural health monitoring, *IEEE Sensors Journal*, 11, 1408, 2011. With permission. Copyright 2010 IEEE.)

has optimal sensitivity and small variations ΔZ_S in the mechanical impedance of the host structure caused by damage can be efficiently detected by corresponding changes ΔZ_E in the electrical impedance.

The analysis of the sensitivity of the transducer as a function of frequency is also important. Figure 28.7 shows the behavior of the electrical impedance of a PZT patch with size (ℓ) of 10 mm as a function of the frequency and the Z_S/Z_T ratio. Once again, the impedance is normalized between 0 and 1. The light areas in Figure 28.7 indicate where a reasonable sensitivity for damage detection is expected. In dark areas, the variations in the electrical impedance are low and the sensitivity of the PZT patch should decrease. In addition, it is seen that the Z_S/Z_T ratio should be lower for higher frequencies.

An experimental investigation was carried out in [25] to confirm this conjecture. Aluminum specimens were analyzed, and the experimental results are in agreement with the theoretical analysis, confirming that this methodology can be an important reference for the correct sizing of PZT patches.

28.6 FREQUENCY RANGE SELECTION

Besides an appropriate measurement system and the analyses of the transducer loading effect to the correct design of PZT patches, it is important to select the suitable frequency range in that the electrical impedance should be measured and the metric indices calculated. This section presents a methodology for the optimal frequency range selection, which was proposed in [29].

The frequency range selection is usually determined by trial and error methods. In [3], for example, the suitable frequency range of 20–40 kHz was determined searching for frequency regions that offered good repeatability under constant conditions, i.e., undamaged cases. In [30], the frequency range of 0–50 kHz was chosen for observing the resonance peaks in the measured impedance signatures on two aluminum plates. Others research studies have proposed more efficient methodologies for frequency range selection. Peairs et al. [31] analyzed the resonance frequencies of the transducer before its installation in the host structure and the variation in undamaged measurements was statically compared to the amount of change in the measurement upon various levels of damage. In [32], the data of structures of steel, aluminum, and concrete were acquired in a wide frequency range. Later, statistical indexes were used to evaluate and compare the sensitivity of various narrower bands within the wide range of frequency in which data were acquired.

The methodology proposed in [29] uses the electromechanical relationship in (28.1) to analyze the sensitivity of the PZT transducer in function of the frequency and mechanical impedance of the host structure. As mentioned before, in SHM systems based on the EMI technique, the damage detection is performed by analyzing the variations in magnitude or shifts in frequency of the electrical impedance of the transducer due to the corresponding change in the mechanical impedance of the host structure. For a general assessment of the sensitivity of the transducer, the changes in the electrical impedance can be estimated assuming a small variation Δ in the mechanical impedance Z_S in Equation 28.1 due to a hypothetical damage. The electrical impedance after damage ($Z_{E,D}$) is given by

$$Z_{E,D} = \frac{1}{j\omega C_0}\left\|jZ_T\left(\frac{s_{11}}{d_{31}\ell}\right)^2\left[\frac{1}{2}\tan\left(\frac{k\ell}{2}\right) - \frac{1}{\sin(k\ell)} + (1+\Delta)\frac{Z_S}{j2Z_T}\right]\right\| \quad (28.5)$$

The transducer sensitivity is obtained by comparing the electrical impedance before damage in Equation 28.1 with the impedance after the damage in Equation 28.5 as follows:

$$\eta = 100\frac{\left\||Z_{E,D}| - |Z_E|\right\|}{|Z_E|} \quad (28.6)$$

where η gives the percentage variation of the electrical impedance of the transducer due to a change Δ in the mechanical impedance of the host structure caused by damage. According to Equations 28.1 and 28.5, the sensitivity η depends on frequency, mechanical impedance Z_S of the host structure and its variation Δ (related to the size of the damage), and the characteristics of the PZT transducer.

Considering a variation $\Delta = 0.05$ (5%) and placing the mechanical impedance Z_S of the host structure in function to the mechanical impedance Z_T of the transducer by the ratio Z_S/Z_T from 1 to 100, the sensitivity η can be calculated for a PZT patch of 20 mm × 20 mm × 0.267 mm using Equation 28.6 and substituting the values of Table 28.1 in Equations 28.1 and 28.5. The sensitivity in a frequency range between 0 and 300 kHz is shown in Figure 28.8.

According to Figure 28.8, the sensitivity of the PZT transducer decreases as the frequency increases or the mechanical impedance of the host structure becomes large compared to mechanical impedance of the transducer. Moreover, throughout the frequency range, there are maximum

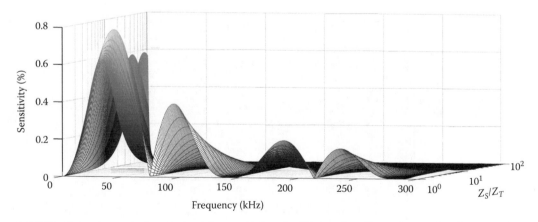

FIGURE 28.8 Theoretical analysis of the sensitivity of a PZT patch to detect damage for the appropriate frequency range selection. (From Baptista, F. and Vieira Filho, J., Optimal frequency range selection for PZT transducers in impedance based SHM systems, *IEEE Sensors Journal*, 10, 1300, 2010. With permission. Copyright 2010 IEEE.)

and minimum local points. These points can be determined graphically or calculated by solving the following equation for the angular frequency ω:

$$\frac{\partial}{\partial\omega} \frac{\left\| \left| Z_{E,D}(\omega) \right| - \left| Z_E(\omega) \right| \right\|}{\left| Z_E(\omega) \right|} = 0 \tag{28.7}$$

It is expected that the damage detection is more efficient in frequency ranges around the maximum points and deficient near to the minimum points.

There is a concentration of maximum global points in the frequency range of approximately 0–50 kHz. In this range, the sensitivity of the PZT transducer is optimal and almost constant in function of the mechanical impedance of the host structure. This theoretical result explains the good repeatability between the measurements and efficient damage detection in the frequency range of 20–40 and 0–50 kHz obtained in [3] and [30], respectively. Although the sensitivity is higher at frequencies below 50 kHz, this range is more susceptible to external disturbances; moreover, higher frequencies allow the detection of smaller damages. Therefore, appropriate ranges around maximum local points at higher frequencies must be found.

In order to verify the usefulness of the sensitivity of PZT transducers given by Equation 28.6 in selecting the appropriate frequency range to detect structural damages, tests were carried out on structures with different sizes and mechanical impedances [29].

Figure 28.9 shows the comparison between the theoretical sensitivity of the PZT patch bonded to an aluminum beam of 500 mm × 30 mm × 2 mm and an adapted version of RMSD metric index obtained for structural damage simulated at a distance of 100 mm from the patch. Observing Figure 28.9, the magnitude of the metric index varies according to the sensitivity of the transducer. It is evident that the higher magnitude of the index occurs at frequencies near to the maximum point, which is 7.1 kHz. On the other hand, the index is approximately zero at frequencies around 74 kHz, which is the minimum point. Although the selection of the appropriate frequency range for damage detection depends on the characteristics of each structure and type of damage, the results show that the range should be selected around the frequencies where the sensibility of the transducer is optimal and the ranges around the minimum points should be avoided.

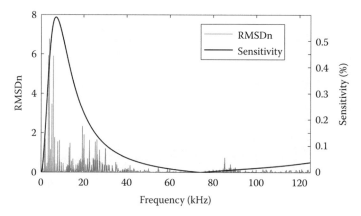

FIGURE 28.9 Comparison between the theoretical sensitivity of the PZT patch and experimental metric indices. (From Baptista, F. and Vieira Filho, J., Optimal frequency range selection for PZT transducers in impedance based SHM systems, *IEEE Sensors Journal*, 10, 1301, 2010. With permission. Copyright 2010 IEEE.)

28.7   CONCLUSIONS

In this chapter, we have presented the basic principle of SHM based on the EMI technique and some issues related to its applications in real-world structures, such as the electrical impedance measurement, transducer loading effect, design of PZT patches, and the frequency range selection.

Methodologies that were recently proposed in the literature to overcome these practical issues were succinctly presented along with some theoretical and experimental results. We hope that these methodologies can be good references in the application of the EMI technique or to lead future research.

REFERENCES

1. Kessler S. S., Spearing S. M., Atala M. J., Cesnik C. E. S., and Soutis C. Damage detection in composite materials using frequency response methods. *Composites Part B: Engineering, Oxford*, 33(1), 87–95, 2002.
2. Sohn H., Farrar C. R., Hemez F. M., Shunk D. D., Stinemates D. W., Nadler B. R., and Czarnecki J. J. A review of structural health monitoring literature: 1996–2001. Los Alamos National Laboratory Report, LA-13976-MS, 2004. Available at: http://www.lanl.gov (Accessed: February 7, 2007).
3. Gyekenyesi A. L., Martin R. E., Sawicki J. T., and Baaklini G. Y. *Damage Assessment of Aerospace Structural Components by Impedance Based Health Monitoring*. Hanover: NASA Technical Memorandum TM— 2005–213579, GLTRS, 2005. Available at: http://gltrs.grc.nasa.gov/ (Accessed: April 3, 2007).
4. Cawley P. The impedance method of non-destructive inspection, *NDT International, Ann Arbor*, 17(2), 59–65, 1984.
5. Giurgiutiu V. and Rogers C. A. Electro-mechanical (E/M) impedance method for structural health monitoring and nondestructive evaluation, in: *Proceedings of the International Workshop on Structural Health Monitoring*, Stanford, CA, pp. 433–444, 1997.
6. Park G., Sohn H., Farrar C., and Inman D. J. Overview of piezoelectric impedance-based health monitoring and path forward, *The Shock and Vibration Digest, Thousand Oaks, CA*, 35(6), 451–463, 2003.
7. Ballato A. Modeling piezoelectric and piezomagnetic devices and structures via equivalent networks, *IEEE Transactions on Ultrasonics, Ferroelectrics, and Frequency Control*, 48(5), 1189–1240, May 2001.
8. Mason W. P. *Electromechanical Transducers and Wave Filters*, 2nd edn., Princeton, NJ: Van Nostrand-Reinhold, 1948.
9. *IEEE Standard on Piezoelectricity*, IEEE Standard 176, 1987.
10. Kossoff G., The effects of backing and matching on the performance of piezoelectric ceramic transducers, *IEEE Transactions on Sonics and Ultrasonics*, SU-13(1), 20–30, Jan. 1966.
11. Giurgiutiu V. and Rogers C. A. Recent advancements in the electro-mechanical (E/M) impedance method for structural health monitoring and NDE, in: *Proceedings of the 5th Annual International Symposium on Smart Structures and Materials*, San Diego, CA, pp. 536–547, 1998.
12. Marqui C. R., Bueno D. D., Baptista F. G., Vieira Filho J., Santos R. B., and Lopes Junior V. External disturbance effect in damage detection using electrical impedance, in: *Proceedings of International Modal Analysis Conference*, Orlando, FL, paper 286, 2008.
13. Peairs D. M., Park G., and Inman D. J. Improving accessibility of the impedance-based structural health monitoring method, *Journal of Intelligent Material Systems and Structures*, 15(2), 129–139, February 2004.
14. Peairs D. M., Park G., and Inman D. J. Low cost impedance monitoring using smart materials, in: *Proceeding of the First European Workshop on Structural Health Monitoring*, Ecole Normale Superieure, Cachan (Paris), France, July 10–12, 442–449, 2002.
15. Xu B. and Giurgiutiu V., A low-cost and field portable electromechanical (E/M) impedance analyzer for active structural health monitoring, in: *5th International Workshop on Structural Health Monitoring*, Stanford University, Stanford, CA, September 15–17, 634–644, 2005.
16. Kim J., Grisso B. L., Ha D. S., and Inman D. J., A system-on-board approach for impedance-based structural health monitoring, in: *Sensors and Smart Structures Technologies for Civil, Mechanical, and Aerospace Systems, Proceedings of SPIE*, Vol. 6529, San Diego, CA, March 18, 652900, 2007.
17. Overly T. G., Park G., Farinholt K. M., and Farrar C. R. Development of an extremely compact impedance-based wireless sensing device. *Smart Materials and Structures* 2008; 17(6): 065011. DOI: 10.1088/0964-1726/17/6/065011.

18. Min J., Park S., Yun C. B., and Song B. Development of multi-functional wireless impedance sensor nodes for structural health monitoring, in: *Proceedings of SPIE Sensors and Smart Structures Technologies for Civil, Mechanical, and Aerospace Systems 2010*, San Diego, CA, 2010, Vol. 7647, 764728-1–764728-8. DOI: 10.1117/12.847458.
19. Park S., Shin H. H., and Yun C. B. Wireless impedance sensor nodes for functions of structural damage identification and sensor self-diagnosis. *Smart Materials and Structures*, 18(5), 055001, 2009. DOI: 10.1088/0964-1726/18/5/055001.
20. Baptista F. G. and Vieira Filho J. A new impedance measurement system for PZT based structural health monitoring. *IEEE Transactions on Instrumentation and Measurement*, 58(10), 3602–3608, Oct. 2009.
21. Baptista F. G., Vieira Filho J., and Inman D. J. Influence of the excitation signal on impedance-based structural health monitoring. *Journal of Intelligent Material Systems and Structures*, 21, 1409–1416, 2010. DOI: 10.1177/1045389X10385032.
22. Park G., Cudney H. H., and Inman D. J., Feasibility of using impedance-based damage assessment for pipeline structures, *Earthquake Engineering Structural Dynamics*, 30(10), 1463–1474, 2001.
23. Baptista F. G. and Vieira Filho J., Transducer loading effect on the performance of PZT-based SHM systems, *IEEE Transactions on Ultrasonics, Ferroelectrics, and Frequency Control*, 57(4), 933–941, Apr., 2010.
24. Park G., Cudney H. H., and Inman D. J., Impedance-based health monitoring of civil structural components, *Journal of Infrastructure Systems*, 6(4), 153–160, 2000.
25. Baptista F. G., Vieira Filho J., and Inman D. J., Sizing PZT transducers in impedance-based structural health monitoring, *IEEE Sensors Journal*, 11(6), 1405–1414, 2011.
26. Park G. and Inman D. J. Structural health monitoring using piezoelectric impedance measurements, *Philosophical Transactions. Series A, Mathematical, Physical, and Engineering Sciences*, 365, 373–392, 2007.
27. Yan W. and Chen W. Q. Structural health monitoring using high-frequency electromechanical impedance signatures, *Advances in Civil Engineering*, 2010, 11 p. 2010.
28. Liu X. and Jiang Z. Design of a PZT patch for measuring longitudinal mode impedance in the assessment of truss structure damage, *Smart Materials and Structures*, 18(12), 125017, 2009.
29. Baptista F. G. and Vieira Filho J. Optimal frequency range selection for PZT transducers in impedance-based SHM systems, *IEEE Sensors Journal*, 10(8), 1297–1303, August 2010.
30. Yang Y., Liu H., Annamdas V. G. M., and Soh C. K. Monitoring damage propagation using PZT impedance transducers, *Smart Materials and Structures*, 18(4), 045003, 2009.
31. Peairs D. M., Tarazaga P. A., and Inman D. J. Frequency range selection for impedance-based structural health monitoring, *Journal of Vibration and Acoustics*, 129(6), 701–719, 2007.
32. Annamdas V. G. M. and Rizzo P. Influence of the excitation frequency in the electromechanical impedance method for SHM applications, in: *Proceedings in Smart Sensor Phenomena, Technology, Networks, and Systems*, San Diego, CA, Vol. 7293, 72930V, 2009.

29 Microwave Sensors for Non-Invasive Monitoring of Industrial Processes

B. García-Baños, Jose M. Catalá-Civera,
Antoni J. Canós, and Felipe L. Peñaranda-Foix

CONTENTS

The rapid development of industrial automation processes has created a growing need for improved sensors for process monitoring and control. With the introduction of ISO 9000, fast and accurate measurements of dielectric properties of materials have grown to a great importance in different applications of industry and agricultural engineering.

Microwave sensors, which allow determining the dielectric properties of materials, have found industrial and scientific applications in the non-destructive testing of materials, identification of surface defects, spectroscopy, medical diagnosis, etc. Their applicability in industrial processes monitoring tasks has allowed new control functionalities, which directly improve the efficiency in the use of resources, and thus, the overall competitiveness.

29.1   DIELECTRIC PROPERTIES OF MATERIALS

29.1.1   Interaction of Microwaves and Matter

When an electric field is applied across a dielectric material, the atomic and molecular charges in the dielectric are displaced from their equilibrium positions and the material is said to be polarized. Dielectric analysis involves the determination of this polarization in materials subjected to a time varying electric field.

The main dielectric parameter, the permittivity, represents the measurement of the maximum dipolar polarization that can be attained by the material under specific conditions (temperature, chemical state, etc.). The permittivity is a frequency-dependent complex number with the following expression:

$$\varepsilon(f) = \varepsilon_0 \left(\varepsilon_r'(f) - j\varepsilon_r''(f) \right) \qquad (29.1)$$

where
 f is the frequency
 ε_0 is a constant, which represents the vacuum permittivity
 the real part $\varepsilon_r'(f)$ is known as the relative dielectric constant, which characterizes the material's ability to store and release electromagnetic energy
 the imaginary part $\varepsilon_r''(f)$ is known as the relative loss factor, which characterizes the material's ability to absorb (attenuate) electromagnetic energy to create heat

Electromagnetic fields and waves are very sensitive to the dielectric properties of the material in which they exist. Microwave sensors are based on the fact that the interaction between microwaves and the medium of propagation is completely determined by the medium dielectric properties. By monitoring the electromagnetic fields/waves that are interacting with the material, it is possible to simultaneously and independently determine the dielectric parameters ($\varepsilon_r'(f)$ and $\varepsilon_r''(f)$), and/or two independent physical parameters (e.g., moisture content and density), while all other physical properties are held constant.

29.1.2   Why Measure Dielectric Properties?

Permittivity is the main factor that determines how the material interacts with an electromagnetic field. In communications and radar devices, it is necessary to know the dielectric properties of the materials involved both in the design and analysis processes [1].

In the present era of worldwide communications, international regulations require rigorous studies about how the electromagnetic fields are absorbed or affect the human body. To this end, specific absorption rate measurements are performed using equivalent dielectric materials whose permittivities have to be previously accurately determined.

If the electromagnetic fields are applied for thermal processing of materials, dielectric characterization is necessary to model, predict, design, and understand these processing systems. For example, dielectric properties can give an answer to questions such as "is the material loss factor high enough?," "how does the permittivity change with temperature?," and "is a thermal runaway possible?"

The range of applicability is vastly broadened because dielectric properties are also sensitive indicators of numerous physical properties of the material, such as moisture content, biomass, density, bacterial content, chemical reaction, viscosity, concentration of components, and most other physical properties. Correlation between dielectric properties and other parameters of interest can be generally done through empirically modeled relationships, which is of crucial interest in industrial applications [2].

29.1.3 Techniques for Dielectric Properties Measurement

Due to the design flexibility, after decades of microwave devices development, there are many different fixtures available to perform dielectric characterization of materials [3,4].

The different techniques can be classified depending on many criteria: size of the measurement cell, resonant/non resonant structures, open/closed cells, etc. One of the possible classifications distinguishes between lumped-element circuits, travelling wave methods, and resonance/cavity methods.

When the size of the measurement cell is smaller than the wavelength (frequency is sufficiently low), the measurement cell can be considered as a lumped circuit. The method usually consists of placing a slab of dielectric material between two conductors. The material permittivity can be recovered from measurements of the circuit impedance, which can be performed with an impedance analyzer.

If the measurement cell has a size comparable to or larger than the wavelength, propagation parameters of the wave have to be considered. These techniques for dielectric characterization are known as travelling-wave methods. The material is inserted or placed in contact with transmission lines (rectangular waveguide, coaxial line, etc.) or between two antennas in a free-space configuration. Measurements of the complex transmission (S_{21}) and/or reflection (S_{11}) scattering parameters are performed by vector network analyzers (VNAs). From these measurements and the electromagnetic model of the measurement cell and the propagating waves, it is possible to calculate the permittivity of the dielectric material.

A widely used group of traveling-wave methods is based on open measurement cells, which are placed near or in contact with the dielectric material at one side. The waves are emitted from the open end and penetrate the material. Their propagation characteristics (transmission, reflection, and attenuation) are thus modified depending on the material dielectric properties. VNA measurements of the reflection coefficient (S_{11}) allow for complex permittivity determination, which can cover several decades of frequency. Some of the most popular fixtures are the open-ended coaxial probe [5] and the coplanar waveguide [6].

If the wavelength is much shorter than the sample size, free space methods are used in which the sample is placed between two antennas (or in front of only one antenna). Quasi-optical methods are used to model the propagation characteristics. Again, VNA measurements of complex scattering parameters allow for permittivity determination of the sample.

Finally, almost any of the previous techniques can be converted into resonance or cavity methods by modifying the propagation conditions in the measurement cell in such a way that a resonant mode is established at a certain frequency. Instead of scattering parameters, resonance frequency and quality factor of the cavity are the measured quantities. The permittivity of the sample can be recovered from these measurements only at the discrete frequencies at which the fixture resonates. The advantage of this method is the great accuracy in the determination of material permittivity, especially the loss factor for low-loss materials.

29.1.4 Examples of Measurements

Due to the wide range of applicability of dielectric characterization of materials, numerous studies and also deep review papers and books have been published by expert dielectric metrologists [3,4,7].

Some illustrative results obtained at ITACA with in-house measurement cells are shown in the subsequent sections. Figure 29.1 shows several open-ended coaxial cells developed for specific applications. In [8], it is demonstrated how a careful selection of measurement frequency and coaxial dimensions can drastically affect the cell performance. Despite the fact that there are some commercially available general-purpose open coaxial cells, it is clear that many applications require a specific design to match with the application requirements. For example, bigger coaxial cells may mitigate the error caused by sample roughness, or may improve the penetration of electromagnetic

FIGURE 29.1 Examples of coaxial cells developed at ITACA for dielectric characterization of materials at microwave frequencies.

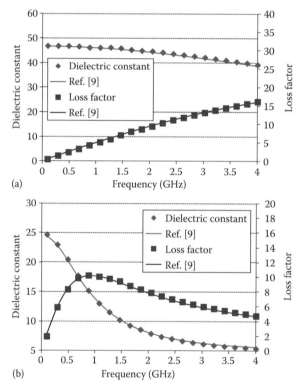

FIGURE 29.2 Dielectric characterization of some liquid samples performed with an open-ended coaxial probe at microwave frequencies. (a) Dimethyl sulfoxide (at 23.3°C) and (b) ethanol (at 23.5°C).

fields in thick samples. This may be necessary if dielectric properties exhibit gradients at the sample surface, and it is of interest to obtain dielectric values representative of the bulk material.

Figure 29.2 shows the dielectric properties of some liquid samples (ethanol and dimethyl sulfoxide) performed in the frequency range from 100 MHz to 4 GHz with an open-ended coaxial probe. This kind of measurement cell is particularly convenient for liquids characterization, because

TABLE 29.1

Dielectric Characterization of Materials with an Open-Ended Coaxial Resonator

Material	Freq. (GHz)	e'_r	e'_r	e'_r (References)	e''_r (References)	References
PTFE	2.37	2.049	6E-4	2.050	6E-4	[10]
PVC	2.34	2.898	0.023	2.919	0.023	[11]
Nylon	2.33	3.092	0.047	3.082	0.044	[11]
Acetal	2.34	2.977	0.124	2.968	0.129	[11]
Crosslinked polystyrene	2.35	2.532	0.002	2.540	0.002	[12]
Methanol	1.83	24.461	11.963	25.738	12.700	[13]

in the absence of air bubbles, a good contact between the sample and the probe surface is assured. Permittivity values obtained by other authors [9] are in good agreement with the presented results.

Table 29.1 shows permittivity results obtained for some common materials with open-ended coaxial cavity cells. In this case, the dielectric characterization is performed at discrete points of frequency, which coincide with the frequencies of the coaxial cavity resonant modes. As the table shows, the resonant coaxial structure gives accurate results for materials with low and high losses. This is possible only if an accurate model of the resonator coupling network is applied, and thus accurate measurements of resonance frequency and quality factor are made. One of these methods will be discussed in next sections. Again, results obtained with coaxial resonators are in good agreement with measurements performed by other authors and with other techniques.

Measurements with an open-coaxial resonator cell assume that the electromagnetic signal emitted from the probe is confined within the sample volume. However, there are applications in which, due to the cost or the material nature, only small sample volumes are available. If the sample is not big enough, a portion of the signal arrives to the sample edges and reflects back to the probe, inducing measurement errors. For these cases, a specific measurement cell has been designed at ITACA, which is based on a single-post coaxial reentrant cavity with a partially dielectric filled gap (see Figure 29.3). With this measurement cell, liquid or solid samples of 1 mL volume can be accurately characterized at microwave frequencies.

Table 29.2 shows permittivity results obtained for some common materials with the single-post coaxial reentrant cavity cell. It should be noted that this fixture allows convenient measurements of liquid, granular, and powder samples. The results show good agreement with measurements performed by other authors and with other techniques.

FIGURE 29.3   Single-post coaxial reentrant cavity sensor developed at ITACA for dielectric materials characterization (1 mL samples in standard vials).

TABLE 29.2
Dielectric Characterization of Materials with a Single-Post Coaxial Reentrant Cavity Cell

Material	Freq. (GHz)	e'_r	e'_r	e'_r (References)	e''_r (References)	References
Water	1.97	77.18	6.537	77.83	6.789	[9]
Dimethyl Sulfoxide	2.10	45.39	8.820	45.02	8.966	[9]
2-Propanol	2.47	4.06	2.925	3.89	2.724	[9]
Quartz sand	2.50	2.49	0.003	2.27	0.012*	[*]
Granular paraffin	2.52	1.71	0.001	1.77	0.011*	[*]
Powdered milk	2.51	2.06	0.072	2.12	0.068	[*]
SiC	2.41	8.57	1.289	8.49	1.335	[*]
Perlite	2.49	2.68	0.018	2.73	0.011	[*]

[*] Measurements performed with Agilent coaxial probe HP-85070B (not suitable for low losses materials).

29.2 MICROWAVE SENSORS FOR DIELECTRIC PROPERTIES MONITORING

29.2.1 BASIC PRINCIPLES OF MICROWAVE SENSORS

Microwave measurement cells can be applied to the characterization of dielectric properties of materials but can also be used to perform a continuous monitoring of these properties if the material under test is undergoing a certain process. To this end, measurements of the cell response have to be performed at a sufficient rate to allow for real-time monitoring of changes in the material state.

This kind of applications is of notable interest in industry, since it permits to implement new control functionalities, which assure the quality of final products by assessing and adapting in real-time the process conditions [2].

In a typical example of a microwave sensor system, the material is subjected to a process (curing, heating, drying, etc.), the microwave sensor is placed in contact or near the material, and its response is dependent on the material state. A microwave transducer comprising a microwave source, and a receiver, together with a network are necessary to separate the incident waves (from the source) from the reflected waves (from the sensor). Everything is controlled by a processing unit, having implemented the additional control functionalities (displays, thresholds, alarms, etc.).

As explained in previous sections, microwave measurement cells have been traditionally used with VNAs, which perform broadband measurements of the complex scattering parameters. This equipment includes the functionalities of microwave transducer and control system. VNAs can be considered as commercially available sophisticated equipment to cover dielectric measurements in the whole microwave spectrum (when used in combination with microwave measurement cells). This equipment is designed to run in the laboratory environment, with specialized operators, and has low portability. On the other hand, the high cost of commercial VNAs, which might be acceptable for scientific institutions, is often prohibitive for industrialists.

The practical nature of monitoring applications has forced researchers to develop microwave equipment according to industrial conditions: simple, affordable, and robust, while retaining the necessary accuracy. In this context, microwave reflectometers have been developed to tackle with this demand [14–16].

In [14,16], detailed descriptions of microwave reflectometers developed at ITACA can be found. The microwave source is designed as a phase-locked-loop-based synthesizer (AD8314) with voltage-controlled oscillator ranging from 1.5 to 2.6 GHz. The separation network has been implemented with two directional couplers. The receiver is designed from the commercial gain and phase detector (AD8302). The reflectometer system is connected to a PC through the USB port to perform the required calculations and to transform the outputs into the desired display.

The reflectometer can be integrated with different microwave measurement cells (sensors) for specific applications, giving rise to robust, compact, and portable systems to perform dielectric materials characterization both in industrial or laboratory environments.

29.2.2 ADVANTAGES AND DRAWBACKS OF MICROWAVE SENSORS

The main advantages of microwave sensors for materials measurement can be summarized in the following points [2]:

- Contact between the material under test and the sensor is not necessary.
- Microwave energy penetrates inside dielectric materials (from mm to several cm). The resulting dielectric properties are representative of a volume of that material, not only of the surface.
- In the absence of sample, the response of a resonator sensor is only dependant on its dimensions (very stable).
- Microwave sensors are not sensitive to common industrial ambient conditions (dust, vibration, etc.) or static charges.
- At microwave frequencies, the influence of direct current conductivity is small and material state is not masked by its ion content.
- The sensors are completely safe for the operator (contrary to ionizing radiation such as x-rays) and do not affect the material.
- The response of the microwave sensor to a change in the material under test is immediate. For this reason, several measurements per second are feasible.
- Manufacturing cost of microwave sensors is not high.

However, there still may be some disadvantages, which have prevented a wider industrial use of microwave sensors:

- These sensors are usually designed and optimized for specific applications.
- If there are several variables changing at the same time (for example, pressure, temperature, and degree of cure), one microwave sensor is not enough to quantify the contribution of each variable to the final response. Additional sensors (microwave or other kind of sensors) may be necessary to solve the ambiguities.
- Due to the cells size and the wavelength of the microwave signals, the spatial resolution with microwave sensors is low.
- The relationship of dielectric properties with process parameters needs to be calibrated for each specific application (i.e., moisture content, degree of curing, etc.).

29.3 MICROWAVE RESONATOR SENSORS

29.3.1 BASICS OF MICROWAVE RESONATORS

A microwave resonator is formed by a section of a transmission line bounded by impedance discontinuities that cause reflections. At a frequency, where the multiple reflections are in phase, constructive interference occurs in the form of a resonance. Two main parameters are necessary to characterize a resonator response: the frequency (f) at which the resonance occurs and the quality factor (Q) of the cavity at that frequency.

The stationary fields in resonators can be considered as a superposition of forward and backward travelling waves; therefore, a dielectric brought in contact with those fields interacts continuously with them, resulting in high sensitivity of the measurement. For this reason, open resonators represent an attractive solution for non-destructive sensing of dielectric materials, since they combine the convenience of one-sided configuration with the high sensitivity of resonance structures [17].

As an example, a transmission line can be converted to a resonator if a coupling network is connected to the line at one side, and the other side is left open (this end is placed in contact with the material under test). Depending on the type of transmission line, the microwave resonator is called, for example, coaxial, microstrip, stripline, slotline, or coplanar resonator.

When a resonator is loaded with a material, its response is related to the permittivity of the material under test. Electromagnetic models of the resonant structure provide a quantitative relation between the resonator parameters and the material dielectric properties. Using them, a measurement of the resonant frequency and quality factor of a resonator gives the necessary information to calculate the complex permittivity of the dielectric material.

29.3.2 COUPLING OF MICROWAVE RESONATORS

In practice, any resonator is not an isolated structure. It has to be fed through a network, which couples the energy from the source. There are some examples of feeding networks: electrical probes, apertures, or magnetic loops. The use of a feeding network to launch energy into the resonator modifies its theoretical (unloaded) response: the resonance frequency is shifted to the loaded resonant frequency, and the quality factor is lowered to the loaded quality factor [18].

The effect of the coupling network can be neglected if the energy is weakly coupled into the resonator, then loaded resonant frequencies are practically equal to the unloaded resonances. However, the undercoupling condition involves an important limitation: materials exhibiting medium or high losses cannot be measured because they absorb and attenuate most of the energy [19]. Then, it is a challenge to develop microwave resonators to cope with any kind of material losses. One possible solution is to design a strongly coupled (overcoupled) cavity, and, consequently, to develop an adequate modeling of the effect of the coupling network to correct the disturbance effect [18].

29.3.3 MEASUREMENT OF RESONANT FREQUENCY AND QUALITY FACTOR

A one-port microwave resonator, connected to a measurement system by means of a transmission line, can be represented by an equivalent circuit in the vicinity of the resonance. This representation is referred to as the first Foster's form. The coupling network is modeled by an ideal transformer of ratio n, and by the complex impedance $r_e + jx_e$. Ohmic losses within the coupling mechanisms are represented by an equivalent resistance r_e. The reactance x_e represents the extra energy storage introduced by the coupling structure. A more detailed description of this representation can be found in [18].

According to this model, the detuning of a resonator due to the coupling elements, namely, the relation between loaded (f_L, Q_L) and unloaded resonator parameters (f_U, Q_U) is given by the following expressions:

$$f_L = f_U \left(1 + \frac{kx_e}{2Q_U} \right) \tag{29.2}$$

$$Q_L = \frac{Q_U}{1 + k} \tag{29.3}$$

where k is the coupling factor. It can be appreciated that, for highly undercoupled resonators ($k \to 0$), the loaded parameters, f_L and Q_L, can be approximated by the unloaded ones, f_U and Q_U. The values of k, f_L, Q_L, and Q_U of Equations 29.2 and 29.3 can be directly derived from the measurement of the magnitude and phase of the reflection factor of the sensor around the resonance, following the procedure described in [20]. However, the reactance x_e cannot be directly extracted from the measurements [20].

This parameter can be obtained by simulating the structure with different coupling levels with the aid of electromagnetic simulators. For example, [21] shows these results in the case of a coaxial resonator with an electrical feeding probe. Once the feeding network is modeled (x_e is known for each f_L), f_U and Q_U can be calculated from measurements (f_L, and Q_L) by using Equations 29.2 and 29.3. With the unloaded parameters, dielectric properties of materials can be accurately determined from electromagnetic models of the resonator.

29.3.4 EXAMPLES

The unique characteristics of microwave sensors, together with the advance of high frequency circuits and processors, are promoting that microwave sensors can take the leap from laboratory into industrial environments with high possibilities of success. In this section, several microwave sensors developed for specific applications are shown. Each of them has been developed with the characteristics required to be used in industrial applications: robust, accurate, and simple to use.

29.3.4.1 Microwave Sensor for Characterization of Liquid Mixtures

Water in oil (W/O) emulsions appear in the sludges of many industries. These pollutants must be treated before discharge, and microwave heating is considered as an environmental-friendly technique to separate these emulsions. The amount of heat generated in the emulsion greatly depends on the emulsion dielectric properties. Precise characterization of emulsions is essential to predict their interaction with microwave radiation.

Previous studies on emulsions dielectric properties have been conducted at radio frequencies below 300 MHz, which do not include the standard heating frequency (2.45 GHz). Now, with a single-post coaxial reentrant cavity sensor (see Figure 29.4), accurate characterization of different

FIGURE 29.4   Single-post coaxial reentrant cavity sensor developed at ITACA for dielectric materials characterization (8 mL samples in standard vials).

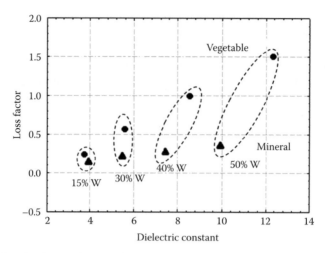

FIGURE 29.5 Dielectric characterization of water-in-oil emulsions (with vegetable and mineral oils) performed with a single-post coaxial reentrant cavity sensor.

W/O emulsions (mineral and vegetable) at microwave frequencies has been performed. Also, an open-ended coaxial sensor has been designed for dielectric characterization of this type of emulsions while flowing through a pipe. These and other related results can be found in [22].

Two types of W/O emulsions were considered: mineral and vegetable oil emulsions. The concentrations of water in samples were 15%, 30%, 40%, and 50% by volume. Figure 29.5 shows that dielectric properties of emulsions depend not only on the water volume fraction but also on the chemical properties of the oils. The information given by the sensor can be used to simultaneously determine type of emulsion, influence of oil composition (vegetal or mineral), and water content. In contrast with other methods to characterize emulsions (such as conductivity measurements, gravity separation, centrifugation, etc.), dielectric measurements are a simple, convenient, nondestructive, and real-time alternative.

29.3.4.2 Microwave Sensors for Monitoring Curing Processes

Cure is a chemical reaction, which converts a certain mixture of liquids into a solid piece. There are many products, which need some kind of cure process during their manufacturing. Typical examples of curing processes are adhesives bonding or polymers shaping into solid pieces inside a mould. Usually, the quality of products obtained after the curing process is highly dependent on variables such as components dosage, ageing or contamination, temperature, or ambient conditions. In the industry, it is very difficult to control these variables and thus, to obtain repetitive cure processes, which assure the adequate final product quality. The common practice is to apply recommended specifications or apply conservative estimations. As a result, cure is less efficient than it could be, and materials waste can be up to 10%, increasing costs and environmental problems [23].

Broadband dielectric relaxation spectroscopy has been widely used in the bibliography to reveal details about curing processes, with the aid of VNAs or impedance analyzers. Most of these measurements are performed with coplanar sensors with interdigitated-comb-like electrode design [24], in the frequency range between 1 kHz and 1 MHz, conductivity being the main parameter of interest. However, cure monitoring through conductivity changes presents many drawbacks [21]: lack of sensitivity, conductivity masked by ionic changes, erroneous readings due to electrode polarization, etc.

One alternative for cure monitoring is to use microwave frequencies (above 1 GHz) and follow the changes in dielectric properties instead of conductivity. At these frequencies, the mobility of molecules (indication of the material viscosity) plays the main role and it is not masked by conductivity. Despite the advantages presented by microwave sensors, industrial application of the technique has been inhibited by a lack of basic knowledge of the relationships between molecular

FIGURE 29.6   Open-ended coaxial resonator sensor developed at ITACA for monitoring the curing process of thermoset samples.

structure and the macroscopic dielectric behavior and by limited availability of robust sensors and instrumentation suitable for the industrial environment [25,26].

Figure 29.6 shows an example of microwave sensor system developed for the noninvasive monitoring of the curing process of thermoset materials. The thermoset material is placed inside a mould, and the microwave sensor is designed with a curved surface adapted to the mould's inner shape. In particular, an open-ended coaxial resonator has been designed as the microwave sensor head.

Figure 29.7 shows some measurements of the sensor response during the cure process of a polyurethane sample. The figure shows how the effect of the polyurethane formation can be seen in the change of the sensor response during time. The figure also shows the complex permittivity during the curing process. Both the dielectric constant and the loss factor decrease with the reaction progress. This means that, during cure, the increase of viscosity of the liquid components cause a drop of the molecular mobility, and thus, the permittivity of the material is drastically decreased. Thus, the rate of reaction can be followed by simple inspection of these curves. The system allows not only monitoring the evolution of the curing process but also assessing the process variables, such as components ageing or contamination [26].

Adhesives represent another interesting example of materials, which undergo cure in numerous manufacturing processes. In the case of adhesives, cure monitoring allows to reliably determine the time of cure and avoid the dependence on the apparent "hardness" of the bond or manufacturer values of cure time which do not take specific conditions into account.

Thermocouples and pressure transducers have been traditionally almost exclusively used for in-situ monitoring of the adhesives cure [27]. For practical reasons, these methods are difficult to incorporate into industrial production.

In [16], a microwave sensor developed for cure monitoring of adhesives samples is shown. In this case, it is based on an open-ended coaxial resonator. The sample is placed on a protective layer, and with the aid of a metallic lid, a slight pressure is applied to get a uniform adhesive layer of the desired thickness and rid the bond of air bubbles.

Figure 29.8 shows the evolution of the shift of resonance frequency during the first 30 min of the curing process for different adhesive samples. A clear dependence of the adhesive type on the curing response of the sample is visible from the figure. As in the previous case with polymer samples, the effect of the adhesive curing can be followed with the change of the resonant frequency of the sensor, showing typical curing curves. For example, Figure 29.8 shows a fast variation of the sensor response for the Loctite 9455 during the first 700 s and a practically flat response in the rest of 1800 s of monitoring. This is a clear indicator that most of the curing of Loctite 9455 falls in these first 700 s. From these measurements, more information can be obtained about the

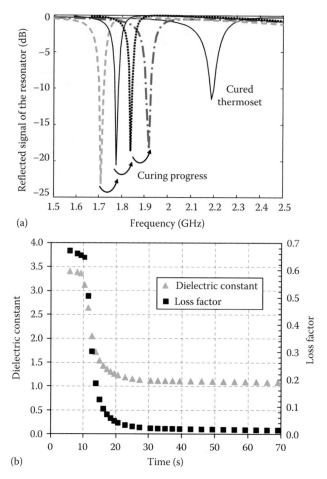

FIGURE 29.7   (a) Microwave sensor response during the cure process of a polyurethane sample. (b) Dielectric properties of the sample during cure.

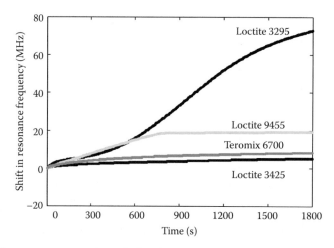

FIGURE 29.8   Microwave sensor response during the cure process of some adhesive samples. (Reproduced from García-Baños, B. et al., Microwave sensor for continuous monitoring of adhesives curing, in Tao, J. ed., *Proc. of the 13th International Conference on Microwave and RF Heating, AMPERE 2011*, September 5–9, 2011, Toulouse, France, pp. 145–149. With permission.)

reaction, since the sensor also provides related results such as the reaction kinetics, cure time, etc. This allows the system to be used as a production monitoring and control tool as well as for laboratory studies [28].

29.3.4.3 Microwave Moisture Sensors for Powder Materials

Knowing the moisture content has enormous economic value in the manufacture and processing of materials. Such information is useful for determining the value of raw materials, for in-process control, and for output quality control.

Moisture determination has been traditionally done by analyzing the loss of weight of a sample during drying, which is very time consuming and labor intensive. When fast and non-destructive moisture measurements are required, two main techniques are available: infrared (IR) methods and microwave dielectrometry. One of the differences between them is that IR methods are only sensitive to the moisture at the material surface, while microwave dielectrometry provides the *bulk* moisture of the material. This is of particular importance if moisture gradients exist in the material.

To illustrate this application, several samples of quartz sand were moistened and measured with a single-post coaxial reentrant cavity sensor (see Figure 29.3). Quartz sand is a granular material often used as representative because its properties are very similar to other materials employed in chemical or pharmaceutical industries. The moisture content of each sample was determined as the mass of water added (ranging from 0% to 5%) to the total mass of dry material.

Figure 29.9 shows the relation between dielectric properties (dielectric constant and loss factor) and moisture content of quartz sand samples at 2.45 GHz. The results show how dielectric properties are strongly influenced by water content at this frequency, as expected, due to the high dielectric constant and loss factor of water drops. Typical measurement deviations (see error bars in the figure) are more than adequate for practical moisture determination.

29.3.4.4 Microwave Sensors for Monitoring Liquids under Flow

In many industrial processes, it is very important to control the state of liquid components to ensure the final product quality. For instance, polyol is a highly reactive basic monomer commonly used for obtaining polymers in the footwear industry. It tends to absorb water, thus modifying the proper stoichiometric relationship in the reaction. Since changes in the composition of a reagent involve variations in its dielectric properties, instantaneous and non-destructive measurements with a microwave sensor can be used to monitor the purity and composition of the material.

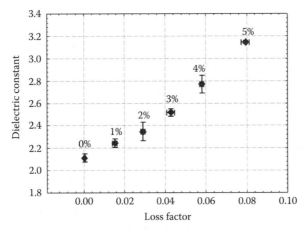

FIGURE 29.9 Dielectric characterization of quartz sand samples with different moisture content (in % of dried weight) performed with a single-post coaxial reentrant cavity sensor.

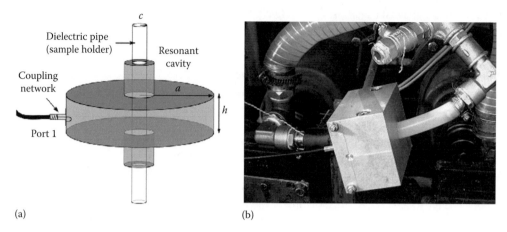

(a) (b)

FIGURE 29.10 (a) Microwave sensor design (cylindrical cavity). (b) Non-intrusive installation of the sensor in the production line. (Reproduced with kind permission from Springer Science+Business Media: *Adv. Microw. Radio Frequency Process.*, Microwave non-destructive evaluation of moisture content in liquid composites in a cylindrical cavity at a single frequency, 2006, 138–148, Catalá-Civera, J.M., Canós, A.J., Peñaranda-Foix, F., and de los Reyes, E., Figure 29.7.)

Figure 29.10 shows the cylindrical cavity designed for this application. The liquid flows inside a pipe, passing through the resonant cavity allowing dynamic measurements.

To follow the evolution of the degradation process in the liquid, only measurements of phase at a fixed frequency are performed. It yields to a cheaper and robust system for the industry and simplified measurements from the operator's point of view.

The designed sensor has been tested in the laboratory, where a continuous line with the liquid flowing through a polytetrafluoroethylene (PTFE) pipe from and to a tank was prepared. A small

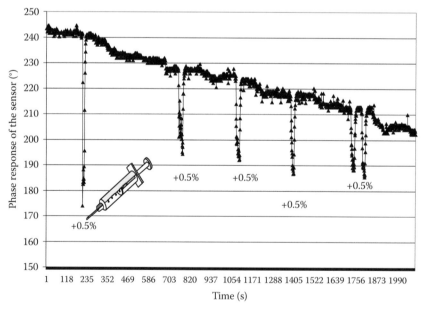

FIGURE 29.11 Phase response of the microwave sensor. The water content added in each step is displayed with percentage in volume. (Reproduced with kind permission from Springer Science+Business Media: Adv. Microw. Radio Frequency Process., Microwave non-destructive evaluation of moisture content in liquid composites in a cylindrical cavity at a single frequency, 2006, 138–148, Catalá-Civera, J. M. et al., Figure 29.8.)

pump is used to maintain constantly the movement of the liquid. The degradation process of the polyol was produced artificially by injecting water into an initially pure polyol sample.

The evolution of the sensor response in time is plotted in Figure 29.11 where each procedure of adding water to the pure material is indicated with the corresponding percentage in volume. As shown, the phase response of the sensor is very sensible to water content. Since water is directly injected into the pipe, a high concentration of water goes through the sensor during each of the adding processes and therefore sharp and momentary changes on the response can be seen. After the water injection, the phase takes its original value and then, as polyol is mixed with water in the tank, the response follows the degradation degree. It is possible to identify variations lower than 0.25% of moisture.

The use of phase monitoring at a single-frequency in the resonator during the degradation process was found as a simple, cheap, robust, and accurate mechanism of control, showing the unique capabilities of microwaves to follow changes in materials properties.

REFERENCES

1. B. Clarke, Dielectric measurements and other measurements relevant to RF & microwave processing, *RF Industrial Processing Club Meeting*, National Physical Laboratory, Middlesex, U.K., 2006.
2. E. Nyfors, Industrial microwave sensors. A review, *Subsurf. Sens. Technol. Appl.*, 1(1), 23–43, 2000.
3. R. N. Clarke et al. A guide to characterization of dielectric materials at RF and microwave frequencies, Institute of Measurement and Control/National Physical Laboratory, Teddington, 2003.
4. J. Krupka, Frequency domain complex permittivity measurements at microwave frequencies, *Meas. Sci. Technol.*, 17, R55–R70, 2006.
5. A. P. Gregory and R. N. Clarke, Dielectric metrology with coaxial sensors, *Meas. Sci. Technol.*, 18, 1372–1386, 2007.
6. A. Raj, W. S. Holmes, and S. R. Judah, Wide bandwidth measurement of complex permittivity of materials using coplanar lines, *IEEE Trans. Instrum. Meas.*, 50(4), 905–909, 2001.
7. J. C. Anderson, *Dielectrics, Science Paperbacks, Modern Electrical Studies, Chapman & Hall*, London, U.K., 1964.
8. B. García-Baños et al., Design rules for the optimization of the sensitivity of open-ended coaxial microwave sensors for monitoring changes in dielectric materials, *Meas. Sci. Technol.*, 16, 1186–1192, 2005.
9. A. P. Gregory and B. Clarke, Tables of the complex permittivity of dielectric reference liquids at frequencies up to 5 GHz, NPL Report, MAT 23, 2009.
10. Rodhe&Schwartz, Measurement of dielectric material properties, Application Note.
11. F. L. Penaranda-Foix and J. M. Catala-Civera, Circuital analysis of cylindrical structures applied to the electromagnetic resolution of resonant cavities, *Passive Microwave Components and Antennas*, ed. V. Zhurbenko, INTECH, 2010.
12. J. Krupka et al., A dielectric resonator for measurements of complex permittivity of low loss dielectric materials as a function of temperature, *Meas. Sci. Technol.*, 9, 1751–1756, 1998.
13. D. V. Blackham, An improved technique for permittivity measurements using a coaxial probe, *IEEE Trans. Instrum. Meas.*, 46(5), 1093–1099, 1997.
14. D. Polo, P. Plaza-Gonzalez, B. García-Baños, and A. J. Canós, Design of a low cost reflectometer coefficient system at microwave frequencies, *10th International Conference on Microwave and High Frequency Heating*, Modena, Italy, p. 235, 2005.
15. G. C. Hock et al. Super-heterodyne interferometer for non-contact dielectric measurements on millimeter wave material, *IEEE International RF and Microwave Conference Proceedings*, Selangor, Malaysia, 2008.
16. B. García-Baños, J. M. Catalá-Civera, F. L. Peñaranda-Foix, A. J. Canós, and O. Sahuquillo-Navarro, Microwave sensor system for continuous monitoring of adhesive curing processes, *Meas. Sci. Technol.*, 23, 1–8, 2012.
17. L. F. Chen, *Microwave Electronics: Measurement and Materials Characterization*, John Wiley & Sons, Ltd., Chichester, U.K., 2004.
18. A. J. Canós, J. M. Catalá-Civera, F. L. Peñaranda-Foix, and E. de los Reyes-Davó, A novel technique for deembedding the unloaded resonance frequency from measurements of microwave cavities, *IEEE Trans. Microw. Theory Tech.*, 54(8), 3407–3416, 2006.
19. D. Xu, L. Liu, and Z. Jiang, Measurement of the dielectric properties of biological substances using an improved open-ended coaxial line resonator method, *IEEE Trans. Microw. Theory Tech.*, MTT-35(12), 1424–1428, 1987.

20. D. Kajfez, *Q Factor*, Vector Fields, Oxford, U.K., 1994.
21. B. García-Baños, A. J. Canós, F. L. Peñaranda-Foix, and J. M. Catalá-Civera, Noninvasive monitoring of polymer curing reactions by dielectrometry, *IEEE Sensor J.*, 11(1), 62–70, 2011.
22. B. García-Baños, R. Perez-Paez, J. M. Catala-Civera, and F. L. Penaranda-Foix, Dielectric characterization of water in oil emulsions flowing through a pipe, *Proceedings of the 12th International Conference on Microwave and High Frequency Heating—AMPERE*, Karlsruhe, Germany, 2009.
23. G. Oertel, *Polyurethane Handbook*, 2nd edn., Hanser, Munich, Germany, 1994.
24. M. C. Hegg, A. Ogale, A. Mescher, A. V. Mamishev, and B. Minaie, Remote monitoring of resin transfer molding processes by distributed dielectric sensors, *J. Compos. Mater.*, 39(17), 1519–1538, 2005.
25. R. Casalini, S. Corezzi, A. Livi, G. Levita, and P. A. Rolla, Dielectric parameters to monitor the crosslink of epoxy resins, *J. Appl. Polym. Sci.*, 65(1), 17–25, 1997.
26. J. M. Catala-Civera, F. Peñaranda-Foix, B. Garcia-Baños, and A. J. Canos, Microwave sensor system for monitoring the curing of polymer thermosets, *Eighth International Conference on Electromagnetic Wave Interaction with Water and Moist Substances (ISEMA)*, Espoo, Finland, 2009.
27. M. J. Lodeiro et al., Cure monitoring techniques for polymer composites, adhesives and coatings, NPL, Measurement Good Practice Guide No. 75, February 2005.
28. B. García-Baños, F. L. Peñaranda-Foix, P. J. Plaza-Gonzalez, and A. J. Canós, Microwave sensor for continuous monitoring of adhesives curing, *Proceedings of the 13th International Conference on Microwave and High Frequency Heating—AMPERE*, Toulouse, France, 2011.
29. J. M. Catala-Civera, A. J. Canós, F. Peñaranda-Foix, and E. de los Reyes, Microwave non-destructive evaluation of moisture content in liquid composites in a cylindrical cavity at a single frequency, *Adv. microw. Radio Frequency Process.*, 138–148, 2006.

30 Microwave Reflectometry for Sensing Applications in the Agrofood Industry

Andrea Cataldo, Egidio De Benedetto, and Giuseppe Cannazza

CONTENTS

30.1 INTRODUCTION

The present chapter describes some of the most promising applications of microwave reflectometry (MR) for monitoring and sensing purposes in the agrofood industry. The interest toward MR originates from the fact that this technique can satisfy several contrasting requirements, such as low implementation cost, real-time response, possibility of remote control, reliability, and adequate measurement accuracy. Typically, in MR measurements, an electromagnetic (EM) signal is propagated into the system under test (SUT): the analysis of the reflected signal along with a specific data processing is used to retrieve the desired information on the SUT [1]. Applications of MR are numerous and cover a wide range of fields; in fact, thanks to the versatility of the approach, this technique has proven useful for several applications, such as

- Localization of faults along cables [2–4]
- Geotechnical engineering for assessing distributed pressure profiles [5–10]
- Measurement of liquid levels (also in stratified liquids) [11–13]

- Monitoring applications in civil engineering for sensing crack/strain in reinforced concrete structures [14], for fault location on concrete anchors [15], for monitoring cement hydration [16], etc.
- Water-leak detection in underground metal pipes [17]

A comprehensive overview of the most important and promising applications of MR can be found in [18]. On the other hand, the present chapter focuses on three specific applications in which MR is employed for monitoring purposes in the agrofood industry. In the following paragraphs, first the basic theoretical principles behind MR are recalled and the adopted measurement strategies are discussed. Finally, some interesting test-cases related to MR-based monitoring of agrofoods are presented; in particular, the following applications are considered:

1. Moisture measurement of granular agrofood materials [19]
2. Quality control of vegetable oils [20,21]
3. Monitoring of dehydration process of fruit and vegetables [22]

30.1.1 MEASUREMENT TECHNIQUES

A MR measurement system generally consists of two main elements: (1) the unit for generating and receiving the EM signal and (2) the measurement probe (or sensing element).

MR measurements can be performed either in time domain (time domain reflectometry [TDR]) or in frequency domain (frequency domain reflectometry [FDR]): the choice of using either approach depends on the intended application. In general, TDR instruments are less expensive and often also portable; as a result, TDR is often considered strategically suitable for in situ application (e.g., in a production line). On the other hand, employing FDR instruments may often ensure a better measurement accuracy. An optimal trade-off of instrument portability/low cost and measurement accuracy can be achieved through a combined time domain/frequency domain (TD/FD) approach (i.e., by performing TDR measurements and extrapolating, through suitable processing techniques, the corresponding FD information) [23].

In TDR measurements, a step-like EM signal is propagated, along a probe, through the SUT. The output of TDR measurements is a reflectogram, i.e., the reflection coefficient $\rho(t)$ along the sections of the probe is displayed as a function of time (or as a function of the traveled electric distance). The behavior of $\rho(t)$ is strictly associated with the impedance variations along the electrical path traveled by the EM signal. In turn, the electric distance traveled by the EM signal, L_{app}, is related to the dielectric characteristics of the SUT:

$$L_{app} = \sqrt{\varepsilon_{app}}L = \frac{ct_t}{2} \qquad (30.1)$$

where
L is the length of the used probe
$c \cong 3 \times 10^8$ m/s is the velocity of light in free space
t_t is the travel time
ε_{app} is the apparent relative dielectric permittivity of the material in which the probe is inserted [24]

The quantity ε_{app} depends on the relative complex dielectric permittivity of the considered material [25]; for low-loss, low-dispersive materials, ε_{app} is approximately equal to the real part of the dielectric permittivity $\left(\varepsilon_r'\right)$ and can be considered constant with frequency [23].

The analysis of TDR waveform directly leads to the evaluation of L_{app}, whose value is used in the subsequent data-processing for retrieving the desired information.

In FDR measurements, generally, the quantity to be measured is the reflection scattering parameter, $S_{11}(f)$. Traditionally, measurements performed directly in FD rely on vector network analyzers (VNAs): in this case, the excitation stimulus is a sinusoidal signal whose frequency is swept over the desired range of analysis. One of the advantages of FD measurements is that there are well-established error correction models that, through appropriate calibration procedures [such as the short–open–load (SOL) calibration], can be used to reduce the influence of systematic errors [26]. The FDR approach is often used for the dielectric spectroscopy of materials and for impedance characterization of electronic devices and components.

Finally, the adoption of the combined TDR/FDR approach allows taking advantage of the benefits of both the approaches. In particular, estimating the $S_{11}(f)$ from TDR measurements can help disclose useful information that is masked in TD (e.g., multiple dielectric relaxation) [27]. This strategy is regarded as a powerful tool for guaranteeing simultaneously low cost and portability of the instruments and measurement accuracy [28–30]. However, there are some crucial aspects that need to be considered in order to perform an accurate TD/FD transformation, thus reducing errors in the assessment of the spectral response of the system and in the extraction of $S_{11}(f)$ [31].

30.1.2 THE SENSING ELEMENT

The sensing element (or probe) is responsible for the interaction of the stimulus signal with the SUT, and it is the ultimate factor that influences the accuracy of results. Since MR senses the changes in impedance, it is extremely important to employ a probe with a well-known impedance profile; in this way, it is easier to discriminate and interpret the impedance variations due to the SUT.

Several probe configurations can be used: multifilar, coaxial, planar, etc. In general, the design of coaxial probes is quite simple and, for this configuration, the impedance profile (in the TEM propagation mode) can be determined from the transmission line (TL) theory [27,32]:

$$Z(f) = \frac{60}{\sqrt{\varepsilon_r^*(f)}} \cdot \ln\left(\frac{b}{a}\right) \qquad (30.2)$$

where
   $Z(f)$ is the impedance of the probe filled with the considered material
   $\varepsilon_r^*(f)$ is the complex relative dielectric permittivity of the material filling the probe
   b is the inner diameter of the outer conductor
   a is the outer diameter of the inner conductor

Coaxial probes are widely used for monitoring and diagnostics of liquids [20,22].

On the other hand, for soil monitoring and for granular materials monitoring in general, the multirod configuration is a widely used solution. In fact, this configuration allows an easier insertion of the probe into the granular material. In particular, three-rod probes have become a widespread solution since their electric behavior resembles that of coaxial cells [33,34]. Furthermore, when a noninvasive approach must be preserved, it is possible to use surface probes [35] or antennas [36]. A numerical analysis of some probe configurations can be found in [37].

30.1.3 DIELECTRIC CHARACTERIZATION AND QUALITY CONTROL OF MATERIALS

The dielectric characteristics of materials are intrinsic features that can be associated to other characteristics of the considered SUT (also to non-electric characteristics); therefore, dielectric spectroscopy has been employed as a non-destructive diagnostic method for a wide range of materials. The combination of TDR measurement with a suitable TL modeling of the measurement cell can provide accurate results on the dielectric properties of materials. With regards, for example, to the adoption of the TDR/FDR for the accurate evaluation of the Cole–Cole parameters, the typical procedure would include three major steps:

- First, the TDR waveforms are acquired and processed in order to obtain the corresponding $S_{11}(f)$
- Second, the measurement cell (i.e., probe and material under test) is accurately modeled as a TL, in which the dielectric characteristics of the considered liquid are parameterized through the Cole–Cole formula [38]. The modeled scattering parameter, $S_{11,MOD}\left(f, \varepsilon_r^*(f)\right)$, is evaluated
- Finally, the Cole–Cole parameters of the SUT are evaluated by minimizing the deviations between (1) the measured reflection scattering parameter, $S_{11}(f)$, obtained from the TD/FD approach and (2) the modeled reflection scattering parameter obtained from the implemented TL model of the probe plus SUT, $S_{11,MOD}(f)$ [27]

As a matter of fact, dielectric properties of foods are a useful tool for controlling the quality of food products [39–41]. Also, knowing the dielectric characteristics of food products is important for allowing the optimization of several industrial processes, for example, in terms of power absorption during microwave drying process [42]; or in terms of effectiveness of radio-frequency heating for disinfestation of fruit [43]; for controlling the salting process of pork meat [44], and so on.

30.2 MOISTURE CONTENT MEASUREMENT OF GRANULAR AGROFOOD MATERIALS

In the agrofood industry, monitoring water content of granular materials (such as coffee, flour, etc.) is extremely important; in fact, water content of these materials is strictly associated to their quality status: a high amount of water may lead to the growth of microorganisms and bacteria and could compromise the storage of the material.

A wide variety of methods can be used for estimating water content of porous media, ranging from destructive (gravimetric) to non-destructive methods (gamma radiation probe, neutron probe, etc.). The gravimetric sampling, for example, is a highly accurate method; nevertheless, since it requires medium samples to be removed from the medium mass, it is extremely time-consuming and difficult to implement on a process line. As for non-destructive methods, such as the neutron scattering method [45] and the gamma ray attenuation method [46], they are accurate as well; nevertheless, their higher costs and caution to avoid possible health hazards limit their adoption [19].

On the other hand, the adoption of TDR can be an excellent candidate for monitoring water content of agrofood materials. TDR is a highly flexible technique; it can guarantee low measurement uncertainty and continuous measurements. Additionally, measurements have only a small dependence on temperature. Furthermore, TDR allows the possibility of controlling several signals through multiplexing.

Basically, in TDR, the value of the volumetric water content, θ, is inferred from dielectric permittivity measurements. In fact, the relative permittivity of water (approximately 80) is considerably higher than the typical relative permittivity of agrofood materials; therefore, the presence of water increases the overall permittivity of the "moistened" material [47]. In TDR-based method, moisture content is determined from measurements of ε_{app}, which can be easily evaluated from the TDR waveform, by applying Equation 30.1; the value of ε_{app} is then substituted in a mathematical model, which gives the value of moisture as a function of ε_{app}. In the literature, there are a number of models that can be used; however, most of them are material-specific and none has a universal validity [48]. A broad classification of the models that describe the relationship between ε_{app}–θ can be made by distinguishing into empirical and/or partly deterministic approaches. The former category simply fits mathematical expressions (i.e., polynomial whose coefficients must be determined for each specific material) to measured data; the latter considers dielectric mixing models, which take into account the dielectric characteristic and the volume fractions of each constituent (e.g., solid, water, and air) of the considered material.

30.2.1 Empirical Models

One of the most widely used empirical model is described by Topp's equation, which is mostly used for soils [49,50]:

$$\theta = 4.3 \times 10^{-6} \varepsilon_{app}^3 - 5.5 \times 10^{-4} \varepsilon_{app}^2 + 2.92 \times 10^{-2} \varepsilon_{app} - 5.3 \times 10^{-2}. \tag{30.3}$$

In this equation, the coefficients are already determined, and the expression is considered generally valid for materials with low content of organic materials. However, when dealing with a more complex system, specific calibration curves must be individuated for each considered material. These models have some shortcomings (such as the limited possibility of extrapolation outside the moisture range of the original set of experiments [48]); nevertheless, they are routinely used.

As aforementioned, the empirical approach simply fits mathematical expressions to measured data; polynomial regression equations are typically used to individuate the relationship for different types of materials. In this way, the TDR-measured ε_{app} can be related to θ; these curves are called calibration curves. Once the calibration curve (for the specific material) is individuated, to determine the unknown moisture level, it is enough to measure the dielectric constant and to retrieve the moisture level from the calibration curve [33].

Typically, the ε_{app}–θ calibration curve can be assessed as follows. Samples of the considered material are moistened at prefixed values of moisture (θ), and the corresponding dielectric constant (ε_{app}) is measured through the TDR method. The (ε_{app}, θ) points are fitted through a non-linear regression method [33].

30.2.2 Dielectric Mixing Models

A more accurate θ–ε_{app} relationship can be obtained through dielectric mixing models, which take into account the volume fraction and the dielectric constant of each single constituent that is present in the material under test (MUT), the porosity, the effect of the geometrical arrangement of the medium components, etc. Therefore, the dielectric mixing models are suitable for the evaluation of the ε_{app}–θ relationship for non-homogeneous granular materials, with low dry bulk density, or with large amount of bound water, or with relatively large permittivity of the solid phase. Despite the extensive scientific efforts that have been devoted to obtain data on the dielectric properties of agrofoods, the development of specific dielectric models for each considered material, relating the dielectric properties of agrofood materials to the corresponding moisture content, is still an open issue [19].

30.2.2.1 Partly Deterministic and Probabilistic Models

Several dielectric mixing models have been proposed in the literature. In the present case study, two families of models are considered: partly deterministic models and probabilistic models.

With regards to the first category, two main models can be considered: (1) the so-called α-model, based on the semi-empirical equations proposed by Birchak et al. [51] and (2) the Maxwell-De Loor (MD) model, a theoretical model proposed by De Loor and based on Maxwell's equation [52]. In these two models, the water fraction is considered as a single phase. To overcome this simplification and to take into account the effect of the water portion close to the medium surface (which is water whose permittivity value is lower than that of the free water), Dobson and Hallikainen [53] extended the two aforementioned models, thus deriving a four-component system that suitably considers the effects of free water, bound water, solid phase, and air.

Finally, the theoretical model of Ansoult et al. [54] belongs to a particular family of the partly deterministic models, the so-called probabilistic model. This model is based on the random propagation of the pulse in a porous medium that is represented as an array of capacitors. In this approach, the granular material is schematized as a set of three capacitors, each associated to the dielectric constants (ε_{ij}) of the different sample components; eventually, a computational algorithm is used to evaluate the overall equivalent dielectric constant [19].

30.2.3 EXPERIMENTAL COMPARISON OF DIFFERENT MODELS FOR MOISTURE CONTENT MEASUREMENTS ON GRANULAR MATERIALS

In this section, a comparative approach between different empirical and deterministic models for TDR-based estimation of water content is described. For each considered material, the functional relations between dielectric properties and the corresponding moisture levels are also assessed.

The following models are considered: (i) a third-degree polynomial with four "adjusted" parameters obtained by calibration procedures on the materials under test; (ii) the polynomial equation of Topp with its original parameters; (iii) the partly deterministic simplified models based on the theory of mixtures of three components; (iv) the partly deterministic simplified models based on the theory of mixtures of four components; (v) the MD model; and (vi) the probabilistic model of Ansoult et al., based on the random propagation of the EM pulse in the soil.

The aforementioned models were tested on three feedstuff materials: corn, corn flour, and bran. TDR measurements were performed through the Hyperlabs 1500 (which is a portable TDR unit); the used probe was a 30 cm long three-wire metallic probe (Campbell Scientific CS610).

Materials were progressively moistened at prefixed (known) levels of moisture in the ranges reported in Table 30.1. For each moisture level, the corresponding ε_{app} was measured. As a result, for each material, the corresponding empirical calibration curve ε_{app}–θ was obtained.

The ε_{app}–θ curves obtained for all the considered models were derived by superimposing the TDR-measured ε_{app} values, and the different models were compared with the empirical calibration curve. Results are shown in Figure 30.1. Repeated measurements showed that the maximum percentage relative standard uncertainty in ε_{app} is less than 8%.

The root mean square error (*rmse*) between the actual values of moisture and the values obtained from the models was taken as a figure of merit of the different ε_{app}–θ relations. The *rmse* values are reported in Table 30.2. It can be seen that the empirical Topp's equation is the least accurate model; conversely, the best fitting is observed for the partly deterministic models. This is attributable to the fact that the effect of porosity and of the additional physical components is explicitly taken into account. More specifically, the MD model predicts the ε_{app}–θ behavior for bran and corn flour better than it does for the other materials; this is due to the fact that bran and corn flour contain the highest percentage of hygroscopic water. On the other hand, for corn samples, α-model succeeds better than the other models in characterizing the ε_{app}–θ relation [19].

TABLE 30.1
Moisture Content Range of the Considered Agrofood Materials

Material	Moisture Levels Range (cm³/cm³)
Corn	0–0.25
Corn flour	0–0.22
Bran	0–0.21

Source: From Cataldo, A., et al., *Computer Standards & Interfaces*, 32(3), 86–95, 2010.

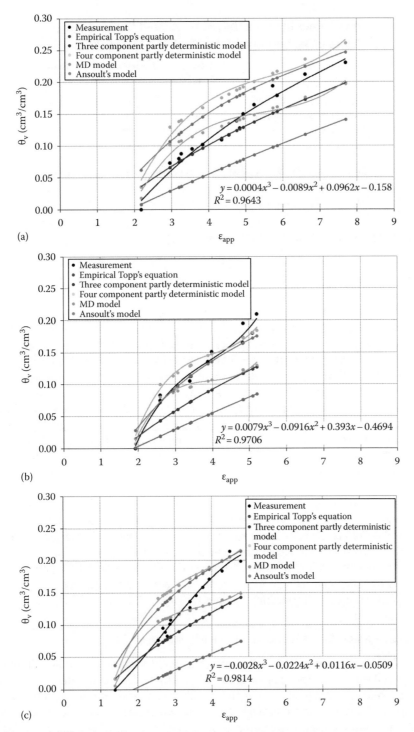

FIGURE 30.1 Comparison among experimental measurements and different empirical and, partly deterministic models for (a) corn flour, (b) corn, and (c) bran. Interpolation functions are also reported. (From Cataldo, A., et al., *Computer Standards & Interfaces*, 32(3), 86–95, 2010.)

TABLE 30.2
***rmse* Values between the Measured and the Modeled**
Values of Water Content, for Feedstuff Materials

Model	Corn	Corn Flour	Bran
Topp's equation	0.08	0.07	0.09
α model (three component)	0.05	0.02	0.04
α model (four component)	0.02	0.05	0.04
MD model	0.04	0.02	0.03
Ansoult's model	0.02	0.03	0.03

Source: From Cataldo, A., et al., *Computer Standards & Interfaces*, 32(3), 86–95, 2010.

30.3 QUALITY CONTROL OF VEGETABLE OILS

Typically, the in-situ evaluation of the Cole–Cole parameters for industrial quality control purposes is not an easy task, especially when reference data are missing. This is the case, for example, with vegetable oils, for whose dielectric characteristics only scarce reference data are available [55,56]. In fact, the complexity of the intrinsic characteristics of vegetable oils, such as density and viscosity, makes the investigation of this kind of material rather challenging, nonetheless interesting. Additionally, since quality control of vegetable oils is becoming more stringent (especially for avoiding adulteration), the estimation of these dielectric parameters may be used as an indicator for certifying the product quality.

Although highly sophisticated methods for the analysis of edible oils are available, many of them are not easily applicable for routine or continuous monitoring (e.g., gas chromatography [57] and liquid chromatography [58]), whereas others (e.g., Fourier transform infrared [59], Raman spectroscopy [60], and nuclear magnetic resonance [61]) are rather complex and highly expensive. On the other hand, other simpler methodologies that are commonly adopted for routine analysis (such as those based on chemical titrations) do not provide much information. As a matter of fact, all of aforementioned methods are affected by some limitations: in particular, they cannot be performed on the process line and they are very laborious.

The following experimental procedure describes a method for evaluating the Cole–Cole parameters of several types of edible oils. Starting from traditional TDR measurements, the evaluation of the Cole–Cole parameters was carried out following the TL-based procedure described in Section 30.1. This is done in view of possible practical applications, which may be useful, for example, in the on-line monitoring of the characteristics of oils throughout a production process. Furthermore, affordable and reliable TDR instruments are readily available on the market, thus making the proposed approach appealing for industrial applications.

30.3.1 TRANSMISSION LINE MODELING OF THE MEASUREMENT CELL

Figure 30.2 shows the picture and the schematic diagram of the used probe. The TL model of the measurement cell was implemented in the AWR Microwave Office® simulator. The basic TL model includes a length of short-circuited coaxial probe, with the physical dimensions of the actual probe, filled with a material characterized through its relative permittivity and loss tangent. These parameters, in turn, vary with frequency according to the Cole–Cole model, described by four parameters.

The implemented TL model includes not only the Cole–Cole parameters of the liquid under test as minimization variables but also some lumped circuit elements that account for additional parasitic effects that cannot be compensated for through traditional SOL calibration procedures (the values and the nature of the lumped circuit elements were assessed through preliminary

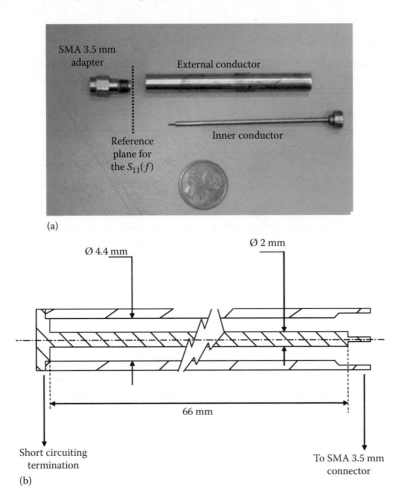

(a)

(b)

FIGURE 30.2   (a) Picture of the used coaxial probe. (b) Schematic configuration of the used coaxial probe. (From Cataldo, A., et al., *Measurement*, 43(8), 1031–1039, 2010.)

TDR measurements on well-referenced materials: these measurements allowed the optimization of the TL model).

30.3.2 DESCRIPTION OF THE PROCEDURE AND DISCUSSION OF RESULTS

The probe, filled with the liquid under test, was connected to the TDR instrument and the corresponding waveform was acquired. For each TDR measurement, also the waveforms corresponding to the SOL calibration standards were acquired. The collected TD data were processed through the fast Fourier transform (FFT)-based algorithm described in [31], thus extracting the corresponding calibrated $S_{11}(f)$ [20].

Finally, the Cole–Cole parameters are extracted by minimizing (over the Cole–Cole parameters) the difference between the FFT-extracted S_{11} (f) and the modeled $S_{11,MOD}(f)$. However, to obtain a more accurate evaluation of the Cole–Cole parameters, it is advisable to reduce the number of variables in the final minimization routine. To this purpose, the values of the static permittivity were assessed through an alternative method, and the minimization routine was performed over the remaining three Cole–Cole parameters (i.e., ε_∞, f_r, and β).

The evaluation of the ε_s of the oils was performed through capacitive measurements carried out with an LCR meter. LCR measurements were performed using the same coaxial probe, but removing the short circuit at the distal end. The probe (filled with the liquid under test) was considered as

TABLE 30.3

Averaged Cole–Cole Parameters for the 10 Types of Vegetable Oil (at 20.0°C)

Type of Oil	ε_s	$\sigma_{\varepsilon s}$	ε_∞	$\sigma_{\varepsilon\infty}$	f_r (MHz)	σ_{fr} (MHz)	β	σ_β
Olive (ac. = 0.3%)	3.08	0.01	2.39	0.01	315	11	0.33	0.01
Olive (ac. = 1.2%)	3.14	0.01	2.38	0.01	288	9	0.36	0.02
Olive (ac. = 1.6%)	3.19	0.03	2.36	0.01	259	4	0.40	0.01
Olive (ac. = 4.0%)	3.19	0.03	2.34	0.02	249	5	0.42	0.02
Peanut	3.05	0.01	2.40	0.01	334	5	0.28	0.01
Sunflower	3.12	0.04	2.40	0.01	292	5	0.31	0.01
Corn	3.11	0.02	2.41	0.01	309	6	0.29	0.02
Castor	4.69	0.01	2.56	0.01	122	2	0.42	0.01
Various seeds	3.10	0.02	2.43	0.01	371	5	0.27	0.01
Soybean	3.09	0.01	2.41	0.02	390	7	0.30	0.02

Source: From Cataldo, A., et al., *Measurement*, 43(8), 1031–1039, 2010.
Note: Standard deviation values are also reported.

a cylindrical capacitor whose capacitance ideally depends on the liquid filling the space between the cylindrical plates. The values of the static permittivity were considered as known (fixed) values in the subsequent minimization routines carried out on FFT transformed TDR data, employing the procedure described in Section 30.1: as expected, since the successive minimizations were performed only on three Cole–Cole parameters, the accuracy of the final results of the minimizations is enhanced (electrical conductivity is negligible for edible oils) [20].

Table 30.3 summarizes the Cole–Cole parameters and the corresponding standard deviations for ten different types of vegetable oils.

The results obtained for the Cole–Cole parameters through the proposed combined procedure are in good agreement with the scarce data available in the literature. Therefore, the proposed technique can be useful for quality monitoring of oils. Results show that the relaxation frequency appears to be the Cole–Cole parameter that differs the most among the considered oils. On this basis, a reliable, real-time, and in-situ quality control procedure may rely on relaxation frequency as the key parameter for identifying different kinds of oil. To this purpose, results may be collected in a database (together with the associated confidence intervals) and used as reference values for future "offline" measurements on unknown oils. More advances and additional details on the assessment of the procedure can be found in [21].

30.4 MONITORING OF THE DEHYDRATION PROCESS OF FRUIT AND VEGETABLES

Another industrial application of MR relates to the monitoring of the osmotic dehydration (OD) process of fruit and vegetables. OD is a process that is employed for the partial removal of water from fruit and vegetables, thus obtaining a significant increase of their shelf-life. In fact, by reducing water content, the growth of microorganisms is inhibited, and the rate of degradation reactions is reduced.

In the OD process, the product is immersed in a hypertonic solution, thus promoting two counter-current flows: (i) water motion from biological tissues to hypertonic solution and (ii) osmotic agents flow toward the product. OD has been extensively used for predehydration of pineapples [62], mushrooms [63], melons [64], bananas [65], and many other perishable goods in general. Nevertheless, as reported in [66], the adoption of OD in industrial production processes is hindered by problem with the overall management of the concentrated solutions. In fact, some practical aspects make managing and controlling the osmotic solution the bottleneck of the OD process. In particular, there are still some major concerns regarding the appropriate individuation of the loss in dewatering capacity and the possibility of reusing the spent solution, so as to make the process more economically advantageous [67].

On such bases, problems related to the OD control and osmotic solution management are still actual and, currently, no methodologies can successfully and rapidly represent the solution to the aforementioned needs. As a matter of fact, the osmotic pressure (π) might represent the primary parameter whose estimation would lead to the identification of the quality status of the osmotic solution. However, many food-industry processes typically involve osmotic solutions with high value of π (up to 250–300 bar), which, in practice, seriously limit the possibility of realizing specific devices for measurements. Additionally, the high uncertainties and the long times needed for measurement are other drawbacks that make the direct measurement of π extremely difficult.

On the other hand, water activity (a_w) is a reliable indicator of the status of the osmotic solution, since it is strictly related to the de-watering capacity of the solution. In fact, the OD proceeds until the osmotic solution and the food product have reached the same a_w value [68]. Nevertheless, although this parameter can be a suitable indicator in the OD management, not only the instrumentation for measuring it is quite costly but also measurements usually have to be performed in laboratory (i.e., samples of the solutions should be taken and subsequently analyzed off-line), thus limiting the possibility of continuous and in-situ monitoring.

To overcome this limitation, an alternative procedure, based on MR, for identifying the status of the industrial osmotic solutions for control and management purposes has been developed. In particular, it was demonstrated that there is a specific linear relationship between characteristic parameters of the solution (i.e., a_w, degree Brix, and consequently, π) and the corresponding measured reflection coefficient. As a direct practical consequence, once the functional relationship is individuated for a specific OD process, it can be used for the subsequent online monitoring in the process line, or for the optimization of the involved process parameters. The clear advantage of resorting to MR is that measurements can be performed in-situ and, most importantly, can provide results in real-time, thus allowing to promptly intervene in the process [22].

The proposed methodology was adopted also in a real test-case involving the OD of tomatoes, which is a typical application in the food industry. In this case, the most appropriate reflectometric parameter was found to be the so-called steady-state reflection coefficient, ρ_∞, which is the value of the reflection coefficient at long times (i.e., when multiple reflections caused by the signal traveling back and forth along the probe have died out). The quantity ρ_∞ is associated to the corresponding variation of a_w and of degree Brix. The experimental results confirm the suitability of the proposed method as a powerful tool for controlling the status of the osmotic solution, thus paving the way for a prompt management of the hypertonic solution in OD processes in the field of food industry.

Sliced tomatoes were immersed in the solution (mass ratio 5:1) of glucose syrup, and the dehydration process was observed for 24 h. TDR measurements were carried out at eight prefixed intervals; results are reported in Figure 30.3. The overall range of variation of the reflection coefficient (from −0.91 to −0.96), corresponding to different dehydration intervals, is rather narrow, even though some well distinguishable differences can be clearly detected. Referring to Figure 30.3, the long-distance traces present a slightly increased sensitivity range, anticipating a better performance of the method around the steady-state conditions. In fact, due to the elevated presence of polar species, such as the high concentration of NaCl, the resultant dielectric response is strongly attenuated; hence, significant variations of the reflected signal can be located prevalently around the static condition. In other words, the solution at the initial time has a high electrical conductivity (approximately 3080 μS/cm); as the dehydration proceeds, the electrical conductivity value increases up to approximately 5600 μS/cm, due to the agents' flow between tomatoes and solutions. However, thanks to the broadband frequency sensitivity of the TDR method, also little variation in static electrical conductivity of osmotic solutions can be detected.

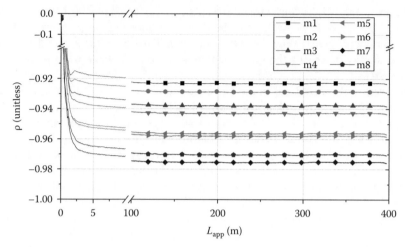

FIGURE 30.3 TDR measurements performed during the 24 h long cycle on the solution of the process line at long distance. (From Cataldo, A., et al., *Journal of Food Engineering*, 105(1), 186–192, 2011.)

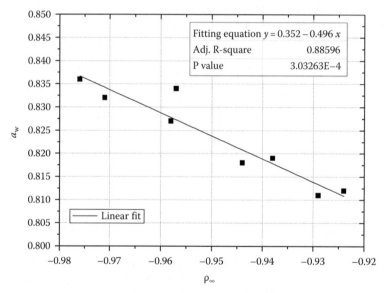

FIGURE 30.4 Behavior of a_w as a function of the steady-state reflection coefficient value, ρ_∞. (From Cataldo, A., et al., *Journal of Food Engineering*, 105(1), 186–192, 2011.)

To individuate the relation between the reference values and the measured data, in Figures 30.4 and 30.5, the long-distance reflection coefficients are plotted against a_w and degree Brix (both evaluated through dedicated methods), respectively, and a linear fitting is imposed. A good linear proportionality is observed since the values of the adjusted R-square indicate the correctness of the linearity assumption for the $\rho_\infty - a_w$ and for the $\rho_\infty -°$Bx relations. As a result, differently from the sugar-based solutions previously considered (which presented a linear trend at higher frequency measured data), for this kind of industrial solutions and for the intended control purposes, the most suitable measurement parameter is ρ_∞. Therefore, the adoption of the specific calibration curves ($\rho_\infty - a_w$ or $\rho_\infty -°$Bx) is the key for controlling the status of high-conductivity osmotic solutions.

On a side note, it is worth mentioning that the osmotic pressure values can be predicted through appropriate models, starting from the inferred values of a_w. This implies that, should there be the specific requirement to calibrate TDR measurements against the corresponding π values, the

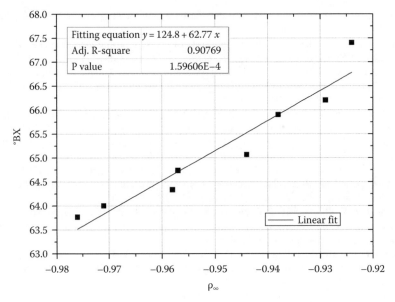

FIGURE 30.5 Behavior of Bx as a function of the steady-state reflection coefficient value, ρ_∞. (From Cataldo, A., et al., *Journal of Food Engineering*, 105(1), 186–192, 2011.)

proposed method would still be successfully applicable even without the difficult direct measurement of the osmotic pressure (given that the independently measurement of a_w and the consideration of a robust model can be suitably considered).

30.5 CONCLUSIONS

In this chapter, a brief overview of current achievements of the use of MR for monitoring and diagnostics purposes was given. The different approaches (TDR, FDR, and TDR/FDR) were addressed, and the pros and cons were described. Additionally, the different strategies for enhancing the final measurement accuracy were presented. Finally, the major applications (and the obtained results) of MR in the field of agrofood monitoring were discussed, thus providing a global picture of the open issues and of the advances of this measurement technique. Results demonstrate that MR is a powerful technique that can be effectively employed in a number of practical applications; in particular, the high versatility and other intrinsic characteristics (such as real-time response and possibility of remote control) have fostered the interest toward the adoption of MR for monitoring purposes. An additional major strength of MR lies in the high adaptability of the technique also to preexisting conditions (i.e., MR-based monitoring systems can be easily implemented even on the already-operating production line). Finally, also the implementation costs of MR-based solutions are highly competitive (especially when resorting to TDR instruments), thus making such systems extremely appealing not only for large-scale applications, but also for small-scale monitoring purposes.

REFERENCES

1. A. M. O. Mohamed, *Principles and Applications of Time Domain Electrometry in Geoenvironmental Engineering*. Great Britain: Taylor & Francis Group, 2006.
2. K. O'Connor and C. H. Dowding, *Geomeasurements by Pulsing TDR Cables and Probes*. Boca Raton, FL: CRC Press, January 1999.
3. D. E. Dodds, M. Shafique, and B. Celaya, TDR and FDR identification of bad splices in telephone cables, in: *Proceedings of IEEE Canadian Conference on Electrical and Computer Engineering*, May 2006, Niagara Falls, Ontario, Canada, pp. 838–841.

4. S. Wu, C. Furse, and C. Lo, Noncontact probes for wire fault location with reflectometry, *IEEE Sensors Journal*, 6(6), 1716–1721, 2006.

5. A. Scheuermann and C. Huebner, On the feasibility of pressure profile measurements with time-domain reflectometry, *IEEE Transactions on Instrumentation and Measurement*, 58(2), 467–474, 2009.

6. A. Scheuermann, C. Hubner, H. Wienbroer, D. Rebstock, and G. Huber, Fast time domain reflectometry (TDR) measurement approach for investigating the liquefaction of soils, *Measurement Science and Technology*, 21, 025104.1–025104.11, 2010.

7. A. Corsini, A. Pasuto, M. Soldati, and A. Zannoni, Field monitoring of the Corvara landslide (Dolomites, Italy) and its relevance for hazard assessment, *Geomorphology*, 66(1–4), 149–165, 2005.

8. W. F. Kane, T. J. Beck, and J. Hughes, Applications of time domain reflectometry to landslide and slope monitoring, in: *Proceedings of the Second International Symposium and Workshop on Time Domain Reflectometry for Innovative Geotechnical Applications*, 2001, Evanston, IL, pp. 305–314.

9. A. M. O. Mohamed and R. A. Said, Detection of organic pollutants in sandy soils via TDR and eigende-composition, *Journal of Contaminant Hydrology*, 76, 235–249, 2005.

10. A. McClanahan, S. Kharkovsky, A. R. Maxon, R. Zoughi, and D. D. Palmer, Depth evaluation of shallow surface cracks in metals using rectangular waveguides at millimeter-wave frequencies, *IEEE Transactions on Instrumentation and Measurement*, 59(6), 1693–1704, 2010.

11. A. Cataldo, L. Catarinucci, L. Tarricone, F. Attivissimo, and E. Piuzzi, A combined TD-FD method for enhanced reflectometry measurements in liquid quality monitoring, *IEEE Transactions on Instrumentation and Measurement*, 58(10), 3534–3543, 2009.

12. A. Cataldo, L. Tarricone, F. Attivissimo, and A. Trotta, Simultaneous measurement of dielectric properties and levels of liquids using a TDR method, *Measurement*, 41(3), 307–319, 2008.

13. C. P. Nemarich, Time domain reflectometry liquid levels sensors, *IEEE Instrumentation and Measurement Magazine*, 4(4), 40–44, 2001.

14. S. Sun, D. J. Pommerenke, J. L. Drewniak, G. Chen, L. Xue, M. A. Brower, and M. Y. Koledintseva, A novel TDR-based coaxial cable sensor for crack/strain sensing in reinforcedconcrete structures, *IEEE Transactions on Instrumentation and Measurement*, 58(8), 2714–2725, 2009.

15. C. Furse, P. Smith, and M. Diamond, Feasibility of reflectometry for nondestructive evaluation of prestressed concrete anchors, *IEEE Sensors Journal*, 9(11), 1322–1329, 2009.

16. N. E. Hager, III and R. C. Domszy, Monitoring of cement hydration by broadband time-domain-reflectometry dielectric spectroscopy, *Journal of Applied Physics*, 96(9), 5117–5128, 2004.

17. A. Cataldo, G. Cannazza, E. De Benedetto, and N. Giaquinto, A new method for detecting leaks in underground water pipelines, *IEEE Sensors Journal*, 12(6), 1660–1667, 2012.

18. A. Cataldo, E. De Benedetto, and G. Cannazza, *Broadband Reflectometry for Enhanced Diagnostics and Monitoring Applications*. Springer, Berlin, Germany, 2011.

19. A. Cataldo, L. Tarricone, M. Vallone, G. Cannazza, and M. Cipressa, TDR moisture measurements in granular materials: From the siliceous sand test-case to the applications for agro-food industrial monitoring, *Computer Standards & Interfaces*, 32(3), 86–95, 2010.

20. A. Cataldo, E. Piuzzi, G. Cannazza, and E. De Benedetto, Dielectric spectroscopy of liquids through a combined approach: Evaluation of the metrological performance and feasibility study on vegetable oils, *IEEE Sensors Journal*, 9(10), 1226–1233, 2009.

21. A. Cataldo, E. Piuzzi, G. Cannazza, and E. De Benedetto, Classification and adulteration control of vegetable oils based on microwave reflectometry analysis, *Journal of Food Engineering*, 112(4), 338–345, 2012.

22. A. Cataldo, G. Cannazza, E. De Benedetto, C. Severini, and A. Derossi, An alternative method for the industrial monitoring of osmotic solution during dehydration of fruit and vegetables: A test-case for tomatoes, *Journal of Food Engineering*, 105(1), 186–192, 2011.

23. A. Cataldo and E. De Benedetto, Broadband reflectometry for diagnostics and monitoring applications, *IEEE Sensors Journal*, 11(2), 451–459, 2011.

24. Hewlett Packard, Time domain reflectometry theory, USA, May 2006, Application Note 1304-2, 1–17.

25. D. A. Robinson, S. B. Jones, J. M. Wraith, D. Or, and S. P. Friedman, A review of advances in dielectric and electrical conductivity measurement in soils using time domain reflectometry, *Vadose Zone Journal*, 2, 444–475, 2003.

26. Hewlett Packard, Applying error correction to network analyzer measurements, 2002, Application Note 1287-3, pp. 1–15.

27. R. Friel and D. Or, Frequency analysis of time-domain reflectometry (TDR) with application to dielectric spectroscopy of soil constituents, *Geophysics*, 64(3), 707–718, 1999.

28. T. J. Heimovaara, W. Bouten, and J. M. Verstraten, Frequency domain analysis of time domain reflectometry waveforms: 2. A four-component complex dielectric mixing model for soils, *Water Resources Research*, 30, 201–209, 1994.

29. T. J. Heimovaara, Frequency domain analysis of time domain reflectometry waveforms: 1. Measurement of the complex dielectric permittivity of soils, *Water Resources Research*, 30(2), 189–199, 1994.

30. S. B. Jones and D. Or, Frequency domain analysis for extending time domain reflectometry water content measurement in highly saline soils, *Soil Science Society of America Journal*, 68, 1568–1577, 2004.

31. A. Cataldo, L. Catarinucci, L. Tarricone, F. Attivissimo, and A. Trotta, A frequency-domain method for extending TDR performance in quality determination of fluids, *Measurement Science and Technology*, 18(3), 675–688, 2007.

32. T. J. Heimovaara, J. A. Huisman, J. A. Vrugt, and W. Bouten, Obtaining the spatial distribution of water content along a TDR probe using the SCEM-UA Bayesian inverse modeling scheme, *Vadose Zone Journal*, 3, 1128–1145, 2004.

33. A. Cataldo, G. Cannazza, E. De Benedetto, L. Tarricone, and M. Cipressa, Metrological assessment of TDR performance for moisture evaluation in granular materials, *Measurement*, 42(2), 254–263, 2009.

34. S. J. Zegelin, I. White, and D. R. Jenkins, Improved field probes for soil water content and electrical conductivity measurement using time domain reflectometry, *Water Resources Research*, 25, 2367–2376, 1989.

35. J. S. Selker, L. Graff, and T. Steenhuis, Noninvasive time domain reflectometry moisture measurement probe, *Soil Science Society of America Journal*, 57, 934–936, 1993.

36. A. Cataldo, G. Monti, E. De Benedetto, G. Cannazza, and L. Tarricone, A noninvasive resonance-based method for moisture content evaluation through microstrip antennas, *IEEE Transactions on Instrumentation and Measurement*, 58(5), 1420–1426, 2009.

37. P. A. Ferré, J. H. Knight, D. L. Rudolph, and R. G. Kachanoski, A numerically based analysis of the sensitivity of conventional and alternative time domain reflectometry probes, *Water Resources Research*, 36(9), 2461–2468, 2000.

38. K. S. Cole and R. H. Cole, Dispersion and absorption in dielectrics: I. Alternating current characteristics, *Journal of Chemical Physics*, 9(4), 341–351, 1941.

39. N. Miura, S. Yagihara, and S. Mashimo, Microwave dielectric properties of solid and liquid foods investigated by time-domain reflectometry, *Journal of Food Science*, 68(4), 1396–1403, 2003.

40. S. O. Nelson, W. Guo, S. Trabelsi, and S. J. Kays, Dielectric spectroscopy of watermelons for quality sensing, *Measurement Science and Technology*, 18, 1887–1892, 1999.

41. S. O. Nelson and S. Trabelsi, Dielectric spectroscopy of wheat from 10 MHz to 1.8 GHz, *Measurement Science and Technology*, 17, 2294–2298, 2006.

42. V. Changrue, V. Orsat, G. Raghavan, and D. Lyew, Effect of osmotic dehydration on the dielectric properties of carrots and strawberries, *Journal of Food Engineering*, 88, 280–286, 2008.

43. S. L. Birla, S. Wang, J. Tang, and G. Tiwari, Characterization of radio frequency heating of fresh fruits influenced by dielectric properties, *Journal of Food Engineering*, 89, 390–398, 2008.

44. M. Castro-Giráldez and P. J. Fito, Application of microwaves dielectric spectroscopy for controlling pork meat (*Longissimus dorsi*) salting process, *Journal of Food Engineering*, 97, 484–490, 2010.

45. W. Gardner and D. Kirkham, Determination of soil moisture by neutron scattering, *Soil Science*, 73, 391–401, 1951.

46. R. Reginato and C. van Bavel, Soil water measurement with gamma attenuation, *Proceedings—Soil Science Society of America*, 28, 721–724, 1964.

47. A. Cataldo, M. Vallone, L. Tarricone, G. Cannazza, and M. Cipressa, TDR moisture estimation for granular materials: an application in agro-food industrial monitoring, *IEEE Transactions on Instrumentation and Measurement*, 58(8), 2597–2605, 2009.

48. R. Cerny, Time-domain reflectometry method and its application for measuring moisture content in porous materials: A review, *Measurement*, 42(3), 329–336, 2009.

49. K. Noborio, Measurement of soil water content and electrical conductivity by time domain reflectometry: a review, *Computers and Electronics in Agriculture*, 31(3), 213–237, 2001.

50. G. C. Topp, J. L. Davis, and A. P. Annan, Electromagnetic determination of soil water content: Measurements in coaxial transmission lines, *Water Resources Research*, 16, 574–582, 1980.

51. J. R. Birchak, D. G. Gardner, J. E. Hipp, and J. M. Victor, High dielectric constant microwave probes for sensing soil moisture, *Proceedings of the IEEE*, 62, 93–98, 1974.

52. G. P. De Loor, Dielectric properties of heterogeneous mixtures, PhD dissertation, University of Leiden, Leiden, the Netherlands, 1956.

53. M. C. Dobson and M. T. Hallikainen, Microwave dielectric behavior of wet soil-part II: Dielectric mixing models, *IEEE Transactions on Geoscience and Remote Sensing*, 23(1), 35–46, 1985.
54. M. Ansoult, L. W. De Backer, and M. Declercq, Statistical relationship between apparent dielectric constant and water content in porous media, *Soil Science Society of America Journal*, 49, 47–50, 1985.
55. H. Lizhi, K. Toyoda, and I. Ihara, Dielectric properties of edible oils and fatty acids as a function of frequency, temperature, moisture and composition, *Journal of Food Engineering*, 88, 151–158, 2008.
56. T. S. Ramu, On the high frequency dielectric behavior of castor oil, *IEEE Transactions on Electrical Insulation*, 14, 136–141, 1979.
57. G. Morchio, A. DiBello, C. Mariani, and E. Fedeli, Detection of refined oils in virgin olive oil, *Rivista Italiana Sostanze Grasse*, 66, 251–257, 1989.
58. A. H. El-Hamdy and N. K. El-Fizga, Detection of olive oil adulteration by measuring its authenticity factor using reversed-phase high-performance liquid chromatography, *Journal of Chromatography*, 708, 351–355, 1995.
59. A. Tay, R. K. Singh, S. S. Krishnan, and J. P. Gore, Authentication of olive oil adulterated with vegetable oils using Fourier transform infrared spectroscopy, *Lebensmittel Wissenschaft und-Technologie*, 35, 99–103, 2002.
60. F. Guimet, J. Ferré, and R. Boque, Rapid detection of olive-pomace oil adulteration in extra virgin olive oils from the protected denomination of origin Siurana using excitation-emission fluorescence spectroscopy and three-way methods of analysis, *Analytica Chimica Acta*, 544, 143–152, 2005.
61. R. Sacchi, F. Adeo, and L. Polillo, 1H and 13C NMR of virgin olive oil. An overview, *Magnetic Resonance in Chemistry*, 35, 133–145, 1997.
62. G. E. Lombard, J. C. Oliveira, P. Fito, and A. Andres, Osmotic dehydration of pineapple as a pre-treatment for further drying, *Journal of Food Engineering*, 85(2), 277–284, 2008.
63. E. Torringa, E. Esveld, R. van den Berg, and P. Bartels, Osmotic dehydration as a pre-treatment before combined microwave-hot-air drying of mushrooms, *Journal of Food Engineering*, 49, 185–191, 2001.
64. S. Rodrigues and F. A. N. Fernandes, Dehydration of melons in a ternary system followed by air-drying, *Journal of Food Engineering*, 80, 678–687, 2007.
65. F. A. N. Ferndandes, S. Rodrigues, O. C. P. Gaspareto, and E. L. Oliveira, Optimization of osmotic dehydration of bananas followed by air-drying, *Journal of Food Engineering*, 77, 188–193, 2006.
66. M. Dalla Rosa and F. Giroux, Osmotic treatments (OT) and problems related to the solution management, *Journal of Food Engineering*, 49, 223–236, 2001.
67. J. Warczok, M. Ferrando, F. Lopez, A. Pihlajamaki, and C. Guell, Reconcentration of spent solutions from osmotic dehydration using direct osmosis in two configurations, *Journal of Food Engineering*, 80, 317–326, 2007.
68. G. V. Barbosa-Canovas and H. Vega-Mercado, *Dehydration of Food*. New York: Chapman & Hall, 1996.
69. A. Cataldo, E. Piuzzi, G. Cannazza, E. De Benedetto, and L. Tarricone, Quality and anti-adulteration control of vegetable oils through microwave dielectric spectroscopy, *Measurement*, 43(8), 1031–1039, 2010.

31 Wearable PTF Strain Sensors

Sari Merilampi

CONTENTS

Sensors are an essential part of a smart system. In many applications, it would be especially useful if the sensing element could be embedded as a part of the structure that is monitored. In this chapter, embedded strain sensors are discussed. Especially, their use in wearable applications and sensing of relatively large displacement is examined. Printing is used to fabricate the sensors, and the strain sensitivity of the prototype sensors is based on the materials used. Two approaches are taken; strain sensitive stretchable conductors and strain sensitive stretchable radio frequency identification (RFID) tags. Both the performance and the modification possibilities of the prototypes are evaluated and their potential applications are discussed. The chapter is divided into following sectors: introduction and background information are first given in Section 31.1 after which the theoretical foundation of polymer thick film (PTF) and RFID technology in strain sensing is introduced in Section 31.2. The performance of simple prototype structures is investigated through case studies in Section 31.3, and their potential applications are then discussed in Section 31.4.

31.1 INTRODUCTION AND BACKGROUND

31.1.1 PRINTED ELECTRONICS AND PTF TECHNOLOGY IN WEARABLE APPLICATIONS

Printing technology can be used to fabricate conductive patterns, passive components, and passive microwave circuits. Screen printing, gravure printing, flexography, ink jet, and lithography are examples of different printing techniques [1–8]. Many kinds of inks are available on the market, most of which are thick film materials. In novel printed electronics applications such as in printed RFID tag antennas, polymer thick film inks are used. Compared to traditional thick film materials, PTF inks do not contain glass and they do not need to be fired but are cured at lower temperatures (100°C–200°C). This makes it possible to use unconventional substrates such as paper and fabric. This creates an opportunity to develop novel light-weight, flexible, and wearable applications, one of which is wearable large area embedded strain sensor. Printed conductive PTF structures can be utilized in such sensors due to their strain sensitive nature [9].

Electrically conductive PTFs are polymer matrix composites. The PTF ink consists of metallic filler, binder material (polymer matrix), solvents, and additives. The ingredients, as well as their relative amounts, are selected according to the printing method. The *filler* of the ink is metal powder, commonly silver, which is also used in the structures discussed in this chapter. Typically, the particle size in screen printable silver inks is a few microns. The particles are usually in flake form, and the size distribution is relatively large [9–12].

The particles are suspended in *vehicle* for printing use. The vehicle normally consists of a non-volatile and a volatile portion, which will evaporate during curing. *Solvents* form the volatile part of the vehicle and they are normally used to impart the desired characteristics of flow, viscosity, and ink density. The non-volatile portion of the vehicle is called *binder*. Binder is needed to hold the particles together once the ink is transferred onto a substrate, to affix them to the surface, and to protect the film from being damaged. Binder materials in screen printing inks are *resins* [10,12–15]. PTF inks also contain *modifiers* or *additive* ingredients, which are used to affect the characteristics of the ink [9,12,15].

The electrical conductivity of composites similar to the silver ink used in the case studies in this chapter is discussed in, for example, articles [4–6,10,16,17]. The total conductivity is a function of the conductivity of the polymer matrix, the quantity and conductivity of conductive filler material, and the tunneling and hopping conductivity [4–6,10,16,17]. The conductivity of the matrix (polymer) is negligible in our case, and the role of tunneling and hopping conductivity is assumed minor due to high particle content. In wearable applications, the substrate material also has an effect on the effective conductivity of the printed film. For example, fabrics typically absorb ink, which affects the ink film resistance. When the printed film is exposed to tensile load, structural and microstructural changes occur. These changes have an effect on the effective conductivity of PTF structures during straining and make them strain sensitive. The strain sensitivity is discussed further in Section 31.2.

31.1.2 RFID Technology

When sensors are part of other structures, wireless passive sensors are preferred. Passive sensors need no battery or maintenance, so the sensors can be embedded in such structures as the wall, packaging, or in clothing [2,7]. The lifetime of a sensor is then required to be the same as the lifetime of the structure in which it is embedded in. RFID technology can be utilized in passive sensors.

Although the technology can be used for sensing, the original purpose of RFID is automatic identification of products. RFID systems consist of tags, a reader, and a data processing system. The tags are located on objects that are identified. The data processing system is connected to the reader in order to process the data and to provide additional information on the identified object.

There are many different types of RFID systems. In this chapter, a passive ultra high frequency (UHF) RFID system is discussed. Passive tags are light-weight, simple in structure, inexpensive, generally resistant to harsh environmental conditions, and they offer virtually unlimited operational lifetime. The passive technology shows promise in embedded applications since passive tags require very little maintenance. A UHF RFID system is used since the read ranges of passive UHF RFID systems are longer compared to other used frequencies. Thus, they also offer promising opportunities in remote sensing. The typical operation frequency band of the UHF RFID systems is 860–960 MHz [9,18–21].

In passive UHF RFID systems, which support the EPC Class 1 Gen 2 standard, communication begins with the reader, which emits a continuous wave (power) to activate the tag. The continuous wave is followed by a command. After the command, the reader starts to send a continuous wave, which is then modulated and backscattered by the tag, according to its identification code. The modulation is executed by changing the scattered field from the tag [9,18,19,22–24].

The passive UHF RFID tags are actually antennas loaded with a microchip. All the electronics that the tag needs are integrated in the chip [9,25–28]. The antenna's function is to receive and back-scatter the electromagnetic wave, which is transmitted by the reader. The wave is led through the tag antenna to the microchip and backscattered to the reader according to modulation. In backscattering systems, the modulation of the scattering from the tag is realized by switching the input impedance of the chip between two different values (typically a matched impedance state and a mismatched impedance state). In the case of poor matching, almost all power is backscattered to the reader. In the well-matched state, the power is transferred from the tag antenna to the chip and a fraction of the transmitted power is backscattered. In addition to the amplitude of the backscattering wave, also the phase may be changed (depending on the chip impedance states). These changes in the backscattering wave can be interpreted as bits 1 and 0 [9,18,29].

31.2 PTF AND RFID TECHNOLOGY IN STRAIN SENSING

31.2.1 ELECTRICAL CONDUCTIVITY OF PRINTED PTF UNDER TENSILE LOAD

The effect of tensile strain on conductivity of PTF composites has been investigated by several scientists. When the polymer matrix composite is strained, several different phenomena are reported to occur. These include the breaking of the 3D network formed by the conductive filler particles, the loss of the contact between particles, the increase in the inter filler distance, the delamination and the reorientation of particles, and the decrease of the volume fraction of the filler material as the material is extended [3–6,30]. Sevkat et al. [4] have also reported the decrease of resistance in the tensile test due to the Poisson effect: more contacts in the lateral direction and fewer contacts in the longitudinal direction are formed. Sevkat et al. [4] reported that the resistance change and the tensile strain follow an exponential or power law. Hu et al. [5] mentioned that the change in the resistance is most sensitive to the strain level at the percolation threshold. At large strain levels, the conditions resemble the state where the content of silver particles is close to the critical volume fraction. The aspect ratio affects the critical volume fraction due to a wider contact area between the particles with increasing aspect ratio, according to Lin et al. [31]. In this way, the aspect ratio may also affect the electrical performance of an ink film under strain. In articles [32–34], the conductivity of the particle-reinforced composites was observed to depend on the frequency near the percolation threshold; but with higher particle loadings, the conductivity was found to be practically independent of the frequency. If the particles move away from each other under large strain, the conditions are similar to the near percolation threshold and the hopping conductivity may be involved in the total conductivity [5]. These aforementioned phenomena make the conductive PTF films strain sensitive [16].

31.2.2 SENSING STRAIN WITH PTF TAG ANTENNA

The tag's response can be manipulated according to the prevailing circumstances, such as mechanical changes in the tag. This is utilized in self-sensing tags [8]. The changes in the antenna affect wirelessly measureable parameters, such as the threshold power and the backscattered power of the tag. The transmitted threshold power is the minimum sufficient power required from the reader to activate the microchip (in matched state). Threshold measurement is performed by increasing the transmitted power until the tag can respond to the query command of the reader [9]. Considering that the power P_t is transmitted, the power P_{tag} received by the tag antenna at distance d, can be calculated as

$$P_{tag} = L_{pol} A_{e,tag} S_t = L_{pol} G_t G_{tag} \left(\frac{\lambda}{4\pi d} \right)^2 P_t \qquad (31.1)$$

where
   L_{pol} is the polarization loss factor
   $A_{e,tag}$ is the effective aperture of the tag in the direction of the incoming wave
   S_t is the power density (equal to the power density of an isotropic radiator multiplied by the gain of the transmitting antenna)

λ is the wavelength

G_t is the reader's transmitting antenna gain

G_{tag} is the tag antenna gain

Gain of an antenna is the product of directivity and radiation efficiency of the antenna ($G = e_{rad}D$) [9,24,35].

The power received by the tag antenna, given in Equation 31.1, is the available power for the on-tag microchip, but the delivered power to the chip depends on the impedance matching between the tag antenna and the chip. It is well known that the maximum power transfer between two complex impedances is achieved under the complex conjugate-matched conditions, and the quality of the power matching between a tag antenna (source) and the microchip (load) can be evaluated in terms of the power reflection coefficient:

$$\Gamma_{tag} = \left|\rho_{tag}\right|^2 = \left|\frac{Z_{ic} - Z_{tag}^*}{Z_{ic} + Z_{tag}}\right|^2 \tag{31.2}$$

where Z_{tag} and Z_{ic} are antenna and chip impedance, respectively, and the impedance ratio denoted by ρ_{tag} is the power wave reflection coefficient [36]. In terms of power, Γ_{tag} is the ratio of the reflected power from the antenna-microchip-interface, due to impedance mismatch, to the available power to the microchip. Thus, combining Equations 31.1 and 31.2, the power delivered to the chip is

$$P_{ic} = (1 - \Gamma_{tag})P_{tag} = (1 - \Gamma_{tag})L_{pol}G_tG_{tag}\left(\frac{\lambda}{4\pi d}\right)^2 P_t. \tag{31.3}$$

Obviously, P_{ic} in Equation 31.3 is maximized by minimizing Γ_{tag}, i.e., by tuning the tag antenna impedance, so that $Z_{tag} = Z_{ic}^*$ [9,24,35].

On the other hand, the power reflection coefficient is increased, when the tag antenna impedance changes from the "matched" state. This happens when the antenna is stretched. In addition, the radiation efficiency and the directivity of the tag antenna change due to the changed effective conductivity of the ink [16] and the changes in antenna dimensions. Since G_{tag} is the product of the directivity and radiation efficiency of the tag, G_{tag} in Equation 31.3 changes during straining. Since the aforementioned parameters affect the amount of power received by the chip, the minimum power required to activate the tag (threshold power) is changed, and the deformation (strain) can therefore be observed wirelessly by measurement of the threshold power.

Another parameter that can be used to measure the strain in self-sensing stretchable tags is the backscattered signal power. The backscattered signal power is the time-average power detected from tag response at the receiver [9]. The power of the tag signal at the receiver $P_{received}$ is [37]

$$P_{received} = \alpha\left|\rho_1 - \rho_2\right|^2\left(\frac{\lambda}{4\pi d}\right)^4 G_t^2 G_{tag}^2 P_t. \tag{31.4}$$

where

G_t is the gain of the reader (transmit/receive) antenna (G_t^2 is the product of transmitting and receiving reader antenna gain in the case of a bi-static reader)

ρ_1 and ρ_2 are the power wave reflection coefficients [36] of the tag in matched and mismatched chip impedance states

α is a coefficient which depends on the specific modulation details [37]

Equation 31.4 shows that the received backscattered signal power depends on the power wave reflection coefficients in both chip impedance states as well as on the tag antenna gain. As discussed earlier, the gain and the impedance of the tag change when the tag is strained and thus parameters G_{tag}, ρ_1 and ρ_2 are all functions of the strain and the backscattered signal power could be used in strain measurements in theory [35].

31.3 CASE STUDIES

The usability of PTF conductors and printed tags in practice is discussed in this section. The performance and modification of the sensors are discussed. PTF conductors are first examined. The effect of the substrate material and ink material are introduced. Similar investigation is then performed with RFID tags. In addition, the effect of the geometry is evaluated.

31.3.1 STRETCHABLE PTF CONDUCTORS

As discussed in Section 31.2, conductive PTFs are strain sensitive. In addition to the PTF ink material, the strain sensitivity depends on other factors. The role of the substrate in strain sensitive screen printed PTF conductors was examined in publications [16,38]. Two substrate materials, elastic polyvinyl chloride (PVC) sheet and fabric, were used. The idea was to select two significantly differently behaving materials. The stretchability of the PVC substrate is based on the material itself, and the elastic behavior of the fabric substrate is based on the texture of the fabric. There are elastane rubber bands inside the woven fabric texture. When the fabric is stretched, the individual fibers do not stretch but the woven structure does [16]. PTF conductors, 97 mm in length and 8 mm in width, were fabricated on the substrates by screen printing. The change in DC resistance of the conductors on the aforementioned substrates during straining was measured. The measured change in resistance is presented in Table 31.1 [9].

It was found that the resistance changes during stretching on both PVC and fabric substrates. Figure 31.1 illustrates the printed film cross section on a PVC substrate in unloaded conditions

TABLE 31.1
The Resistance of Printed Film
on Fabric and PVC during Straining

Substrate	Strain (%)	DC Resistance (Ω)
Fabric	0	1
Fabric	5	7
Fabric	10	12
Fabric	20	40
Fabric	28	70
Fabric	38	200
Fabric	48	1,200,000
Elastic PVC	0	0.5
Elastic PVC	5.7	0.75
Elastic PVC	10.8	1.2
Elastic PVC	20.5	2.4
Elastic PVC	29.2	4.1
Elastic PVC	40.3	7.1
Elastic PVC	48.1	9.4

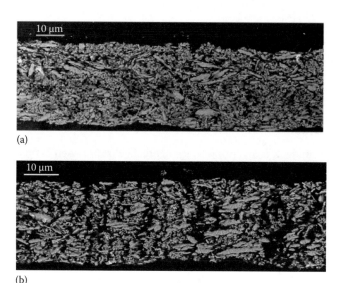

(a)

(b)

FIGURE 31.1 SEM micrograph from a cross section of printed film on PVC under 0% (a) and 50% (b) strain. (From Merilampi, S. et al., *Microelectron. Reliab.*, 50(12), 2001, 2010.)

and during stretching. The effect of tensile strain on the resistance of the ink film on PVC can be explained by microstructural changes in the printed film. Poisson's effect, found in [4], was not observed in the samples because of the non-elastic matrix polymer cracking. Instead, matrix micro-cracking is shown in Figure 31.1 [16,28].

The cracking of the matrix leads to partial breaking of the 3D network and reduction of the contact area between the particles. The interfiller distance also increases and the delamination of particles occurs. It was estimated in [16] that about 20% volume fraction reduction under ≈50% strain occurs. The volume fraction of particles is recovered almost entirely after straining. Only particle delamination from the matrix could be seen after recovery. The particle size distribution of the ink used in the samples is large and strain levels are high. Thus, instead of the movement of the particles relative to each other, matrix cracking and its consequences are believed to be the main cause for the increase in resistance. It should be noted that in addition to microstructural changes, change in the physical dimensions (width, length, and thickness) affects resistance.

The resistance of the printed film on fabric changes more rapidly during straining compared with samples on PVC. It can be seen from Figure 31.2 that the decrease in the conductivity of the ink on fabric is due more to the structural change of the fabric than to microstructural changes of the printed film. The pictures in the lower left corners in Figure 31.2a and b are taken from above the sample conductors and they illustrate the behavior of the fabric. The magnification in Figure 31.2b shows that no microstructural changes occur in the printed film during straining [9,16].

The effect of the ink material on the behavior of printed PTF structures was examined in article [38]. Two inks "A" and "B" were selected as the conductor material. The particles in the two inks are similar in shape and size, but there is a significant difference in the particle content. Ink "A" contains 72 wt% silver flakes, whereas ink "B" contains 52 wt%. Similar conductors as those used previously (of length and width of 97 and 8 mm) were printed by screen printing method. The resistance of the conductors was first measured in unloaded conditions after which strain was applied [38].

Because of the lower particle content, ink "B" has higher resistance already without strain, being 2.4 Ω, whereas the resistance of sample printed with ink "A" is 0.7 Ω. Resistance of the sample conductors as a function of strain is introduced in Figure 31.3. It is seen that the difference in the measured DC resistance between inks "A" and "B" increases as the conductor is strained. The resistance of ink "A" changed during straining (from 0% to 51.5%) from 0.7 to 7.3 Ω and the resistance of ink "B" from 2.4 to 190 Ω.

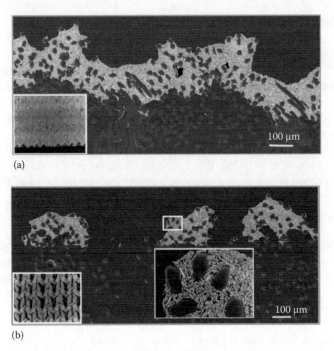

(a)

(b)

FIGURE 31.2   SEM micrograph of a cross section of printed film on fabric under 0% strain (a) and 50% strain (b). (From Merilampi, S. et al., *Microelectron. Reliab.*, 50(12), 2001, 2010.)

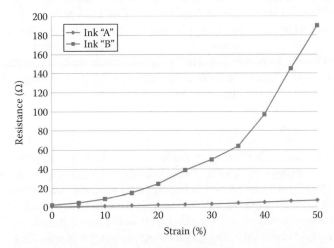

FIGURE 31.3   The resistance of sample conductors as a function of strain. (From Merilampi, S. et al., Modification of printed wearable strain sensors by PTF ink particle content adjustment, in *Fourth International Symposium on Applied Sciences in Biomedical and Communication Technologies*, Isabel, Barcelona, Spain, 2011.)

The strain sensitivity of the ink films is assumed to be caused by the same phenomena as in the case of the sample conductors on PVC discussed earlier (Figure 31.1). Ink "B" has a larger portion of polymer, which is assumed to lead to increased amount of matrix cracking that would explain the more significant increase in resistance. In addition, there may be differences between the inks in the increase in inter filler distance as well as the loss of contact between particles and the particle orientation. The difference due to the particle shape and size is assumed to be insignificant since the particles of both inks are similar. The difference between the change in dimensions of conductors between ink "A" and ink "B" is also assumed insignificant [38].

It can be concluded that in addition to the substrate material, the particle content of the ink plays a very important role in the behavior of the printed conductors during straining. These can be utilized in sensing different strain levels [38].

In articles [6,10,11,31,39], it was discussed that the size and shape of the particles both affect the percolation threshold and thus also affect the behavior of the printed film electrical performance under tensile load. Also the behavior of the polymer is important. This means that in addition to particle content of the ink, the behavior of a printed PTF strain sensor can be modified with the size and shape of the conductive particles as well as by polymer matrix choice [38].

It is also worth remembering that the geometry of the printed structures also affects the resistance change during straining [6,10,16]. This means that both material choices and geometry can be used to adjust the printed sensor according to the requirements set by the application [38].

31.3.2 STRETCHABLE RFID TAGS

The strain sensitive behavior of PTFs was exploited in a wireless strain sensor whose operation is based on passive UHF RFID technology in articles [28,35,38]. The tag antenna used as a strain sensor is a rectangular short dipole, as shown in Figure 31.4. The dimensions of the tag are $L = 97$ mm, $W = 8$ mm, $s = 17$ mm, $t = 0.5$ mm, $u = 2$ mm, and $v = 5$ mm. The geometry is selected because it is simple and the outer dimensions are the same as those of the sample conductors introduced earlier in this chapter [9].

EPC Gen 2 protocol-based measurements were performed to study tag performance. The transmitted threshold power and the backscattered signal power were measured as a function of frequency. The measurements were performed without and during straining [9,28].

The threshold power showed an ambiguous response to strain at certain frequencies. As discussed, the power the chip receives is, in addition to tag antenna gain (G_{tag}), proportional to the power reflection coefficient (Γ_{tag}) of the tag antenna (Equation 31.3). When the tag antenna is stretched, the power reflection coefficient is increased because the tag antenna impedance changes from the "matched" state. Impedance mismatch and decreased radiation efficiency increase the required threshold power, whereas the increasing directivity decreases it. The ambiguous behavior of the tag threshold power is caused by impedance matching, which changes the frequency of optimal impedance matching. However, with some form of calibration, the threshold power could be used in strain sensing [9,28].

The backscattered signal power showed far less ambiguous behavior and thus seems a promising method in strain measuring. The backscattered signal power of the samples on PVC monotonically increases during stretching, although the conductor loss resistance increases. This increase in backscattered signal power is presumed to be caused by the increased antenna directivity as the length of the dipole increases toward half wavelength. Due to changed antenna impedance, the power wave reflection coefficients ρ_1 and ρ_2, are also affected by antenna deformation. However, it is unclear how $|\rho_1 - \rho_2|$ is affected. Also, the frequency dependency of the backscattered signal power is affected by the modulation properties of the chip, described by the quantity $\rho_1 - \rho_2$ in (31.4).

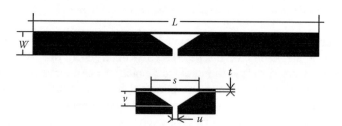

FIGURE 31.4 Strain sensor tag geometry called simple dipole. (From Merilampi, S. et al., *Sensor Rev.*, 31(1), 32, 2011.)

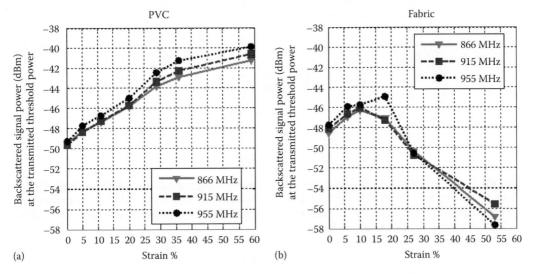

FIGURE 31.5 The backscattered signal power of the tag on (a) PVC and on (b) fabric as a function of strain at 866, 915, and 955 MHz. (From Merilampi, S. et al., *Sensor Rev.*, 31(1), 32, 2011.)

However, the measured responses at three different frequencies (Figure 31.5) are very similar, suggesting that the frequency dependence of the modulation efficiency of the chip is not strong within the frequency range studied. Also, in our measurements, the power delivered to the chip remains constant (which is the threshold power) so that the power dependence of its modulation properties does not affect the measured backscattered signal power as a function of strain [9,28].

In the case of PTF conductors, it was found that the substrate affected the resistance change of the conductor during stretching. The conductor loss resistance increase of the conductors on PVC was significantly smaller than that of conductors on fabric (Table 31.1). Since backscattered power showed promising behavior, the study of the effect of changing the substrate on the backscattered power was performed in article [28,35]. Similar to the PTF conductors, stretchable fabric was used as the substrate (in addition to the PVC) also in the case of the RFID tags. Figure 31.5 shows the backscattered power as a function of strain at 866, 915, and 955 MHz for samples on PVC and on fabric.

We can conclude that the response from the tag on PVC is fairly linear until 30% strains after which the role of the ohmic losses becomes more significant and the increase in backscattered power as a function of strain slows down.

In the case of the tag on fabric, the backscattered signal power first increases as a result of increasing directivity, but when the ohmic losses increase significantly from 10% to 20% strain levels up (see Table 31.1), they start to dominate and backscattered power starts to decrease. The difference in the ohmic losses is assumed to be the main reason for the differences in behavior of the tag on PVC and that on fabric [9,28].

In addition to the substrate, in the case of the conductors, also the ink material affected the behavior of the printed structure during stretching. In article [38], two inks with different particle content, ink "A" and ink "B," were also used to print the tag geometry shown in Figure 31.4. The inks are the same as were used in the case of stretchable conductors introduced in Figure 31.3. The effect of the particle content on the backscattered signal power was investigated. The backscattered signal power of the sample tags under strain is presented in Figures 31.6 and 31.7. The backscattered signal power of the prototype tags printed with ink "A" first starts to increase compared to the reference value, which is the backscattered signal power in the unloaded conditions. This is assumed to be caused by the increasing directivity of the tag antenna. After approximately 27% strain, the backscattered signal power starts to decrease at 866.6 and 915 MHz. Also at 954.2 MHz, the increase of the backscattered signal power as a function of strain remarkably slows down. This contributes

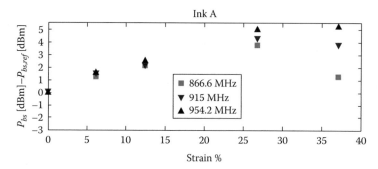

FIGURE 31.6 The backscattered signal power of a prototype tag printed with ink "A" as a function of strain. (From Merilampi, S. et al., Modification of printed wearable strain sensors by PTF ink particle content adjustment, in *Fourth International Symposium on Applied Sciences in Biomedical and Communication Technologies*, Isabel, Barcelona, Spain, 2011.)

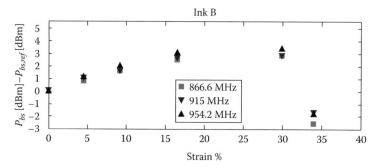

FIGURE 31.7 The backscattered signal power of a prototype tag printed with ink "B" as a function of strain. (From Merilampi, S. et al., Modification of printed wearable strain sensors by PTF ink particle content adjustment, in *Fourth International Symposium on Applied Sciences in Biomedical and Communication Technologies*, Isabel, Barcelona, Spain, 2011.)

to the increase of the ohmic loss of the tag antenna conductor, reducing the tag antenna gain and backscatter. It can be seen from Figure 31.6 that the frequency also affects the behavior of the tag during straining. It is assumed that the radiation efficiency of the tag antenna at 866.6 MHz is lower compared with 915 and 954.2 MHz to begin with (due to smaller electrical length), and thus the tag antenna gain and backscattered power slow down at a smaller strain level at 866.6 MHz, compared with 915 and 954.2 MHz [38].

The backscattered signal power of the prototype tag printed with ink "B" is presented in Figure 31.7. Generally, the backscattered signal power is higher in the case of tags printed with ink "A," but this does not show in Figures 31.6 and 31.7 since they present the difference between the measured value and the reference level. The reason for the higher back scattered signal power is the higher conductivity of ink "A" and thus the lower ohmic losses [38].

The behavior of the tag printed with ink "B" as a function of strain seems similar to the tag printed with ink "A." There are though small differences. As is seen from Figure 31.3, the resistance of the sample conductor printed with ink "B" increases faster compared with a sample conductor printed with ink "A." The faster increasing resistance also increases the ohmic losses more rapidly. It is seen from Figure 31.7 that the backscattered signal power measured under about 15% strain is almost the same as that below 27% strain. This means that the maximal backscattered signal power lies between 15% and 27%; and after 27% strain, the backscattered signal power decreases rapidly. More measurements are required to define the maximum of the curve more exactly. Nevertheless, it is clear that the increase of the backscattered signal power of the tag printed with ink "B" starts to slow down at lower strain levels compared with the tag printed with ink "A" (see measurements at 954.2 MHz, for example). It is noted that for all frequencies tested, the response from the tag printed with ink "A" is monotonic until 27% strain and from the tag printed with ink "B" at least until 15% strain [38].

It is concluded from the results that the particle content also affects the printed tags behavior during straining, although the effect is more pronounced for the printed conductor resistance. Compared with the results in Figure 31.5, the difference in behavior between the printed tags with different particle content is smaller than the difference in behavior between tags printed on fabric and PVC.

In addition to ink and substrate material, the effect of the tag geometry on strain behavior has been investigated. The tag geometry in Figure 31.4 is now called simple dipole. In article [35], another geometry called folded dipole (Figure 31.8) was examined. The folded dipole tags were printed on elastic PVC with the same ink that was used in printing the simple dipole tags in articles [28] and [35]. Backscattered signal power was measured in unloaded conditions and during straining.

The measured backscattered signal power of the simple and folded dipole tags is presented in Figure 31.9. The measured response of the folded dipole to strain is more ambiguous than the response of the simple dipole. As the antenna geometry is a decisive factor for the antenna impedance, it is

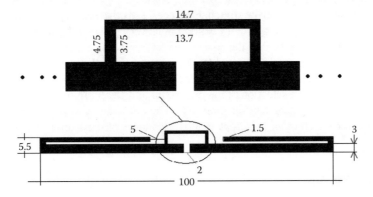

FIGURE 31.8   The folded dipole tag geometry. Dimensions are in mm. (From Merilampi, S. et al., Modification of printed wearable strain sensors by PTF ink particle content adjustment, in *Fourth International Symposium on Applied Sciences in Biomedical and Communication Technologies*, Isabel, Barcelona, Spain, 2011.)

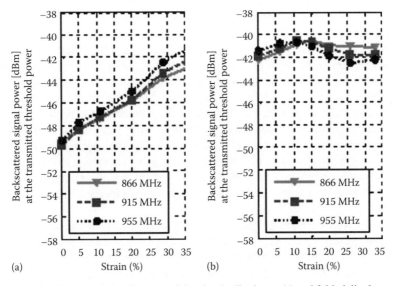

FIGURE 31.9   The backscattered signal power of the simple dipole tag (a) and folded dipole tag (b) on PVC. (From Merilampi, S. et al., Printed passive UHF RFID tags as wearable strain sensors, in *Third International Symposium on Applied Sciences in Biomedical and Communication Technologies*, Isabel, Rome, Italy, 2010.)

expected to be the main reason for the difference observed. The folded dipole geometry on PVC substrate behaves in a manner similar to a simple dipole tag on fabric. The increase of backscattered signal power of the folded dipole on PVC starts to slow down at lower strain levels than in the case of the simple dipole geometry on PVC. Finally, the backscattered signal power begins to decrease. One reason for this might be the narrower traces in the folded dipole geometry, which accelerates the increase in the ohmic losses of the ink film on PVC as a function of strain [35].

The folded dipole tag could be used in measuring strains from 0% to 10% (backscattered signal power increases as a function of strain) and also for larger strains from 10% up (backscattered signal power decreases as a function of strain) at least between 866 and 915 MHz. However, simple dipole tag geometry offers a more practical and useful option. The results suggest that, at least in the case of the simple dipole geometry on PVC, the increase of the backscattered signal power is monotonic until very large levels of strain are applied, mostly because of the relatively slowly increasing ohmic losses. In general, the slowly growing ohmic losses of the tag are achieved by using wide conductors in the tag geometry. The increase of the area of the wider tag geometry under strain is also larger than in the case of thinner tag geometry. This affects the directivity increase. If the tag design contains very narrow stretchable traces, sensitive sensors for measuring relatively small strains are possible, but the geometry would not be very practical for measuring large strains because of the high ohmic losses (and the consequent short reading distances).

We have seen in this chapter that printed PTFs on stretchable substrate can be utilized in wearable strain sensing, and there are many ways to modify their sensitivity such as material and geometry selection. In addition, the selection of the microchip of different impedance offers possibilities in RFID tag-based strain sensor modification [35]. Thus, it is concluded that printed stretchable PTF structures can be used in various applications. Examples of these are briefly discussed in the next chapter.

31.4 APPLICATIONS

Since the sensitivity of the printed conductor-based strain sensor is adjustable by ink choice, substrate material, and geometry of the conductive pattern, the conductors offer various application possibilities. Passive UHF RFID-based sensors are wireless, they do not need a battery or any maintenance, they are inexpensive, and they can be remotely read, also through some materials. Thus, passive UHF RFID tag-based strain sensors are of most interest in application where the sensors are embedded in other structures. Printed conductors and tags introduced in this chapter were fabricated from materials which can be integrated as a part of clothing. Thus, they offer interesting possibilities, especially in wearable strain sensing applications. Large human body movements could be monitored by attaching the strain sensitive structure in such places on the human body that experience strain (elbows, knees, etc.). Thus, the strain sensors could be used to monitor that not too large a movement is performed with a healing body limb. In addition, the trajectory of a certain joint could be monitored, for example, after a surgery. The strain sensors could also be used to count how many times a certain movement is made during an exercise. The strain sensitive structures might as well be integrated into exercising equipment to count the movements. The automatic counting could motivate people to move more in similar manner as the pedometer, for example. The sensor could also be used in force measurements, since strain is proportional to force [9,35,38].

In addition, the strain sensors could be used as a game controller in exercising games (exergames). The idea of exergames is to prevent (childhood) obesity and diseases related to it. Recently, game consoles such as Nintendo Wii have become popular. Nintendo Wii games are controlled by the body movements of the players. In a similar manner, also strain sensitive structures introduced in this chapter could be used as game controllers. The strain sensitive conductors could be used as a separate controller or they could be integrated into the clothing of the players. This enables new possibilities to develop more versatile games and thus elongate their life time and attractiveness.

In gaming applications, the passive UHF RFID tag-based sensors offer wireless game controllers, which need no charging due to their passive nature. Strain sensors might also be helpful in the prevention of cheating in exercise games. By adding a strain sensor on joints (knees, for example), it is possible to monitor that the player actually moves the part of the body that is meant to move [9,35,38,40,41].

In the future, RFID could be integrated into smart phones. Near field communication equipment, which utilizes RFID technology, can already now be found in some smart phones. There is yet no UHF RFID reader in game consoles or in cell phones, but the functionality of the equipment has increased and researchers are already working with miniaturizing the UHF readers. Games and other applications mentioned earlier offer end user applications for the RFID technology. End user applications are important since they have a huge market potential [9,35,38,40,41].

Although self-sensing tags offer various possibilities, the design of a proper "sensor component" in the tag might be sometimes problematic [8]. In these cases, there is also another way to utilize passive UHF tags in strain sensing. An RFID tag can be connected to a traditional sensor and used only for power supply and data transfer [2]. This means even more versatile sensing possibilities.

Despite the promising results discussed in this chapter, a lot of work is still to be done. Further work is required to investigate the sensors in their real use environment, the long-term reliability of the sensors and the recovery from cyclic straining. In terms of practical applications for the tag-based sensors, the permitted power levels and read ranges should be determined as well as the effect of the human body and other proximate materials. Promising results from electromagnetic responses independent of the position and orientation of the reader and the nearby environment have already been published [42,43]. Further work also needs to be done to develop the background systems for the sensors.

REFERENCES

1. R. Bhattacharyya, C. Floerkemeier, and S. Sarma, Towards tag antenna based sensing—An RFID displacement sensor, *IEEE International Conference on RFID*, Orlando, FL, 2009, pp. 95–102.
2. S. Suzuki, H. Okamoto, H. Murakami, H. Asama, S. Morishita, T. Mishima, X. Lin, and H. Itoh, Force sensor system for structural health monitoring using passive RFID tags, *Sensor Review*, 29(2), 2009, 127–136.
3. K. Kure, T. Kanda, K. Suzumor, and S. Wakimoto, Flexible displacement sensor using injected conductive paste, *Sensors and Actuators A*, 143, 2008, 272–278.
4. E. Sevkat, J. Li, B. Liaw, and F. Delale, A statistical model of electrical resistance of carbon fiber reinforced composites under tensile loading, *Composite Science and Technology*, 68, 2008, 2214–2219.
5. N. Hu, Y. Karube, C. Yan, Z. Masuda, and H. Fukunaga, Tunneling effect in a polymer/carbon nanotube nanocomposite strain sensor, *Acta Materialia*, 56, 2008, 2929–2936.
6. G. Hay, D. Southee, P. Evans, D. Harrison, G. Simpson, and B. Ramsey, Examination of silver-graphite lithographically printed resistive strain sensors, *Sensors and Actuators A: Physical*, 135(2), 2007, 534–546.
7. K. Loh, J. Lynch, and N. Kotov, Passive wireless sensing using SWNT based multifunctional thin film patches, *International Journal of Applied Electromagnetics and Mechanics*, 28(1–2), 2008, 87–94.
8. J. Gao, J. Siden, and H.-E. Nilsson, Printed temperature sensors for passive RFID tags, *Progress in Electromagnetics Research Symposium Proceedings*, Xi'an, China, 2010.
9. S. Merilampi, The exploitation of polymer thick films in printing passive UHF RFID dipole tag antennas on challenging substrates, TUT Publication 967, Tampere, Finland, 2011.
10. S. Merilampi, T. Laine-Ma, and P. Ruuskanen, The characterization of electrically conductive silver ink patterns on flexible substrates, *Microelectronics Reliability*, 49(7), 2009, 782–790.
11. M. Prudenziati (ed.), *Thick Film Sensors*, Elsevier, Amsterdam, the Netherlands, 1994.
12. S. B. Hoff, *Screen Printing—A Contemporary Approach*, Delmar Publishers, Albany, NY, 1997.
13. I. Wheeler, *Metallic Pigments in Polymers*, Smithers Rapra Technology, Shrewsbury, UK, 1999.
14. J. W. Gooch, *Analysis and Deformulation of Polymeric Materials: Paints, Plastics, Adhesives and Inks*, Kluwer Academic Publishers, New York, 1997.
15. H. Ujiie, *Digital Printing of Textiles*, Woodhead Publishing Limited, Sawston, Cambridge, UK, 2006.

16. S. Merilampi, T. Björninen, V. Haukka, P. Ruuskanen, L. Ukkonen, and L. Sydänheimo, Analysis of electrically conductive silver ink on stretchable substrates under tensile load, *Microelectronics Reliability*, 50(12), 2010, 2001–2011.

17. L. Jylhä and A. Sihvola, Equation for the effective permittivity of particle-filled composites for material design applications, *Journal of Physics D: Applied Physics*, 40(16), 2007, 4966–4973.

18. K. Finkenzeller, *RFID Handbook*, 2nd edn., John Wiley & Sons, New York, 2003.

19. J. Nummela, Studies towards utilizing passive UHF RFID technology in paper reel supply chain, Tampere University of Technology doctoral thesis, TUT Publication 876, Tampere, Finland, 2010.

20. L. Ukkonen, Development of passive UHF RFID tag antennas for challenging objects and environments, Tampere University of Technology doctoral thesis, TUT Publication 621, Tampere, Finland, 2006.

21. Z. N. Chen, *Antennas for Portable Devices*, John Wiley & Sons, Chichester, UK, 2007.

22. L. Ukkonen, L. Sydänheimo, and M. Kivikoski, Read range performance comparison of compact reader antennas for a handheld UHF RFID reader, *IEEE International Conference on RFID*, Grapevine, TX, 2007, pp. 63–70.

23. EPC Global, *EPC Radio-Frequency Identity Protocols, Class 1 Generation-2 UHF RFID Protocol for Communications at 860 MHz–960 MHz, Version 1.2.0*, 2008, Available: http://www.epcglobalinc.org/standards/uhfc1g2/uhfc1g2_1_2_0-standard-20080511.pdf (accessed on October 24, 2011).

24. C.-H. Loo, K. Elmahgoub, F. Yang, A. Elsherbeni, D. Kajfez, A. Kishk, T. Elsherbeni, L. Ukkonen, L. Sydänheimo, M. Kivikoski, S. Merilampi, and P. Ruuskanen, Chip impedance matching for UHF RFID tag antenna design, *Progress in Electromagnetics Research*, 81, 2008, 359–370.

25. S. Merilampi, L. Ukkonen, L. Sydänheimo, P. Ruuskanen, and M. Kivikoski, Analysis of silver ink bow-tie RFID tag antennas printed on paper substrates, *International Journal of Antennas and Propagation*, 2007, 2007, 9 pp.

26. S. Merilampi, V. Haukka, L. Ukkonen, P. Ruuskanen, L. Sydänheimo, M. Kivikoski, C.-H. Loo, F. Yang, and A. Z. Elsherbeni, Printed RFID tag performance with different materials, *The Fifth International New Exploratory Technologies Conference (NEXT Conference)*, Turku, Finland, 2008, pp. 265–274.

27. S. Merilampi, T. Björninen, L. Ukkonen, P. Ruuskanen, and L. Sydänheimo, Characterization of UHF RFID tags fabricated directly on convex surfaces by pad printing, *The International Journal of Advanced Manufacturing Technology*, 53(5), 2011, 577–591.

28. S. Merilampi, T. Björninen, L. Ukkonen, P. Ruuskanen, and L. Sydänheimo, Embedded wireless strain sensors based on printed RFID tag, *Sensor Review*, 31(1), 2011, 32–40.

29. D. Dobkin, *RF in RFID—Passive UHF RFID in Practice*, Elsevier, Burlington, MA, 2008.

30. F. Meraghnia and M.L. Benzeggagha, Micromechanical modelling of matrix degradation in randomly oriented discontinuous-fibre composites, *Composites Science and Technology*, 55(2), 1995, 171–186.

31. Y.-S. Lin and S.-S. Chiu, Effects of oxidation and particle shape on critical volume fractions of silver-coated copper powders in conductive adhesives for microelectronic applications, *Polymer Engineering and Science*, 44(11), 2004, 2075–2082.

32. K.S. Deepa, S. Kumari Nisha, P. Parameswaran, M.T. Sebastian, and J. James, Effect of conductivity of filler on the percolation threshold of composites, *Applied Physics Letters*, 94, 2009, 142902.

33. H.B. Brom, L.J. Adriaanse, P.A.A. Teunissen, J.A. Reedijk, M.A.J. Michels, and J.C.M. Brokken-Zijp, Frequency and temperature scaling of the conductivity in percolating fractal networks of carbon-black/polymer composites, *Synthetic Metals*, 84(1–3), 1997, 929–930.

34. Z. Rimska, V. Kresalek, and J. Spacek, AC conductivity of carbon fiber-polymer matrix composites at the percolation threshold, *Polymer Composites*, 23(1), 2004, 95–103.

35. S. Merilampi, T. Björninen, L. Ukkonen, P. Ruuskanen, and L. Sydänheimo, Printed passive UHF RFID tags as wearable strain sensors, *Third International Symposium on Applied Sciences in Biomedical and Communication Technologies*, Isabel, Rome, Italy, 2010.

36. K. Kurokawa, Power waves and the scattering matrix, *IEEE Transactions on Microwave Theory and Techniques*, 13(2), 1965, 194–202.

37. P.V. Nikitin and K.V.S. Rao, Antennas and propagation in UHF RFID systems, *IEEE RFID 2008 Conference Proceedings*, Las Vegas, NV, 2008, pp. 277–288.

38. S. Merilampi, T. Björninen, A. Koivisto, L. Ukkonen, and L. Sydänheimo, Modification of printed wearable strain sensors by PTF ink particle content Adjustment, *Fourth International Symposium on Applied Sciences in Biomedical and Communication Technologies*, Isabel, Barcelona, Spain, 2011.

39. X. Jing, W. Zhao, and L. Lan, The effect of particle size on electric conducting percolation threshold in polymer/conducting particle composites, *Journal of Materials Science Letters*, 19(5), 2000, 377–379.

40. K. Kiili and S. Merilampi, Developing engaging exergames with simple motion detection, *Mindtrek Conference*, Tampere, Finland, 2010, pp. 103–110.

41. A. Koivisto, S. Merilampi, and K. Kiili, Mobile exergames for preventing diseases related to childhood obesity, *Fourth International Symposium on Applied Sciences in Biomedical and Communication Technologies*, Isabel, Barcelona, Spain, 2011.

42. G. Marrocco, RFID grids: Part I—Electromagnetic theory, *IEEE Transactions on Antennas and Propagation*, 59(3), 2011, 1019–1026.

43. S. Caizzone and G. Marrocco, RFID grids: Part II—Experimentations, *IEEE Transactions on Antennas and Propagation*, 59(8), 2011, 2896–2904.

32 Application of Inertial Sensors in Developing Smart Particles

Ehad Akeila, Zoran Salcic, and Akshya Swain

CONTENTS

This chapter describes the design of a smart particle (SP) used to monitor the behavior of particles inside river beds, which is used to study the forces and pressures of water flows that cause the particles to be lifted up. In developing this application, low-cost inertial sensors (i.e., accelerometers and gyroscopes) have been utilized to make the SP capable of measuring its own accelerations in all three coordinates. Considering the design constraints (size, power consumption, and accuracy), the SP has gone through several stages of developments. The error models of the sensors as well as the overall system have been derived and analyzed in order to evaluate the performance of SP and identify the necessary calibration procedures to reduce these errors.

32.1 INTRODUCTION

The past few years have witnessed an increasing interest in deploying smart sensors in different applications, which requires monitoring of several physical properties such as temperature, light, and motion. The progress in manufacturing integrated circuits has made such sensors possible in low cost small packages. These sensors combined with microelectronic circuits are becoming more intelligent in terms of functionality and more efficient in terms of power consumption and cost. Strapdown inertial navigation systems (INSs) are among those systems, which rely on mechanical sensors in tracking of moving objects. These systems have been applied in several fields including consumer electronics and automotive applications. The popularity of using such systems comes from the fact that they utilize passive sensors, which make inertial systems entirely self-contained and independent of external sources in determining acceleration, velocity, and position of moving objects. Furthermore, those sensors are commercially available at low cost in small-size packages, which makes them suitable for a wide range of applications, which require portability and minimum cost [1].

A motivating example of our research was to apply such systems in studying the behavior of particles inside riverbeds. The granular materials (sands and gravels) comprising the beds of rivers are typically in motion during periods of high river flow and stationary at other times. A crucial aspect related to those particles is the cause of the initial motion of these particles, which lifts them up in water. This process is often called *entrainment* or *pickup*. The effects of the entrainment process become critical when it progresses over time leading to disastrous consequences, especially when considering buildings and other structural sites in nearby harbors. For that reason, the study of the particle entrainment has attracted the attention of researchers for many years [2–4]. However, the actual mechanics of this process has not yet been fully understood. One of the major reasons is due the limitation of the measurement systems used to monitor this process. Most of the experimental systems used in this field rely on high-speed video cameras for monitoring the motions of the particles inside riverbeds [5]. However, there are several limitations related to these cameras when they are used for such applications such as capability to capture a particle under water, cost, complexity, as well as the computational effort to analyze recorded video clips.

An alternative solution was the design of a smart pebble or smart particle (SP) based on use of low-cost mechanical sensors. The SP is smart in a way that it is able to measure directly applied forces and monitor its own motions while inside the water. Thus, the entrainment process can be studied more appropriately by finding out the magnitude and the direction of forces, which cause this process. Figure 32.1 shows the typical test arrangement of the SP when it is being used to monitor its own motion by placing it amongst a bed of similar fixed particles.

The following section shows some of the technologies used in monitoring movements of particles inside rivers and discusses their advantages and drawbacks.

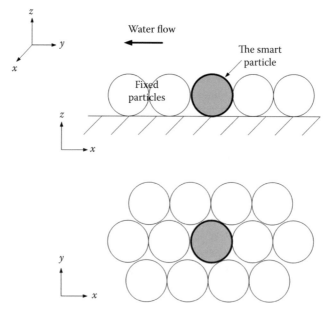

FIGURE 32.1   Test arrangement in using a smart particle while it is placed among fixed particles in a river.

32.2 TECHNOLOGIES USED IN MONITORING SEDIMENT TRANSPORT

As mentioned in the previous section, high-speed video cameras are one of the widely used devices, which are being used in monitoring and tracking movements of particles inside rivers [5–7]. Figure 32.2 shows the typical experimental setup when using this technology. Studying particles transport usually requires that the particles are in a controlled laboratory environment called *flume*.

Video cameras are useful in this field since they are able to monitor the motion of multiple particles at the same time when they move within the image range of the camera. Using these cameras, it is possible to track the position of particles as well as the timing during which the entrainment process occurs. However, there are some limitations when using cameras in this field. Some of these limitations are like: (1) video cameras are only capable of measuring the position of particles, with no information about the forces and pressures applied from the water flow or the other surrounding particles and (2) these cameras usually monitor motions of particles in two dimensions, while it is necessary to study these motions in all three dimensions. Studying the entrainment of particles in all directions would significantly contribute to the knowledge of the factors, which cause this process. (3) Particles are only monitored when they move within a limited range (i.e., within the camera view as shown in Figure 32.2). It is sometimes interesting to monitor the behavior of particles after their entrainment.

The other technology, which is being used in this field for long-term particle tracking, is the radio frequency identification (RFID) [8–10]. Passive RFID devices do not require batteries since they rely on the external power received from an external transmitter. For that reason, the size of such devices can be small and comparable to the actual moving particles, which are to be tracked. Furthermore, the low cost of this technology has a major impact on using the RFID devices in tracking a large number of particles.

This technology is used by encapsulating RFID receivers in a number of particles and placing these particles in certain area within the actual river flow. Unlike the video cameras, this technology makes it possible to track particles moving in actual rivers. These particles can then easily be located using an RFID transmitter after the river flows over these particles. Hence, it becomes possible to draw a map showing how these particles have been moved as a result of the river flow.

The drawback of RFID technology when applied to this field is that it is only applicable for locating particles in shallow water environments and therefore cannot be used for deep rivers since radio signals do not propagate through water in such environments. Furthermore, the accuracy achieved from this technology is very low and it does not accurately measure the internal forces, which act on the particles, which makes it difficult to monitor the entrainment process. This kind of measurement is very critical in studying the nature of the particles' movements.

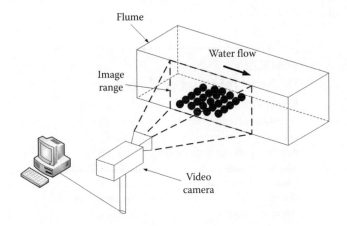

FIGURE 32.2 The experimental setup of a video imaging of particles monitoring.

The SP described in this chapter has the ability to accurately monitor the forces and accelerations of the particles in all three directions while using low cost and small size components. The following sections describe the evolution of the smart particle through different stages of developments, as well as the calibration and testing procedures done to verify the performance of the SP.

32.3 SMART PARTICLE DEVELOPMENT STAGES

The main objective was to develop a SP, which is capable of detecting the entrainment event and measuring its accelerations accurately inside river beds. In developing the SP for this particular application, the following key features were essential in addition to the basic motion-related capabilities: packaging, power supply, relative physical placement of the micro-electro-mechanical system (MEMS) sensors, signal conditioning system, and a compact and simple data communication interface. Packaging has direct influence on the size of the pebble. Since elements involved in the sediment entrainment study are ping pong balls of 40 mm diameter, the diameter of the pebble should not exceed 40 mm. This is possible only with proper compact packaging of the embedded electronics. Since the entrainment is an instantaneous event, the time at which a particle will be entrained can vary from few seconds to few minutes from the moment the particle is placed into the river bed. This directly puts constraint on the power of the SP. Experience in the present application suggests that SP should be powered for at least 15 min in order to capture meaningful data before entrainment, during and after the pebble entrainment. The target design requirements are listed in Table 32.1. As shown in the table, the main challenge is to achieve good performance within constrained size and low overall cost. The electronic components have to be carefully selected to satisfy the requirement of low power consumption and fit into the target package size.

The accuracy in measuring the entrainment accelerations is determined by the type of the used sensors. This means that the sensors have to be selected based on their low signal to noise ratio (SNR), acceleration sensitivity, and range. In addition, the sampling rate must be such that it matches the water flow motions, which are expected to be within 20 Hz. However, the exact motions of the pebble have frequencies within 5 Hz range, and the other higher frequencies are related to the water flow turbulences and can be considered as noise.

In meeting all the target specifications, the SP has evolved through several stages of implementation and development, which are described as follows:

Stage 1: The first work done in this project is described in [11]. The work has focused on proving the feasibility of using low cost inertial sensors in monitoring the motions of particles instead of meeting the target requirements and achieving the goals regarding accuracy, size, and power consumption.

Stage 2: The focus in this stage was on reducing the size of the SP as well as the power consumption, which enabled longer operation in more realistic environment, but still packaged within an 8 cm metallic ball shown in Figure 32.3a.

TABLE 32.1
The Smart Pebble Initial Project Requirements

Size of the pebble	4 cm diameter
Time of operation	>5 min
Acceleration measurement accuracy	90%–95%
Acceleration range	1–2 g (9.8–16 m/s^2)
Expected input frequency	5–20 Hz
Overall cost	As low as possible
Operating temperature	25°C–30°C

FIGURE 32.3 The SP hardware versions. (a) Stage 2 hardware prototype and (b) the current version. (From Akeila, E. et al., *Sensors J. IEEE*, 10, 1705, 2010. With permission.)

Stage 3: In this stage, the size of the SP has been significantly reduced and more work has been focused on the accuracy and processing of the system's outputs. Figure 32.3b shows the actual 4 cm package with the printed circuit boards (PCBs) inside [12].

Stage 4: Further work has been carried out in the past couple of years in terms of evaluating the performance of the sensors used within the SP and analyzing the error sources, which affect the accuracy of the SP as described in [13]. In terms of improving the power consumption as well as reducing the size of the SP, some of the designs have been documented in [14,15]. One of these designs suggests using the technology of *inductive power transfer* (IPT) for charging the SP battery wirelessly without the need for replacing the battery from time to time by opening the actual pebble.

32.4 INERTIAL NAVIGATION SYSTEM

The basic working principle of the SP in monitoring its motions is based on the implementation of INS [1]. The main functionality of the INS-based devices is to measure the accelerations of moving objects. This is performed by measuring two types of motions: (1) linear accelerations and (2) rotational motions. Linear accelerations are typically measured using accelerometers, while rotational motions are measured using gyroscopes or compass sensors. Typical applications that employ basic INS sensors require measurement of accelerations in all three directions. The configuration of

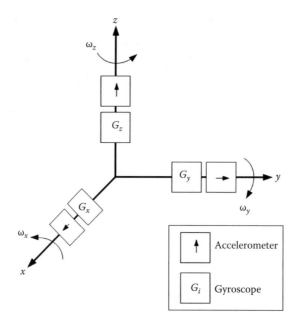

FIGURE 32.4   The typical configuration of the INS sensors based on Gyro-based methods.

the INS sensors generally takes the one shown in Figure 32.4. In the figure, the gyroscopes measure the rotational velocities ω_x, ω_y, and ω_z around each axis of the object to be tracked.

In real applications, accelerometers normally measure the directly applied accelerations, which express the accelerations of moving objects. However, when objects accelerate in a certain direction while having rotational motions, the acceleration along this direction changes in a way that accelerometers will no longer be able to measure the actual linear acceleration of the object. For that reason, it is necessary to measure the rotational motions along with the linear accelerations in order to transform the measurements from the accelerometers to the initial coordinate system of the object before it starts moving. In that light, two frames are defined when using INS-based methods for measuring accelerations: (1) reference frame and (2) body frame. The reference frame is usually defined as the initial frame of the object before it starts moving. This frame consists of fixed axes, which all the other measurements should be related or transformed to. The body frame is defined as the actual axes of the object (i.e., where the INS sensors located), and this frame consists of moving axes as the object moves [16].

The transformation of the measurements from the body frame to the reference frame is based on the algorithm of *Euler angle transformation* [1]. In this algorithm, the displacement of a moving object is determined relative to initial starting axes (i.e., the reference frame). Accelerometers are used to measure the linear accelerations along the three axes of the body frame. Gyroscopes are employed to measure the rotational changes of the body frame in order to correct for changes in directions of the accelerometers' axes and make them measure the accelerations along the original reference frame axes. All sensors are orthogonally placed relative to each other inside the body frame as shown in Figure 32.4. Therefore, to transform the related motion parameters from the body frame to the reference frame, three angular rotations, ψ, θ, and ϕ are performed around the z, y, and x axes, respectively. When the three rotations are performed, the transformation from the body frame to the reference frame is obtained using the following rotation matrix:

$$C_n^b = \begin{bmatrix} \cos\theta\cos\psi & -\cos\phi\sin\psi + \sin\phi\sin\theta\cos\psi & \sin\phi\sin\psi + \cos\phi\sin\theta\cos\psi \\ \cos\theta\sin\psi & \cos\phi\cos\psi + \sin\phi\sin\theta\cos\psi & -\sin\phi\cos\psi + \cos\phi\sin\theta\sin\psi \\ -\sin\theta & \sin\phi\cos\theta & \cos\phi\cos\theta \end{bmatrix} \tag{32.1}$$

Therefore, the conversion of the accelerometers' measurements from the body frame to the reference frame is done by using the following equation:

$$
\begin{bmatrix} Ax_r \\ Ay_r \\ Az_r \end{bmatrix} = C_n^b \cdot \begin{bmatrix} Ax_b \\ Ay_b \\ Az_b \end{bmatrix}
\tag{32.2}
$$

where
  Ax_r, Ay_r, Az_r are the accelerations measured relative to the reference frame (starting point)
  Ax_b, Ay_b, Az_b are the accelerations measured relative to the body frame

32.5 SYSTEM DESIGN AND IMPLEMENTATION

This section describes the hardware design of the SP as well as the necessary offline data processing algorithms, which are being used to process the data acquired from the sensors.

32.5.1 HARDWARE DESIGN

Figure 32.5 shows the block diagram of the SP hardware components. In this application, commercial off-the-shelf type accelerometers and gyroscope MEMS from analog devices were selected based on their sensitivity, accuracy, noise behavior, offset, as well as the total cost of the system. ADXL202 dual axis accelerometers and ADXRS150 yaw rate gyroscopes were found suitable for this application. These devices were placed in three orthogonal sensor modules with suitable signal conditioning circuitry. Figure 32.6 shows a 3D model of the physical placement of the PCBs inside the SP. The overall power consumption of the system was measured to be around 0.25 W. For that reason, the unit is powered from a 6 V alkaline battery, which generally powers the circuit for up to about 9 min continuously. A Texas Instruments TPS60132 charge pump has been utilized to regulate the supply voltage of the SP at 5 V. Additionally, it boosts the battery voltage as it dips below 5 V

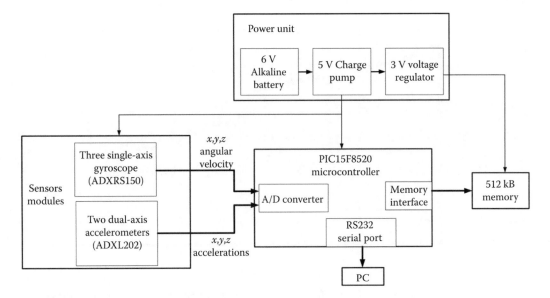

FIGURE 32.5 Hardware block diagram.

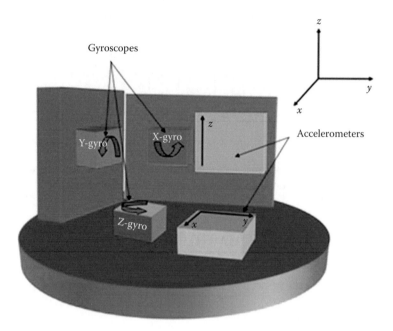

FIGURE 32.6   3D model of the smart particle hardware configuration. (From Akeila, E. et al., *Sensors J. IEEE*, 10, 1705, 2010. With permission.)

to maintain the accuracy of the sensor readings. Figure 32.7 shows the performance of the battery measured, as well as the output from the charge pump. The entire system works at 5 V, except for the external flash memory used to store the acquired data, which requires 3 V supplied through a REG102 3 V regulator.

Since dual axis accelerometers are used, it was possible to take all the necessary measurements with two chips to keep the real estate of the PCBs to a minimum. Each sensor produces an analogue voltage proportional to the acceleration or gyration measured. These outputs are periodically sampled at a rate of 128 Hz, and averaged out to minimize noise, through the analogue-to-digital-converter channels of the Microchip PIC18 F8520 microcontroller, which has a 16 channel 10 bit

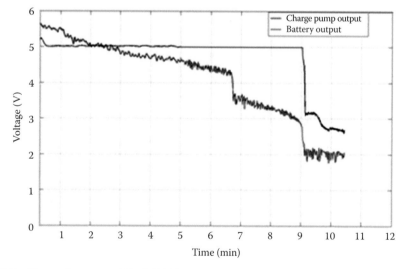

FIGURE 32.7   The performance of the L1016 battery used in the SP along with the output of the charge pump. (From Akeila, E. et al., *Sensors J. IEEE*, 10, 1705, 2010. With permission.)

A/D converter suitable for this application. The digitized sensor data are then stored in an external flash memory, which has a capacity of 512 kBs. This memory is large enough to record data from the six sensors (three accelerometers + three gyroscopes) for up to 5 min. After finishing the measurements, the data are uploaded to a PC via the serial port.

32.5.2 Offline Data Processing

After collecting the data from the sensors, all necessary processing is done off-line using MATLAB®. The outline of offline processing of data is shown in Figure 32.8. The first step in this processing is to filter out all frequencies above 5 Hz, thus being able to focus on the slow motions, such as the entrainment, which has a frequency of around 5 Hz. Other high frequencies due to water flow turbulences are considered as noise. The raw data processed in MATLAB are the digitized voltages of the outputs of the six sensors (three accelerometers and three gyroscopes), which are converted to the actual physical measurements. Once that is done, the misalignment errors are corrected as described in Section 32.8. After that, the outputs from gyroscopes are integrated once to calculate the final rotational angles from which the rotation matrix C_n^b is obtained. The final step is to compensate for the effects of gravity from accelerometers followed by applying Euler's angles formulae on the accelerations and angular rotations to get the final measured accelerations along the x, y, and z axes. The gravity compensation algorithm is described in [17].

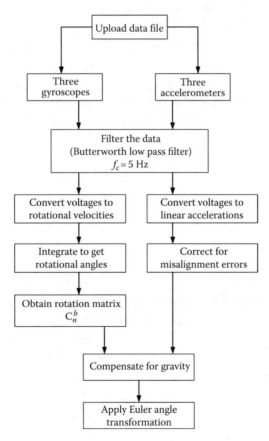

FIGURE 32.8 Offline software data processing. (From Akeila, E. et al., *Sensors J. IEEE*, 10, 1705, 2010. With permission.)

32.6   ANALYSIS OF THE SP ERRORS

The purpose of analyzing the errors of the INS system in this application is to have a clear view about the expected system performance and required calibration procedures. The analysis is done in two steps: (a) analyzing the separate sources of errors and (b) analyzing the combined error sources affecting the overall system performance. The following sections describe these errors in detail.

32.6.1   SOURCES OF ERRORS

The sources of errors in this system can be classified as follows:

Errors in sensors: These types of errors are related to the properties of the materials used in making the sensors and the imperfections resulted from the construction process. The following are the major sources of errors in the accelerometers and gyroscopes:

Sensitivity (scale factor) error: This is defined as the nonlinear relationship between the input motion and the output voltage. This type of error is a result of the material properties and can be affected by temperature changes.

Bias (offset) error: This represents the change in initial output from the sensors when no motion is applied.

Nonlinearity: This is a measurement of deviation from a perfectly constant sensitivity, specified as a percentage with respect to full-scale range.

Alignment error: This is specified as the angle between the true and indicated axis of sensitivity.

Cross-axis sensitivity: It is a measure of how much output is seen on one axis when motion is imposed on a different axis, typically specified as a percentage. The cross-sensitivity is one of the major errors included in multi-axis accelerometers as the one used in this project. In the single-axis gyroscopes, the cross sensitivity is defined as the response of the sensing axis to linear accelerations in addition to the rotational motions.

Table 32.2 shows the estimated error values for the accelerometers and gyroscopes according to the sensors' datasheets. It is noticed from the table that the nonlinearity and temperature drifts are relatively small and can be ignored when modeling the errors of the sensors. Moreover, the operating temperatures required for this application are within a small range, which limits the variations in the sensitivity of the sensors due to changes in temperatures.

TABLE 32.2

The Estimated Error Values in the Accelerometers and Gyroscopes Used in the SP Design

	ADXL202 Accelerometer	ADXRS150 Gyroscope
Nonlinearity	0.2% of F.S	
Alignment error	0.01°	
Cross-axis sensitivity	±2%	
Sensitivity change due to temperature drift	±0.5%	±0.5%
Offset vs. temperature	2 mg/°C	
Linear acceleration effect		0.2°/s/g

Errors due to A/D converter: The errors expected from this part of the SP should be due to the signal to quantization noise ratio, which is modeled according to [18]

$$SNR(dB) = 20\log_{10}(2^n) \approx 6.02n \tag{32.3}$$

where n is the number of bits of the A/D converter. The A/D converter selected in this project has 10 bit resolution, hence the SNR resulting from this component is 60.2 dB. This value should be appropriate for this system, considering the SNR of the other components used in the system [19].

32.6.2 ERROR ANALYSIS OF THE OVERALL SYSTEM

The full error analysis of the SP is shown in Figure 32.9. In the figure, ε_{accel} and ε_{gyros} correspond to the errors in sensors due to noise, misalignments, and biases. The error in the measurements obtained from the accelerometers A_e is the sum of the errors from the A/D converter $\varepsilon_{A/D}$ and the imperfections included in the accelerometers ε_{accel}. On the other hand, the errors in calculating the final rotational angles, θ_e, from the gyroscopes are similar to those in accelerometers except that they have a nature of accumulation due to the integration process. Such errors, if not properly compensated, normally affect the estimation of the rotational matrix C_n^b and lead to poor performance when computing the final measured acceleration A_{final}.

In order to reduce such errors from the system, calibrations of the sensors have to be performed to estimate the bias and scale-factor of the sensors and compensate for the misalignment errors. This can help reduce the inaccuracies in the final acceleration measurement in the SP which is due to the imperfection of sensors.

The initial experiments of the movements of the particles inside river beds indicate that the amount of rotations applied to the particles during the entrainment process is very small compared to the linear accelerations. This means that the errors from calculating the rotational angles obtained from gyroscopes are expected to be small because of the small values of the rotational motions and relatively short time of operation (3 min). The small operation time leads to the small error accumulation due to the integration process. Moreover, it is noticed that the linear accelerations occur mainly along the x and z axes of the flume according to Figure 32.1, while the motions on the y axis can be ignored. The following section describes in detail the calibration procedures of the sensors and the overall system.

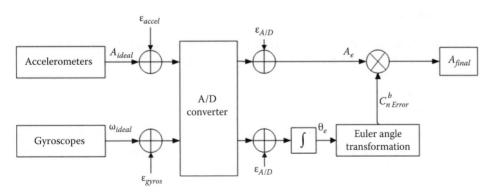

FIGURE 32.9 The error flowchart in the SP design. (From Akeila, E. et al., *Sensors J. IEEE*, 10, 1705, 2010. With permission.)

32.6.3 Effect of the Sampling Frequency

The sampling frequency is one of the major error factors which needs to taken into account when designing any measurement system. The amplitude error of the final measured acceleration signals is closely related to the ratio between the sampling rate f_s and the full bandwidth of the signal BW. The amplitude error due to the sampling frequency is expressed according to the following equation [20]:

$$\varepsilon = \frac{\sqrt{2}\pi V_{s\_pk} BW}{\sqrt{5} f_s V_{FS\_pk}} \tag{32.4}$$

where
  V_{S_pk} is the peak value of the actual input signal to be measured
  V_{FS_pk} is the analog-to-digital converter (ADC) full-scale range
  BW is the bandwidth of the analog signal
  f_s is the sampling frequency

Since the maximum BW of the signal to be measured is 20 Hz in this application (i.e., the frequency of the water flow motions), it is necessary to find out the best sampling rate suitable for measuring the applied accelerations. The calibration procedure and essential testing were carried out and results are shown in Section 32.8.

32.7 TESTING EQUIPMENT

In the real environment, the SP is exposed to forces, which may cause to move the SP randomly inside water. These motions can be classified into two main types: (a) linear accelerations and (b) rotational motions. The following devices have been utilized to simulate each type of motion and calibrate the sensors as well as the whole system:

1. *Shake-table*: This machine is generally used to simulate and test the effect of earthquakes on structures and buildings. The shake-table has a highly sensitive single axis accelerometer, which measures the accelerations generated by the table. Figure 32.10a shows the shake-table and the direction of motion produced by it.
2. *2D rotational motors*: This device was designed and built to generate rotational motions in two dimensions as shown in Figure 32.10b. It is manually controlled to rotate the SP at certain desired angles. The motors are then placed on the shake table such that linear accelerations are combined with rotational motions at the same time. Figure 32.10c shows the final setup.

32.8 CALIBRATION

Calibration is a critical process in INS-based applications as it significantly affects the final performance of the device. In this application, the calibration process was done as described in the following sections.

32.8.1 Calibration of Sensors

The accelerometer sensors (ADXL202) were calibrated against gravity by orienting their sensing axes toward and against the gravitation direction to get $+g$ and $-g$ outputs, respectively. From those outputs, the gain and the offset of the sensor are calculated using the following equations:

$$offset(V) = \frac{V_{+g} + V_{-g}}{2} \tag{32.5}$$

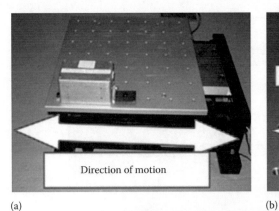

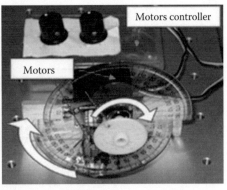

(a)   (b)

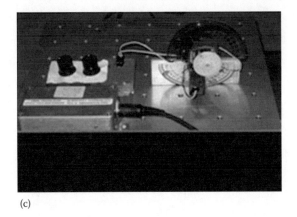

(c)

FIGURE 32.10 The testing devices used to evaluate the SP performance. (a) The shake table used to produce a single-axis linear motion, (b) 2D rotational motion motors, and (c) the complete SP testing devices setup. (From Akeila, E. et al., *Sensors J. IEEE*, 10, 1705, 2010. With permission.)

$$Gain(V/g) = V_{+g} - offset \tag{32.6}$$

where V_{+g} and V_{-g} are the voltages of the outputs of the accelerometers at $+g$ (toward the direction of gravity) and $-g$ (against the direction of gravity), respectively.

The gyroscopes (ADXRS150) were calibrated by placing each sensor on a servo motor, which produces known angular velocities. Figure 32.11a shows a block diagram of the gyroscope calibration setup. The motion of the motor is controlled by a microcontroller, which is programmed to produce periodic rotational motions. The rotational speed of the motor was adjusted so that it does not exceed the maximum sensitivity (150°/s) of the gyroscope. This was done by controlling the voltage level that supplies the motor (at 3 V the speed is 83°/s). The output of the gyroscope is shown in Figure 32.11b.

With reference to Figure 32.11a and b, the gyroscope gives a high pulse (V_+) when the motor rotates in a positive direction relative to the sensing axis (+) of the sensor and a low pulse (V_-) when moving in the opposite direction. The offset voltage (V_{offset}) is obtained by measuring the gyroscope output when no motion is applied as indicated by the red circles in Figure 32.11b. The sensitivity (gain) of the gyroscope is calculated using

$$Gain(V/°/s) = \frac{(V_+ - V_{offset})\Delta t}{180} \tag{32.7}$$

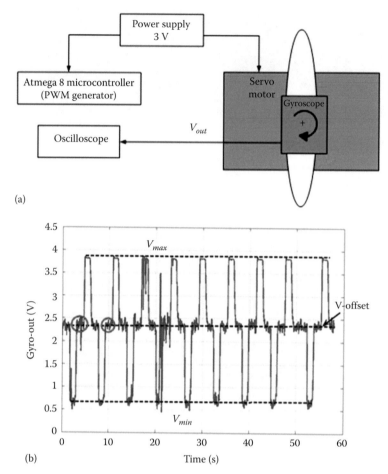

FIGURE 32.11 Gyroscope calibration setup. (a) Hardware block diagram and (b) gyroscope signal output that resulted from the servo motor rotation. (From Akeila, E. et al., *Sensors J. IEEE*, 10, 1705, 2010. With permission.)

32.8.2 MISALIGNMENT CORRECTION

Another major source of errors in most of the INS systems is the misalignment between the axes of the sensors. These errors lead to corresponding errors when deriving the INS Euler angle equation described in Section 32.4, which assumes that the three axes of the system are perfectly orthogonal. However, due to the imperfections in the sensors and the placement of the PCBs, this orthogonality is not guaranteed. In the SP application, such misalignments are estimated and corrected as illustrated in Figure 32.12. The first step is to choose one of the axes as a reference axis, (the y-axis in this case), and orientate it to be perpendicular to the direction of gravity (i.e., parallel to the earth's surface). After that, the tilt angle of the z-axis θ_{z_tilt} is measured from which the misalignment angle θ_{z_error} is obtained. Since the x and y axes are within one accelerometer, the misalignment angle error of the x-axis, θ_{x_error}, can be obtained from the datasheet of the sensor. Once these angles are known, the actual axes are transformed to the corrected orthogonal axes $x_i - y_i - z_i$.

32.8.3 SAMPLING FREQUENCY

The essence of this calibration process is to establish a relation between the amplitude error of the sampled signal and the ratio between the sampling rate and the signal bandwidth as described earlier in Section 32.6.3. The effects of the sampling rate were analyzed first by sampling a 1 Hz sine wave at different sampling rates using a 10 bit ADC. The peak-to-peak amplitude of the sine wave was

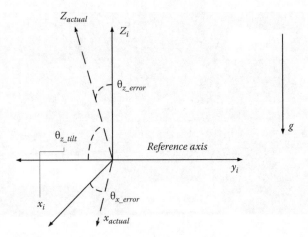

FIGURE 32.12 The misalignment error correction in the SP. (From Akeila, E. et al., *Sensors J. IEEE*, 10, 1705, 2010. With permission.)

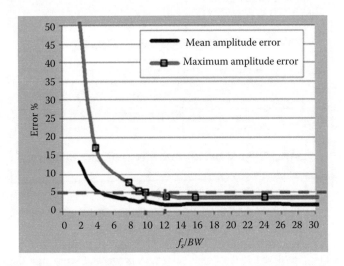

FIGURE 32.13 Amplitude percentage error at different sampling frequencies. (From Akeila, E. et al., *Sensors J. IEEE*, 10, 1705, 2010. With permission.)

adjusted to match the full scale range of the ADC analog input. Figure 32.13 shows the maximum and mean error versus the ratio f_s/BW. From the figure, it is clear that a sampling ratio (f_s/BW) between 10 and 12 is necessary to keep the amplitude errors below 5% as it is required in the SP system.

According to Nyquist sampling theorem, the full recovery of a sampled analog signal can be obtained by sampling the analog signal at a rate of at least twice the maximum bandwidth of the system. From the aforementioned experiment with the two sampling rates, it is clear that the sampling rate of 2 in Nyquist sampling theorem guarantees only recovery of the spectrum of the original input signal; however, this not sufficient to recover the amplitude of signals in time domain.

32.9 FINAL EXPERIMENTAL RESULTS

The SP was tested inside a real flume in order to evaluate its final performance. Measurements performed by a video camera are used as a reference to verify the outputs of the SP. The camera normally takes a video of the SP for a period of time, which ends when the entrainment event occurs. The video is then processed to monitor the final positions of the SP in two-dimensional space.

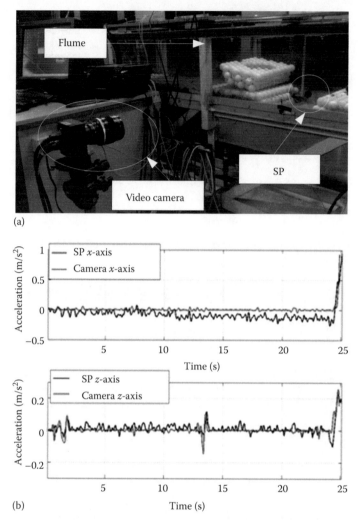

FIGURE 32.14 Testing the SP inside the flume. (a) The experimental setup with the video camera and (b) the acceleration output of the SP compared to the acceleration calculated from the video camera taken at the entrainment event. (From Akeila, E. et al., *Sensors J. IEEE*, 10, 1705, 2010. With permission.)

After studying the entrainment event, the effects of the horizontal and vertical forces that lift up the particles are investigated.

Since the outputs from the camera are the positions of the SP, it was necessary to mathematically differentiate this output twice to obtain its accelerations. By doing so, it is possible to efficiently compare the SP outputs with those taken from the camera.

Figure 32.14 shows the experimental setup as well as the final results from both the camera and the SP. From the figure, it is clear that the SP can accurately measure the accelerations of the SP's movements during the initial small vibrations, which occur before the entrainment as well as during the actual entrainment. However, some errors seem to be accumulating especially on the x-axis prior to and during the entrainment. This error is mainly due to integration of the outputs of the gyroscopes.

32.10 FUTURE IMPROVEMENTS

The current version of the SP has relatively high power consumption (about 0.5 W). This requires a powerful battery (such as the alkaline battery used in this application), which is normally large in size and weight. This causes a restriction on the overall size of the SP, which could be reduced when

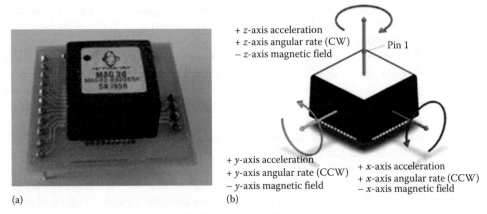

+ z-axis acceleration
+ z-axis angular rate (CW)
− z-axis magnetic field

Pin 1

+ y-axis acceleration
+ y-axis angular rate (CCW)
− y-axis magnetic field

+ x-axis acceleration
+ x-axis angular rate (CCW)
− x-axis magnetic field

(a)   (b)

FIGURE 32.15 (a) The MAG3D sensor hardware and (b) a schematic of the INS system based on the MAG3D sensor system. (From Akeila, E. et al., A new algorithm for direct gravity estimation and compensation in gyro-based and gyro-free INS applications, in S.C. Mukhopadhyay, G.S. Gupta, and R.Y.M. Huang, eds., *Recent Advances in Sensing Technology*, Lecture Notes in Electrical Engineering, LNEE, Springer, Berlin, Germany, 2009, pp. 203–219. With permission.)

reducing the power consumption. One way of reducing the power consumption is by replacing the microcontroller (which consumes most of the power in the SP) by a low power consumption one, such as Texas Instruments MSP430 [21]. This can have a direct impact on the size of the SP as it would be possible in this case to use one of the lithium coin cell batteries, which are normally used in watches and low power operated devices.

Another factor that affects the size of the SP is the placement of the three orthogonal PCBs of sensors in order to form the INS system. In addition to the space taken by these PCBs, there is an issue with the accuracy of the SP caused by the misalignments between the INS sensors. One solution for this problem can be using one of the fully packed INS sensors units as the one shown in Figure 32.15. MAG3D sensor [22] is a full INS solution consisting of three types of sensors; accelerometers, gyroscopes, and magnetometers. This INS unit comes in a 20 mm × 20 mm package, which has the advantage of low sensors noise in addition to the high degree of orthogonality between the sensors.

Another drawback in the current version of the SP is in the measurement procedure, which requires that the SP has to be opened several times in order to replace the battery and upload the data. This can be inconvenient for the users and some of the components and connections can be broken after the several times of opening of the SP. Some of the future work in solving such problems can be as follows:

- Wireless charging using the IPT technology as in the design described in [14,15]. With this technology, it would be possible to charge the SP wirelessly without opening it.
- The current version of the SP communicates with the PC through a serial port for initiating the measurements as well as downloading the data stored in the memory. An alternative solution can be using a wireless communication between the SP and the PC using one of the RF-based technologies, such as Bluetooth or ZigBee.

32.11 SUMMARY

This chapter describes the design of a smart particle used to monitor the behavior of particles inside river-beds. Considering the design constraints (size, power consumption, and accuracy), the SP has gone through several stages of developments. The error models of the sensors as well as the overall system have been derived and analyzed in order to evaluate the performance of SP and identify the necessary calibration procedures, which may help to reduce these errors. The calibration process was divided into two independent steps; calibration of sensors and overall system calibration.

The effect of the sampling rate has been studied by testing a sinusoidal wave sampled under different rates. Accordingly, it was discovered that in order to efficiently recover the amplitude of signals in time domain, the minimum sampling rate required using an ADC is 10 times the maximum frequency component inside the signal, assuming that the peak amplitude matches the full-scale range of the ADC component.

The misalignments of the sensors have been corrected using the system's outputs under stationary conditions. After that, the gravitational changes have been compensated from the outputs of the accelerometers used in the SP.

Final experimental results of the SP show that it is able to accurately measure its own accelerations in all three directions when the entrainment process occurs. However, there is some error accumulation caused by the integration of the outputs of the gyroscopes.

REFERENCES

1. D. H. Titterton and J. L. Weston, *Strapdown Inertial Navigation Technology*. Stevenage, U.K.: Institution of Electrical Engineers, 2004.
2. Y. Nino, F. Lopez, and M. Garcia, Threshold for particle entrainment into suspension, *The Journal of the International Association of Sedimentologists*, 50, 247–263, 2003.
3. B. P. K. Yung, H. Merry, and T. R. Bott, The role of turbulent bursts in particle re-entrainment in aqueous systems, *Chemical Engineering Science*, 44, 873–882, 1989.
4. A. J. Grass, Initial instability of fine bed sand, *Journal of the Hydraulics Division*, 96, 619–632, 1970.
5. Y. Niro and M. H. Garcia, Experiments on particle-turbulence interactions in the near-wall region of an open channel flow: Implications for sediment transport, *Journal of Fluid Mechanics*, 326, 285–319, 1996.
6. K. G. Heays, H. Freidrich, and B. W. Melville, Advanced particle tracking for sediment movement on river beds: A laboratory study, in: *17th Australian Fluid Mechanics Conference*, Auckland, New Zealand, 2010.
7. R. Middletona, J. Brasingtonb, B. J. Murphya, and L. E. Frosticka, Monitoring gravel framework dilation using a new digital particle tracking method, *Computers & Geosciences*, 26, 329–340, 2000.
8. H. L. N. Lamarre, B. Macvicar, and A. G. Roy, Using passive integrated transponder (Pit) tags to investigate sediment transport in gravel-bed rivers, *Journal of Sedimentary Research*, 75, 736–741, 2005.
9. M. H. Nichols, A radio frequency identification system for monitoring coarse sediment particle displacement, *Applied Engineering in Agriculture*, 20, 783–787, 2004.
10. J. Schneider, R. Hegglin, S. Meier, J. M. Turowski, M. Nitsche, and D. Rickenmann, Studying sediment transport in mountain rivers by mobile and stationary RFID antennas, *River Flow*, Braunschweig, Germany, 2010, pp. 1723–1730.
11. N. Kularatna, C. Wijeratne, and B. Melville, Mixed signal approach for rapid prototyping of a compact smart pebble for sediment transport monitoring in river beds, in *Sensors, 2005 IEEE*, Irvine, CA, 2005, p. 5.
12. N. Kularatna, B. Melville, E. Akeila, and D. Kularatna, Implementation aspects and offline digital signal processing of a smart pebble for river bed sediment transport monitoring, in *Fifth IEEE Conference on Sensors,*, Daegu, Korea, 2006, pp. 1093–1098.
13. E. Akeila, Z. Salcic, and A. Swain, Smart pebble for monitoring riverbed sediment transport, *Sensors Journal, IEEE*, 10, 1705–1717, 2010.
14. D. K. Abeywardana, A. P. Hu, and N. Kularatna, Design enhancements of the smart sediment particle for riverbed transport monitoring, in *4th IEEE Conference on Industrial Electronics and Applications, 2009. ICIEA 2009*, Xi'an, China, 2009, pp. 336–341.
15. N. Kularatna and D. K. Abeywardana, Use of motion sensors for autonomous monitoring of hydraulic environments, in *Sensors, 2008 IEEE*, Lecce, Italy, 2008, pp. 1214–1217.
16. D. H. Titterton and J. L. Weston, *Strapdown Inertial Navigation Technology*, vol. 5. London, U.K.: Institution of Electrical Engineers, 2004.

17. E. Akeila, Z. Salcic, and A. Swain, A new algorithm for direct gravity estimation and compensation in gyro-based and gyro-free INS applications, in: S. C. Mukhopadhyay, G. S. Gupta, and R. Y. Min Huang (eds.), *Recent Advances in Sensing Technology. Lecture Notes in Electrical Engineering*, LNEE, Springer, Berlin, Germany, 2009, pp. 203–219.
18. S. V. Vaseghi, *Advanced Digital Signal Processing and Noise Reduction*, 3rd edn. Chichester, England: Wiley, 2006.
19. N. Gray, Selecting high-speed A/D converters, in *Electronic Products*. National Semiconductor, Santa Clara, CA, 1999. Retrieved from http://www2.electronicproducts.com/Selecting_high_spead_A_D_converters-article-NOVNAT1-nov1999-html.aspx.
20. P. H. Garret, *High Performance Instrumentation and Automation*. Boca Raton, FL: Taylor & Francis, 2005.
21. Texas Instruments, MSP430C33x, MSP430P337A mixed signal microcontrollers datasheet.
22. Memsense, MAG3, triaxial magnetometer, accelerometer & gyroscope analog inertial sensor, in: MAG3 datasheet.

Index